MTP International Review of Science

Volume 9

Chemical Kinetics

Edited by **J. C. Polanyi, F.R.S.**
University of Toronto

Butterworths · London
University Park Press · Baltimore

THE BUTTERWORTH GROUP

ENGLAND
Butterworth & Co (Publishers) Ltd
London: 88 Kingsway, WC2B 6AB

AUSTRALIA
Butterworth & Co (Australia) Ltd
Sydney: 586 Pacific Highway 2067
Melbourne: 343 Little Collins Street, 3000
Brisbane: 240 Queen Street, 4000

NEW ZEALAND
Butterworth & Co (New Zealand) Ltd
Wellington: 26–28 Waring Taylor Street, 1

SOUTH AFRICA
Butterworth & Co (South Africa) (Pty) Ltd
Durban: 152–154 Gale Street

ISBN 0 408 70270 2

UNIVERSITY PARK PRESS

U.S.A. and CANADA
University Park Press Inc
Chamber of Commerce Building
Baltimore, Maryland, 21202

Library of Congress Cataloging in Publication Data

Polanyi, J   C   1929–
  Chemical kinetics.

  (Physical chemistry, series one, v. 9) (MTP
international review of science)
  Includes bibliographies.
  1. Chemical reaction, Rate of. I. Title.
QD453.2.P58 vol. 9 [QD501] 541'.3'08s [541'.39]
ISBN 0–8391–1023–5   72–39496

First Published 1972 and © 1972
MTP MEDICAL AND TECHNICAL PUBLISHING CO. LTD.
Seacourt Tower
West Way
Oxford, OX2 0JW
and
BUTTERWORTH & CO. (PUBLISHERS) LTD.

Filmset by Photoprint Plates Ltd., Rayleigh, Essex
Printed in England by Redwood Press Ltd., Trowbridge, Wilts
and bound by R. J. Acford Ltd., Chichester, Sussex

MTP International Review of Science

Chemical Kinetics

MTP International Review of Science

Publisher's Note

The MTP International Review of Science is an important new venture in scientific publishing, which we present in association with MTP Medical and Technical Publishing Co. Ltd. and University Park Press, Baltimore. The basic concept of the Review is to provide regular authoritative reviews of entire disciplines. We are starting with chemistry because the problems of literature survey are probably more acute in this subject than in any other. As a matter of policy, the authorship of the MTP Review of Chemistry is international and distinguished; the subject coverage is extensive, systematic and critical; and most important of all, new issues of the Review will be published every two years.

In the MTP Review of Chemistry (Series One), Inorganic, Physical and Organic Chemistry are comprehensively reviewed in 33 text volumes and 3 index volumes, details of which are shown opposite. In general, the reviews cover the period 1967 to 1971. In 1974, it is planned to issue the MTP Review of Chemistry (Series Two), consisting of a similar set of volumes covering the period 1971 to 1973. Series Three is planned for 1976, and so on.

The MTP Review of Chemistry has been conceived within a carefully organised editorial framework. The over-all plan was drawn up, and the volume editors were appointed, by three consultant editors. In turn, each volume editor planned the coverage of his field and appointed authors to write on subjects which were within the area of their own research experience. No geographical restriction was imposed. Hence, the 300 or so contributions to the MTP Review of Chemistry come from many countries of the world and provide an authoritative account of progress in chemistry.

To facilitate rapid production, individual volumes do not have an index. Instead, each chapter has been prefaced with a detailed list of contents, and an index to the 13 volumes of the MTP Review of Physical Chemistry (Series One) will appear, as a separate volume, after publication of the final volume. Similar arrangements will apply to the MTP Review of Organic Chemistry (Series One) and to subsequent series.

Butterworth & Co. (Publishers) Ltd.

**Physical Chemistry
Series One**
Consultant Editor
A. D. Buckingham
*Department of Chemistry
University of Cambridge*

Volume titles and Editors

1 **THEORETICAL CHEMISTRY**
Professor W. Byers Brown, *University of Manchester*

2 **MOLECULAR STRUCTURE AND PROPERTIES**
Professor G. Allen, *University of Manchester*

3 **SPECTROSCOPY**
Dr. D. A. Ramsay, F.R.S.C.,
National Research Council of Canada

4 **MAGNETIC RESONANCE**
Professor C. A. McDowell, *University of British Columbia*

5 **MASS SPECTROMETRY**
Professor A. Maccoll, *University College, University of London*

6 **ELECTROCHEMISTRY**
Professor J. O'M Bockris, *University of Pennsylvania*

7 **SURFACE CHEMISTRY AND COLLOIDS**
Professor M. Kerker, *Clarkson College of Technology, New York*

8 **MACROMOLECULAR SCIENCE**
Professor C. E. H. Bawn, F.R.S.,
University of Liverpool

9 **CHEMICAL KINETICS**
Professor J. C. Polanyi, F.R.S.,
University of Toronto

10 **THERMOCHEMISTRY AND THERMODYNAMICS**
Dr. H. A. Skinner, *University of Manchester*

11 **CHEMICAL CRYSTALLOGRAPHY**
Professor J. Monteath Robertson, F.R.S.,
University of Glasgow

12 **ANALYTICAL CHEMISTRY –PART 1**
Professor T. S. West, *Imperial College, University of London*

13 **ANALYTICAL CHEMISTRY – PART 2**
Professor T. S. West, *Imperial College, University of London*

INDEX VOLUME

Physical Chemistry
Series One

Consultant Editor
A. D. Buckingham

Consultant Editor's Note

The MTP International Review of Science is designed to provide a comprehensive, critical and continuing survey of progress in research. The difficult problem of keeping up with advances on a reasonably broad front makes the idea of the Review especially appealing, and I was grateful to be given the opportunity of helping to plan it.

This particular 13-volume section is concerned with Physical Chemistry, Chemical Crystallography and Analytical Chemistry. The subdivision of Physical Chemistry adopted is not completely conventional, but it has been designed to reflect current research trends and it is hoped that it will appeal to the reader. Each volume has been edited by a distinguished chemist and has been written by a team of authoritative scientists. Each author has assessed and interpreted research progress in a specialised topic in terms of his own experience. I believe that their efforts have produced very useful and timely accounts of progress in these branches of chemistry, and that the volumes will make a valuable contribution towards the solution of our problem of keeping abreast of progress in research.

It is my pleasure to thank all those who have collaborated in making this venture possible — the volume editors, the chapter authors and the publishers.

Cambridge   A. D. Buckingham

Preface

Though this volume is entitled 'Chemical Kinetics', it is symptomatic of a significant current development that a substantial portion of the material could be described as lying in the field of 'Reaction Dynamics'. The earliest concern of the kineticist was the overall rate of chemical reactions, and the dependence of these rates on such macroscopic parameters as concentration and temperature. Half a century ago these same interests became focused on the more fundamental question of the rates of elementary reactive steps. Today these individual reactive encounters are coming under a new type of scrutiny designed to uncover the atomic and molecular motions (the 'dynamics') of the reactive event. In these dynamical studies the variables include such parameters as reagent and product energies, as well as angles of approach and separation.

The benefits that accrue from the study of reaction probability at this level of detail are twofold. In the first place the theory of thermal reaction rates can be placed on a firmer basis than has been possible heretofore. The second consequence is likely to be even more notable. There are many environments in which chemical reactions occur, or can be made to occur, at rates which differ enormously from their thermal equilibrium rates. This happens naturally, for example, in the upper atmosphere, and perhaps in the awesome reaction-vessel of inter-stellar space. It happens also in environments where there is a large and specific energy-flux, as is the case, for example, within a laser or a plasma. In these environments we shall be totally dependent for guidance concerning the rates of reaction on the application of information culled from the new field of reaction dynamics.

It will be a long time, however, before our knowledge of the detailed dynamics of reactive encounters extends to more than a fraction of the huge number of reactions which fall within the domain of chemistry. There will continue to be a pressing need for more studies, and more effective studies, of the rates of a wide variety of reactions occurring under conditions of thermal equilibrium. Moreover, these two types of study, macroscopic and (more or less) microscopic, illuminate one another. It is a source of satisfaction, therefore, that both are well represented in this volume.

I should like to extend my sincere thanks to the authors, all of whom are very actively engaged in research in the fields covered by their chapters, for devoting their time and talent to this enterprise. They have aimed for, and I believe achieved, a certain liveliness that comes from being contemporary and critical. They were not asked to cover their subject area exhaustively – it is well-nigh impossible to do this and still remain readable. Nor does this volume attempt to cover the entire field of reaction kinetics under its ten subject headings. There will soon be an opportunity in the next biennial review to see that important omissions are made good. Meanwhile it is the hope of all involved that the compilation of a somewhat personal album of recent developments in a rapidly growing branch of chemistry will prove of value to workers in this, and in neighbouring, fields.

Toronto   J. C. Polanyi

Contents

1
Unimolecular Reactions of Polyatomic Molecules, Radicals and Ions

D. W. SETSER
Kansas State University, Manhattan

1.1 INTRODUCTION

1.1.1 General scope

During the last decade a quantum statistical theory of unimolecular reactions, which includes the activated complex of absolute rate theory, has been applied to a variety of unimolecular reactions of polyatomic molecules, radicals and ions with a high measure of success. The central objective will be to review some quantitative experimental work that can be related to theoretical predictions. Attention will be focused upon experimental methods providing non-equilibrium energy distributions. Although often considered separately, unimolecular reactions of polyatomic ions should be a natural part of a discussion concerned with non-equilibrium unimolecular reactions. Space limitations prevent discussion of many interesting studies and the examples chosen reflect the author's current interests and knowledge of the literature. The coverage for activated molecules and radicals is more comprehensive than for ions. The 1970 Annual Review of Physical Chemistry[1] contains material on unimolecular reactions and overlap with that review was minimised. Formulation of the statistical theory of unimolecular reactions is included in two recently published textbooks[2, 3].

Although some elementary reactions of polyatomic molecules in excited electronic states have been shown to be truly unimolecular[4, 5], these processes differ fundamentally from unimolecular reactions of ground electronic states because the rate-limiting step for the former often involve crossing to a different electronic state of the parent molecule. Unless explicitly stated otherwise, reactions considered in this review occur on the ground-state potential-surface. Polyatomic, as used in the title, denotes four or more atoms. Unimolecular reactions of triatomics[6, 7] were excluded because electronically excited states often are involved and because the basic postulates of the statistical theory may not be applicable. The situation is improved slightly for tetra-atomic molecules and, in fact, such cases may offer the possibility of ascertaining the small-molecule limit for the statistical theory. An important development for small molecules is the measurement of rate constants at pressures near the high-pressure limit[7, 8].

1.1.2 Equilibrium versus non-equilibrium methods of activation

According to the statistical theory, molecules with internal energy, E, have specific rate constants which can be computed[2, 3, 9] from a knowledge of the internal energy and the threshold energy, E_0, of the reaction. If the transition state of absolute rate theory is incorporated into the statistical formulation, the result commonly is known as the RRKM[10] (Rice–Ramsperger–Kassel–Marcus) theory for molecules and radicals or as the QET (Quasi Equilibrium Theory)[11] for ions. Unimolecular reactions traditionally were studied by use of thermal collisional activation at high pressure. For such experiments the Boltzmann distribution function, $F(E)dE$, characterises the internal energy distribution of the activated molecules.

$$F(E)dE = N(E)e^{-E/RT}dE/Q_1 \qquad (1.1)$$

Q_1 is the partition function of the internal degrees of freedom at temperature, T, and $N(E)$ is the density of internal energy states at energy, E. Figure 1.1 shows $F(E)dE$ and k_E values for HF elimination and dissociation of 1,1,1-trifluoroethane[12]. The measured high-pressure first-order rate-constant obtained from pyrolysis studies is the textbook[2, 3] form of the absolute rate-theory rate-constant for a unimolecular reaction.

$$k_{uni}^{\infty} = \int_{E_0'}^{\infty} k_E F(E)\,dE = \langle k_E \rangle = \frac{\ell kT}{h}\left[\frac{I_A^\dagger I_B^\dagger I_C^\dagger}{I_A I_B I_C}\right]^{\frac{1}{2}}\frac{Q_1^\dagger}{Q_1}e^{-E_0/RT} \qquad (1.2)$$

The † symbol denotes quantities associated with the activated complex. The quantity in brackets is the ratio of the moments of inertia and l is the reaction path degeneracy, which continues to present problems of definition[13].

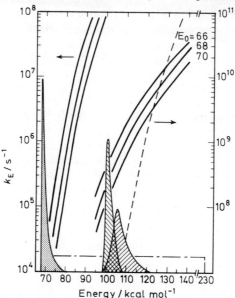

Figure 1.1   Rate constants and distribution functions for $CH_3CF_3^*$. (a) Rate-constants; —for HF elimination with E_0 as indicated; the model[12] has a pre-exponential factor of 5.3×10^{13} s^{-1} at 800 K; —for dissociation to $CF_3 + CH_3$ with $E_0 = 99$ kcal mol^{-1} and a pre-exponential factor[12] of 1.7×10^{17} s^{-1}. (b) Four distribution functions of activated molecules: the closely spaced dotted area is the thermal distribution at 800 K; cross-hatched areas are chemical-activation distributions $(CH_3 + CF_3)$ at 300 and 600 K and the area below the flat dotted line is for hot atom displacement $(F^* + CH_3CF_3)$. The first three distributions are normalised to unity on a linear scale; the last one is not. The distribution functions of *reacting* molecules for the chemically activated distributions are similar to those shown for the formed molecules. However, a drastic change occurs for thermal activation at high pressure since the product, $k_E F(E)$, extends to much higher energy than $F(E)$. This is illustrated by the average energy of reacting $CH_3CF_3^*$ molecules at high pressure, which is approximately $E_0 + 10$ kcal mol^{-1} at 800 K.

Equilibrium pyrolysis studies provide thermodynamic information, which is summarised by Arrhenius pre-exponential factors (or the equivalent entropies of activation) and activation energies. Alternatively the partition function pre-exponential factor and threshold energy of equation (1.2) may be used. The pre-exponential factor is very useful and the threshold energy is *essential* for more advanced dynamical treatment of unimolecular reactions. Equilibrium thermal-activation studies are not the subject of this review; however, the lack of reliable threshold energies often prevents, or at least confuses, attempts at more advanced treatment. A comprehensive evaluation of Arrhenius parameters was published in 1970 [14] and there are several earlier reviews[15-17]. Empirical rules and tabulations provide useful methods[18, 19] for estimating entropies of activation, threshold energies or even fall-off curves[20]. In fact such estimates by experienced investigators often are about as good as the experimental pyrolysis data. Reactions that have exceptionally large[17, 21, 22] or exceptionally small pre-exponential factors[23-26] still present difficulties in terms of rationalising the structure of the transition state relative to the reactant molecule.

A unimolecular reaction may occur whenever a molecule, radical or ion acquires energy in excess of its threshold energy. In the gas phase the number of activation processes which produce non-Boltzmann distributions, $f(E)dE$, far exceeds those which yield thermal equilibrium distributions. Two possibilities for CH_3CF_3, (i) combination of CH_3 and CF_3 and (ii) hot atom, F^* for F, displacement[27], are shown in Figure 1.1. The whole field of mass spectrometry, whether using electron impact, photoionisation or some other technique of ion production, is concerned with non-Boltzmann distributions of activated molecular ions. Most non-equilibrium activation studies are concerned with measurement of a rate constant for a specific level of excitation energy; however, the energy and momentum released to the products are becoming increasingly important experimental quantities[28-30]

1.2 EXPERIMENTAL METHODS OF ACTIVATION WITH EXAMPLES

1.2.1 Fall-off results and isotope effects in thermal collisional activation

1.2.1.1 General description

According to a Lindeman mechanism with unit collisional efficiency for the bath gas, M, the first-order rate-constant is given by:

$$k_{uni} = k_M[M] \int_{E_0}^{\infty} k_E/(k_E + k_M[M])F(E)dE \qquad (1.3)$$

with k_M being the collision number associated with a collision diameter S namely: $k_M = S^2(8\pi kT/\mu)^{\frac{1}{2}}$. The experimental measurements are first-order rate-constants over a range of pressure, and plots of $\log(k_{uni}/k_{uni}^{\infty})$ versus $\log(\text{pressure})$ are termed fall-off plots. Whenever $k_{uni}/k_{uni}^{\infty} < 1$, the distribution of the reacting molecules is not the equilibrium case and, thus, fall-off studies qualify for this review. If (i) the unit deactivation formulation is appropriate and if (ii) the collision diameter is known, it is apparent from equation (1.3) that the degree of fall-off at a given pressure depends upon the competition between k_E and $k_M[M]$ and provides a measure of the specific rate-constant. Tardy and Rabinovitch[31] examined the consequences of multi-step activation and deactivation for several thermal-reaction systems. The criterion for unit deactivation, in the second-order region, can be summarised by the relation, $\langle \Delta E \rangle / \langle E^{\dagger} \rangle \gtrsim 6$, where $\langle \Delta E \rangle$ is the average energy transferred per collision and $\langle E^{\dagger} \rangle$ is the average thermal energy above E_0 for the reacting molecules in the low-pressure limit. Measurements of $\langle \Delta E \rangle$ in chemical-activation studies[32-36] and of the collisional efficiency for CN_3NC isomerisation in the low-pressure region have shown that large and/or polar polyatomic bath-gas molecules have virtually unit efficiency. Although the interpretation of 'collisional efficiency' is more complex[31b, 37c] in the inter-mediate fall-off region of pressure, the unit deactivation assumption for the parent molecule would seem appropriate for the rather large polyatomic

Table 1.1 Summary of fall-off studies

Molecule	Comments on data	Comments on RRKM calculation	Reference
	Isomerisation of isocyanides $(RNC \rightarrow RCN)$		
CH_3NC	Pressure: 10^5–10^{-2} Torr; $k/k_\infty = 1.00$–1.6×10^{-3} which covered the complete transition from first-order to second-order kinetics; E_a measured as a function of pressure.	Good fit to fall-off and variation of E_a with pressure; best model included one active overall rotation and anharmonicity effects. Absolute pressure fit was better than a factor of 1.5 for $S = 0.45$ nm.	44
CDH_2NC	Internal comparison versus CH_3NC from 7×10^3–5×10^{-2} Torr; $k_H/k_D = 1.02$–0.7; $k/k_\infty = 1.0$–4×10^{-3}.	Good agreement, using the same model as for CH_3NC, for k_H/k_D as a function of pressure and the low pressure limiting value of $c. 0.7$.	45
CD_3NC	Internal and external comparison versus CH_3NC from 10^4–10^{-2} Torr; $k_H/k_D = 1.07$ to 0.28; $k/k_\infty = 1.0$ to $c. 2 \times 10^{-3}$.	Good agreement, by use of the same model as for CH_3NC, for k_H/k_D as a function of pressure and the low pressure limiting value of 0.3.	46
C_2H_5NC	Pressure: 10^3–10^{-3} Torr; $k/k_\infty = 1$ to 3×10^{-3} for three temperatures, heterogeneity problems were found at lowest pressure; $k/k_\infty = 0.2$ at $c. 0.1$ Torr.	Good fit to curvature; absolute pressure fit to within a factor of 1.5 for $S = 0.5$ nm; adequate fit to change in E_a with pressure. Best model did not include an active overall-rotation (compare with CH_3NC).	47
C_2D_5NC	Internal comparison from 10^3–10^{-3} Torr; $k_H/k_D = 1.0$ to 0.21; $k/k_\infty = 0.2$ at $c. 0.025$ Torr.	Good fit for k_H/k_D over the whole pressure range, with quantitative account of inverse isotope effect.	48
	Structural isomerisation of cyclopropane $(\nabla \rightarrow = \diagup)$		
Cyclopropane	Pressure 10^3–10^{-1} Torr; $k/k_\infty = 1.0$–0.1; $k/k_\infty = 0.2$ and 0.4 at $c. 0.25$ and $c. 1.2$ Torr, respectively, at 718 K.	Good fit of both shape and absolute pressure of the fall-off curve using $S = 0.42$ nm.	51
Cyclo-propane-d_6	Extension of isotope-effect studies to low pressure: 10^{-1}–10^{-4} Torr; wall complications enter at $c. 5 \times 10^{-4}$ Torr.	Calculations agree with experiment until onset of wall effect; $k_H/k_D = 1.96$ to $c. 0.8$ at 3×10^{-2} torr.	53

Table 1.1 — *continued*

Molecule	Comments on data	Comments on RRKM calculation	Reference
Methyl-cyclopropane	Pressure: 200–0.05 Torr; $k/k_\infty = 1.0$–$c.\ 0.3$; $k/k_\infty = 0.4$ at 0.06 Torr.	No RRKM calculation, but it was noted that addition of the CH_3 group to cyclopropane caused a shift of k/k_∞ to lower pressure in support of a model having free flow of energy.	53
		This recent calculation obtained good fit to the curvature but the collision frequency required multiplication by $c.\ 0.15$ to fit of absolute pressure using $S = 0.47$ nm.	51
Ethyl-cyclopropane	Pressure: 84–0.05 Torr; $k/k_\infty = 1.0$ to $c.\ 0.7$.	The fit of the absolute pressure seems to be poor but the data are not extensive.	54
Dimethyl-cyclopropane	Pressure: 35–10^{-2} Torr; $k/k_\infty = 1.0$–$c.\ 0.4$; E_a measured as function of pressure.	No RRKM calculation, but the fall-off came at lower pressure than for methylcyclopropane in accordance with theory.	55
Dichloro-cyclopropane	Pressure: 70–0.024 Torr; $k/k_\infty = 1.0$–0.27.	Experimental k/k_∞ values decline faster than calculated curves, poor pressure fit. Experimental error seems likely.	62

Isomerisation of cyclobutene

Molecule	Comments on data	Comments on RRKM calculation	Reference
Cyclobutene	Pressure: 23–0.015 Torr; $k/k_\infty = 10$–$c.\ 0.12$ E_a was measured as a function of pressure.	New calculations, taking into account the reverse isomerisation reaction, give good agreement with both the shape and pressure fit of the fall-off.	63 64 65
2-Methyl-cyclobutene	Pressure: 15–0.01 Torr; $k/k_\infty = 1.0$–0.34.	Calculations need improvement, see 64, however, the pressure shift of the fall-off relative to cyclobutene is as expected from theory.	65 66

Isomerisation of cyclobutanes

($\square \rightarrow 2C_2H_4$)

Molecule	Comments on data	Comments on RRKM calculation	Reference
Cyclobutane	Pressure: 1.5×10^3–5×10^{-4} Torr; $k/k_\infty = 1.0$–$c.\ 0.2$ for several temperatures; the degree of fall-off lessens below 10^{-2} Torr for unknown reasons. E_a measured as a function of pressure.	The recent calculations give good agreement with experiment for both shape and pressure fit (to 10^{-2} Torr); the calculated change of E_a with pressure also agrees with experiment.	68 69 51
Cyclo-butane-d_8	Pressure: 100–5×10^{-3} Torr; $k_H/k_D = 1.4$–0.83.	The recent calculations for k_H/k_D give good agreement with experiment; a maximum in $E_a^D - E_a^H$ versus pressure was predicted which is consistent with the limited data.	70 51

molecules that are usually studied (see Table 1.1). In the past, most workers have used Lennard Jones collision diameters. However, good arguments now exist for using $S^2 = \sigma^2 \Omega^{(2.2)*}$ as the collision diameter for energy-transfer processes involving highly vibrationally excited molecules[35, 38–40]. The omega integral[41] is based upon ε values that are usually derived from viscosity measurements.

The evaluation of equation (1.3) is straightforward after the frequencies and moments of inertia of the activated complex are assigned, which is done so that agreement is obtained with the high-pressure pre-exponential factor. A recent development[42, 43] is a formulation which correctly includes centrifugal effects for reactions having transition states with moments of inertia much larger than the moments of inertia of the molecule, e.g. $C_2H_6 \rightarrow 2\,CH_3$. The centrifugal effect, which has the maximum value of $(I_A^\dagger I_B^\dagger I_C^\dagger / I_A I_B I_C)^{\frac{1}{2}}$ at high pressure, declines to 40–60 % of the high pressure at the low-pressure limit. For most unimolecular reactions $I_A^\dagger I_B^\dagger I_C^\dagger \approx I_A I_B I_C$ and the centrifugal effect can be ignored.

The general conclusions from some well-studied reactions, summarised in Table 1.1, are: (i) If all vibrational degrees are taken as active and if the vibrational frequencies of the transition state are adjusted so as to fit the entropy of activation, the calculated fall-off curves are in good agreement with reliable data. (ii) The calculated curves are quite insensitive to individual frequencies selected for the transition state and the fall-off data are not of much aid in choosing a model for the activated complex. (iii) As predicted by the theory, increasing the number of vibrational modes in a homologous series shifts the fall-off region to lower pressure. (iv) The absolute pressure fit has been the most troublesome feature. However, reasonably good agreement has been found for the isocyanide, cyclobutene (after accounting for the reverse reaction), cyclopropane, cyclobutane, and but-2-ene (after accounting for reverse reaction) isomerisation reactions. (v) Although the data are somewhat sketchy, the calculated variation of the activation energy with pressure agrees with experiment.

In addition to the mechanistic details provided by the equilibrium isotope effects, the non-equilibrium secondary statistical isotope-effect strongly affects fall-off curves. The ratio of the second-order rate-constants is:

$$\frac{k_0^H}{k_0^D} = \frac{k_M^H[M^H]\,Q_I^D \int_{E_{OH}}^{\infty} N(E)_H e^{-E/RT} dE}{k_M^D[M^D]\,Q_I^H \int_{E_{OD}}^{\infty} N(E)_D e^{-E/RT} dE} \tag{1.4}$$

The density of states, $N(E)$, for the deuterium-substituted compound is considerably larger than the corresponding term for the hydrogen-substituted compound. The exact magnitude depends upon E_0 and the number of deuterium atoms, but has as a maximum c. 2.3 ± 0.4 per deuterium atom[9]. The predicted inverse secondary isotope-effect, which has been quantitatively confirmed by studies for methyl isocyanide, ethyl isocyanide, cyclopropane and cyclobutane, conclusively showed that quantum statistics must be used to evaluate sums and densities of states.

1.2.1.2 Examination of some fall-off studies

The most definitive work has been with the isocyanide isomerisation reaction[44-48]. Except for a mild reservation[49] about a concurrent free-radical catalysed isomerisation, it can be concluded that studies of isocyanide isomerisation have illustrated more aspects of thermally activated unimolecular reactions than any other example. One of the many interesting features about methyl isocyanide is the better agreement between data and calculations for a model having energy exchange between one overall rotation (an active overall rotation) and the vibrations. This and the decomposition of the ethyl radical[9] are the only well documented examples which have an active overall rotation. The CH_3NC reaction also is one of the few examples for which inclusion of the effects of anharmonicity upon the density of states enhanced agreement between experiment and calculations. In fact, thermally activated unimolecular reaction rates in the low-pressure second-order region provide one of the few ways to measure consequences of anharmonicity in highly vibrationally excited polyatomic molecules. Good fit to the C_2H_5NC data was obtained without recourse to making a rotation active or to inclusion of anharmonicity effects. These two differences between CH_3NC and C_2H_5NC are consistent with the changes in moments of inertia of the complexes relative to the molecules (a factor of c. 3.6 and c. 1.2 for CH_3NC and C_2H_5NC, respectively) and with the lower average energy per mode for C_2H_5NC. This experimental work has stimulated a trajectory calculation[50] for the CH_3NC reaction. Although the final results[50] are not yet available, the mathematical model apparently does not show energy randomisation on a time scale that is sufficiently fast to make the statistical theory applicable to CH_3NC.

Experimental[56, 57] and theoretical[58] interest in the probable intermediate, the trimethylene diradical, for cyclopropane structural isomerisation continues. However except for an excellent summary paper[51], little recent work has been done on this historically important reaction. Good agreement exists between the RRKM calculations, which are independent of the various concerns about trimethylene, and the data for structural and geometric isomerisation[51]. Earlier calculations of Wieder and Marcus[59] for cyclopropane had left the false impression that the absolute pressure fit was poor for the cyclopropane reaction. The pressure fit for other members of the homologous series is rather poor, but the data are not of high quality. The calculations of Lin and Laidler[51] predicted a maximum in the activation energy difference for cyclopropane-d_0 and -d_6. This maximum, which should occur just before the high-pressure region is attained, is sufficiently large (c. 0.6 kcal mol^{-1}) that significant error could arise in using this as the supposed equilibrium isotope-effect to assign structural changes from zero-point-energy considerations. Similar maxima have been calculated for cyclobutane-h_6 and -d_6[51] and for chloroethane-h_5 and -d_5[60]; these predicted maxima merit experimental verification. The variation in $E_a^D - D_a^H$ at relatively high pressures is related to a more general problem. Since both Arrhenius parameters decline with pressure, a unimolecular reaction can be in the fall-off region in regard to either parameter separately before the effect will be noticed upon the rate constant[61]. Before leaving cyclopropanes, the isomerisation (by chlorine-atom migration) of 1,1-dichlorocyclopropane should be noted[62]. The poor agreement with calculations probably arises from experimental complications.

When account[64] was taken of the reverse reaction, the former discrepancy[65] in the pressure fit of the fall-off curves of cyclobutene was resolved. Consideration of the reverse reaction and use of slightly different Arrhenius factors also improved the agreement between experimental[67] and calculated[64] results for cis-trans isomerisation of but-2-ene. Good agreement exists between the calculated and experimentally measured fall-off, change in E_a with pressure, and isotope effects for cyclobutane. Unfortunately the fall-off data, but not the Arrhenius parameters, for HCl elimination from chloroethane are not in agreement from independent studies[71, 72], and fall-off data cannot provide a test of theory[74, 75]. The fall-off region[76] for 1,2-dichloropropane[77] comes at a lower pressure than for chloroethane and argues against an earlier suggestion that some vibrational modes of t-butyl chloride[78] were not active.

The C_2H_5 and C_2H_6 reactions are considered in Sections 1.2.2.4 and 1.4.1.1.

From the extensive investigations listed in Table 1.1, it seems unlikely that additional fall-off studies will contribute in any decisive way to our understanding of unimolecular reactions. This is not to say that equilibrium Arrhenius parameters, isotope effects, and sufficient fall-off data to show that a reaction is truly unimolecular are not useful. Studies in the second-order region using a variety of bath gases can contribute toward an understanding of intermolecular energy transfer[31, 37]. A possibility for such studies may be the decomposition of tetrafluoroethylene oxide[79].

1.2.2   Chemical activation yielding products of known energy

1.2.2.1   Competition between unimolecular reaction and collisional stabilisation

Chemical activation denotes the use of exothermic chemical reactions to generate product molecules with selected, usually high, energy. Only examples giving ground-electronic-state products with known internal energy distributions, $f(E)dE$, will be discussed in this Section, which reduces the scope to bimolecular addition or combination reactions and to unimolecular isomerisation reactions. The latter are difficult to initiate, and they have not been very useful. The general bimolecular case,

$$R_1 + R_2 \rightarrow R_1R_2^*; \Delta H_0^0 = D_0^0(R_1 - R_2),$$

has an internal energy distribution

$$f(E)dE = k_E' F(E)dE / \int_{E_{min}}^{\infty} k_E' F(E)dE, \quad (1.5)$$

$F(E)dE$ is the Boltzmann distribution for $R_1R_2^*$ at the temperature of the experiment and k_E' is the rate constant for dissociation of $R_1R_2^*$; an example is shown in Figure 1.1. The thermochemical values, see Figure 1.2, are very important, and E_{min} is equal to $D_0^0(R_1 - R_2)$ plus the threshold energy, E_0', of the bimolecular step. Variation of temperature can be used to effect small changes in the average thermal energy[80, 35] of $R_1R_2^*$, see Figure 1.1. Reaction possibilities for the activated molecule are:

$$R_1R_2^* \xrightarrow{\quad\quad} \text{unimolecular reaction} \quad (D)$$

$$\xrightarrow{\quad k_M[M] \quad} \text{collisional deactivation} \quad (S)$$

$$\xrightarrow{\quad \tau^{-1} \quad} \text{infrared emission}$$

$$\xrightarrow{\quad k_Q[Q] \quad} \text{bimolecular reaction}$$

The unimolecular channel may be the reverse of the formation reaction or a different reaction (or reactions) with a lower threshold energy. Study of the last two steps may become more frequent in the future[81, 82], but to date most

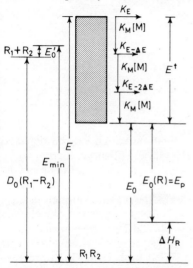

Figure 1.2 Summary of thermochemical terms for chemical activation. The scheme applies to the bimolecular combination of R_1 and R_2 with a slight activation energy, E_0', yielding $R_1R_2^*$, which subsequently undergoes a unimolecular reaction with threshold energy, E_0. The energy, E, of $R_1R_2^*$ is shown as some arbitrary level in the distribution function of formed molecules. The collisional deactivation scheme is a simple stepladder cascade, with an average loss of $\langle \Delta E \rangle$ per collision.

effort has been expended on the first two processes. Experimental observations are D/S ratios as a function of pressure, and some sample data[12] for $CH_3CF_3^*$ are shown in Figure 1.3. Rather than measuring both D and S, an internal standard technique[83, 84] can be used which involves comparing either the D or S yield to the yield of some product which is invariant with pressure.

For comparison with calculations, the approximate mechanism for deactivation must be known. If $E_{min} - E_0$ is not too large and if an efficient polyatomic molecule is the bath gas, the unit deactivation formulation is satisfactory.

$$k_a = k_M[M][D/S] = k_M[M] \cdot \frac{\int_{E_{min}}^{\infty} \frac{k_E}{k'_E + k_E + k_M[M]} f(E)dE}{\int_{E_{min}}^{\infty} \frac{k_M[M]}{k'_E + k_E + k_M[M]} f(E)dE} \quad (1.6)$$

If the only unimolecular channel is the reverse reaction, k'_E is substituted for k_E in the numerator and k_E is dropped from the denominator of (1.6). If another channel exists, k_E is usually, but not always, much larger than k'_E, and the latter does not need to be considered. Depending upon the breadth of $f(E)\mathrm{d}E$ and the difference in $E_{\min} - E_0$, the values of k_a can be somewhat pressure dependent and may decline with decreasing pressure[80]. If unit

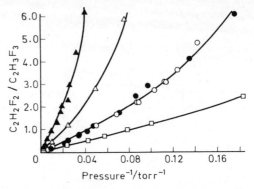

Figure 1.3 Chemical-activation data for $CF_3CH_3^*$ in various bath gases. The lines are calculated results for the following cascade deactivation models: H_2, ▲, exponential with $\langle \Delta E \rangle \approx 1.2$ kcal; CH_3Cl, △, stepladder with $\langle \Delta E \rangle \approx 4$ kcal; $CF_3N_2CH_3$, ●, stepladder with $\langle \Delta E \rangle \approx 8$ kcal; C_2F_6, ○, stepladder with $\langle \Delta E \rangle \approx 7$ kcal; $n\text{-}C_6F_{14}$, □, stepladder with $\langle \Delta E \rangle \approx 10$ kcal.

deactivation is not followed, cascade deactivation, which can be detected from (i) upward curvature in D/S versus $1/P$ plots, and (ii) larger high pressure limiting slopes (see Figure 1.3), must be considered. The treatment can be simple (the stepladder model of Figure 1.2) or more complex[31, 32], as the situation warrants. Complicated (i.e. non-stepladder) cascade models usually are needed only for atomic or diatomic bath gases[32–39]. If the objective is study of unimolecular reactions, a bath gas should be chosen which has a stepladder mechanism or preferably even reduces to unit deactivation. For stepladder cascade with $\langle \Delta E \rangle = 5, 10, 15$ kcal mol^{-1}, the high-pressure rate-constants for $CH_3CF_3^*$ are larger than the unit deactivation rate-constant only by factors of 1.58, 1.14, and 1.03, respectively. Finally a collision diameter must be selected to calculated k_M. As mentioned in Section 1.2.1.1, present evidence[38–40] favours $S = [\sigma^2 \Omega^{(2, 2)*}]^{\frac{1}{2}}$ as the best collision diameter.

For chemical activation having the reverse reaction as the unimolecular step, the experimental measurements are related to fall-off curves of thermal activation.

$$R_1 + R_2 \underset{k'_E}{\overset{k_{bi}}{\rightleftharpoons}} R_1 R_2^* \xrightarrow{k_M[M]} R_1 R_2$$

Measurement of $-\mathrm{d}[R_1]/\mathrm{d}t$ in excess R_2 is often made in discharge-flow experiments.

$$-\frac{\mathrm{d}[R_1]}{\mathrm{d}t} = k_{\exp}[R_1][R_2] = k_{bi}[R_1][R_2] \int_{E_{\min}}^{\infty} \frac{k_M[M]}{k'_E + k_M[M]} f(E)\mathrm{d}E \tag{1.7}$$

At high pressure the bimolecular process is rate limiting and k_{exp} reduces to the equilibrium bimolecular rate-constant, k_{bi}. Substituting for the distribution function yields

$$k_{exp} = k_{bi} \int_{E_{min}}^{\infty} \frac{k_M[M]}{k'_E + k_M[M]} k'_E F(E)dE \Big/ \int_{E_{min}}^{\infty} k'_E F(E)dE \qquad (1.8)$$

The part of the numerator under the integral sign is just k_{uni}, at the temperature of the flow-system experiment, and the denominator is k_{uni}^{∞}. Thus, a plot of log (k_{exp}/k_{bi}) versus log P is the same as a fall-off plot in thermal activation. The low-pressure form is applicable to third-order combination reactions studied in flow systems[85].

$$\lim_{P \to 0} k_{exp} = k_{bi} k_M[M] \int_{E_{min}}^{\infty} \frac{f(E)dE}{k'_E} = k_{bi} k_M[M] \langle 1/k'_E \rangle \qquad (1.9)$$

In many experiments the carrier gas is either He or Ar, and a simple collisional inefficiency factor often is used. However, deactivation by He and Ar follows an exponential type distribution of transition probabilities[31, 37, 39] and a more complete deactivation formulation eventually should be considered[85].

1.2.2.2 Energy partitioning by non-equilibrium unimolecular reactions

Occasionally the energy released to the products can be measured, and such data provide another measure of theory. According to the basic assumption of any statistical theory, the non-fixed energy, $E - E_0 = E^\dagger$, should be released randomly to the degrees of freedom of the products[86-88]. However, the potential energy released after passing through the configuration corresponding to the transition state is released to the products according to the characteristics of the multidimensional potential surface describing the reaction. If $E_0 \sim \Delta H_R^0$, the experimental energy partitioning measurements[89, 90] can be compared to the predictions of the statistical theory. On the other hand if $E^\dagger$ is small and E_p is large ($\Delta H_R^0 < E_0$), the energy partitioning data serve to characterise features of the potential-energy surface in the same way as for bimolecular reactions[91-94].

The statistical partitioning of the non-fixed energy to relative translation and to internal energy of products by a decomposition reaction ($R_1 R_2^* \to R_3 + R_4$) can be calculated from the following assumptions[86]: (i) The total energy of the molecule and, therefore, $E^\dagger$ is randomised in all internal modes before decomposition. (ii) During the course of the decomposition, one vibrational mode disappears and becomes the relative translation along the reaction coordinate[88]. (iii) Five other vibrational modes (if R_3 and R_4 are polyatomics) and the three overall rotational modes become rotations and perpendicular relative translations of the products. The relative kinetic energy along the line of centres[88] follows from (ii) and has a simple form.

$$f(E_t)dE_t = N^\dagger (E^\dagger - E_t) / \sum_{E_v^\dagger \leqslant E^\dagger} P(E_v^\dagger) \qquad (1.10)$$

The denominator is the sum of energy eigen-states of the transition state and

$N^\dagger(E^\dagger - E_t)$ is the density of states in the complex with internal energy, $E^\dagger - E_t$. The internal energy distribution for either fragment has a similar form,

$$f(E_3)dE_3 = N^\dagger(E_3)\Big[\sum_{E_v^\dagger \leqslant E' - E_3} P(E_v^\dagger)_R\Big]\, dE_3 / \sum_{E_v^\dagger \leqslant E'} P(E_v^\dagger) \tag{1.11}$$

The density of states for the vibrational modes associated with R_3 at energy E_3 is $N^\dagger(E_3^\dagger)$; $P(E_v^\dagger)_R$ is the degeneracy of states for the remaining modes of the transition state and $P(E_v^\dagger)$ is the total degeneracy of internal states for the entire activated complex. In order to obtain the rotational energy distributions for R_3 and R_4, explicit conservation of angular momentum must be incorporated into the model[90, 95].

1.2.2.3 Activation by bimolecular combination of radicals

The combination of two doublet free radicals, R_1 and R_2, to form a stable molecular adduct, $R_1R_2^*$, releases a sizeable (50–110 kcal mol^{-1}) quantity of internal energy. Often unimolecular reaction channels with lower threshold than the reverse reaction are available, and such cases are virtually ideal for chemical activation studies.

(a) *Activation of fluoro- chloro- and bromo-alkanes* – In addition to the convenience of activation, the HX(X = F, Cl and Br, but not I because C—I bond rupture is favoured rather than HI elimination[96]) elimination reactions are a good series because threshold energies, pre-exponential factors, equilibrium isotope effects, and the molecular structure of the molecule are known for many halo-alkanes.

The chemical activation studies have been done primarily by three groups, the author's at Kansas State University[12, 35, 61, 70, 74, 97–99], Professor Pritchard's at the University of California (Santa Barbara)[100–102] and Professor Trotman-Dickenson's at Aberystwyth University, Wales[103–109]. Early studies from the other two laboratories used classical RRK theory; the threshold energies based upon those interpretations are not reliable[35] and both groups[100, 108] have revised their earlier conclusions. When the programme was initiated at Kansas State University, chemical activation data[9] indicated that randomisation of energy was sufficiently rapid that RRKM rate constants usually fitted the available data. Since the thermochemistry and molecular structure of the halo-alkanes were known, the objective was to obtain a sufficiently large body of experimental data so that the absolute rate theory features of the RRKM formulation could be more severely tested. Table 1.2 contains a summary of chemical activation results; agreement between calculations and experiment is always within experimental error. As can be seen from Figure 1.1, variation in E_0 by c. 2 kcal mol^{-1} and in $\langle E \rangle$ by 5 kcal mol^{-1} changes k_E by a factor of c. 2. In very few, if any, cases are both E_0 and $\langle E \rangle$ sufficiently reliable to fix k_E better than a factor of 2. Nevertheless, the average energies, the threshold energies, the pre-exponential factors, the molecular structure of the molecules, and the chemical activation rate constants are known better for the HX elimination reactions than for any other set of unimolecular reactions.

The overall results of Table 1.2 strongly support the claim that RRKM calculations give reliable k_E values (within a factor of 2) up to energies such that k_E approaches 10^{12} s^{-1} when the calculations are based upon threshold energies and transition-state models which match the thermal pre-exponential factor. The explicit energy dependence of k_E for the halo-alkanes satisfied the following tests: (i) the variation of k_a with temperature for several halo-alkanes[35]; (ii) the change in k_a for activation of $C_2H_5F^*$ and $C_3H_7F^*$ by radical combination versus methylene insertion[108, 109].

Isotopic labelling studies have shown that, except for molecules[99, 102] containing two halogen atoms on the same carbon, HX elimination from halo-alkanes strictly follows α,β-elimination[60, 73]. A systematic procedure[60, 61, 98] was developed for selecting common transition-state models for the α,β-elimination transition-states by assigning bond orders to the C—H, C—C, H—X and C—X bonds. The vibrational frequencies associated with the atoms attached to the corners of the ring were assigned by analogy with olefins, cyclobutanes and the parent molecule. The remaining frequency, a ring-puckering mode, was adjusted to simultaneously fit all experimental data. The empirical model has five variables (four bond orders and the ring puckering frequency). A unique description was not forced by the data; however, the hydrogen atom must be rather weakly bound and the halogen atom must be rather strongly bound in order to fit the isotope effects and the pre-exponential factor, respectively. In both of these respects the model proposed by Benson and co-workers[14, 18] is deficient. A persistent problem was that the calculated pre-exponential factors tended to be larger and the calculated chemical activation rate-constants to be smaller than the experimental values. Inclusion of anharmonicity for computation of *both* densities and sums[60] increased the disparity between the calculated and experimental values of k_a.

The available energy for partitioning[60, 91] by the CF_3CH_3 reaction is about equally divided between non-fixed and potential energy. The HF* contained[91] c. 13% of the total energy, which is slightly more than a statistical fraction, and the vibrational populations were $N_0:N_1:N_2:N_3:N_4 = 1.0:0.75:0.32:0.10:0.03$. Although vibrational relaxation of the HF* cannot be totally discounted, the results indicate that surprisingly little energy was released to HF. Since CH_2CF_2 from $CH_3CF_3^*$ or C_2H_3F from 1,2-$C_2H_4F_2^*$ contain insufficient energy for further HF elimination[60], their internal energies must[114] be $\leqslant 70$ kcal mol^{-1}. For statistical release of $E^\dagger$ and for release of E_p according to a Gaussian distribution, the centre of the Gaussian must be $\leqslant \frac{2}{3}$ of E_p. The pattern of energy partitioning suggests that a 1.9 bond order is too high for the C—C bond and that the H—X bond order may be somewhat larger than c. 0.1. However, the isotope effects are not consistent with bond orders larger than c. 0.5 for H—X and C—H. The well known polar-effects[16] demonstrated by the lowering of E_0 with α-methyl substitution and the predictions of orbital symmetry considerations have not yet been included in this dynamical model.

(b) *Some other examples* — Combination of CH_3 and NF_2 generates CH_3 NF_2^* which eliminates HF. Analysis[115] by RRKM theory gave $E_0 = 42$ kcal mol^{-1} for an excitation energy of c. 74 kcal mol^{-1}. Laser emission[116] from HF $(N_1/N_0 = 0.33 - 0.37)$ has been observed but the emission may result from

Table 1.2 Chemically activated rate-constants for halo-alkanes*

Molecule	E_0†/kcal mol⁻¹	Pre-exponential factor §/s⁻¹	$\langle E \rangle$‡ /kcal mol⁻¹	Rate-constants‡ Torr	s⁻¹§	Reference†
CH_3CH_2Cl	55.0 ± 0.5	1.2×10^{13}		320 ± 60	3.4×10^9	60
	55.0	1.6×10^{13}	91.0		2.4×10^9	60
CD_3CH_2Cl	56.7 ± 0.5	1.9×10^{13}		140 ± 30	1.6×10^9	60
	56.4	1.9×10^{13}	92.0		1.2×10^9	60
CD_3CD_2Cl	56.9 ± 0.5	1.9×10^{13}		120 ± 20	1.2×10^9	60
	56.4	2.0×10^{13}	92.5		0.81×10^9	60
CH_2ClCH_2Cl	55	2.2×10^{13}	88.3	17 ± 1	1.6×10^8	74, 98
	60	2.2×10^{13}	88.3		9.6×10^8	74
CD_2ClCD_2Cl	56.3	2.5×10^{13}	89.3	5.1 ± 0.5	2.0×10^8	98
	61.3	2.5×10^{13}	89.3		5.0×10^7	74
CH_3CHCl_2	52 ± 1.0	1.3×10^{13}		c. 1000‖	3.4×10^8	74
	52	2.3×10^{13}	92.3		5.8×10^7	39
$CH_2ClCHCl_2$	53.0 ± 1.2	3.2×10^{13}		c. 15	c. 1.2×10^{10}	97
$n\text{-}C_3H_7Cl$	53.0	1.5×10^{13}		6 ± 2	c. $6—10^9$	‖
					c. 1.4×10^8	99‖
$1,3\text{-}C_3H_6Cl_2$	55.0	2.8×10^{13}	89.5	1.5 ± 0.3	7.1×10^7	98
					6.2×10^7	98
					1.7×10^7	98
$1,4\text{-}C_4H_8Cl_2$	53.0	3.2×10^{13}	88.4	0.4 ± 0.1	2.9×10^7	98
					4.5×10^6	98
					2.5×10^6	98
CH_3CH_2Br	52.0 ± 1.0	1.1×10^{13}	88.0	800 ± 200	6.1×10^9	61
	52.0	1.6×10^{13}			5.8×10^9	61
CH_2BrCH_2Br	52	1.7×10^{13}	91.4	220 ± 40	1.7×10^9	61
CH_2BrCH_2Cl	54(HBr)	1.1×10^{13}	88.5	3.8 ± 0.9	1.5×10^9	61
					0.3×10^9	61
CH_2BrCH_2Cl	57(HCl)	1.0×10^{13}	88.4	1.5 ± 1.4	0.5×10^9	61
			88.4		0.12×10^9	61
					0.2×10^9	61
C_2H_5F	57.9 ± 1.0	1.0×10^{13}	91.5	170 ± 10	1.4×10^9	(110), 12
	57	1.6×10^{13}			1.1×10^9	35, 12

Table 1.2 – continued

Molecule	$E_0^{\ddagger}$/kcal mol⁻¹	Pre-exponential factor §/s⁻¹	$\langle E\rangle^{\ddagger}$/kcal mol⁻¹	Rate-constants‡ Torr	s⁻¹§	Reference†
C₂H₅F	57		90	140	1×10^{9}	108, 103
	CH₂ + CH₃F → C₂H₅F*					
C₂H₅F		2.2×10^{13}	c. 106**	2200	1.2×10^{9}	108
CH₂FCH₂F			c. 112**		1.6×10^{10}	109
				20 ± 1	$1.6{-}10^{10}$	35, 108
					1.8×10^{8}	35
					1.2×10^{8}	35
CH₃CHF₂	$\overline{62}$	$\overline{2.0 \times 10^{13}}$	92.5	200‖	2.0×10^{9} ⎫	(112–113) 107
	{60.5 ± 1.8	3×10^{13}		70‖	0.7×10^{9} ⎬	100
	57	3.1×10^{13}	94.1		2.5×10^{9}	35
	α,α-HF elimination/α,β-DF elimination					
CD₃CHF₂	{69.6 ± 1.5	1.2×10^{13}		c. 0.4	50(470 K)	99, 101
CH₃CF₃	67.2 ± 2.4	1.5×10^{13}		29(425 K)	$1.6{-}2.7 \times 10^{8}$	(111), 100
	68	4.2×10^{13}	102.4¶	12–20	3.4×10^{8}	(112b) 12
						12
n-C₃H₇F	56.8 ± 1.0	0.8×10^{13}	87.0	1.0	9×10^{6}	109 (110)
	57.0				4.3×10^{6}	35
	CH₂ + CH₃CH₂F → CH₃CH₂CH₂F					
n-C₃H₇F		2.0×10^{13}	c. 106**	c. 20	c. 1.8×10^{8}	109
			c. 108		1.8×10^{8}	35

Chemical-activation rate-constants for the more highly fluorinated ethanes[100, 104] decline strongly with increasing numbers of F atoms; thermal-activation measurements and RRKM calculations have not been done. Conflicting views[101, 105] exist concerning competitive intramolecular elimination of HCl and HF from chlorofluoroethanes, according to theory $k_t(\text{HCl}) > k_t(\text{HF})$.

*Experimental values are on the first line and RRKM calculated values on the following line or lines. Except for the CH_2 insertion cases, all molecules were activated by radical combination at room temperature; the thermochemistry was obtained from tables of reference 18.

†The experimental Arrhenius parameters were converted to threshold energies and partition function pre-exponential-factors. If the thermal pyrolysis data were reviewed in the paper reporting the chemical-activation results, no reference is given to the thermal pyrolysis data. Most of the experimental pre-exponential factors have uncertainties of c. 2.

‡At room temperature unless specified otherwise; $\langle E\rangle$ is the average energy of the formed molecules. The 'rate-constant' in Torr is the pressure required for half-quenching with an efficient bath gas. For some cases several measurements have been made and a weighted average was chosen for the chemical-activation rate-constant.

§Both the experimental and calculated rate-constants in s⁻¹ are for unit deactivation. The experimental measurements usually have been done in gases which fit an 8 ± 2 kcal mol⁻¹ stepladder cascade deactivation model. For such deactivation models minor correction, about 0.8, gives the unit deactivation rate-constant. For the most part investigators have used Lennard Jones collision diameters to obtain rate-constants in s⁻¹.

‖The experimental rate-constant includes both 1,1- and 1,2-elimination. Earlier values[97] for the HCl-elimination rate-constants from $CH_2ClCHCl_2$ were in error[99].

¶The high $D(CH_3{-}CF_3)$ value is somewhat suspect; however, for this high value E_0 also must be high and shock tube[111–113] studies support the tabulated value. If future work shows $D(CH_3{-}CF_3)$ to be too high, the E_0 value also must be lowered.

**The energy on the basis of thermochemical analysis is 106 kcal; the energy needed to give the listed RRKM rate-constant is 112 kcal mol⁻¹ for C₂H₅F and 108 kcal mol⁻¹ for C₃H₇F.

a second step since the stable products are HCN and 2HF. Chemically activated methylcyclopropane has been generated by combination of cyclopropyl and methyl radicals[117] in a He bath gas. An unusual reaction[118,119],

$$CF_3CH_2SiCX_3^* \longrightarrow CF_2CH_2 + SiFX_3$$

follows combination of CF_3 with $CH_2SiCX_3(X = Cl, F)$. The dissociation of hydrazine has been treated by use of an improved transition state[120]; unfortunately the threshold energy used in that calculation may need revision upwards[121]. Little experimental work has been done for chemically activated RHN—NH$_2$ molecules. Unimolecular reactions of chemically activated alkanes are mentioned in Section 1.4.1.1.

1.2.2.4 Activation by bimolecular addition and insertion reactions of atoms

Studies of chemically activated ethyl and sec-butyl radicals[9] laid the foundations for the chemical-activation technique and addition of atoms, of various multiplicities, to unsaturated molecules give chemically activated radicals of almost endless variety. Singlet $O(^1D)$ and $S(^1D)$ atoms, which insert into C—H bonds, also provide activated molecules. Reactions of $CH_2(^1A_1)$ were important in establishing the chemical-activation technique; however, $CH_2(^1A_1)$ or other carbenes often have non-equilibrium internal energy distributions and are discussed in Section 1.2.3.3.

(a) *Hydrogen atoms + olefins* — An interesting discovery, although adding chemical complications, is the isomerisation of chemically activated alkyl radicals by 5-, 6- and 7-membered cyclic transition states[25,26]. The chemical-activated isomerisation rate-constants can be fitted by models having reasonable E_0 and pre-exponential factors. These isomerisation reactions can be viewed as either unimolecular isomerisations or as bimolecular abstractions of highly vibrationally excited radicals[26]. Isomerisation of n-pentyl and n-hexyl from concurrent chemical activation (the radicals were formed from $n-C_3H_7$ or $n-C_4H_9 + C_2H_4$) and thermal activation may explain the anomalously low pre-exponential factors previously reported from thermal-activation studies[25]. Allowance for isomerisation for sec-octyl and sec-hexyl enhanced the agreement between the RRKM calculations and previously measured decomposition rate-constants[26].

A comprehensive investigation of the intramolecular competitive unimolecular reactions of large alkyl radicals of the following type (R = methyl, ethyl, n-propyl, i-propyl, neopentyl and t-butyl) has been completed[122–126].

Other radicals, e.g. 2-methylhexyl-3, heptyl-3, etc., having several competitive reaction channels which do not fit into the generalised scheme were investigated also. Absolute rate-constants were estimated, and relative rate-constants were accurately measured for these radicals which have $\gtrsim 40$ kcal mol^{-1} of energy. Formation of the *trans*-olefin was favoured over formation of the *cis*-isomer, and the channel giving an R group (especially i-propyl and t-butyl) dominated the methyl channel. The ratio of chemical activation rate-constants at high pressure depends only upon the structure assigned to the transition states and the difference in threshold energies of the competing reaction channels. Results from previously measured absolute reaction rates of alkyl radicals[9] were used to calibrate a model for the transition state. Since the rate-constant ratio is quite sensitive to the difference in critical energy, typically a difference of 1.5 kcal mol^{-1} gives a ratio of c. 3, the ratios were used to assign a difference in critical energies of competitive reaction channels. This assigned difference in critical energy was reduced to the least accurately known thermochemical values, which were the differences in threshold energies for addition of radicals to the olefins and the bond dissociation energies, $D(CH_3—H)$ and $D(R—H)$. The bond energies that best fitted all the experimental data were the same as those commonly accepted in the literature. As direct measurements of the threshold energies for radical addition to the olefins become available, comparison with those tabulated[122-126] should prove interesting. The activation energy difference for addition of CH_3 to *trans*- and *cis*-butene was measured[127] as 0.6 kcal; whereas the chemical activation result is -0.05 kcal. However, the rate ratio for addition of CH_3 to *trans* and *cis*-but-2-ene agreed with the direct measurement[127]. In summary, the consistency between calculations and experiment for these complex alkyl radicals is amazingly good. Although these large radicals have only c. 40 kcal mol^{-1} of energy, typical values of rate-constants are $10^5 - 10^6$ s^{-1}, and the lifetimes are sufficiently long for the statistical theory to be appropriate.

Although the internal consistency obtained for the large alkyl radicals is impressive, more quantitative testing of theory will be found with smaller species having fewer parameters. The first member of the alkyl radical series, ethyl radical, has received much attention in thermal[128, 129] and chemical activation[130-133] (rate of H atom removal by C_2H_4 as a function of pressure) studies. The data provide fall-off curves at 800–900 K and at 300 K. In accordance with early calculations[134], ethyl radical decomposition at high temperature was found to be in the fall-off region at most experimentally convenient pressures. Lin and Laidler[129] recalculated k_E and the high temperature fall-off curves for $C_2H_5^*$, as well as doing new calculations for $CH_3OCH_2^*$. They favoured an active overall rotation for $C_2H_5^*$ as suggested by Rabinovitch and co-workers[9]. Lin and Laidler's model, which gives higher values for k_E than the earlier model[9, 134], fitted the high-temperature experimental fall-off rather well. However, a collision diameter of 0.275 nm was used which, together with the higher k_E values, displaced the fall-off curve to higher pressure by a factor of eight relative to Setser and Rabinovitch's curve[134] (S = 0.495 nm). The collision diameter for C_2H_5 with C_2H_6 surely is larger than 0.275 nm and it seems premature to use ethyl-radical decomposition as a textbook[2] example illustrating RRKM fall-off behaviour.

Plots of k_{exp}^{-1} (see Section 1.2.2.1) versus (pressure)$^{-1}$ for $H + C_2H_4$ have considerable curvature[132] in agreement with theoretical predictions[134, 135]. Various workers[130a, 132, 133] now seem to agree on the high pressure rate-constant, and a value near[132] 13.6×10^{-13} cm^3 mol^{-1} s^{-1}, is favoured. This high value of k_{bi}, when combined with the equilibrium constant, provides better agreement with the unimolecular models[129, 134] than do lower values. For the higher k_{bi} value, the decline in k_{ex} with pressure requires an unusually large inefficiency factor for He in order to fit the theoretical unit deactivation curve[132]. The larger k_E values[129] would give a fit with somewhat more conventional efficiency for He. Since He deactivation follows an exponential transition-probability model, the inefficiency factor may depend on temperature and pressure[85]. As expected, reverse dissociation following terminal addition of H to propene[136] is shifted to much lower pressure than for H + ethene.

Arrhenius energies and pre-exponential factors based upon thermal activation data for alkyl radicals were summarised by Frey and Walsh[15]. New data[137] as well as the tabulation[15] indicate that the transition-state model previously used for n-propyl[9] could be improved by lowering some frequencies. Since equilibrium constants are accurately known for alkyl radicals, *all reliable kinetic and thermodynamic data for both forward and reverse steps should be used when formulating transition-state models.* Relatively good agreement between the pre-exponential factor from the unimolecular model developed from chemical activation data[9] and the product of $K_{eq}k_{bi}$ [127] is obtained for butyl radical.

(b) *Halogen atoms + olefins* – Addition of chlorine atoms to olefins, which has virtually zero activation energy, generates radicals with $c.$ 25 kcal mol^{-1} of energy. The entire series (C$_2$H$_4$Cl—C$_2$Cl$_5$) has been studied[138, 139]; the bimolecular rate constants are all similar, except for C$_2$Cl$_4$. Although theory predicts a tenfold variation in k_a from high to low pressure, experimental difficulties have prevented measurement over a wide range of pressure. A possibility for obtaining more data, which avoids the wide variation in k_a with pressure, is activation by H-atom addition to chlorinated ethenes; the rate-constants[140] should be 10^{10}–10^{11} s^{-1}. The model for the transition states was treated as a Cl atom weakly bound to the olefin. The C—Cl bending frequencies were set at $\frac{1}{8}$ of the values in the radical; the C—Cl stretching coordinate was taken as the reaction coordinate; one overall rotation was taken as active (by analogy with C$_2$H$_5$)[141]; and the motion about the C—C axis was treated as a free rotation in the radical and as a torsion in the activated complex. Agreement between experiment and calculations was poor for both the bimolecular and the non-equilibrium unimolecular rate-constants. Use of a transition state[142] which has a free internal rotation and high C–Cl bending frequencies is not consistent with the zero activation energy for addition of Cl to the double bond. The reactions of chloroethyl radicals present problems that are common for loose transition states; the magnitude of k_E is sensitive to the values of the low bending frequencies and angular momentum considerations and centrifugal effects may be important.

The addition of F atoms to 2,3-dichloroperfluorobutene gives sec-2,3-dichloroperfluorobutyl radicals with $c.$ 54 kcal mol^{-1} of energy which decompose by elimination of a Cl atom. After estimation of the thermo-

and experiment was obtained[143]. The decomposition rate of chemically activated sec-1,4-dichloroperfluorobutyl was 1.3 times faster than for the sec-2,3-dichlorobutyl radical. These data provide strong confirmation of the basic mechanism as well as giving additional evidence for rapid randomisation of energy (within 10^{-9} s) since activation was at the same location for both radicals but the site for Cl rupture was different. The addition of fluorine atoms to vinyl fluoride gives activated CH_2CHF_2 radicals that subsequently decompose to H and CH_2CF_2 [144].

(c) *Hydrogen atoms + acetylenes* — A complete fall-off curve at room temperature, partial curves at other temperatures, and Arrhenius parameters have been obtained[145] for the $H + C_2H_2$ reaction. The fall-off appears to be displaced slightly to lower pressure relative to that from $H + C_2H_4$ [130b]. The mechanism[146] and the isotope effects[146] seem well established and the unimolecular reaction of vinyl radical merits a careful RRKM analysis. Addition of H atoms to alkylacetylenes produces activated RHCCH* radicals which decompose with formation of $R + C_2H_2$ [147]. Isomerisation of chemically activated alkenyl radicals has been demonstrated by the 1,4-hydrogen-atom migration following addition of i-propyl to C_2H_2 [148].

(d) *Insertion and addition reactions of* $O(^1D_2)$ *and* $S(^1D_2)$ *atoms* — Insertion by $O(^1D_2)$ or $S(^1D_2)$ into C—H bonds of alkanes release *c.* 140 and *c.* 83 kcal mol^{-1}, respectively, and subsequent unimolecular decomposition of the activated molecules and laser emission from products[250] has been observed. Activation by $S(^1D)$ has been restricted[149] to CH_3SH* since the energy is only slightly in excess of threshold. With an internal standard technique, the rate-constants of neopentanol, isobutanol and propanol (formed by $O(^1D) +$ alkane) were measured[150-152] as 2.4×10^8, 2.2×10^9 and *c.* 10^{11} s^{-1}. The reaction channels of these activated alcohols have been difficult to identify[150-155]. The addition of $O(^1D_2)$ to olefins also provides high energy (*c.* 130 kcal mol^{-1}) activation system.

1.2.3 Chemical activation yielding products with unspecified energy

1.2.3.1 Energy assignments from measurements of rate-constants

If reactants do not have Boltzmann energy distributions or if two products are formed by the activating reaction, the internal energy of the products cannot be specified. However, if a product species undergoes a unimolecular reaction having calibrated rate-constants, the experimental rate-constant can be used to assign its internal energy. For best results the non-equilibrium rate-constant should be measured over a wide pressure range with an *efficient deactivating gas*. The high pressure value is the average rate constant for that particular distribution function and the variation of the rate-constant with pressure depends upon the shape of the distribution. Although the variation of the rate-constant with pressure is not highly sensitive to the exact shape of the distribution function, the mean energy and the approximate width can be ascertained.

1.2.3.2 Activation by hot-atom displacement and addition reactions

This topic is covered more fully in Chapter 4; examples were chosen here to with large amounts of translational energy may displace either H or F from CF_3CH_3 [27] and either add to the double bond or displace F from C_2F_4 [156]. All of these reactions deposit large amounts of internal energy in the products, which subsequently relax by unimolecular reaction or collisional stabilisation. More than 89% of the $CF_2{=}CF^{18}F^*$ molecules had sufficient ($\geqslant 76$ kcal mol^{-1}) energy for decomposition to $CF_2 + CF^{18}F$. The favoured reaction channel for $C_2^{18}FF_4^*$ radicals was decomposition to $CF_2^{18}F + CF_2$ rather than reverse dissociation[156]. The ^{18}F for F reaction with CF_3CH_3 was studied from a few Torr to c. 40 atm [27]. At $D/S = 1$ the rate-constant was 2.5×10^{11} s^{-1}, which corresponds to c. 180 kcal mol^{-1} of energy. The D/S versus $1/P$ plot had a great deal of curvature from the wide energy distribution[27]. The ^{18}F for H-displacement with CH_3CF_3 gave $CH_2^{18}FCF_3^*$, which decomposed by HF elimination and by dissociation to radicals[27]. According to predictions of the RRKM theory, the dissociation channel should be competitive with HF elimination at high energy (see Figure 1.1). Since dissociation of $CH_3CF_2^{18}F$ was not observed, failure of the energy randomisation hypothesis was suggested[27] for $CH_3CF_2^{18}F^*$, but not for $CH_2^{18}FCF_3^*$. Based upon angular momentum considerations. Bunker[251] has provided an intriguing tentative explanation of these observations. Since the energy range encompasses electronically excited states and since the total chemistry of the system is complicated, some reservation should be maintained. Nevertheless, the work[27] illustrates the potential of hot-atom displacement activation for gaining information about intramolecular relaxation of vibrational energy. Cases having several unimolecular reaction channels seem especially promising.

The average energy of molecules formed by T for H displacement in CF_3CH_3 was estimated[157] as c. 105 kcal mol^{-1} on the basis of the experimental HF-elimination rate-constant, c. 4.5×10^8 s^{-1}. In general, the molecules formed from displacement by hot ^{18}F atoms have more internal energy than those formed by hot T atoms.

Cyclohexene-t formed by hot T atom displacement decomposed to ethene + butadiene[158]. The excitation energy may be either electronic or vibrational. If the latter, the magnitude of the rate-constant suggested an average excitation energy of c. 115 kcal mol^{-1}. The CH_2TNC^* from H displacement by hot T atoms isomerised to CH_2TCN even at pressures of 5 atm [159]; less than 1% of the molecules[159] can have energies below 90 kcal mol^{-1}.

The decomposition rate of $C_2H_4D^*$ produced by krypton-sensitised decomposition of D_2 in the presence of C_2H_4, was enhanced by one–two orders of magnitude relative to the $C_2H_4D^*$ formed by addition of thermal D to C_2H_4. The D atoms were estimated to carry 10–15 kcal mol^{-1} of translational energy into $C_2H_4D^*$.

1.2.3.3 Activation by bimolecular reaction with reactants having excess internal energy

Studies mentioned here have as their objective the generation of activated molecules rather than measurement of photopartitioning of energy (see

Section 1.2.4.3). The excess internal energy of the reactants should be added to the exoergicity of the reaction and the thermal energy of the reactants to obtain the distribution function of an activated product molecule. Obvious examples are the insertion and addition reactions of singlet methylene. The non-equilibrium unimolecular rate-constants of cis-1,2-dimethylcyclopropane, cis-pent-2-ene[84], 1,1-dimethylcyclopropane[83] and 2- and 3-methylbut-1-ene[161] formed by photolysis of diazomethane at 438 and 366 nm with cis-but-2-ene and isobutene are consistent with $\Delta H_{f_0}^0(CH_2) + \langle E^*CH_2) \rangle = 112.6$ and 116.1 kcal mol^{-1}. Since $\Delta H_f^0(CH_2) = 94$ kcal mol^{-1}, $\langle E^*(CH_2) \rangle = 19$ and 22 kcal mol^{-1}. These energy assignments were utilised for the unimolecular reactions of chemically activated ethyltrimethylsilane, ethydimethylsilane and tetramethylsilane formed from methylene and trimethylsilane[162]. Fitting an RRKM analysis to the $Si(CH_3)_4$ data gave a transition-state model with a pre-exponential factor of $10^{15.0 \sim 0.5}$ s^{-1}, which is an order of magnitude smaller than for neopentane[163, 164].

Methylcyclobutene formed from insertion of CH_2 into the CH bonds of cyclobutene[165] gave experimental rate-constants which agreed with the calculated values using the above value $\Delta H_f^0(CH_2) + \langle E^*(CH_2) \rangle$. Since a common energy assignment fits the kinetic data for several activated molecules, the energies of molecules formed by $CH_2(^1A_1)$ reactions are unlikely to be in serious error[252].

Other examples demonstrating the incorporation of excess vibrational energy into product molecules are rare, because several collisions usually intervene between formation and subsequent bimolecular reaction of the species of interest. However in the future one can anticipate preparation of activated molecules via combination of vibrationally excited radicals[165] or from vibrationally excited ions in ion–molecule reaction.

1.2.3.4 Activation by reactions yielding two products

Since more is known about unimolecular rate-constants than about energy partitioning, the unimolecular reaction rates are used to gain information about energy partitioning for reactions yielding two products. The number of examples is small (see Section 1.2.5.4) because the threshold energies of many unimolecular reactions are relatively high and division of the exoergicity between two products reduces the energy below the threshold of the unimolecular reaction channels. The reaction of alkyl radicals with fluorine:

$$R + F_2 \longrightarrow F + RF^* \underset{k_a}{\overset{k_M[M]}{<}} \begin{array}{l} RF \\ HF + \text{olefin,} \end{array}$$

is reported to give chemically activated molecules. To be consistent[35, 108] with the measured rate-constant, more than 90% of the available energy must be retained as internal energy of RF*. This large fraction and the apparent lack[167] of a wide energy distribution is rather surprising[35].

1.2.4   Photochemical activation

1.2.4.1   *Unimolecular chemical reactions of excited electronic states*

In principle this activation method leads to unimolecular *chemical* reaction whenever the potential-energy surface of the excited state formed by absorption of a photon correlates directly with the potential energy surface of the products. However, in practice many excited singlet electronic states of polyatomic molecules relax to triplet or lower singlet states by radiationless nonadiabatic processes. Thus, the problem becomes one of identifying steps that are truly unimolecular *chemical* reactions. Dissociation from the first excited singlet state of ketene[168], hexafluoroacetone[169], azoethane and perfluoroazoethane[170] have been treated as straightforward unimolecular reactions. However, recent data[171, 172] indicate that the first singlet state of hexafluoroacetone should not be treated by RRKM theory because intersystem crossing to the triplet state is important. Competition between dissociation and collisional stabilisation of the first triplet state of ketene has been invoked[173] to explain the pressure dependence for removal of triplet CH_2 by CO.

1.2.4.2   *Activation of ground electronic state molecules by internal conversion*

Internal conversion to the ground electronic state after absorption of a photon gives vibrationally activated molecules with precisely known energy. The photolysis of *cis*-but-2-ene and *cis*-hex-2-ene at 202.6 nm (141 kcal mol^{-1}) appears to be an example[174]. The experimental rate-constants for rupture of H and CH_3 from but-2-ene and C_2H_5 from hex-2-ene were in agreement with calculations based on models developed from chemical activation studies at *c*. 118 kcal mol^{-1}. The difficulties encountered with this activation technique are illustrated by a study[175] with *cis*-pent-2-ene at 185 nm. In an inert bath gas *cis–trans* isomerisation did not occur, and the pressure dependence of the fragmentation products was explained as a competition between dissociation from a Rydberg state and collisional electronic quenching to the ground state[175].

The most ambitious investigation[176] has been with cycloheptatriene-d_0 and $-d_8$, which isomerises to toluene with an Arrhenius rate-constant of $3.5 \times 10^{13} \exp(-51\,100/RT)$. The photon energy, which was varied from 91–125 kcal mol^{-1}, gave rate constants of 4.4×10^6–8.1×10^8 s^{-1}. The agreement between approximate RRKM calculations and the experimental data was unsatisfactory. The authors assigned responsibility for the disagreement to failure of the strong-collision assumption, even for cycloheptatriene or toluene as the bath gas. The data for He, CO_2 and SF_6 were interpreted with a stepladder cascade model having much smaller average energy increments than normally reported from chemical-activation studies. The data are not free from problems and in particular the non-equilibrium kinetic isotope effect is too small and has the wrong dependence on energy. A new interpretation[177] of Atkinson and Thrush's data reduces the discrepancies relative to conclusions from chemical-activation studies.

One of the reaction channels for cyclobutanone(S_1) is internal conversion to the ground state[178] which decomposes (the Arrhenius parameters are $10^{14.56} \exp(-52\,000/RT)$) to ketene and ethene. For photolysis at 313, 302, 289 and 254 nm, the rate-constants increased from $7 \times 10^9 - 5 \times 10^{10}$ s^{-1}. Approximate RRKM calculations were in agreement with experiment at low energy, but were low by a factor of 2 at the highest energy. Studies have been extended to dimethylcyclobutanone and the decline in the rate-constant agrees with estimated RRKM results[179].

1.2.4.3 Activation by photodissociation

Cyclopropane isomerisation has been the most widely used reaction for measuring photopartitioning of energy. Photolysis of 1-pyrazolines[180—181] is a good example.

The cyclopropane isomerisation rate-constants were compared to calculations using known k_E values and trial distribution functions. Since the data do not match a statistical energy partitioning pattern, the authors[180—181] chose to fit a Gaussian function and assigned the most probable energy, E_{mp}, and the dispersion, σ. The portion of the available energy (132 kcal mol^{-1}) associated with the cyclopropanes, $E_{mp} = 73–83$ and $\sigma = 12–18$ kcal mol^{-1} was surprisingly constant for several pyrazolines.

One reaction channel in the photolysis of tetrahydrofuran yields vibrationally excited cyclopropane[182] with c. 75% of the available energy. The second reaction-channel available to cyclobutanone(S_1) is intersystem crossing to cyclobutanone(T_1), which subsequently yields cyclopropane and carbon monoxide[178, 183]. The cyclopropane has a broad energy distribution[184], and c. 85% of the energy was retained by the cyclopropane.

The ratio of propene to cyclopropane from photolysis of cyclobutanone at fixed pressure but with various wavelengths has been used as a calibration scale[183] for the energy in cyclobutanone (T_1). Sensitisation by singlet benzene (109 kcal mol^{-1}) gives cyclobutanone (S_1), which subsequently intersystem crosses to cyclobutanone (T_1) and decomposes. According to the calibration scale, 94 kcal mol^{-1} was deposited in the cyclobutanone (S_1). On the other hand, triplet benzene released nearly all of its energy (84 kcal mol^{-1}) to cyclobutanone(T_1). Application of this technique to the interaction of $Hg(^3P_1)$ and cyclobutanone indicated[185] that 105 ± 2 of the possible 112.7 kcal mol^{-1} were transferred to cyclobutanone(T_1).

The unimolecular reaction of t-butoxy radicals produced by photolysis 313 and 249 nm (available energy = 61 and 84 kcal mol^{-1}) of t-butylperoxide has been observed[186]. The non-equilibrium rate-constants were fitted to a Gaussian distribution with $E_{mp} = 19$ and $\sigma = 3$ and $E_{mp} = 22$ and $\sigma = 3$ kcal mol^{-1} for 313 and 249 nm, respectively. This example[186] shows that small amounts of energy with narrow distributions can be partitioned as internal energy, presumably because dissociation proceeded via a repulsive potential surface.

A variety of photochemical processes yield an excited ethene species which eliminates H_2. Combining experimental data with an RRKM analysis indicated[187] that, if the ethene species was vibrationally excited the ground electronic state, E_0 must be $c.$ 80 kcal mol^{-1}.

A slight variation is to trap a photodissociation product by some bimolecular reaction that yields a product molecule which undergoes a unimolecular reaction. The energy released to CH_2^* following photolysis (123.6 nm) of propane and cyclopropane was estimated[188, 189] from the rate constant[164] of n-butane*, formed by trapping CH_2^* with propane. The C_2O radical from photolysis of C_3O_2 was reacted[190] with CH_3F; the rate-constant for HF elimination from $C_2H_3F^*$ corresponding to a $C_2H_3F^*$ energy of $c.$ 127 kcal mol^{-1}.

1.2.5   Activation methods for positive and negative ions

1.2.5.1   Unimolecular reactions of ions versus neutrals

The application of the statistical theory of unimolecular reactions to polyatomic ions[11] requires only the additional assumption that electronic excited states, which may have been produced in the ionisation step, rapidly relax to the ground potential surface by radiationless processes. Fragmentation then occurs if the internal energy of the ion exceeds the threshold energy for some reaction channel. Since excitation is often to higher energies than for neutrals, the effect of anharmonicity[191] upon the densities and sums of states should be remembered. Photoelectron spectroscopy has shown that many excited electronic states are produced in photon and electron impact processes. Rapid relaxation of electronically excited states may be valid in most cases, but the complexities of molecular photochemistry leads one to expect at least a few instances[192–195] having fragmentation from excited electronic states.

Study of the unimolecular reactions of ions and of neutral molecules developed as virtually independent fields with separate literature, symbolism and nomenclature. The separation has been to the detriment of both fields. For example[11], application of QET theory to polyatomic ions suffered for a time from use of an erroneous equation for computing the density of states and, in fact, the equation commonly used[11] now for computing density of states for polyatomic ions differs from the exact-count result by a factor of nearly 2, even at high energies[196]; several other equations are superior and are just as easy to use (see Section 1.3.3). On the other hand, interest in energy partitioning[88] from fragmentation of ions preceded such studies for neutral molecules. Since experimental techniques for neutrals and ions complement each other and since the QET and RRKM theories are the same, the present trend of merging the two fields is welcome.

Threshold energies, pre-exponential factors and molecular structure of the reactant often are well known for reactions of neutral molecules. For polyatomic ions these advantages are not present; the strong dependence of k_E upon E_0 makes assignment of this quantity especially important. Although pre-exponential factors are unlikely to be measured for reaction of molecular ions, activated-complex models can be conveniently summarised by citing the calculated pre-exponential factor, which would facilitate comparison with similar processes for neutrals. Careful QET calculations for small molecules, such as propane, have been done for some time and these were summarised in 1968 [11]. Combining the thermal distribution of the precursor to the ion with photoelectron spectroscopy results appears[197] to give good energy distribution functions for ions produced by electron impact. For large molecules the thermal distribution of internal energy carried by the precursor can be important[197]. A recent valiant attempt[197] was made to relate the qualitative intuitive reasoning of the organic chemist to the more physical description provided by theory[9, 11]. The arbitrary adjustment of pre-exponential factors in the classical RRK formulation which has been practiced by some investigators[198, 199] studying large organic ions should be avoided.

1.2.5.2 Change transfer and electron and photon impact excitation

The first derivative of relative ion intensities from photoionisation[200] gives the 'breakdown curve' for the molecular ion. Charge transfer from a neutral molecule to a slow ion activates the new ion to a known level of energy[201, 253]. For example[201], 19 different ions, with recombination energies varying from 10.5 to 25 eV, were allowed to exchange charge with NH_3. The subsequent decomposition reactions of NH_3^{*+} give directly the 'breakdown curve' at the various energies. The 'breakdown curve' can be compared directly to theory by use of the decay time that corresponds to the experimental conditions in the following equations[202].

$$A_j(E,t) = \left[k_{E_j}/\sum_i k_{E_i}\right]\left[1 - \exp\left(-\sum_i k_{E_i}t\right)\right]\left[1 - P_j(E,t)\right]$$

$$A_p(E,t) = \exp\left(-\sum_j k_{E_j}t\right)$$

(1.12)

$A_j(E,t)$ and $A_p(E,t)$ are the relative abundance of the jth primary product ion and the parent, respectively, at energy, E, and time, t. The rate-constant at energy E for formation of the ith primary ion is k_{E_i}; $P_j(E,t)$ is the probability for further decomposition of the primary product ion into secondary productions. The mass spectrum obtained with conventional instruments is the product of the 'breakdown curve' and the internal energy distribution of the parent ion from electron impact excitation. Fitting experimental mass spectra or even 'breakdown curves' is not considered[202] a good test of theory because of possible compensating factors, i.e. only relative values of the rate-constants are measured. However, the use of isotopically substituted ions minimises the possible compensating factors and the use of the metastable ions (those ions decomposing during the time of 0.45–2.75×10^{-6} s for the instrument used for the propane studies[202]) provides stronger support for the theory. Careful calculations[202] have been done for the mass spectrum from $C_3H_8^+$, $CH_3CD_2CH_3^+$, $CD_3CH_2CD_3^+$ and $C_3D_8^+$ for an energy distribution obtained from the results of photoelectron spectroscopy and derivatives of photoionisation curves. Comparison also was made to the photoionisation 'breakdown curve' for $C_3H_8^+$, which does not require an energy distribution, and to the metastable ions for all four isomers. Reasonable assumptions regarding structures of the anticipated complexes gave good agreement between theory and experiment.

Some examples show consecutive metastable peaks, i.e. ions which dissociate $(ABC^+ \longrightarrow AB^+ + C)$ in the first field-free region of a double-focusing mass spectrometer and then dissociate further $(AB^+ \longrightarrow A^+ + B)$ in the second field-free region[203, 204]. The agreement between calculated and experimental results were satisfactory for the dissociations $C_7H_7^{*+}$ and $C_5H_5^{*+}$ formed by electron impact on toluene.

It commonly is assumed that the activation energy for the reverse of fragmentation reactions of ions has zero threshold energy. Using this as a starting point, Lin and Rabinovitch[86] calculated the internal energy partitioned to $C_2H_5O^{*+}$ from a series (C_3-C_9) of parent alcohols. Statistical partitioning of energy explained the variation of the metastable (decomposition of $C_2H_5O^{*+}$) intensity with the internal degrees of freedom of the parent alcohol. Franklin and co-workers[87, 205, 206] have directly measured the kinetic energy released in several fragmentation reactions (see Section 1.3.2.5 for more detail). A threshold energy for the reverse reaction was suggested[207] to explain the release of $c.\ 25\%$ of the excess energy as kinetic energy from $[C_6H_6-CH=NOCH_3^*]^+ \rightarrow [C_6H_6OCH_3]^+ + HCN$. The release of a large amount of kinetic energy by decomposition of CH_2OH^+ was explained[208] as crossing to a repulsive state. Thus, interpretation of energy partitioning depends upon the potential surface in question and some judgement should be exercised before assuming statistical partitioning or zero reverse threshold energy[209].

1.2.5.3 Ion–molecule association reactions

Condensation of an ion with a neutral molecule is a form of chemical activation and, with allowance for the excess kinetic or internal energy (if any), the subsequent unimolecular reactions can be treated in the way described for activation by radical association, see Section 1.2.2.3. Since the reaction cross-section for these condensation reactions can be quite large, angular momentum restrictions may need to be considered[28, 95]. In beam studies intermediate complex formation is identified by the symmetry of forward and backward scattering in the centre-of-mass coordinate system[29, 210]. This symmetry can only be achieved if the complex lives for a time equivalent to a few rotational periods, $\gtrsim 10^{-12}$ s, which should be adequate for the statistical theory to describe the breakdown pattern of the adduct ion.

Ion cyclotron resonance spectrometers seem especially well suited for study of unimolecular reactions activated by association of an ion and a neutral molecule. Recent examples[211, 212] are H_2S^+ with C_2H_4 and C_2H_2, $C_2H_4^+$ with C_2H_4, $C_2H_4^+$ with C_2H_2, and $C_2H_2^+$ with C_2H_4. A careful application[190] of theory was made for the $C_4H_6^{*+}$ ion (assumed to be butadiene ion) formed from $C_2H_2^+ + C_2H_4$ and from $C_2H_4^+ + C_2H_2$ at 93 and 71 kcal mol^{-1} of internal energy, respectively. Absolute experimental rate-constants for the H and CH_3 rupture channels were not available; however, the calculations gave excellent agreement with rate-constant ratios for $C_4H_6^{*+}$, $C_4D_6^{*+}$, and $C_4H_2D_4^{*+}$. Thus, QET or RRKM theory was successful, as might have been predicted from analogous chemical-activation studies of neutrals.

Molecular-beam studies can give the translation energy spectrum of products from decomposition of the long-lived complex and this has been achieved for $Cs + SF_6$ [94] and for $O_2^+ + D_2$ [213]. Somewhat surprisingly and perhaps fortuitously, the short lifetime (c. 10^{-13} s) of the $O_2^+D_2$ complex formed in 5.5 eV relative energy collisions was in approximate agreement with the RRKM prediction[28].

1.2.5.4 Ion–molecule transfer reactions

Ion–molecule reactions proceeding by a direct mechanism and yielding two products are similar to the chemical activation systems discussed in Section 1.2.3.4. The internal energy of the product ion depends upon the translational and internal energies of the ion plus the partitioning of the exoergicity by the transfer reaction. It is not always easy[214–216] to identify whether the mechanism is direct or intermediate complex formation and, in fact, the mechanism may change with the relative kinetic energy of the collision[28, 30]. Proton transfer reactions[217, 218] from CHO^+ seem to have dynamics close to a stripping model (but see Ref. 254).

$$CHO^+ + CH_4 \longrightarrow CH_5^{*+} + CO$$

$$CHO^+ + C_2H_6 \longrightarrow C_2H_7^{*+} + CO$$

$$CHO^+ + C_3H_6 \longrightarrow C_3H_7^{*+} + CO$$

The decomposition rate-constants for CH_5^{*+} and $C_2H_7^{*+}$ depended upon the precursor used to obtain CHO^+, which is evidence that CHO^+ contained excess internal energy. Comparison of the experimental decomposition rate of $C_3H_7^{*+}$ to the rate calculated from the model used for interpreting the mass spectrum of propane gave good agreement with the energy expected on the basis of a stripping model for $CHO^+ + C_3H_6$ [217]. The $CH_3^+ + CH_4 \longrightarrow C_2H_5^+ + H_2$ reaction gives no evidence for a $C_2H_7^+$ long-lived complex[217]; however, the condensation releases sufficient energy so that the expected $k_E \geqslant 10^{12}$ s^{-1}, thus explaining the absence of a long-lived complex[218].

Tables of proton affinities are being compiled[214, 219, 220, 254], and studies of ions activated by selective proton transfer soon should be possible. Just as for the condensation reactions, experiments using ion cyclotron spectrometers seem very promising. Proton transfer reactions (Chemical Ionisation Mass Spectrometry) have been used for some time as an alternative ion source for mass analysis[221]; protonating agents have been CH_5^+, $C_2H_5^+$ and t-$C_4H_9^+$.

1.2.5.5 Activation of negative ions

Activated polyatomic negative ions can be generated by electron attachment, charge transfer, ion–molecule reactions, and even three-body association reactions[222]. Only electron-attachment activation has been sufficiently well studied to identify subsequent unimolecular mechanisms. The observations have been explained by use of the same approach as for positive ion break-down[223, 224]. The metastable intensity for $SF_6^{*-} \longrightarrow SF_5^- + F$ was compared to results from statistical theory for $E.A.(SF_6) = 1.49$ eV, and a semi-rigid complex with $E_0 = 1.56$ eV fitted the data[223].

The lifetime of the long-lived auto-ionising component from polyatomic negative ions formed by capture of low energy electrons increases with the number of degrees of freedom of the negative ion[224, 225], as for a chemical unimolecular reaction. This was explained[226] by assuming that the energy released by the combination of the electron and the neutral was converted to vibrational energy. Both auto-ionisation and fragmentation (which has the higher threshold) rate-constants increase with energy; since the latter increases more rapidly, fragmentation dominates at higher energy.

The relative kinetic energy released by the electron attachment-fragmentation[227, 228] process fitted the same empirical relation used for positive ions[205, 206].

1.3 STATISTICAL THEORY OF UNIMOLECULAR REACTIONS

1.3.1 Some of the assumptions

The success of the statistical theory, in RRKM or QET form, in fitting the large variety of reactions, which take place on a single potential energy surface, is evidence of its usefulness, flexibility and basic validity. The assumptions necessary to obtain the expression for the specific rate constant, k_E, can be placed in three categories: (i) The rapid intramolecular randomisation of internal energy, (ii) The assumptions inherent to absolute rate theory, and (iii) The assumptions related to the mechanics of calculating k_E. The first assumption is the most important, and experiments most directly related to it are considered in Section 1.3.2. The assumption in (ii) are not unique to unimolecular reactions and will not be discussed; however, they make the testing of (i) difficult since their effects are difficult to separate. Most of (iii) is considered in Section 1.3.3, although some terms are defined below.

The rate-constant of a rotationless ($J = 0$) molecule with energy, E is given by

$$k_E = \frac{\ell}{h} \frac{\sum\limits_{E_v^\dagger \leqslant E^\dagger} P(E_v^\dagger)}{N(E^\dagger + E_0)} \tag{1.13}$$

$E^\dagger = E - E_0$ is the non-fixed energy of the complex; ℓ is the reaction path degeneracy (including rotational symmetry numbers); h is Planck's constant; $\Sigma P(E_v^\dagger)$ is the sum of all internal energy states up to $E_v^\dagger \leqslant E^\dagger$, and $N(E^\dagger + E_0)$ is the density of internal energy states of the molecule at E. In the use of this equation all internal degrees of freedom, including internal rotations, are to be used in evaluating the density and sum terms.

Rabinovitch and Waage[21, 42, 43] have reviewed the ways of including the effects of overall rotation in equation (1.13). The first RRKM formulation multiplied (1.13) by the ratio of adiabatic (those degrees of freedom staying in the same quantum state throughout the reaction) partition functions; $P_1^\dagger/P_1$. These usually are just the overall rotational partition functions and equation (1.14), which applies to all rotational states, results.

$$k_E = \frac{P_1^\dagger}{P_1} \frac{\ell}{h} \frac{\sum\limits_{E_v^\dagger \leqslant E^\dagger} P(E_v^\dagger)}{N(E^\dagger + E_0)} \tag{1.14}$$

When (1.14) is placed in equation (1.2), the correct equilibrium rate-constant is obtained and for unimolecular reactions having rigid or semi-rigid activated complexes $P_1^\dagger/P_1$ is close to unity and (1.14) is adequate. However (1.14) is not appropriate for loose transition states; in particular, the thermal low-pressure expression is incorrect and $P_1^\dagger/P_1$ overestimates the centrifugal effect for k_E. Therefore, more explicit consideration must be given[42, 43] to each J state.

For loose transition states, two moments of inertia of the transition state differ substantially from the molecule. Thus a rigid-rotor model is used and the rotational energy in the molecule and transition is represented by E_J and $E_J^\dagger$. Since J cannot change during the reaction, $E_J > E_J^\dagger$ and the threshold energy becomes dependent on J: $E_0(J) = E_0(J = 0) + E_J^\dagger - E_J$. The rate-constant depends on J in the following way:

$$k_{EJ} = \frac{\ell}{h} \frac{\sum\limits_{E_v^\dagger \leqslant E^\dagger} P(E_v^\dagger)}{N(E^\dagger + E_0 + E_J^\dagger - E_j)} \tag{1.15}$$

Marcus[10] introduced approximations and averaged equation (1.15) over all J to obtain equation (1.16).

$$k_E = \frac{\ell}{h} \frac{\sum\limits_{E_v^\dagger \leqslant E^\dagger} P(E_v^\dagger)}{N(E^\dagger + E_0)F}; \quad F = \frac{N(E^\dagger + E_0 - nRT[I^\dagger - I]/2I)}{N(E^\dagger + E_0)} \tag{1.16}$$

where n = number of adiabatic rotations associated with $I^\dagger$ and I, which for loose transition states usually are the two of the rigid-rotor model. The F factor is approximately constant with increasing $E^\dagger$. For rigid complexes $I^\dagger \approx I$; F is near unity and the numerical result of equation (1.16) is similar to equation (1.14). For loose transition states F is less than unity ($F \approx 0.8$ for ethane decomposition[21]), but F^{-1} tends to be appreciably smaller than $P_1^\dagger/P_1$. For activated processes giving high values of J, averaging over J may not be appropriate[28, 95].

Other theories, even those having a statistical basis[229], will not be reviewed. However, the relationship between the RRKM absolute rate theory formulation and Light's phase-space theory[230] is worth noting. Mies[231] and, more recently, Klots[232] showed that for the extreme of loose transition states, i.e., the Gorin model, the phase space and RRKM results are identical. Klots' paper, which provides some interesting insights, avoids the transition state by using microscopic reversibility to formulate the unimolecular specific rate-constant in terms of the cross-section of the bimolecular reverse reaction and the density of states of the molecule. Since the Gorin transition state is equivalent to the Langevin cross-sections for ion–molecule reactions, the formulation by Klots also is equivalent to RRKM theory for Gorin type transition states. Klots[232] used the Langevin cross-sections with explicit consideration of angular momentum to obtain rate-constants for three fragmentation channels of CH_4^{*+}.

1.3.2 Experimental tests for intramolecular energy randomisation

1.3.2.1 General comments

Although most data support rapid randomisation of internal energy relative to the time available for unimolecular reaction, randomisation must fail for some circumstances, i.e. for sufficiently extreme experimental tests or for molecules lacking the necessary (but undefined) coupling which facilitates energy transfer. Finding examples illustrating non-randomisation will be necessary before the limitations of a statistical theory can be ascertained.

The listing of topics below is largely from Spicer and Rabinovitch[1]; only results not mentioned by them are discussed. In addition to purely experimental results, computer simulation can be compared to theory or to experiment. Simulated rate-constants for model triatomics[233] agreed with RRKM predictions to better than 50%. However, lack of energy randomisation was encountered in initial efforts[50] to simulate the isomerisation of CH_3NC.

1.3.2.2 Increase in volume of vibrational phase space

Energy randomisation is tested by measuring the rate-constants of a homologous series having the same basic reaction coordinate and level of internal energy. The decline in the rate-constants with increasing size of the molecule has been quantitatively fitted by the theory for both thermal and chemical activation studies. Chloro-[98] and fluoro-alkanes[109, 103] and other examples mentioned in previous sections can be added to the list given by Spicer and Rabinovitch[1]. The large statistical isotope effects[9, 74] observed for chemically activated molecules also provide strong evidence for energy randomisation.

1.3.3.3 Non-random excitation

Any chemically activated molecule which decomposes by a process that differs from the activation reaction, i.e., energy is released at one site and unimolecular reaction occurs at a different site, demonstrates a degree of energy randomisation. However, a much stronger test is to observe the competition between two identical, but independent, reaction channels when the molecule is activated in different ways. This recently was achieved in a very elegant and novel fashion[234, 235]; hexafluorobicyclopropane-d_2 was produced with

c. 110 kcal mol^{-1} of energy by either of two ways:

$$CF_2\!\!-\!\!\underset{\underset{CH_2}{\diagdown\!\diagup}}{CF}\!\!-\!\!CF\!\!=\!\!CF_2 + CD_2$$

or

$$CF_2\!\!-\!\!\underset{\underset{CH_2}{\diagdown\!\diagup}}{CF}\!\!-\!\!\underset{\underset{CD_2}{\diagdown\!\diagup}}{CF}\!\!-\!\!CF_2^*$$

$$CF_2\!\!=\!\!CF\!\!-\!\!\underset{\underset{CD_2}{\diagdown\!\diagup}}{CF}\!\!-\!\!CF_2 + CH_2$$

The activated molecule can decompose by elimination of CF_2 from either cyclopropane ring and deuterium provided a label for identifying the competing reactions. Experiments utilising both activating reactions were done from 1 to 3×10^3 Torr. Even at the lowest pressure complete randomisation was not observed since activation by CD_2 addition favoured elimination at the CD_2 cyclopropane ring; whereas activation by CH_2 favoured the CH_2 cyclopropane ring. As the pressure was increased, i.e. the time interval shortened, the non-random component increased. Analysis of the data gave a rate-constant for intramolecular energy relaxation of c. 1×10^{12} s^{-1}. Since both methods of activation gave the same result, one's confidence is increased in this isolated example which illustrates non-random behaviour.

1.3.2.4 Decrease in mean interval between collisions

The time scale available to an activated molecule can be shortened by doing experiments at super high pressure. The decomposition of butyl radicals has been studied[236] up to 200 atm of H_2, which corresponds to an average collision interval of 2×10^{-13} s. No deviation in the chemical-activation rate-constant was observed. Experiments of this type are inherently difficult and, so far[1], have not shown evidence for non-randomisation.

1.3.2.5 Energy partitioning

In order for energy partitioning to be a good measure of energy randomisation, data should be obtained for a case having a zero threshold energy for the reverse of the unimolecular step. Several studies[86, 89, 90, 99] already have been mentioned which support statistical energy partitioning from a unimolecular reaction. The 1,1-HCl elimination from 1,1,2-trichloroethane should nearly qualify for zero activation energy for the reverse reaction. The molecular-beam experiment[90] with $Cs + SF_6$ is of great interest because for the first time the translational, rotational and vibrational energy, as well as the angular distribution of the CsF product was specified completely. The translational, rotational and vibrational energies of CsF were Boltzmann with $T_t = 1190 \pm 150$, $T_r = 1050 \pm 200$ and $T_v = 1120 \pm 90$ in accordance with a statistical model.

Franklin and co-workers[87, 206, 207, 225] measured the kinetic energy (extrapolated to the appearance potential) released in the decomposition of positive and negative ions having known excess energy, E^* (obtained from the difference between the appearance potential and the known thermochemistry for the reaction). They found an empirical correlation of $E_t = E^*/\alpha N$, where α is a constant, c. 0.44, and N is the number of internal degrees of freedom. This empirical relationship suggests statistical partitioning, which was investigated[205] later by use of equation (1.10). Frequencies used for the transition states were those of the molecules except for deletion of either 1 or 3 modes. Although the authors did not use the following interpretation, the deletion of 3 modes can be viewed as transition-state structure which is tighter than that of the molecule, i.e., a pre-exponential factor $< 10^{13} \text{ s}^{-1}$, while the deletion of one mode corresponds to a pre-exponential factor of c. 10^{13} s^{-1}. The authors concluded that agreement with statistical theory was satisfactory except for three cases, and two of these were rearrangement processes (see Section (1.2.5.2)). Plots of the kinetic energy versus electron energy have a variety of forms[87, 227, 228]. Many do not have the form expected for statistical energy release, which may suggest the importance of small threshold energies for some of the reverse reactions.

The data of Franklin and Haney have been reinterpreted by LeRoy[237] as showing non-randomisation of energy. The plot made by LeRoy is based upon an incorrect definition of the *unimolecular* lifetime and, as noted by the original authors[205], most of the data are adequately represented by statistical partitioning.

1.3.3 Parameters of the statistical transition state theory

In order to evaluate k_E the quantum statistical sums and densities must be calculated. Forst and Prášil[238] compared the results of seven approximate formulae to the exact count for *harmonic* oscillators. Four of the seven were excellent approximations with little difference between them. However, the approximations by Vestal and some other workers using QET were in error by about a factor of 2. In all seven cases the approximations were poor at very low energy but the sums and densities can be obtained by direct counting at the low energies. The steepest descent method, as formulated by Hoare[239, 240] and by Lin and co-workers[241], but appears to be another good approximation[238b]. The steepest descent method[238–242] is applicable to complex cases involving anharmonic oscillators and perhaps even hindered internal rotations.

The general effect of anharmonicity at $E \leqslant 100 \text{ kcal mol}^{-1}$ is to increase the magnitude of the sums of the complex and the densities of the molecule, but the latter is affected more strongly because $E > E^\dagger$, and k_E (anharmonic) $< k_E$ (harmonic). For the same anharmonicity constants in the molecule and transition state and with a Morse oscillator formulation, the lowering usually is a factor of c. 2–3 [60, 239, 241]. The widespread agreement between experiment and calculated results without inclusion of anharmonicity has always puzzled the present author. It may be that some energy states are disallowed[238b, 242]; thus, the harmonic approximation is a fortuitous compromise between inclusion of anharmonicity and the disallowance of certain levels.

Given the facts that all internal modes are to be taken as active and that E_0 and $N(E)$ are known, a frequently encountered response is 'of course an experimental rate-constant can be fitted with $3N - 7$ adjustable parameters'. Actually common sense and a knowledge of molecular structure[9, 14] reduces the unknown transition-state frequencies to a much smaller number. Consider the rupture of an atom (X) from R—X; the only frequencies that can differ appreciably from R—X or R are the *two* bending modes perpendicular to the breaking R⋯X bond. Rearrangement processes are somewhat more difficult, but for elimination of HCl from C_2H_5Cl the frequencies can be divided into two main categories (see Section (1.2.2.3)). The out-of-ring frequencies cannot differ much from analogous modes of ethene or cyclobutane. Thus the unknown frequencies are, at most, only five and these can

Table 1.3   Variation of k_E and pre-exponential factor with transition state frequencies

Transition state frequencies*	k_E (91.5 kcal mol^{-1})	Pre-exponential factor†
(1) 3054(4) 1315(3) 987(4) 888(4) 576(1) 400(1)	2.42×10^9 (1.00)	1.46×10^{13} (1.00)
(2) 3054(4) → 2800(4)	1.08	1.01
(3) 888(4) → 700(4)	2.00	1.54
(4) 987(1) → 250(1)	3.08	2.28
(5) 576(1) → 400(1)	1.34	1.25
(6) 400(1) → 200(1)	1.85	1.65
(7) 400(1) → 75(1)	4.70	4.06

*See reference 74 for a full description of molecular and transition state models; for the calculations of this table only the transition state frequencies shown were altered.
†The pre-exponential factor, in the form of equation (1.2), was calculated at 800 K.

be estimated with aid of empirical rules[18]. Some k_E and pre-exponential factor results for various C_2H_5Cl transition state models are shown in Table 1.3. Two things should be noted: (i) Both k_E and $Q_I^\dagger$ are sensitive to low frequencies (compare entries 3, 5 and 6 in Table 1.3) and both are affected to about the same degree for a given variation in frequency. (ii) Variation of the high and intermediate range frequencies affects k_E somewhat more strongly than $Q_I^\dagger$.

Assignment of torsional modes of large linear molecules or transition states presents the greatest single problem (see Section (1.4.1.1) for an example), which is especially acute for cases involving a change in the torsional modes to some other type of mode. For methyl groups the contribution from the torsional mode to the number of internal states is approximately the same for use of either the upper limit (a free internal rotation) or a lower limit (a harmonic oscillator) for the calculation. However, this is not true for other internal rotors and a convenient method of calculating densities and sums for hindered internal rotors is desirable.

The sensitivity of k_E to the thermochemistry, E_0 and E, should be remembered. Uncertainties in thermochemistry usually overshadow concerns about assignment of specific vibrational frequencies in the transition state.

1.4 SUMMARY

1.4.1 Applications of non-equilibrium techniques

1.4.1.1 Characterisation of unimolecular reactions: the alkane decomposition problem

The usefulness of combining RRKM calculations with non-equilibrium rate-constants for evaluating models of unimolecular reactions (primarily of hydrocarbons) has been emphasised. Fruitful application to ion–molecule reactions and to non-hydrocarbon molecules and radicals can be anticipated. Some problems have been encountered (see below), for reactions having extremely loose transition states and further study of this reaction type is needed.

Alkanes have been activated thermally, by methylene insertion and by radical combination[243]. The reliable data for $C_2H_6^*$ activated by radical combination (see reference 21 for a review) can be reconciled to a transition state model that is consistent with rates of methyl radical combination. However, the new [244] chemical-activation rate-constants for $C_2H_6^*$ activated by $CH_2 + CH_4$ are much lower than the values predicted from this transition state model. Furthermore, the loose model does not fit the high-temperature fall-off data and gives a pre-exponential factor that is nearly an order of magnitude higher than that obtained from C_2H_6 pyrolysis[15, 17]. The situation is somewhat reversed for the larger alkanes. Models having low pre-exponential factors fit the chemical-activation rate-constants for methylene insertion[164, 245]; whereas, except for the shock tube results of Tsang[22], most pyrolysis data[15, 17] (although the error limits are large) favour high pre-exponential factors, which are consistent with rate-constants for radical combinations. By use of loose free-rotor activated-complex models, the highest possible threshold energies for decomposition, and tight free-radical structures, reconciliation[164, 245] of radical combination and the chemical-activation decomposition rate-constants was barely possible. The tight radical structure is especially suspect, and the transition state models for alkane decomposition and radical combination certainly are not satisfactory. Considering the uncertainties in the data, it is premature to fault the theory. Since E is high and $E^\dagger$ is low, anharmonicity may have a significant effect on the chemical-activation rate-constants and inclusion of it would tend to reduce the discrepancy between the models deduced from bimolecular radical combination versus those from the unimolecular chemical-activation data.

1.4.1.2 Vibrational energy transfer probabilities

Much of the knowledge concerning collisional removal of internal energy from highly vibrationally excited polyatomic molecules and radicals has been deduced from chemical-activation studies. Extensive new measurements[12, 39] for $C_2H_4Cl_2^*$ and $CH_3CF_3^*$ generally agree with other chemical-activation data[32–36] in regard to the magnitude of energy removed per collision. However, two new features are suggested: (a) c. 12 kcal mol^{-1} may be the upper limit for the energy removed from the activated molecule by even efficient gases. (b) Significant differences were found for the average amount of energy removed from $C_2H_4Cl_2^*$ and $CH_3CF_3^*$ by the same bath gas. As improved activating reactions are developed, better energy transfer data from chemical-activation studies may be anticipated; some information has even been obtained for ions[246].

1.4.1.3 Assignment of internal energies

The use of non-equilibrium rate-constants to assign internal energies is likely to increase in photochemical and ion–molecule studies. For the technique to be of optimum usefulness, well calibrated rate-constants are needed for a variety of unimolecular reactions and reports dealing primarily with characterisation of a unimolecular reaction should present the rate-constants, k_E, so that they can be used by others.

1.4.2 Some new directions

The obvious question concerning unimolecular reactions is the mechanism for the fast intramolecular energy transfer. Studies of competitive intramolecular reactions from molecules, radicals or ions that are initially activated at a particular site are likely sources of new evidence. There is at least one theoretical attempt[247] in progress to develop a mechanism for explaining fast intramolecular energy transfer. The consequences of failure of fast randomisation have been explored[248] from a stochastic point of view. A great bulk of data support the validity of RRKM theory for calculating rate-constants (an interesting new test[253] has been made for ions prepared at known energies by charge transfer excitation) and further efforts at developing computational models primarily fall into the area of applications. In this review the threshold energy was viewed as a thermochemical quantity; however it seems likely that in the near future more attention will be devoted toward understanding the potential surfaces for unimolecular reactions.

Acknowledgement

I wish to thank the individuals who made their work available to me prior to publication. The assistance of Dr W. H. Duewer, Mr H. W. Chang and Mr B. Holmes in the preparation of this manuscript is appreciated.

Acknowledgment would be incomplete without mention of the debt to Professor Rabinovitch who was responsible for my introduction to the field of unimolecular reactions.

Financial assistance from the National Science Foundation, U.S.A., is acknowledged for preparation of this report and for support of other unimolecular reaction studies done at Kansas State University.

References

1. Spicer, L. D. and Rabinovitch, B. S. (1970). *Ann. Rev. Phys. Chem.*, **21**, 349
2. Laidler, K. J. (1969). *Theories of Chemical Reaction Rates* (New York: McGraw-Hill Co.)
3. Gardiner, W. C., Jr. (1970). *Rates and Mechanisms of Chemical Reactions* (New York: Benjamin Inc.)
4. Parmenter, C. S. and Poland, H. M. (1969). *J. Chem. Phys.*, **51**, 1551
5. Schlag, E. W. and Weyssenhoff, H. V. (1969). *J. Chem. Phys.*, **51**, 2508
6. Belford, R. L. and Strehlow, R. A. (1969). *Ann. Rev. Phys. Chem.*, **20**, 247
7. Troe, Von J. and Wagner, H. Gg. (1967). *Ber. Bunsenges. Phys. Chem.*, **71**, 937
8. Troe, Von J. (1969). *Ber. Bunsenges. Phys. Chem.*, **73**, 188
9. Rabinovitch, B. S. and Setser, D. W. (1964). *Advan. Photochem.*, **3**, 1
10. Marcus, R. A. (1968). *Chemische Elementarprozesse*, 23, 109, ed. by Hartmann, H. (Berlin: Springer-Verlag)
11. Vestal, M. L. (1968). *Fundamental Procession Radiation Chemistry*, ed. by Ausloos, P. (New York: Interscience)
12. Chang, H. W., Craig, N. L. and Setser, D. W. (1972). *J. Phys. Chem.*, **76**, in the press
13. Bishop, D. M. and Laidler, K. J. (1970). *Trans. Faraday Soc.*, **66**, 1685
14. Benson, S. W. and O'Neal, H. E. (1970). *Kinetic Data on Gas Phase Unimolecular Reactions*, NSRDS-NBS 21, Superintendent of Documents, U. S. Goverment Printing Office, Washington
15. Frey, H. M. and Walsh, R. (1969). *Chem. Rev.*, **69**, 103
16. Maccoll, A. (1969). *Chem. Rev.*, **69**, 33
17. Leathard, D. A. and Purnell, J. H. (1970). *Ann. Rev. Phys. Chem.*, **21**, 197
18. Benson, S. W. (1968). *Thermochemical Kinetics* (New York: John Wiley)
19. O'Neal, H. E. and Benson, S. W. (1970). *Int. J. Chem. Kinetics*, **2**, 423
20. Golden, D. M., Solly, R. K. and Benson, S. W. (1971). *J. Phys. Chem.*, **75**, 1333
21. Waage, E. V. and Rabinovitch, B. S. (1971). *Int. J. Chem. Kinetics*, **3**, 105
22a. Tsang, W. (1970). *Int. J. Chem. Kinetics*, **2**, 311
22b. Tsang, W. (1969). *Int. J. Chem. Kinetics*, **1**, 245
23. Frey, H. M. and Solly, R. K. (1968). *Trans. Faraday Soc.*, **64**, 1859
24. O'Neal, H. E. and Frey, H. M. (1970). *Int. J. Chem. Kinetics*, **2**, 343
25a. Watkins, K. W. and Ostreko, L. A. (1969). *J. Phys. Chem.*, **73**, 2080
25b. Watkins, K. W. and Lawson, D. R. *J. Phys. Chem.*, **75**, 1632
26. Hardwidge, E. A., Larson, C. W. and Rabinovitch, B. S. (1970). *J. Amer. Chem. Soc.*, **92**, 3278
27. Krohn, K. A., Parks, N. J. and Root, J. W. (1971). *J. Chem. Phys.*, **55**, 5785
28. Chiang, M. H., Gislason, E. A., Mahan, B. H., Tsao, C. W. and Werner, A. S. (1971). *J. Phys. Chem.*, **75**, 1426
29. Mahan, B. H. (1970). *Accounts Chem. Res.*, **3**, 393 (1970)
30. Herman, Z., Lee, A. and Wolfgang, R. (1968). *J. Chem. Phys.*, **51**, 452
31a. Tardy, D. C. and Rabinovitch, B. S. (1967). *J. Chem. Phys.*, **45**, 3720
31b. Tardy, D. C. and Rabinovitch, B. S. (1968). *J. Chem. Phys.*, **48**, 1282
32. Larson, C. W. and Rabinovitch, B. S. (1969). *J. Chem. Phys.*, **51**, 2293
33. Tardy, D. C. and Rabinovitch, B. S. (1968). *J. Chem. Phys.*, **48**, 5194
34. Rynbrandt, J. D. and Rabinovitch, B. S. (1970). *J. Phys. Chem.*, **74**, 1679
35. Chang, H. W. and Setser, D. W. (1969). *J. Amer. Chem. Soc.*, **91**, 7648
36. Clark, W. G., Setser, D. W. and Siefert, E. E. (1970). *J. Phys. Chem.*, **74**, 1670
37a. Lin, Y. N. and Rabinovitch, B. S. (1970). *J. Phys. Chem.*, **74**, 3151
37b. Chan, S. C., Rabinovitch, B. S., Bryant, J. T., Spicer, L. D., Fujimoto, T., Lin, Y. N. and Pavlou, S. P. (1970). *J. Phys. Chem.*, **74**, 3160
37c. Pavlou, S. P. and Rabinovitch, B. S. (1971). *J. Phys. Chem.*, **75**, 1366, 2171
38. Chan, S. C., Bryant, J. T., Spicer, L. D. and Rabinovitch, B. S. (1970). *J. Phys. Chem.*, **74**, 2058
39. Siefert, E. E. and Setser, D. W. (1972). *J. Chem. Phys.*, to be published
40. Spicer, L. D. and Rabinovitch, B. S. (1970). *J. Phys. Chem.*, **74**, 2445
41. Hirschfelder, J. O., Curtiss, C. F. and Bird, R. B. (1965). *Molecular Theory of Gases and Liquids* (New York: John Wiley)
42. Waage, E. V. and Rabinovitch, B. S. (1970). *J. Chem. Phys.*, **52**, 5581
43. Waage, E. V. and Rabinovitch, B. S. (1970). *Chem. Rev.*, **70**, 377

44. Schneider, F. W. and Rabinovitch, B. S. (1962). *J. Amer. Chem. Soc.*, **84**, 4215
45. Rabinovitch, B. S., Gilderson, P. W. and Schneider, F. W. (1965). *J. Amer. Chem. Soc.*, **87**, 158
46. Schneider, F. W. and Rabinovitch, B. S. (1963). *J. Amer. Chem. Soc.*, **85**, 2365
47. Maloney, K. M. and Rabinovitch, B. S. (1969). *J. Phys. Chem.*, **73**, 1652
48. Maloney, K. M., Pavlou, S. P. and Rabinovitch, B. S. (1969). *J. Phys. Chem.*, **73**, 2756
49. Yip, C. K. and Pritchard, H. O. (1970). *Can. J. Chem.*, **48**, 2942
50. Bunker, D. L. (1971). *Chem. Phys. Lett.*, in the press
51. Lin, M. C. and Laidler, K. J. (1968). *Trans. Faraday Soc.*, **64**, 927
52. Rabinovitch, B. S., Gilderson, P. W. and Blades, A. T. (1964). *J. Amer. Chem. Soc.*, **86**, 2994
53. Chesick, J. P. (1960). *J. Amer. Chem. Soc.*, **82**, 3277
54. Halberstadt, M. L. and Chesick, J. P. (1965). *J. Phys. Chem.*, **69**, 429
55. Flowers, M. C. and Frey, H. M. (1962). *J. Chem. Soc.*, 1157
56a. Carter, W. L. and Bergman, R. G. (1968). *J. Amer. Chem. Soc.*, **90**, 7344
56b. Bergman, R. G. and Carter, W. L. (1969). *J. Amer. Chem. Soc.*, **91**, 7411
57. McGreer, D. E. and McKinley, J. W. (1961). *Can. J. Chem.*, **49**, 105
58. Siu, A. K. Q., St. John, W. M. and Hayes, E. F. (1970). *J. Amer. Chem. Soc.*, **92**, 7249
59. Wieder, G. M. and Marcus, R. A. (1962). *J. Chem. Phys.*, **37**, 1835
60. Clark, W. G., Dees, K. and Setser, D. W. (1971). *J. Amer. Chem. Soc.*, **93**, 5328
61. Johnson, R. L. and Setser, D. W. (1967). *J. Phys. Chem.*, **71**, 4366
62a. Holbrook, K. A., Palmer, J. S. and Parry, K. A. W. (1970). *Trans. Faraday Soc.*, **66**, 869
62b. Holbrook, K. A. and Parry, K. A. W. (1970). *J. Chem. Soc. B*, 1019
63. Hauser, W. P. and Walters, W. D. (1963). *J. Phys. Chem.*, **67**, 1328
64. Lin, M. C. and Laidler, K. J. (1968). *Trans. Faraday Soc.*, **64**, 94
65a. Frey, H. M. and Elliot, C. S. (1966). *Trans. Faraday Soc.*, **62**, 895
65b. Frey, H. M. and Pope, B. M. (1969). *Trans. Faraday Soc.*, **65**, 441
66. Marshall, D. C. and Frey, H. M. (1965). *Trans. Faraday Soc.*, **61**, 1715
67. Jeffers, P. M. and Shaub, W. (1969). *J. Amer. Chem. Soc.*, **91**, 7706
68. Vreeland, R. W. and Swinehart, D. F. (1963). *J. Amer. Chem. Soc.*, **85**, 3349
69. Carr, R. W. and Walters, W. D. (1963). *J. Phys. Chem.*, **67**, 1370
70. Carr, R. W. and Walter, W. D. (1966). *J. Amer. Chem. Soc.*, **88**, 884
71. Holbrook, K. A. and Marsh, A. R. W. (1967). *Trans. Faraday Soc.*, **63**, 643
72. Heydtmann, H. and Volker, G. W. (1967). *Z. Phys. Chem. (Frankfurt)*, **55**, 296
73. Volker, G. W. and Heydtmann, H. (1968). *Z. Naturforsch.*, **23b**, 1407
74. Dees, K. and Setser, D. W. (1968). *J. Chem. Phys.*, **49**, 1193
75. Heydtmann, H. (1968). *Ber. Bunsenges. Phys. Chem.*, **72**, 1009
76. Holbrook, K. A. and Palmer, J. S. (1971). *Trans. Faraday Soc.*, **67**, 80
77. Martens, G. J., Golfroid, M. and Ramoisy, L. (1970). *Int. J. Chem. Kinetics*, **2**, 123
78. Heydtmann, H. (1967). *Chem. Phys. Lett.*, **1**, 105
79. Lenzi, M. and Mele, A. (1965). *J. Chem. Phys.*, **43**, 1974
80. Rabinovitch, B. S., Kubin, R. F. and Harrington, R. E. (1963). *J. Chem. Phys.*, **38**, 405
81. Perona, M. J., Johnson, R. L. and Setser, D. W. (1969). *J. Phys. Chem.*, **73**, 2091
82. Baer, T. and Bauer, S. H. (1970). *J. Amer. Chem. Soc.*, **92**, 4773
83. Taylor, G. W. and Simons, J. W. (1971). *Int. J. Chem. Kinetics*, **3**, 25
84a. Simons, J. W. and Taylor, G. W. (1969). *J. Phys. Chem.*, **73**, 1274
84b. Simons, J. W. and Taylor, G. W. (1970). *J. Phys. Chem.*, **74**, 464
85. Duewer, W. H., White, I., Clyne, M. A. A. and Setser, D. W. (1972). *Int. J. Chem. Kinetics*, to be submitted
86. Lin, Y. N. and Rabinovitch, B. S. (1970). *J. Phys. Chem.*, **74**, 1769
87. Haney, M. A. and Franklin, J. L. (1968). *J. Chem. Phys.*, **48**, 4093
88. Klots, C. E. (1964). *J. Chem. Phys.*, **41**, 117
89. Chang, H. W., Setser, D. W. and Perona, M. J. (1971). *J. Phys. Chem.*, **75**, 2070
90. Freund, S. M., Fisk, G. A., Herschbach, D. R. and Klemperer, W. (1971). *J. Chem. Phys.*, **54**, 2510
91a. Mok, M. H. and Polanyi, J. C. (1970). *J. Chem. Phys.*, **53**, 4588
91b. Mok, M. H. and Polanyi, J. C. (1969). *J. Chem. Phys.*, **51**, 1451
92. Miller, G. and Light, J. C. (1971). *J. Chem. Phys.*, **54**, 1643
93. Clough, P. N., Polanyi, J. C. and Taguchi, R. T. (1970). *Can. J. Chem.*, **48**, 2920

94. Berry, M. J. and Pimentel, G. C. (1968). *J. Chem. Phys.*, **49**, 5190
95. Miller, W. B., Safron, S. A. and Herschbach, D. R. (1967). *Discuss. Faraday Soc.*, **44**, 108, 291
96. Yang, J. H. and Conway, D. C. (1965). *J. Chem. Phys.*, **43**, 1296
97. Hassler, J. C. and Setser, D. W. (1966). *J. Chem. Phys.*, **45**, 3246
98. Dees, K. and Setser, D. W. (1971). *J. Phys. Chem.*, **75**, 2231 (1971)
99. Kim, K. C. and Setser, D. W. (1972). Unpublished results and *J. Phys. Chem.*, **76**, in the press
100. Pritchard, G. O. and Perona, M. J. (1970). *Int. J. Chem. Kinetics*, **2**, 281; this paper contains a summary of the fluoroethane studies from Professor Pritchard's laboratory
101. Pritchard, G. O. and Perona, M. J. (1969). *J. Phys. Chem.*, **73**, 2944
102. Perona, M. J., Bryant, J. T. and Pritchard, G. O. (1968). *J. Amer. Chem. Soc.*, **90**, 4782
103. Kerr, J. A., Kirk, A. W., O'Grady, B. V., Phillips, D. C. and Trotman-Dickenson, A. F. (1967). *Discuss. Faraday Soc.*, **44**, 263
104. Phillips, D. C. and Trotman-Dickenson, A. F. (1968). *J. Chem. Soc. A*, 1667
105. Phillips, D. C. and Trotman-Dickenson, A. F. (1968). *J. Chem. Soc. A*, 1667
106. Cadman, P., Phillips, D. C. and Trotman-Dickenson, A. F. (1968). *Chem. Commun.*, 796
107. Kerr, J. A., Phillips, D. C. and Trotman-Dickenson, A. F. (1968). *J. Chem. Soc. A*, 1806
108. Kirk, A. W., Trotman-Dickenson, A. F. and Trus, B. L. (1968). *J. Chem. Soc. A*, 3058
109. Kerr, J. A., O'Grady, B. V. and Trotman-Dickenson, A. F. (1969). *J. Chem. Soc. A*, 275
110a. Day, M. and Trotman-Dickenson, A. F. (1969). *J. Chem. Soc. A*, 233
110b. Cadman, P., Day, M. and Trotman-Dickenson, A. F. (1970). *J. Chem. Soc. A*, 2498
111. Cadman, P., Day, M., Kirk, A. W. and Trotman-Dickenson, A. F. (1970). *Chem. Commun.*, 203
112a. Tschuikow-Roux, E., Quiring, W. J. and Simmie, J. M. (1970). *J. Phys. Chem.*, **74**, 2449
112b. Tschuikow-Roux, E. and Quiring, W. J. (1971). *J. Phys. Chem.*, **75**, 295
113. Sianesi, D., Nelli, G. and Fontanelli, R. (1968). *Chim. Ind. (Milan)*, **50**, 619
114. Simmie, J. M. and Tschuikow-Roux, E. (1970). *J. Phys. Chem.*, **74**, 4075
115. Ross, D. S. and Shaw, R. (1971). *J. Phys. Chem.*, **75**, 1170
116. Padrick, T. D. and Pimentel, G. C. (1971). *J. Chem. Phys.*, **54**, 720
117. Jakubowski, E., Sandhu, H. S. and Strausz, O. P. (1971). *J. Amer. Chem. Soc.*, **93**, 2610
118a. Bell, T. N. and Zucker, U. F. (1970). *Can. J. Chem.*, **48**, 1209
118b. Bell, T. N. and Zucker, U. F. (1970). *J. Phys. Chem.*, **74**, 979
119. Bell, T. N. and Platt, A. E. (1970). *Int. J. Chem. Kinetics*, **2**, 299
120. Setser, D. W. and Richardson, W. G. (1969). *Can. J. Chem.*, **47**, 2593
121. Solly, R. K., Gac, N. A., Golden, D. M. and Benson, S. W. (1970). 160th American Chemical Society Meeting, Chicago, Illinois, Abstract 99 of the Physical Division
122. Larson, C. W., Rabinovitch, B. S. and Tardy, D. C. (1967). *J. Chem. Phys.*, **47**, 4570
123. Larson, C. W., Tardy, D. C. and Rabinovitch, B. S. (1968). *J. Chem. Phys.*, **49**, 299
124. Larson, C. W. and Rabinovitch, B. S. (1969). *J. Chem. Phys.*, **50**, 871
125. Larson, C. W., Hardwidge, E. A. and Rabinovitch, B. S. (1969). *J. Chem. Phys.*, **50**, 2769
126. Larson, C. W. and Rabinovitch, B. S. (1970). *J. Chem. Phys.*, **52**, 5181
127. Cvetanović, R. J. and Irwin, R. S. (1967). *J. Chem. Phys.*, **46**, 1694
128. Louks, L. F. and Laidler, K. J. (1967). *Can. J. Chem.*, **45**, 2795
129. Lin, M. C. and Laidler, K. J. (1968). *Trans. Faraday Soc.*, **64**, 79
130a. Barker, J. R., Keil, D. G., Michael, J. V. and Osborne, D. T. (1970). *J. Chem. Phys.*, **52**, 2079
130b. Cowfer, J. A., Keil, D. G., Michael, J. V. and Yeh, C. (1971). *J. Phys. Chem.*, **75**, 1584
130c. Michael, J. V. (1971). Private communication. New measurements from this laboratory give $k_{bi}(C_2H_4) = 14.8 \pm 2.0 \times 10^{-13}$ cm^3 molecule^{-1} s^{-1}
131. Halstead, M. D., Leathard, D. A., Marshall, R. M. and Purnell, J. H. (1970). *Proc. Roy. Soc. A*, **316**, 575
132. Kurylo, M. J., Peterson, N. C. and Braun, W. (1970). *J. Chem. Phys.*, **53**, 2776
133. Eyre, J. A., Hikada, T. and Dorfman, L. M. (1971). *J. Chem. Phys.*, **54**, 3422
134. Setser, D. W. and Rabinovitch, B. S. (1964). *J. Chem. Phys.*, **40**, 2427
135. Oref, I. and Rabinovitch, B. S. (1968). *J. Phys. Chem.*, **72**, 4488
136. Kurlo, M. J., Peterson, N. C. and Braun, W. (1971). *J. Chem. Phys.*, **54**, 4662
137. Papic, M. M. and Laidler, K. J. (1971). *Can. J. Chem.*, **49**, 549
138. Knox, J. and Waugh, K. C. (1969). *Trans. Faraday Soc.*, **65**, 1585
139. Franklin, J. A., Goldfinger, P. and Huybrechts, G. (1968). *Ber. Bunsenges. Phys. Chem.*, **72**, 173

140. Beadle, P. C., Knox, J. H., Placido, F. and Waugh, K. C. (1969). *Trans. Faraday Soc.,* **65,** 1571
141. Tardy, D. C. and Rabinovitch, B. S. (1968). *Trans. Faraday Soc.,* **64,** 1844
142. Huy, L. K., Forst, W., Franklin, J. A. and Huybrechts, G. (1969). *Chem. Phys. Letts.,* **3,** 307
143a. Rodgers, A. S. (1968). *J. Phys. Chem.,* **72,** 3400
143b. Rodgers, A. S. (1968). *J. Phys. Chem.,* **72,** 3407
144. Kirk, A. W. and Tschuikow-Roux, E. (1970). *J. Chem. Phys.,* **53,** 1925
145a. Von Hoyermann, K., Wagner, H. Gg., Wolrum, J. and Zellner, R. (1970). *Ber. Bunsenges. Phys. Chem.,* **75,** 22
145b. Von Hoyermann, K., Wagner, H. Gg. and Wolfrum, J. (1968). *Ber. Bunsenges. Phys. Chem.,* **72,** 1004
146. Michael, J. V. and Niki, H. (1967). *J. Chem. Phys.,* **46,** 4969
147. Stief, L. J., De Carlo, V. J. and Payne, W. A. (1971). *J. Chem. Phys.,* **54,** 1913
148. Watkins, K. W. and O'Deen, L. A. (1971). *J. Phys. Chem.,* **75,** 2665
149. Fowles, P., de Sorgo, M., Yarwood, A. J., Strausz, O. and Gunning, H. E. (1967). *J. Amer. Chem. Soc.,* **89,** 1352
150. Paraskevopoulos, G. and Cvetanović, R. J. (1970). *J. Chem. Phys.,* **52,** 5821
151. Paraskevopoulos, G. and Cvetanović, R. J. (1969). *J. Chem. Phys.,* **50,** 590
152. Yamazaki, H. and Cvetanović, R. J. (1964). *J. Chem. Phys.,* **41,** 3703
153. Bradley, J. N., Edwards, A. D. and Gilbert, J. R. (1971). *J. Chem. Soc. A,* 327
154. Scott, P. M. and Cvetanović, R. J. (1971). *J. Chem. Phys.,* **54,** 1440
155. Paraskevopoulos, G. and Cvetanović, R. J. (1969). *J. Amer. Chem. Soc.,* **91,** 7573
156. Smail, T., Miller, G. E. and Rowland, F. S. (1970). *J. Phys. Chem.,* **74,** 3464
157. McKnight, C. F., Parks, N. J. and Root, J. W. (1969). *J. Phys. Chem.,* **74,** 217
158. Weeks, R. W., Jr. and Garland, J. K. (1971). *J. Amer. Chem. Soc.,* **93,** 2381
159. Ting, C. T. and Rowland, F. S. (1970). *J. Phys. Chem.,* **74,** 4080
160. Tewarson, A. and Lampe, F. W. (1968). *J. Phys. Chem.,* **72,** 3261
161. Taylor, G. W. and Simons, J. W. (1971). *Int. J. Chem. Kinet.,* **3,** 453
162. Hase, W. L. and Simons, J. W. (1970). *J. Chem. Phys.,* **52,** 4004
163. Johnson, R. L., Hase, W. L. and Simons, J. W. (1970). *J. Chem. Phys.,* **52,** 3911
164. Hase, W. L. and Simons, J. W. (1971). *J. Chem. Phys.,* **54,** 1277
165. Elliot, C. S. and Frey, H. M. (1968). *Trans. Faraday Soc.,* **64,** 2353
166. Callear, A. B. and Van Den Berg, H. E. (1970). *Chem. Phys. Lett.,* **5,** 23
167. Kirk, A. W. (1969). Private communication
168. Bowers, P. G. (1967). *J. Chem. Soc. A,* 466
169. Bowers, P. G. (1968). *Can. J. Chem.,* **46,** 307
170. Bowers, P. G. (1970). *J. Phys. Chem.,* **74,** 952
171. Gandini, A., Whytock, D. A. and Kutschke, K. O. (1968). *Proc. Roy. Soc. A,* **306,** 541
172. Halpern, A. M. and Ware, W. R. (1970). *J. Chem. Phys.,* **53,** 1969
173. Dees, K. and Setser, D. W. (1971). *J. Phys. Chem.,* **75,** 2240
174. Chesick, J. P. (1966). *J. Chem. Phys.,* **45,** 3934
175. Borrell, P. and Cashmore, P. (1969). *Trans. Faraday Soc.,* **65,** 2412
176. Atkinson, R. and Thrush, B. A. (1970). *Proc. Roy. Soc. A,* **316,** 123, 131, 143
177. Rabinovitch, B. S. and Thrush, B. A. (1971). *J. Phys. Chem.,* **75,** 3376
178. Lee, E. E. and Lee, E. K. C. (1969). *J. Chem. Phys.,* **50,** 2094
179a. Carless, H. A. J. and Lee, E. K. C. (1970). *J. Amer. Chem. Soc.,* **92,** 4482
179b. Carless, H. A. J. and Lee, E. K. C. (1970). *J. Amer. Chem. Soc.,* **92,** 6683
180a. Dorer, F. H. (1969). *J. Phys. Chem.,* **73,** 3109
180b. Dorer, F. H. (1970). *J. Phys. Chem.,* **74,** 1142
180c. Dorer, F. H., Brown, E., Do, J. and Rees, R. (1971). *J. Phys. Chem.,* **75,** 1640
181. Cadman, P., Meunier, H. M. and Trotman-Dickenson, A. F. (1969). *J. Amer. Chem. Soc.,* **91,** 7640
182. Roquitte, B. C. (1969). *J. Amer. Chem. Soc.,* **91,** 7664
183. Denschlag, H. O. and Lee, E. K. C. (1968). *J. Amer. Chem. Soc.,* **90,** 3628
184. Campbell, R. J. and Schlag, E. W. (1967). *J. Amer. Chem. Soc.,* **89,** 5103
185. Montague, D. C. and Rowland, F. S. (1969). *J. Amer. Chem. Soc.,* **91,** 7230
186. Dorer, F. H. and Johnson, S. H. (1971). *J. Phys. Chem.,* **75,** 3651
187. Kirk, A. W. and Tschuikow-Roux, E. (1969). *J. Chem. Phys.,* **51,** 2247
188. Dhingra, A. K., Vorachek, J. H. and Koob, R. D. (1971). *Chem. Phys. Lett.,* **9,** 17

189. Vorachek, J. H. and Koob, R. D. (1971). *J. Chem. Phys.*, submitted for publication
190. Tschuikow-Roux, E. and Kodama, S. (1969). *J. Chem. Phys.*, **50**, 5297
191. Prášil, Z. and Forst, W. (1969). *J. Phys. Chem.*, **71**, 3166
192a. Lifshitz, C. and Long, F. A. (1965). *J. Phys. Chem.*, **69**, 3746
192b. Lifshitz, C. and Long, F. A. (1965). *J. Phys. Chem.*, **69**, 3737
193. Lindholm, E. (1968). *Arkiv Fysik.*, **37**, 37
194. Rowland, C. G., Eland, J. H. D. and Danby, G. J. (1968). *Chem. Commun.*, 1535; but see reference 195
195. Bogan, D. J. and Hand, C. W. (1971). *J. Phys. Chem.*, **75**, 1532
196. Buttrill, S. E., Jr. (1970). *J. Chem. Phys.*, **52**, 6174
197. McLafferty, F. W., Wachs, T., Lifshitz, C., Innorta, G. and Irving, P. (1970). *J. Amer. Chem. Soc.*, **92**, 6867
198. Howe, I. and Williams, D. H. (1969). *J. Amer. Chem. Soc.*, **91**, 7137
199. Yeo, A. N. H. and Williams, D. H. (1971). *J. Amer. Chem. Soc.*, **93**, 395
200. Chupka, W. A. and Berkowitz, J. (1967). *J. Chem. Phys.*, **47**, 2921
201. Sahlstrom, G. and Szabo, I. (1968). *Arkiv. Fysik.*, **38**, 145; the interested reader should consult numerous other publications on charge exchange from Lindholm's laboratory
202. Vestal, M. and Futrell, J. H. (1970). *J. Chem. Phys.*, **52**, 978
203. Hills, L. P., Futrell, J. H. and Wahrhaftig, A. L. (1969). *J. Chem. Phys.*, **51**, 5255
204. Tou, J. C. (1970). *J. Phys. Chem.*, **74**, 3076
205. Spotz, E. L., Seitz, W. A. and Franklin, J. L. (1969). *J. Chem. Phys.*, **51**, 5142
206. Haney, M. A. and Franklin, J. L. (1969). *J. Chem. Phys.*, **50**, 2028
207. Cooks, R. G., Setser, D. W., Jennings, K. and Jones, S. (1972). *Int. J. Mass Spectrom. Ion Phys.*, in the press
208. Smyth, K. C. and Shannon, T. W. (1969). *J. Chem. Phys.*, **51**, 4633
209. Marcotte, R. E. and Tierman, T. O. (1971). *J. Chem. Phys.*, **54**, 3385
210. Wolfgang, R. (1970). *Accounts Chem. Res.*, **3**, 48
211. Buttrill, S. E., Jr. (1970). *J. Amer. Chem. Soc.*, **92**, 3560
212. Bowers, M. T., Elleman, D. D. and Beauchamp, J. L. (1968). *J. Phys. Chem.*, **72**, 3599
213. Diang, A. and Henglein, A. (1969). *Ber. Bunsenges. Phys. Chem.*, **73**, 562
214. Holtz, D., Beauchamp, J. L. and Eyler, J. R. (1970). *J. Amer. Chem. Soc.*, **92**, 7045
215. Bowers, M. T. and Elleman, D. D. (1970). *J. Amer. Chem. Soc.*, **92**, 1847
216. Futrell, J. H., Abramson, F. P., Bhattacharya, A. K. and Tiernan, T. O. (1970). *J. Chem. Phys.*, **52**, 3655
217. Futrell, J. H. and Tiernan, T. O. (1970). *J. Chem. Phys.*, **52**, 2366
218. Herman, Z., Hierl, P., Lee, A. and Wolfgang, R. (1969). *J. Chem. Phys.*, **51**, 454
219. Bowers, M. T. and Elleman, D. D. (1970). *J. Amer. Chem. Soc.*, **92**, 7258
220. Beauchamp, J. L. and Buttrill, S. E., Jr. (1968). *J. Chem. Phys.*, **48**, 1783
221a. Field, F. H. (1968). *Accounts Chem. Res.*, **1**, 42
221b. Field, F. H. (1970). *J. Amer. Chem. Soc.*, **92**, 2672
222. Ferguson, E. E. (1970). *Accounts Chem. Res.*, **3**, 902
223. Lifshitz, C., Peers, A. M., Grajower, R. and Weiss, M. (1970). *J. Chem. Phys.*, **53**, 4605
224. Naff, W. T., Compton, R. N. and Cooper, C. D. (1971). *J. Chem. Phys.*, **54**, 212
225. Naff, W. T., Cooper, C. D. and Compton, R. N. (1968). *J. Chem. Phys.*, **49**, 2784
226. Klots, C. E. (1967). *J. Chem. Phys.*, **46**, 1197
227. De Corpo, J. J., Bafus, D. A. and Franklin, J. L. (1971). *J. Chem. Phys.*, **54**, 1592
228. De Corpo, J. J. and Franklin, J. L. (1971). *J. Chem. Phys.*, **54**, 1885
229. Keck, J. and Kalelkar, A. (1968). *J. Chem. Phys.*, **49**, 3211
230. Light, J. C. (1967). *Discuss. Faraday Soc.*, **44**, 14
231. Mies, F. H. (1969). *J. Chem. Phys.*, **51**, 798
232. Klots, C. E. (1971). *J. Phys. Chem.*, **75**, 1526
233. Bunker, D. L. and Pattengill, M. (1968). *J. Chem. Phys.*, **48**, 772
234. Rynbrandt, J. D. and Rabinovitch, B. S. (1971). *J. Phys. Chem.*, **75**, 2164
235. Rynbrandt, J. D. and Rabinovitch, B. S. (1971). *J. Chem. Phys.*, **54**, 2275
236. Oref, I., Schuetzle, D. and Rabinovitch, B. S. (1971). *J. Chem. Phys.*, **54**, 575
237a. LeRoy, R. L. (1970). *J. Chem. Phys.*, **53**, 846
237b. LeRoy, R. L. (1971). *J. Chem. Phys.*, **55**, 1476
238a. Forst, W. and Prášil, Z. (1969). *J. Chem. Phys.*, **51**, 3006
238b. Forst, W. (1971). *Chem. Rev.*, **71**, 339
239. Hoare, M. R. and Ruijgrok, Th. W. (1970). *J. Chem. Phys.*, **52**, 113

240. Hoare, M. R. (1970). *J. Chem. Phys.,* **52,** 5695. (See comment in footnote 31 of reference 232 regarding misprints.)
241. Lau, K. H. and Lin, S. H. (1971). *J. Phys. Chem.,* **75,** 2458
242. Forst, W. and Prášil, A. (1970). *J. Chem. Phys.,* **52,** 113
243. Kobrinsky, P. C., Pritchard, G. O. and Toby, S. (1971). *J. Phys. Chem.,* **75,** 2225
244. Hase, W. L., Growcock, F. W. and Simons, J. W. (1971). Private communication
245. Hase, W. L., Johnson, R. L. and Simons, J. W. (1971). *J. Chem. Phys.,* in the press
246. Gill, P. S., Inel, Y. and Meisels, G. G. (1971). *J. Chem. Phys.,* **54,** 2811
247. Kay, K. and Rice, S. A. (1972). Private communication
248. Gelbert, W. M., Rice, S. A. and Freed, K. F. (1970). *J. Chem. Phys.,* **52,** 5718
249. Hofmann, R. (1971). Private communication
250. Lin, M. C. (1971). *J. Phys. Chem.,* **75,** 3642
251. Bunker, D. L. (1972). *Chem. Phys. Lett.,* in the press
252. Hase, W. L., Phillips, R. J. and Simons, J. W. (1971). *Chem. Phys. Lett.,* **12,** 6
253. Andlauer, B. and Ottinger, Ch. (1971). *J. Chem. Phys.,* **55,** 1471 and *Z. Naturforsch.,* submitted
254. Roche, A. E., Sutton, M. M., Bohme, D. K. and Schiff, H. I. (1971). *J. Chem. Phys.,* **55,** 5480

2
Atomic and Bimolecular Reactions

I. M. CAMPBELL and D. L. BAULCH
University of Leeds

2.1 INTRODUCTION

The ever increasing literature on the reaction kinetics of atomic and diatomic species makes selectivity inevitable in this review. A recent survey of reactions of excited atoms[1] largely pre-empts the need to cover these, and for the rest emphasis is placed upon quantitative aspects of publications during 1969 and 1970, where the increasing sophistication of techniques has made most impact. Despite restriction to areas of particular interest, we hope to convey a general impression of this intensely active field. In the review all quoted data are for room temperature and all energies are in $J\ mol^{-1}$ unless otherwise stated. ($4.184\ J\ mol^{-1} = 1\ cal\ mol^{-1}$)

2.2 EFFECTS OF INTERNAL ENERGY ON REACTIVITY

Polanyi *et al.*[2-4] have attempted a theoretical rationalisation of the roles of translational and vibrational energy in determining the rates of atom–

diatomic and diatomic–diatomic transfer reactions. Calculations, locating the activation energy barrier positions on potential energy surfaces (L.E.P.S. or B.E.B.O.) for linear and rectangular co-planar complexes respectively, distinguished two extreme cases. Translational energy was dominant for activation when the barrier was in the 'entry' valley (Case I) whereas vibrational energy was most important for an 'exit' valley barrier (Case II); the underlying key factor appeared to be the alignment of directional characteristics of these modes with the reaction coordinate in the barrier region. For atom–diatomic reaction, Cases I and II corresponded to substantially exothermic and endothermic reactions respectively. As yet, no detailed calculations on the role of rotational energy have been performed.

Experimental evidence provided the impetus for these calculations. Jaffe and Anderson[5], in a molecular beam study, showed that relative translational energies of up to 899 kJ mol^{-1} did not induce the reaction

$$HI + DI \longrightarrow HD + I_2 \quad (\Delta H = -19.6 \text{ kJ mol}^{-1}) \quad (2.1)$$

As this occurs readily under conditions of thermal equilibrium, internal excitation must be critical. Mok and Polanyi[4] interpret this as an example of Case II, a view not inconsistent with an alternative treatment by trajectory analysis[6].

Application of the Principle of Microscopic Reversibility to exothermic reactions, yielding significant vibrationally excited products, indicates that the reverse endothermic reaction will be favoured by vibrational excitation of reactants. Anlauf et al.[7] and Polanyi and Tardy[8] measured vibraluminescence intensities at low pressures from the reactions

$$Cl + HI \rightarrow HCl + I \quad (2.2)$$

$$F + H_2 \rightarrow HF + H \quad (2.3)$$

The results suggested that the rate constants for the reverse endothermic reactions would increase by orders of magnitude for even small vibrational excitation of HCl and HF. Similar conclusions for reaction (2.3) come from classical trajectory calculations by Anderson[9].

More problematical is the postulate by Bauer et al. that critical vibrational excitation of one reactant is rate determining for isotope exchange systems in shock tubes, e.g. H_2/D_2 [10]. Mok and Polanyi[4] note that such systems will have symmetrically placed energy barriers and therefore could be intermediate between Cases I and II. Further discussion is reserved until Section 2.10.

Most recently, Stedman et al.[11] have found that vibrational excitation of H_2 enhances its reactivity towards Cl (Section 2.4.2), O and possibly OH, all interpretable within Case II. However, Wray et al.[12] have found one example of an atom–diatomic endothermic reaction where vibrational excitation does not enhance the rate. Dilute O_3/N_2 mixtures were shock heated to induce the reaction

$$O + N_2 \longrightarrow NO + N \quad (\Delta H = 318.4 \text{ kJ mol}^{-1}) \quad (2.4)$$

monitored by NO infrared thermal emission. Microscopic reversibility then suggests that the reverse does not yield vibrationally excited nitrogen directly.

2.3 REACTIONS OF CARBON ATOMS C(^{3}P), C(^{1}D) AND C(^{1}S)

Only recently has a suitable source of carbon atoms been found; Braun et al.[13] have shown that vacuum u.v. photolysis of carbon suboxide yields C(^{3}P) and C(^{1}D) as primary, and C(^{1}S) as minor secondary, products. Flash photolysis in the presence of added substrates produced chemical reaction as evidenced by the appearance of product absorption spectra. With excess additives, at large dilution in argon, pseudo-first-order decays were observed by atomic absorption spectrometry. High rate constants between 10^{10} and 10^{11} l mol^{-1} s^{-1} for reactions of C(^{3}P) with NO and O$_2$ and for C(^{1}D) with NO, H$_2$ and CH$_4$ showed these to have little activation energy. For C(^{1}D) reaction with O$_2$ and C(^{1}S) with H$_2$, lower rate constants ($\leqslant 3 \times 10^9$ l mol^{-1} s^{-1}) were found; the latter accords with a value of 1.2×10^8 l mol^{-1} s^{-1} obtained by pulse radiolysis[14] but in neither case was a chemical interaction proved.

Braun et al.[13] also found that the rates for reaction of C(^{3}P) with N$_2$, CO and H$_2$ were strongly pressure dependent, indicating three-body association to yield CNN, C$_2$O and CH$_2$, respectively, for which absorption spectra were observed. However, no quantitative information was given.

2.4 REACTIONS OF HALOGEN ATOMS (^{2}P$_{\frac{3}{2}}$)

2.4.1 Three-body recombinations

2.4.1.1 Chlorine atoms

In contrast to iodine and bromine much less attention has been paid to the three-body recombination of chlorine atoms. However, Burns and Browne[15] have used the existing discharge flow data in the range 195–500 K to derive Lennard–Jones interaction potentials of $\varepsilon_{(Cl-Ar)} \sim 9.6$ and $\varepsilon_{(Cl-Cl_2)} \sim 8.4$ kJ mol^{-1}. They consider that these findings favour the lower values of the recombined rate constants found in shock-tube studies over those up to 20 times larger obtained in other work. Despite the lack of a specific Cl–Cl$_2$ interaction indicated above, in low temperature matrices a Cl$_3$ species has been detected[16]. With krypton no Cl–Kr complexing was found, suggesting, by extrapolation along the rare gas series, a lower value of $\varepsilon_{(Cl-Ar)}$ than for $\varepsilon_{(Cl-Cl_2)}$ in contrast to the above kinetic analysis result.

2.4.1.2 Bromine atoms

Unlike the situation for other atoms, rate constant data for the recombination

$$Br + Br + M \rightarrow Br_2 + M \tag{2.5}$$

are available over the whole range 300–3000 K. This provides what is at present a unique opportunity for comparison of overlapping sets of results from various techniques.

Most of the lower temperature (300–1250 K) values of $k_{2.5}$ have been obtained by flash photolysis[17]. For the series M = He, Ne, Ar, Kr, N_2, O_2 the values of $k_{2.5}$ at 300 K increased in that order from 1.1×10^9 to 5.8×10^9 1^2 mol^{-2} s^{-1}. All had similar temperature coefficients approximating to $T^{-1.4}$, but showed significant variations in the detailed expressions summarising the reaction rate constants.

Burns and Ip[17, 18] found that flash photolysis values of $k_{2.5}$ at ~ 1250 K were lower than those derived from shock-tube studies where the rate of dissociation of Br_2 was followed absorptiometrically. The error in the latter arose from coupling between vibrational excitation and dissociation[18], precluding the use of the equilibrium constant to obtain $k_{2.5}$. The existence of such effects was confirmed experimentally by Boyd et al.[19] where $k_{-2.5}$ was measured in a shock tube using bromine atom two-body luminescence intensities to follow the dissociation of Br_2. Such values were independent of coupling effects, and combination with the equilibrium constant yielded $k_{2.5}$ (M = Ar) in good agreement with the flash-photolysis results. The temperature coefficient of $k_{2.5}$ became more negative with increasing temperatures[19] from 1300–2000 K. This was interpretable in terms of vibrational non-equilibrium in Br_2 under these conditions (cf. Section 2.6.1). All existing shock-tube data can be rationalised only by considering a time-dependent vibrational temperature for Br_2 even in the very early stages of dissociation[18].

Blake et al.[20] have calculated Lennard–Jones potentials from available values of $k_{2.5}$ assuming the radical–molecule complex mechanism (Section 2.4.1.3). The indications are that Br–M species may be regarded as bound dimers with $\varepsilon_{(Br-Ne)} = 2.5$, $\varepsilon_{(Br-Ar)} = 5.9$, $\varepsilon_{(Br-K)} = 8.4$ and $\varepsilon_{(Br-N_2)} = 6.3$ kJ mol^{-1}. Several third bodies appear to have relatively strong interactions with Br as indicated by high values of $k_{2.5}$ and larger temperature coefficients. In contrast to the chlorine system, for M = Br_2 analysis of limited data indicated a specific interaction[17]; for the series M = Ne, SF_6, CCl_4 an increasingly strong interaction was evident[21] in results from kinetic spectroscopy. An even higher but unconfirmed value of $k_{2.5}$ for M = C_3H_8 comes from use of the rotating sector method on the acetone-photosensitised decomposition of HBr in the presence of propane[22].

For the first time, using a technique other than the shock tube, recombination with the atom as third body has been detected[11]. The results, which were consistent with shock-tube data, were summarised between 300 and 3000 K by $k_{2.5}$ (M = Br) = $10^{12.2}(T/300)^{-4.3}$ 1^2 mol^{-2} s^{-1} with no indication of a maximum value (cf. Section 2.5.1).

2.4.1.3 Iodine atoms

An interesting recent result is the demonstration by Burns et al.[23] that measurement of $k_{2.6}$ for the recombination written formally as

$$I + I + M \rightarrow I_2 + M \tag{2.6}$$

must usually take into account a significant concentration of IM complexes. In flash photolysis of dilute I_2/Ar mixtures constant values of $k_{2.6}$ (M = Ar) = $(3.0\pm0.2)\times10^9$ and $k_{2.6}$ (M = I_2) = $(1.0\pm0.1)\times10^{12}$ l^2 mol^{-2} s^{-1}, independent of composition, could only be obtained if corrections for IAr and I_3 concentrations were applied. Similar considerations also explained an apparent transition of I_2 dissociation in shocked argon from bimolecular to unimolecular with increasing argon pressure[24].

On the basis of the radical complex mechanism for reaction (2.6):

$$I+M \rightleftharpoons IM \tag{2.6a}$$

$$I+IM \rightarrow I_2+M \tag{2.6b}$$

the temperature variation of $k_{2.6}$ (M = Ar, I_2) indicated an interaction potential $\varepsilon_{(I-Ar)} = 6.3$ kJ mol^{-1} and a much larger value for the I_3 complex. The increased stability of trihalogen complexes from Cl to I has also been demonstrated in crossed molecular-beam experiments[25]. For $Cl+Br_2$, trajectory analysis indicated formation and dissociation of the complex in a time shorter than the rotational period. However, $Br-I_2$ complexes in similar experiments had considerable stability with $\varepsilon_{(Br-I_2)} \geqslant 40$ kJ mol^{-1}.

In contrast to the situation for bromine atoms, the values of $k_{2.6}$ (M = Ar) from high-temperature flash photolysis almost match those from absorptiometric measurement of I_2-dissociation in shock waves[26], implying here relatively minor coupling between vibrational excitation and dissociation.

2.4.2 Bimolecular reactions of halogen atoms

2.4.2.1 Chlorine atoms

A reaction currently attracting great interest is

$$Cl+H_2 \longrightarrow HCl+H \tag{2.7}$$

In the range 250–600 K there is reasonable agreement in the values of $k_{2.7}$ from widely different techniques. A direct discharge flow study[27], based on e.s.r. measurement of atom decay in excess H_2, gave $k_{2.7} = (8.1\pm0.4)\times10^6$ l mol^{-1} s^{-1} with Arrhenius activation energy of 18.0 ± 0.9 kJ mol^{-1} over the range 251–453 K. A short extrapolation yields results 25–45% lower from 476 to 594 K than those determined by Benson et al.[28], from thermal decomposition of ICl in the presence of excess I_2 as the source of $Cl(^2P)$. The room-temperature result above is confirmed by $k_{2.7} = (8.4\pm0.6)\times10^6$ l mol^{-1} s^{-1} obtained by Davis et al.[29], by use of vacuum ultraviolet photolysis of CCl_4 as the source of $Cl(^2P_{\frac{3}{2}})$ in excess H_2. The resonance-fluorescence detection employed showed that concentrations of $Cl(^2P_{\frac{1}{2}})$ were negligible in these experiments.

The problem in this system is that complementary measurements of $k_{-2.7}$ in the discharge flow system[27] revealed a value of $k_{2.7}/k_{-2.7}$ smaller than the thermodynamic equilibrium constant by at least a factor of two. No obvious explanation is available for this as yet, although postulates have been made that effects of internal excitation (rotational since $\Delta H_{2.7} \sim 5$ kJ mol^{-1}) on reactivity[27], or enhanced reactivity of a less than equilibrium concentration of $Cl(^2P_{\frac{3}{2}})$[30] may be significant.

In a discharge flow study of the rate of formation of HCl when Cl_2 was added to discharged hydrogen–inert gas mixtures, Stedman et al.[11] found that $H_2(v' = 1)$ reacted some hundred times more rapidly than $H_2(v = 0)$ in reaction (2.7), demonstrating the importance of internal excitation. However, no vibrationally excited H_2 could be present in the above studies of reaction (2.7).

2.4.2.2 Bromine ($^2P_{\frac{3}{2}}$) and iodine ($^2P_{\frac{3}{2}}$) atoms

A large number of absolute rate constants for $I(^2P_{\frac{3}{2}})$ reactions have been measured by following spectrophotometrically the disappearance of I_2, in thermal equilibrium with atoms, in the presence of added substrate[31]. It is only recently that the same method has been applied to $Br(^2P_{\frac{3}{2}})$ reactions to provide urgently required *absolute* rate data[32, 33].

Another useful development leading to the determination of absolute rates for $I(^2P_{\frac{3}{2}})$ reactions is their production for the first time under controlled conditions in a discharge-flow system[34]. Addition of HI to flowing Cl (or Br) atoms in an argon-carrier induced reaction (2.2) (Section 2.2) yielding $I(^2P_{\frac{3}{2}})$. The downstream removal of added ozone, followed by u.v. absorption, through reaction (2.8)

$$I + O_3 \rightarrow IO + O_2 \qquad (2.8)$$

occurred under pseudo-first-order conditions due to rapid IO disproportionation regenerating atoms. A value of $k_{2.8} = 10^{8.7 \pm 0.3} \ 1 \ mol^{-1} \ s^{-1}$ was obtained from the early stages of reaction.

2.5 REACTIONS OF HYDROGEN ATOMS (^2S)

2.5.1 Three-body recombination of H(^2S) and D(^2S)

Despite the large number of attempts to measure $k_{2.9}$

$$H + H + M \rightarrow H_2 + M \qquad (2.9)$$

there is still doubt regarding the precise values for 300 K. Table 2.1 lists recent results from fast flow systems; the discrepancies are outside reasonable error limits, and are particularly disturbing in that all workers have claimed rigorous control over impurity and surface effects. Moreover, a trend between measurement technique and magnitude of $k_{2.9}$ is evident for $M = H_2$.

In this respect, the results of Ham et al.[40] (calorimeter) and Bennett and Blackmore[36] (e.s.r.), on the isotope recombination

$$D + D + D_2 \rightarrow D_2 + D_2 \qquad (2.9D)$$

may be significant, since ratios of $k_{2.9D}/k_{2.9}$ ($M = H_2$) = 0.75 ± 0.06 and

1.25 ± 0.25 respectively were obtained. The agreement between the absolute values of $k_{2.9D}$ may imply that absolute values of $k_{2.9}$ (M = H$_2$) differ because of calibration errors, and hence for the latter a comparative study by use of both e.s.r. and calorimetric detection in the same laboratory could be valuable.

Quantum mechanical calculations on reaction (2.9), by use of a model where two hydrogen atoms form orbiting resonance or quasi-bound states

Table 2.1 $10^{-9} \times k_{2.9}$ (l^{2}mol^{-2}s^{-1}) at 300 K ($k_{2.9} = \dfrac{d[H_2]}{d}/[H]^2[M]$)

Technique	M = H$_2$	Ar	He	H$_2$O
E.S.R.	1.4(35)*, 2.1(36)	5.6(37)	7.2(37)	(45 ± 10) (38)
Isothermal calorimeter	3.4(39), 3.0(40)	2.3(39)	2.4(39)	≤25(39)
Theoretical[41]	4.6	3.2	3.3	

*Reference numbers are in parentheses.

before critical rotational relaxation by M, have been performed by Roberts et al.[41]. The resultant $k_{2.9}$ (Table 2.1) and predicted ratio[42] $k_{2.9D}/k_{2.9} = 0.7 \pm 0.2$ are in sufficient agreement with the experimental results to emphasise further the discrepancies in the latter. However, the model is less successful in another respect; it predicts a maximum in $k_{2.9}$ for c. 100 K whereas experiment[40] shows only a smooth variation as $T^{-0.6}$ for 77–300 K.

In the range 2500–7000 K, shock-tube measurements by Hurle et al.[43] of the rate of H$_2$ dissociation in argon, when combined with the equilibrium constant, yielded values of $k_{2.9}$, represented by

$$\log_{10}k_{2.9} (M = H_2) = 9.24 - 1.95 \times 10^{-4}T$$

$$\log_{10}k_{2.9} (M = Ar) = 9.79 - 2.75 \times 10^{-4}T$$

Comparison with Table 2.1 suggests that $k_{2.9}$ may have slightly different temperature coefficients for different M. A recent H$_2$–O$_2$ flame study[44] supports the results of Hurle et al.[43], and further indicates a reduced relative efficiency for M = H$_2$O at 1900 K compared to 300 K.

Above 2500 K, the hydrogen atom becomes an important third body[43, 45]. As $k_{2.9}$ (M = H) $\leqslant 5 \times 10^9$ l^2 mol^{-2} s^{-1} at 300 K [35] but is $\sim 10^{10}$ l^2 mol^{-2} s^{-1} at 3000 K [43] and decreases above that, it is probable that a maximum value is reached between 2000 and 3000 K, particularly since no effects of M = H were found in H$_2$–O$_2$ flames at 1900 K [44]. To account for this Hurle et al.[45] propose an orbiting-resonance theory very similar to that of Roberts et al.[41], but with hydrogen-atom exchange now responsible for relaxation. Balance between the orbiting-resonance equilibrium and a postulated small positive temperature coefficient for relaxation quantitatively yielded the correct $k_{2.9}$ (M = H) for 3000 K and a maximum near 2000 K. However, it was difficult to select reasonable physical parameters to match the fall-off above 3000 K.

2.5.2   Simple bimolecular reactions of $H(^2S)$

2.5.2.1   Isotopic exchange

Prospects for accurate comparison between theory and experiment rightly continue to stimulate interest in the set of reactions

$$H + p\text{-}H_2 \rightarrow o\text{-}H_2 + H \tag{2.10}$$

$$H + D_2 \rightarrow HD + D \tag{2.11}$$

$$D + H_2 \rightarrow HD + H \tag{2.12}$$

$$D + o\text{-}D_2 \rightarrow p\text{-}D_2 + D \tag{2.13}$$

The largely inconclusive situation in 1969 was reviewed by Kaufman[46]. The best absolute values of $k_{2.11}$ and $k_{2.12}$ appear to be those derived in a discharge-flow study by Westenberg and de Haas[47]; values measured by LeRoy et al.[48] were too high due to back-diffusion effects at lower temperatures, although those for $k_{2.13}$ are free of these effects[49]. However, Quickert and LeRoy[48] consider that rate-constant ratios will afford more accurate comparison with theory. Transition-state calculations matched their values of $k_{2.10}/k_{2.11}$ above 350 K but deviation below was ascribed to tunnelling effects which are difficult to assess theoretically.

The validity of the transition-state approach is supported by largely classical calculation of energy profiles along the reaction coordinate[50]. The single maximum required by theory was found on imposition of the condition that the zero-point energy of reactants could pass smoothly into that of the intermediate complex. Further confirmation came from trajectory calculations[50] showing the threshold energies to be equal to the barrier energies in each case. Such considerations are specific to reactions (2.10)–(2.13) where relatively large zero-point energies and vibrational-level separations are favourable to this vibrationally adiabatic model.

In molecular-beam experiments, Geddes et al.[51] found the angular distribution and velocity of the HD product from reaction (2.12) consistent with the formation and rapid dissociation of linear D—H—H complexes. This provides experimental justification for a further assumption of transition-state theory as applied to these reactions.

2.5.2.2   Reaction with halogens and hydrogen halides

Several recent investigations, using photolytically generated $H(^2S)$, of reaction pairs,

$$H + X_2 \longrightarrow HX + X \tag{2.14}$$

$$H + HX \longrightarrow H_2 + X \tag{2.15}$$

with X = Cl, Br, I, have shown $k_{2.14}$ and $k_{2.15}$ to be very sensitive to 'hot' atom effects; consequently, any study which does not thoroughly investigate these must be treated with caution[52].

Fass[53] measured the change in Br_2 concentration when admixed HBr was photo-dissociated at 184.9 and 248.0 nm. 'Hot' atom effects were evident in the variation of the apparent $k_{2.14}/k_{2.15}$ ratio with pressure of inert-gas diluent. High-pressure limiting values of $k_{2.14}/k_{2.15} = (6.8 \pm 2.0)$ exp(3340 $\pm 1260/RT$) over 300–523 K were obtained with a pressure of 180 kN m^{-2} of helium moderator. For the analogous X = I reactions, Penzhorn and Darwent[54] found $k_{2.14}/k_{2.15} = 4.95 \text{ exp} (2680/RT)$ for the temperature range 303–533 K by photo-dissociating HI and measuring H_2 production in the presence of I_2 and excess CO_2 moderator. A similar expression comes from Sullivan's work[55] on the thermal and photo-initiated H_2–I_2 reaction over the range 418–738 K. With unmoderated $H(^2S)$, $k_{2.14}/k_{2.15} = 5.3 \pm 0.4$ (X = Br)[53] and ~ 4 (X = I)[54]. Both are close to the ratio of pre-exponential factors for moderated atoms, consistent with the simple collision-theory interpretation that pre-exponential factors are proportional to the probability that a collision involving energy above the reaction threshold will result in chemical reaction.

The reactivity of $H_2(v' \neq 0)$ towards chlorine atoms (Section 2.4.2.1) means that $k_{2.14}$ (X = Cl) can only be measured in discharge-flow systems by the rate of removal of Cl_2 in excess hydrogen atoms. Stedman et al.[11] found $k_{2.14}$ (X = Cl) $= (2.1 \pm 0.7) \times 10^{10} \text{ l mol}^{-1} \text{ s}^{-1}$ on this basis. A relatively large isotope effect was indicated by a rate constant of $(7.2 \pm 3.0) \times 10^9 \text{ l mol}^{-1} \text{ s}^{-1}$ for the analogous $D(^2S)$ reaction.

Westenberg and de Haas[56], from e.s.r. spectral measurements of hydrogen-atom decay rates with excess HCl in inert carriers in an acid-poisoned flow-tube, found $k_{2.15}$ (X = Cl) $= (2.3 \pm 0.7) \times 10^{10}$ exp $((-14\,600 \pm 800)/RT)$ $1 \text{ mol}^{-1} \text{ s}^{-1}$ for nine temperatures from 195 to 500 K. Clyne and Stedman[57] confirmed these values to within 30% at four temperatures; they also used the discharge-flow technique but with HNO chemiluminescence to follow atom decay. If significant $H_2(v' \neq 0)$ had been present in these experiments, it should have reacted rapidly with chlorine atoms to give apparently low values of $k_{2.15}$ (X = Cl). This is in the wrong direction to be an alternative explanation for the discrepancy between $k_{2.15}/k_{-2.15}$ (X = Cl) (i.e. $k_{-2.7}/k_{2.7}$) and the equilibrium constant (Section 2.4.2.1), but could explain the curvature of the Arrhenius plot of Clyne and Stedman[57].

2.5.3 Complex reactions with organic molecules

An underlying controversy in this section concerns the importance of heterogeneous combination of hydrogen atoms with methyl radicals, invoked in some cases but apparently absent in others.

2.5.3.1 Alkanes (CH_4, C_2H_6, C_3H_8)

This uncertainty exists in two recent quantitative studies of reactions with alkanes. Kurylo et al.[58, 59] measured $k_{2.16} = (6.3 \pm 2.0) \times 10^{10}$ exp$(-48\,500/RT) 1 \text{ mol}^{-1} \text{ s}^{-1}$ over the range 424–732 K from e.s.r. absorption measurements of hydrogen atom decay rates in excess methane in a discharge-flow system,

$$H + CH_4 \longrightarrow CH_3 + H_2 \qquad (2.16)$$

This agrees reasonably with $k_{2.16} = (1.3 \pm 0.3) \times 10^{11}$ exp$(-49\,800/RT)$ 1 mol^{-1} s^{-1} evaluated by Walker[60] from literature prior to 1968. Kurylo et al.[58, 59] considered that effects of the secondary reaction

$$H + CH_3 \longrightarrow CH_4 \qquad (2.17)$$

were negligible; a homogeneous component was precluded by the low helium-carrier pressures (0.2–0.8 kN m^{-2}) and the lack of effect of addition of 1% O_2, a known methyl scavenger. Moreover, since heterogeneous occurrence of reaction (2.17) would lower the measured pre-exponential factor, comparison of the above expressions suggests this was negligible also.

In contrast Azatyan and Fillipov[61] using a similar technique and conditions in a quantitative study of the reaction

$$H + C_2H_6 \longrightarrow C_2H_5 + H_2 \qquad (2.18)$$

assumed the importance of secondary reactions (2.19) *and* (2.17)

$$H + C_2H_5 \longrightarrow CH_3 + CH_3 \qquad (2.19)$$

From the implied stoichiometry $k_{2.18} = 1.1 \times 10^{11}$ exp$(-38\,800/RT)$ 1 mol^{-1} s^{-1} for the temperature range 290–579 K was obtained, which agrees well with $k_{2.18} = 1.3 \times 10^{11}$ exp$(-40\,500/RT)$ 1 mol^{-1} s^{-1} evaluated by Baldwin and Melvin[62] from literature prior to 1964. A confirmatory value of $k_{2.18} = 2 \times 10^8$ 1 mol^{-1} s^{-1} for 773 K has been derived[63] from product concentration-profile analysis slowly reacting $H_2 + O_2 + C_2H_6$ mixtures.

An important general reactivity formula for higher alkanes may be implied by the work of Campbell et al.[64]. Yields of isomeric hexanes (from propyl radical combination), which result from Hg-photosensitised H(^{2}S) production in the presence of propane, indicated a ratio $k_{2.20}/k_{2.21} = 10^{0.11 \pm 0.11}$ exp$((-8400 \pm 800)/RT)$ for the temperature range 310–450 K.

$$H + C_3H_8 \longrightarrow n\text{-}C_3H_7 + H_2 \qquad (2.20)$$

$$H + C_3H_8 \longrightarrow i\text{-}C_3H_7 + H_2 \qquad (2.21)$$

A very similar result was deduced[64] from available absolute values of $k_{2.18}$ and $(k_{2.20} + k_{2.21})$, assuming that the reactivity of primary C—H bonds was independent of the particular alkane. A general reactivity formula appears to exist for analogous O atom (Section 2.7.3.1) and OH reactions (Section 2.9.1.4).

2.5.3.2 Ethylene

Problems of rate and mechanism, apparent in the earlier literature and arising largely from reversibility of the initial step

$$H + C_2H_4 \rightleftharpoons C_2H_5^* \qquad (2.22)$$

now appear substantially resolved. $C_2H_5^*$ is formed with $\leqslant 168$ kJ mol^{-1} excitation energy and will re-dissociate unless it is stabilised by a collision with a third body

$$C_2H_5^* + M \longrightarrow C_2H_5 + M \qquad (2.23)$$

Under low-pressure conditions, the interplay of these reactions affects the overall stoichiometry because of secondary reactions of H atoms.

Three recent studies, with good limiting conditions, leave little doubt that $k_{2.22}$ lies between 5×10^8 and 9×10^8 l mol^{-1} s^{-1}. Eyre et al.[65] have isolated $k_{2.22}$, following pseudo-first-order H-atom decay by Lyman α-absorptiometry, after pulse radiolysis of high pressure H_2(100–240 kN m^{-2}) with at least a hundredfold excess of ethylene (13–267 N m^{-2}) over atoms. Hence $k_{2.22} = (5.5 \pm 0.5) \times 10^8$ l mol^{-1} s^{-1} was obtained directly. Also $k_{2.22} \sim 5 \times 10^8$ l mol^{-1} s^{-1} can be deduced by combining the ratio $k_{2.22}/k_{2.24} = 0.83$ found by Woolley and Cvetanovic[66] (Section 2.5.3.3) with the absolute $k_{2.24} = (6.0 \pm 0.4) \times 10^8$ l mol^{-1} s^{-1} measured by Mihelcic and Schindler[67] in discharge-flow experiments with e.s.r. detection.

$$H + H_2S \longrightarrow H_2 + HS \qquad (2.24)$$

Finally Kurylo et al.[68] have used Lyman α-fluorescence and absorptiometry to follow H-atom decay in vacuum u.v. flash-photolysed ethylene (6–14 N m^{-2})–helium(0.67–6.7 kN m^{-2}) mixtures. The relatively low total pressures meant that only an apparent bimolecular rate constant, rather than the high-pressure limit, was obtained by reiterative computer matching of decay profiles to obtain optimum values of rate constants. Extrapolation to infinite pressure on the basis suggested by Oref and Rabinowitch[69] yielded $k_{2.22} = (8.2 \pm 1.1) \times 10^8$ l mol^{-1} s^{-1}.

The conditions of low pressure and comparable H-atom and ethylene concentrations used in discharge-flow studies are conducive to elucidation of secondary reactions in this system. Mass spectrometric[70] and gas chromatographic[71] analyses of products have established a generally agreed overall mechanism, an interesting aspect of which is the role of reaction (2.17). Halstead et al.[71] find by reiterative computer matching of concentration–time profiles that methane formation is entirely homogeneous with $k_{2.17} = 5.4 \times 10^{12}$ l^2 mol^{-2} s^{-1} for an argon carrier, and clearly third order in the range 1.1–2.2 kN m^{-2}. A similar conclusion, but higher $k_{2.19} = (3 \pm 1.5) \times 10^{13}$ l^2 mol^{-2} s^{-1} for M = He, was postulated by Barker et al.[70]. No indications of heterogeneous methane formation were apparent in these studies, supported by the fact that methyl radicals were actually detectable[70], in contrast to the situation when this reaction does occur. (Section 2.5.3.4).

2.5.3.3 Higher olefins trans-but-2-ene cis-but-2-ene, isobutene

Here the situation is simpler than for ethylene since the initial adduct does not re-dissociate to original reactants. The measured rate constants are, therefore, pressure independent. For trans-but-2-ene rate constants of $(6.0 \pm 0.9) \times 10^8$ and $(5.4 \pm 0.4) \times 10^8$ l mol^{-1} s^{-1} have been measured by vacuum u.v. flash photolysis[72] and mass-spectrometric measurement of olefin removal in a large excess of H atoms[73] respectively. Other rate constants relative to this are set out in Table 2.2 which incorporates some of the data of Cvetanovic *et al.*[66], in which H atoms were generated in the presence of olefins either by Hg-photosensitised dissociation of H_2 or photo-dissociation of H_2S in the presence of excess moderator.

Table 2.2   Rate constants for H+olefins at 300 K relative to H+*trans*-but-2-ene*

	$k/l\,mol^{-1}\,s^{-1}$		$k/l\,mol^{-1}\,s^{-1}$
Ethylene	1.11 (66), 1.07(65), 1.60(68)	*cis*-But-2-ene	0.80(66), 0.86(73)
Isobutene	4.08(66), 4.0(72)		

*Reference numbers are in parentheses

Agreement is good between different determinations and some support is given to lower values of $k_{2.22}$ for ethylene. Only small isotope effects were found in analogous D-atom reactions[73].

2.5.3.4   H+CH₃ *(heterogeneous)*

The importance of a recent discharge-flow study by Davis *et al.*[74] of the sequence of reactions initiated by

$$D + CH_3Br \longrightarrow DBr + CH_3 \qquad (2.25)$$

lies in the information obtained on subsequent methyl radical reactions. E.S.R. and gas chromatographic measurements showed that yields of H atoms were much less than CH_3Br consumption or yields of methanes. The reaction

$$D + CH_3 \longrightarrow CH_3D^* \longrightarrow CH_2D + H \qquad (2.26)$$

is strongly favoured by the intramolecular isotope effect. With argon-carrier pressures of 0.13–1.33 kN m^{-2} significant stabilisation of the adduct was not expected. Accordingly heterogeneous occurrence of reaction (2.17), limited by the diffusion of methyl radicals to the walls, was postulated. On this basis a rate constant of 8.3×10^9 l mol^{-1} s^{-1} for homogeneous association of CH_3 was derived at 0.26 kN m^{-2}, not inconsistent with a high-pressure limiting value of $(2.6 \pm 0.3) \times 10^{10}$ l mol^{-1} s^{-1} [75]. Less satisfactory is $k_{2.26} = 2 \times 10^8$ l mol^{-1} s^{-1} for 0.13 kN m^{-2} of argon; this should approximate to the high-pressure limit of $k_{2.17}$ (homogeneous) but it falls an order of magnitude below values quoted in Section 2.5.3.2. Also unexpected is the positive activation energy of 9.1 ± 1.5 kJ mol^{-1} found for $k_{2.26}$ over the temperature range 297–480 K.

A similar situation of high methane yields compared to ethane was found by Carr *et al.*[76] where CH_3 was produced by the rapid primary reaction between H atoms and ketene. A time-of-flight mass spectrometer failed to detect CH_3 (in contrast to the ethylene system (Section 2.5.3.2)) and 1 % O_2 addition did not reduce methane yields. Again heterogeneous reaction (2.17) was postulated.

The clear differences in the roles of reaction (2.17) in the various systems discussed in Section 2.5.3 cannot be rationalised as yet; no systematic variation with wall treatment or gas-phase composition is apparent.

2.6 REACTIONS OF NITROGEN ATOMS N(^{4}S)

2.6.1 Three-body recombination

Discharge-flow and shock-tube studies have given values of $k_{2.27}$ in the ranges 90–610 K and 6000–15 000 K respectively.

$$N + N + M \longrightarrow N_2 + M \qquad (2.27)$$

A promising application of modified phase-space theory has been made by Shui et al.[77] to develop the form of temperature dependence reconciling all results. For M = Ar, N–N and N–Ar interaction potentials were critical parameters; the former was derived from spectroscopic tables while the latter was treated as a variable parameter, except that it was realistic to incorporate in it a Morse parameter derived from scattering of a fast beam of argon by N_2 gas[78]. The resultant absolute expression exhibited a complex temperature-dependence; comparison with an Arrhenius plot showed $k_{2.27}$ increased, primarily by the N–Ar interaction, at low temperatures and decreased by vibrational non-equilibrium in N_2 at high temperatures (cf. Section 2.4.1.2). A further prediction, that c. 70% of recombinations populate $N_2(A^3\Sigma_u^+)$ directly, agreed with the chemiluminescence result[79].

More recently Shui et al.[80] have demonstrated the success of their method for other recombinations with M = Ar. Further extension is awaited with interest.

2.6.2 Bimolecular transfer reactions

2.6.2.1 O_2 ($^3\Sigma_g^-$), $O_2(^1\Delta_g)$

The rate constant for the ground-state reaction

$$N + O_2(^3\Sigma_g^-) \longrightarrow NO + O \qquad (2.28)$$

is well established, a recent evaluation[81] yielding $k_{2.28} = 10^{6.8 \pm 0.1}$ exp $((-26\,150 \pm 800)/RT)$ 1 mol^{-1} s^{-1} for 300–1700 K. Westenberg et al.[82], in a discharge-flow study with e.s.r. detection, have shown that the corresponding $O_2(^1\Delta_g)$ reaction is negligibly slow at 300 K. The conflicting conclusions of Clark and Wayne[83] are explained if physical quenching of $O_2(^1\Delta_g)$ by N(^{4}S) is moderately fast in the range 195–431 K, although an activation energy of 5 kJ mol^{-1} appears anomalous for such a process.

2.6.2.2 $SO(^3\Sigma_g^-)$

Closely related to reaction (2.28), the reaction

$$N + SO \longrightarrow NO + S \qquad (2.29)$$

is now known to be rapid and stoichiometric at 300 K.

Luiti[84] generated $SO(^3\Sigma_g^-)$ by discharge of flowing SO_2; after hetero-geneous removal of O atoms, downstream mixing with N atoms resulted in rapid deposition of sulphur on the walls, together with the blue glow characteristic of co-existing N and O atoms. Similar observations, confirming reaction (2.29) followed by rapid reaction of N atoms with the NO product, have been made by Jacob and Winkler[85]. Quantitative generation of SO by reaction of O atoms with COS and measurement of N-atom decay in its presence by NO titration, gave $k_{2.29} = (3.8 \pm 0.1) \times 10^8$ l mol^{-1} s^{-1} at 300 K. Comparison with $k_{2.28}$ shows that $k_{2.29}$ has a larger pre-exponential factor and suggests a lower activation energy also.

2.6.2.3 CN

From the CN and C_2 emission intensity profiles produced by shock heating of BrCN diluted in argon (5000–10 000 K), Slack and Fishburne[86] postulate a chain mechanism

$$N + CN \longrightarrow N_2 + C \qquad (2.30)$$

$$C + CN \longrightarrow C_2 + N \qquad (2.31)$$

A value of $k_{2.30} = 2 \times 10^{11}$ l mol^{-1} s^{-1} was deduced on the basis that the activation energy of reaction (2.31) cannot be less than the endothermicity (142 kJ mol^{-1}). Despite the high temperatures, the near identity of $k_{2.30}$ with the collision frequency implies that $k_{2.30}$ cannot have a large temperature coefficient. Extrapolation suggests that reaction (2.30) could be significant in the acetylene system at lower temperatures (Section 2.6.3.2).

2.6.3 Complex reactions with acetylene

By use of discharge-flow systems with coupled mass spectrometers both Arrington et al.[87] and Safrany and Jaster[88] found that reaction between N atoms and acetylene was slow and bimolecular with excess acetylene. Major products were C_2N_2 and cyanoacetylene but not HCN. A slow reaction was also encountered with over 20-fold excess of N atoms. Between these two regimes a rapid chain reaction took place.

2.6.3.1 The slow reaction

In excess acetylene Arrington et al.[87] found a rate constant which increased with total pressure to an extrapolated limit of $10^{(8.82 \pm 0.13)}$ exp$((-8400 \pm 1650)/RT)$ l mol^{-1} s^{-1} over 303–468 K. In a large excess of N atoms, Herron and Huie[89] found a rate constant for acetylene disappearance of $< 4 \times 10^6$ l mol^{-1} s^{-1} at 500 K with a total pressure of 330 N m^{-2}; this limit agrees with the prediction of the above expression, after compensating for the low pressure. Thus it seems likely that the same initial reaction is operative in both slow-reaction regimes. The identity of this reaction is disputed; Arrington et al.[87] postulate a mechanism

$$N + C_2H_2 \rightleftharpoons C_2H_2N^* \tag{2.32}$$

$$C_2H_2N^* + M \longrightarrow C_2H_2N + M \tag{2.33}$$

with C_2H_2N, stabilised by collision, reacting further to yield the products. Safrany and Jaster[88] prefer a direct bimolecular step

$$N + C_2H_2 \longrightarrow C_2HN + H \tag{2.34}$$

rejected by Arrington et al.[87] as endothermic, although this may be questionable. Little isotopic scrambling results however from reaction of a C_2D_2–C_2H_2 mixture[87], which seems inconsistent with reaction (2.34), and combined with the pressure dependence of the rate constant favours reaction (2.32).

2.6.3.2 The chain reaction

Both groups[87, 88] agree that CN is the chain carrier and its formation appears to involve an energy transfer from $N_2(A^3\Sigma_u^+)$ resulting from N-atom recombination. Small additions of acetylene in this regime were consumed almost instantaneously but N atoms continued to be removed further downstream. Arrington et al.[87] propose that this is due to the reactions

$$N + CN + M \longrightarrow NCN + M \tag{2.35}$$

$$N + NCN \longrightarrow N_2 + CN \tag{2.36}$$

Chain termination was ascribed to combination

$$CN + CN + M \longrightarrow C_2N_2 + M \tag{2.37}$$

The CN concentration was deduced from the rate of C_2N_2 production in conjunction with $k_{2.37} = 1.8 \times 10^{13}$ l mol^{-2} s^{-1} at 300 K; this value came from extrapolation of shock-tube results for the rate of C_2N_2 dissociation at 2500 K. Such a long extrapolation may involve substantial error and a direct measurement[90] of $k_{2.37} \approx 1.7 \times 10^{10}$ l^2 mol^{-2} s^{-1} at 300 K supports this contention. Accordingly $k_{2.35}$ (M = N_2) = $(5.4 \pm 1.8) \times 10^{12}$ l^2 mol^{-2} s^{-1} obtained by Arrington et al.[87] must be treated with caution at present.

The above difficulty may lend support to the alternative hypothesis of Safrany and Jaster[88], invoking reaction (2.30) followed by a reaction of carbon atoms, which may be heterogeneous, to regenerate CN. As noted above (Section 2.6.2.3), present evidence suggests that reaction (2.30) may be quite fast at 300 K.

2.7 REACTIONS OF OXYGEN ATOMS, O(^{3}P)

2.7.1 Three-body association (with O_2, SO_2 and CO)

From the pulse radiolysis of high pressure O_2–M mixtures, Sauer[91] has evaluated $k_{2.38} = 7 \times 10^8$ l mol^{-1} s^{-1} for the initial step of ozone formation:

$$O + O_2 \rightleftharpoons O_3^* \tag{2.38}$$

$$O_3^* + M \longrightarrow O_3 + M \tag{2.39}$$

This agrees well with $k_{2.38} = 9 \times 10^8$ l mol^{-1} s^{-1} from a mass-spectrometric study[92] of exchange of ^{18}O with $^{16}O_2$. However Sauer's values for the overall rate constants for M = He, Ar are a factor of c. 2 lower than those from discharge-flow[93] or flash-photolysis[94] experiments, where O-atom decays were followed. Hochnanadel et al.[95] have probably resolved this problem in showing that vibrationally excited O_3 absorption above 250 nm does reduce the apparent rate constant from absorptiometric measurement of O_3 formation as used by Sauer[91]; compensation for this resulted in good agreement.

Mulcahy et al.[96] show that surface-conditioning effects by SO_3 must be taken into account in measuring $k_{2.40}$

$$O + SO_2 + M \longrightarrow SO_3 + M \tag{2.40}$$

Adsorption of the relatively stable SO_3 on the walls of the stirred-flow reactor appeared to accelerate heterogeneous O-atom removal. Without compensation for this, apparent values of $k_{2.40}$ for O-atom removal similar to those available in the literature were found. However these decreased with decreasing O-atom concentrations to a zero limit of $k_{2.40}$ (M = Ar) = $(1.1 \pm 0.3) \times 10^9$ l^2 mol^{-2} s^{-1}, which could be four times lower.

As anticipated from its overall spin-forbidden nature, reaction (2.41) is much slower than most other associations

$$O(^3P) + CO(^1\Sigma) + M = CO_2(^1\Sigma) + M \tag{2.41}$$

Recent work has yielded acceptable values of $k_{2.41}$. Slanger and Black[97] obtained $k_{2.41}$ (M = He) = $(2.1 \pm 0.5) \times 10^6$ l^2 mol^{-2} s^{-1}, following photodissociation of O_2 by pulsed vacuum u.v. radiation in the presence of high pressures ($\geqslant 3.3$ kN m^{-2}) of 10% CO in He mixtures. Resonance-fluorescence detection had the high sensitivity required for the very small concentrations of O atoms. The above value is well within the upper limit of $k_{2.41}$ (M = He) $\leqslant 1.7 \times 10^7$ l^2 mol^{-2} s^{-1} derived by Donovan et al.[98] in vacuum u.v. flash photolysis studies. The results of Slanger and Black[97] also establish that reaction (2.41) is third order, resolving a further controversy. They found $k_{2.41}$ (M = Ar) = $(2.5 \pm 1.3) \times 10^6$ l^2 mol^{-2} s^{-1} and $k_{2.41}$ (M = N$_2$) = $(5.0 \pm 1.4) \times 10^6$ l^2 mol^{-2} s^{-1} on substitution of these diluents for helium.

Lin and Bauer[99] have measured $k_{2.41}$ between 1600 and 3000 K in a shock tube with thermal dissociation of N_2O as the source of O atoms. Unfortunately the expression used for the dissociation rate constant had an erroneously high temperature-dependence (Section 2.7.2.3) but its absolute value was correct at c. 2000 K where $k_{2.41}$ (M = Ar) = 1.1×10^9 l^2 mol^{-2} s^{-1} was obtained. Combination with the above room temperature value gives $k_{2.41}$ (M = Ar) $\approx 3 \times 10^9$ exp($-18\,000/RT$) l^2 mol^{-2} s^{-1}. Significantly this positive activation energy is close to that found for the luminescent association[100], while the pre-exponential factor is typical of three-body reactions.

For M = $O_2(^3\Sigma_g^-)$, reaction (2.41) is formally spin allowed; however, Mulcahy and Williams[101] found $k_{2.41}$ (M = O_2) $\approx 10^8$ l^2 mol^{-2} s^{-1} at 456 K, about a factor of 4 larger than the predicted value of $k_{2.41}$ (M = Ar), suggesting that no significance is attached to this. Kondratiev and Intezarova[102] studied the thermal decomposition of flowing O_3 in O_2–CO carriers at atmospheric pressure and in the range 409–503 K; they derived $k_{2.41}$ from the rates of removal of O_3 and formation of CO_2. Although the values appear to be of the correct order of magnitude, the negative temperature coefficient proposed appears unacceptable since, as well as disagreeing with the above, it predicts $k_{2.41}$ (M = O_2) to be much larger at 300 K than an established upper limit[100] of $\leqslant 3 \times 10^7$ 1^2 mol^{-2} s^{-1}. It is perhaps notable that the flowing O_3 system will be highly sensitive to hydrogeneous impurities which could establish a catalytic cycle based on

$$H + O_3 \longrightarrow OH + O_2 \tag{2.42}$$

$$CO + OH \longrightarrow CO_2 + H \tag{2.43}$$

2.7.2 Bimolecular reactions of O(^{3}P) with inorganic molecules

2.7.2.1 *Ammonia*

The primary step which is generally accepted is

$$O + NH_3 \longrightarrow NH_2 + OH \tag{2.44}$$

Albers *et al.*[103] have provided strong evidence from discharge-flow experiments with mass-spectrometric and e.s.r. detection that, with excess O atoms,

$$O + NH_2 \longrightarrow HNO + H \tag{2.45}$$

occurs with HNO being very rapidly removed probably by

$$O + HNO \longrightarrow OH + NO \tag{2.46}$$

$$OH + HNO \longrightarrow H_2O + NO \tag{2.47}$$

Such reactions accounted for the rapid production of both NO and H_2O, while the alternative to reaction (2.45) yielding NH radicals could not. This alternative is, moreover, probably c. 9 kJ mol^{-1} endothermic, and hence probably slow at 300 K. With excess NH_3, three oxygen atoms were consumed per NH_3 molecule to yield 1.5 hydrogen atoms, as required by the mechanism above, while the pseudo-first-order decay yielded $k_{2.44} = 1.5 \times 10^9$ $\exp(-25\,000/RT)$ 1 mol^{-1} s^{-1} from 300–1010 K. A similar study by Kurylo *et al.*[104] gave $k_{2.44} = (4.0 \pm 0.9) \times 10^9$ $\exp((-27\,600 \pm 540)/RT)$ 1 mol^{-1} s^{-1} over 361–763 K by use of a similar but assumed stoichiometry factor. Further work is required to resolve the significant divergences of these values at higher temperatures.

2.7.2.2 *Hydrazine*

A highly complex reaction occurs between O atoms and hydrazine in a discharge-flow system[105]; Becker and Bayes[106] detected luminescence demonstrating the involvement of OH, NH, NH_2 and $N_2(A^3\Sigma_u^+)$. Mass-spectrometric measurement of the decay rate of N_2H_4 in excess flowing atoms gave $k_{2.48} = 8.5 \times 10^{10}$ $\exp(-5020/RT)$ 1 mol^{-1} s^{-1} from 230–500 K. A rather unusual major primary step, involving simultaneous abstraction of two H atoms, has been established by two independent studies[105, 107] by use of crossed molecular-beams and mass-spectrometric detection.

$$O + N_2H_4 \longrightarrow N_2H_2 + H_2O \tag{2.48}$$

Only 4 % of the overall reaction went by an alternative step to yield N_2H_3 [107].

2.7.2.3 Isotopic exchange reactions (SO_2, CO_2)

Clark et al.[108] have measured $k_{2.49}$ and $k_{2.50}$

$$^{18}O + C^{16}O_2 \longrightarrow C^{16}O^{18}O + ^{16}O \tag{2.49}$$

$$^{18}O + S^{16}O_2 \longrightarrow S^{16}O^{18}O + ^{16}O \tag{2.50}$$

from 2900–3425 K and 1600–2100 K respectively by in situ mass-spectrometric analysis of products in reflected shock-waves. Both reactions appear to occur by a simple atom-displacement mechanism. Combination of the results with values for lower temperatures from earlier work yielded $k_{2.49}$ $= 10^{7.65 \pm 0.30} \exp((-15\,200 \pm 2900)/RT) \, 1 \, mol^{-1} \, s^{-1}$ and $k_{2.50} = 10^{7.71 \pm 0.30}$ $\exp((-540 \pm 2900)/RT) \, 1 \, mol^{-1} \, s^{-1}$. In the experiments with SO_2, the source of atomic oxygen was the thermal dissociation of added N_2O and the value of $k_{2.50}$ is very sensitive to the dissociation rate constant used. In Section 2.7.1 a similar situation was described for Lin and Bauer's work[99] on reaction (2.41). When the dissociation rate constant used by Lin and Bauer was applied to the SO_2 exchange system[108], it led to the absurd activation energy for reaction (2.50) of $-187 \, kJ \, mol^{-1}$.

2.7.3 Bimolecular reactions of $O(^3P)$ with organic molecules

2.7.3.1 Alkanes

Herron[109] and Herron and Huie[110, 111] have obtained rate-constant data for a series of alkanes and halogenated alkanes by mass-spectrometric measurement of their removal rate when added to a large excess of flowing oxygen atoms. For the alkanes Herron[109] has attempted to derive a general reactivity formula based on the assignment of a rate expression to each type of C—H bond attacked, namely, primary, secondary or tertiary, followed by summation over the total number of bonds in the molecule. This was capable of predicting the rate constants to within a factor of three.

2.7.3.2 Acetylene

It is now agreed that the overwhelmingly dominant path is

$$O + C_2H_2 \longrightarrow CH_2 + CO \quad \Delta H = -200 \, kJ \, mol^{-1} \tag{2.51}$$

$$O + CH_2 \longrightarrow CO + 2H \quad \Delta H = -306 \, kJ \, mol^{-1} \tag{2.52}$$

with a further step significant in the presence of excess acetylene

$$CH_2 + C_2H_2 \longrightarrow C_3H_4 \tag{2.53}$$

Two discharge-flow studies[112, 113] of CO infrared emission from this system at low pressures have found an excitation limit of $CO(X^1\Sigma)$ $v' = 13$ or 14; this limit corresponds closely with the exothermicity of reaction (2.52), and thus supports its occurrence. There are other minor reactions; at 298 K reactions (2.54) and (2.55)

$$O + C_2H_2 \longrightarrow C_2O + H_2 \tag{2.54}$$

$$O + CH_2 \longrightarrow OH + CH \tag{2.55}$$

occur to the extent of c. 0.3% of reaction (2.51)[114] and c. 6% of (2.52)[115] respectively.

Values of $k_{2.51}$ at 298 K are in reasonable agreement; six independent studies give values in the range (7 ± 3) to $(10.7 \pm 0.8) \times 10^7$ l mol^{-1} s^{-1} recently. The higher values are favoured because the more precise of the lower values were obtained under conditions of excess C_2H_2 without taking account of reaction (2.53); compensation for this raises the values significantly[115]. This may also explain some of the difficulty in obtaining concordant temperature coefficients for $k_{2.51}$. Of these experiments[116-118], only those of James and Glass[118], who obtained $k_{2.51} = (1.43 \pm 0.50) \times 10^{10}$ exp $((-13\ 180 \pm 840)/RT)$ l mol^{-1} s^{-1} over 273–729 K for excess O-atom conditions, appear to be immune from complications induced by reaction (2.53).

2.8 REACTIONS OF SULPHUR ATOMS, S(^{3}P) and S(^{1}D)

Discharge-flow studies by Fair and Thrush[119] and Mihelcic and Schindler[67] have shown that the fast reaction of H atoms with H_2S (reaction (2.24)) may serve as a source of S(^{3}P) for absolute rate measurements. Only isolated rate constants of 7.2×10^8 l mol^{-1} s^{-1} for reaction with O_2 [119] and 2.7×10^{10} l mol^{-1} s^{-1} for reaction with SH [67] have emerged so far.

In virtually all studies of S(^{1}D) reactions, the source has been photodissociation of carbon oxysulphide at 225–265 nm. Re-investigation of this by Breckenridge and Taube[120] has shown that the ratio S(^{1}D):S(^{3}P) produced in the primary photochemical act is 3:1 in contrast to the assumption of sole production of the former made in previous studies. On this basis, relative rate constants for reactions of S(^{1}D) with COS, alkanes, olefins, N_2O and CS_2 were derived[120]; all were between 1.0 and 2.0. Donovan et al.[121], in a flash-photolysis study, set a lower limit for S(^{1}D) + COS of one collision in four, indicating very low activation energies for all the above S(^{1}D) reactions. No accurate absolute rate data are available.

2.9 REACTIONS OF DIATOMIC SPECIES

2.9.1 Hydroxyl radicals

2.9.1.1 Disproportionation (homogeneous and heterogeneous)

Until recently the simple gas-phase reaction

$$OH + OH \longrightarrow H_2O + O \tag{2.56}$$

was considered solely responsible for OH decay in flow systems at low pressures. However, Breen and Glass[122], by use of e.s.r. detection, found evidence of a concurrent first-order heterogeneous decay ($\gamma = 3 \times 10^{-3}$ for boric acid coated quartz) and hence a lower value of $k_{2.56} = (5.0 \pm 1.6) \times 10^8$ l mol^{-1} s^{-1} ($-d[OH]/dt = 2 \, k_{2.56}[OH]^2$) than $k_{2.56} = (1.55 \pm 0.12) \times 10^9$ l mol^{-1} s^{-1} postulated by Dixon–Lewis et al.[123], from a similar study. The latter detected no heterogeneous contribution but found good second-order behaviour with similar initial concentrations to those used by Breen and Glass[122] but in a *smaller*-diameter quartz tube which should have emphasised heterogeneous effects. The nature of the walls could exert a marked effect on the heterogeneous decay rate[122] but further work is required to resolve the true value of $k_{2.56}$.

2.9.1.2 Reaction of OH with NO₂

In a flow-discharge system with e.s.r. detection, Wilson and O'Donovan[124] observed first-order decay of OH in the presence of excess NO_2. The rate related to the concentration of the latter, but it could not be established whether the reaction was homogeneous or heterogeneous. From a recent static photolysis study[125] of HNO_3–NO_2 mixtures diluted with krypton, a rate constant $k_{2.57}$ (M = Kr) $\approx 1.3 \times 10^9$ l^2 mol^{-2} s^{-1} has been derived for the reaction

$$OH + NO_2 + M \longrightarrow HNO_3 + M \tag{2.57}$$

Calculations using this value suggest that reaction (2.57) might be significant for Wilson and O'Donovan's system where it could be followed by reaction (2.58).

$$OH + HNO_3 \longrightarrow H_2O + NO_3 \tag{2.58}$$

However reaction (2.59) is about 10^4 times faster than reaction (2.58) at 300 K [125]

$$O + HNO_3 \longrightarrow OH + NO_3 \tag{2.59}$$

and could reduce the net effect.

2.9.1.3 Reactions of OH with inorganic molecules (CO, H₂, SO, N₂H₄)

The rate constants for reactions with carbon monoxide and hydrogen are still controversial.

$$CO + OH \longrightarrow CO_2 + H \tag{2.60}$$

$$OH + H_2 \longrightarrow H_2O + H \tag{2.61}$$

Both have been studied over 300–500 K by Greiner[126] by use of flash photolysis and $k_{2.60}$ and $k_{2.61}$ for 300 K agree with the evaluations of Drysdale and Lloyd in a recent review[127]. However Greiner's measurements indicate smaller temperature coefficients with values of $k_{2.60}$ and $k_{2.61}$ for 500 K about a factor of two lower than the predictions of the review but at the same time values of the ratio $k_{2.60}/k_{2.61}$ are in good agreement. It is difficult to pinpoint any deficiency in the flash photolysis technique but from material in Section 2.9.1.4 it appears to yield consistently lower temperature co-efficients for a series of rate constants than do other methods. Therefore at present the results must be treated with caution. Values of $k_{2.60}$ and $k_{2.61}$ have also been measured by Wong and Belles[128] by use of a discharge-flow stirred reactor and mass-spectrometric analysis. Their results favour the higher temperature coefficient for $k_{2.60}$ but the lower one for $k_{2.61}$, hence disagreeing with both Drysdale and Lloyd[127] and Greiner[126]. It is likely that any error here is in the determination of $k_{2.61}$ since further con-firmation of Drysdale and Lloyd's assessment of $k_{2.60}$ comes from the work of Kochubei and Moin[129]. They determined the reverse rate constant, $k_{-2.60}$, over 1023–1523 K in a flow reactor; combination with the equilibrium constant yielded $k_{2.60}$.

A very fast reaction occurs between OH and SO

$$SO + OH \rightarrow SO_2 + H \tag{2.62}$$

A value of $k_{2.62} = (7 \pm 3) \times 10^{10}$ 1 mol^{-1} s^{-1} was derived from a discharge-flow study[119] where SO and OH were generated by addition of O_2 and NO_2 respectively to the H atom–H_2S system and the resultant kinetics followed by use of chemiluminescence. Some confirmation of this high value comes from its successful incorporation into a computer matching of profiles in the O atom–H_2S system[130].

The reaction of OH with N_2H_4 has an unusual primary step,

$$OH + N_2H_4 \longrightarrow NH_3 + NH_2O \tag{2.63}$$
$$cf.\ O + C_2H_4 \longrightarrow CH_3 + CHO$$

demonstrated by crossed molecular-beam experiments[131]. As indicated this is analogous to the initial step in the O atom–ethylene reaction.

2.9.1.4   Reactions of OH with hydrocarbons (alkanes, ethylene, acetylene)

The rate measurements up to 1969 on reactions (2.64) and (2.65), covering the ranges 300–3000 K and 300–1500 K respectively, have been evaluated by Drysdale and Lloyd[127].

$$OH + CH_4 \rightarrow CH_3 + H_2O \tag{2.64}$$

$$OH + C_2H_6 \longrightarrow C_2H_5 + H_2O \tag{2.65}$$

Recent results have been obtained by use of the flash photolysis technique by Greiner[132] in the range 300–500 K. As for $k_{2.60}$ and $k_{2.61}$, the pattern of Greiner's values of $k_{2.64}$ and $k_{2.65}$ is agreement with the evaluation at 300 K but lower temperature coefficients, with the values at 500 K lower by a factor of around two than those of the evaluation[127].

The only available absolute data on the analogous abstraction reactions of higher alkanes comes from the flash-photolysis work of Greiner[132]. Investigation for a range of alkanes enabled the derivation of a general reactivity formula for abstraction of primary, secondary and tertiary hydrogens. However for the last, a small negative temperature coefficient for the rate constant was obtained, which seems unlikely to be correct, and probably points once again to the tendency of the flash-photolysis technique to underestimate positive temperature coefficients. This view is reinforced by values of the rate constants for propane, n-butane and iso-butane reactions relative to $k_{2.61}$ at 753 K obtained by Baker et al.[133] from measurements of the rate of pressure change and removal of these reactants added to slowly reacting H_2-O_2 systems in turn. Even the ratios were up to three times larger than predicted by extrapolation of Greiner's results[126, 132]. By use of the evaluated $k_{2.61}$ for 753 K and making the reasonable postulate that Greiner's rate constants are probably correct for 300 K, a general rate constant $k(OH + RH) = 1.5 \times 10^{10} \exp(-14\,700/RT)N_1 + 2.2 \times 10^{10} \exp(-10\,670/RT)N_2 + 5.3 \times 10^{10} \exp(-10\,000/RT)N_3$ l mol^{-1} s^{-1} was derived, where N_1, N_2 and N_3 are the respective numbers of primary, secondary and tertiary hydrogen atoms in RH. Further confirmation of this potentially very useful general form is required.

Greiner[134] has also studied the reaction of OH with ethylene over 300–500 K and found a rate constant at 300 K an order of magnitude larger than those for O- and H-atom reactions with ethylene. Less acceptable is a negative activation energy of -3770 ± 570 J mol^{-1} for the reaction, particularly in view of our previous comments on the flash-photolysis technique.

Investigations of the reaction between OH and acetylene suggest that the reaction mechanism varies from low to high temperatures. At 300 K, crossed molecular-beam studies[131] indicate a pathway

$$OH + C_2H_2 \longrightarrow CO + CH_3 \qquad (2.66)$$

Bradley and Tse[115] detected minor formation of OH in O atom–acetylene systems (Section 2.7.3.2) by e.s.r. spectroscopy; its removal could be ascribed to reaction (2.66), and from concentration profiles $k_{2.66} = (7.8 \pm 1.5) \times 10^6$ l mol^{-1} s^{-1} was derived at 298 K. However this is probably an upper limit (although of the correct order of magnitude) since calculation suggests that enough carbon monoxide was formed prior to this reaction zone by the major reactions to make reaction (2.60) competitive for hydroxyl radicals. In acetylene–O_2–N_2 flames[135] at temperatures of c. 1600 K, on the other hand, it is suggested that the course of the reaction is different

$$OH + C_2H_2 \longrightarrow H_2O + C_2H \qquad (2.66a)$$

2.9.2 Oxyhalogen radicals

2.9.2.1 ClO

In low-pressure (0.1–1.0 kN m^{-2}) flow experiments, Clyne and Coxon[136] have detected ClO radicals absorptiometrically following their production by reaction of Cl atoms with chlorine dioxide or thermal decomposition of the latter. The decay mechanism

$$ClO + ClO \longrightarrow Cl + ClOO \qquad (2.67)$$

$$Cl + ClOO \longrightarrow Cl_2 + O_2 \qquad (2.68)$$

$$ClOO + M \longrightarrow Cl + O_2 + M \qquad (2.69)$$

was established with $k_{2.67} = (1.1 \pm 0.1) \times 10^7$ l mol^{-1} s^{-1} at 298 K with an activation energy of $10\,450 \pm 1250$ J mol^{-1} over 294–495 K. Johnston *et al.*[137] have irradiated static Cl$_2$–O$_2$–M mixtures with oscillating u.v. light with phase-sensitive absorptiometry to detect intermediates. For ClO, it was found that, although reactions (2.67)–(2.69) were important at the lower total pressures, an alternative termolecular process, effectively written as

$$ClO + ClO + M \longrightarrow Cl_2 + O_2 + M \tag{2.70}$$

became important at higher pressures of the range 6.5–101 kN m^{-2}. Values of $k_{2.67} = 3.8 \times 10^6$ l mol^{-1} s^{-1}, $k_{2.70}$ (M = Ar) $= 1.2 \times 10^{10}$ l^2 mol^{-2} s^{-1} and $k_{2.70}$ (M = O$_2$) $= 1.8 \times 10^{10}$ l mol^{-1} s^{-1} were determined, the first in encouraging agreement with the result from the flow system considering the vastly different techniques.

2.9.2.2 BrO, IO

BrO and IO can be generated by reacting ozone with Br and I respectively[34, 138] in a flow system. Unlike ClO, these do not react in a scheme of the type (2.67)–(2.69) because of the low stability of BrOO and IOO radicals, but rather disproportionate.

$$BrO + BrO \longrightarrow 2Br + O_2 \tag{2.71}$$

$$IO + IO \longrightarrow 2I + O_2 \tag{2.72}$$

Accordingly, over 293–573 K, the large pre-exponential factor and low activation energy of $k_{2.71} = 10^{10.5 \pm 0.3} \exp((-1880 \pm 1250)/RT)$ l mol^{-1} s^{-1}, compared to $k_{2.67}$, reflects the more weakly bound transition state expected for reaction (2.71) [138].

2.9.3 Chemical reactions of $O_2(a^1\Delta_g)$ and $O_2(b^1\Sigma_g^+)$

The situation up to 1969 has been reviewed by Wayne[139].

The only established dissociative reactions for both of these species are those with ozone.

$$O_2(a^1\Delta_g) \text{ or } O_2(b^1\Sigma_g^+) + O_3 \longrightarrow 2O_2 + O \qquad \text{(2.73a) (2.73b)}$$

In a discharge-flow study Clark *et al.*[140] found $k_{2.73a} = (4.0 \pm 1.5) \times 10^8$ $\exp(-13\,000/RT)$ l mol^{-1} s^{-1} over 195–439 K. The pre-exponential factor is typical of a diatomic–triatomic reaction while the activation energy merely reflects the endothermicity of reaction (2.73a). Reaction (2.73b) is exothermic and a flow-system study[141] and flash-photolysis study[142] agree on a value of $k_{2.73b} = (4.1 \pm 0.6) \times 10^9$ l mol^{-1} s^{-1} at 300 K.

Stuhl and Niki[143] have followed the decay of $O_2(b^1\Sigma_g^+)$ in the presence of H$_2$, D$_2$ and HD, obtaining rate constants of 6×10^8, 1.1×10^7 and 1.9×10^8 l mol^{-1} s^{-1} respectively. The large isotope effects may imply four-centre chemical reaction but at present this is purely speculative.

Addition across olefinic double-bonds is typical of $O_2(a^1\Delta_g)$. However,

Herron and Huie[144] have shown that significant chemical reaction is limited to olefins above C_4. The reaction with tetramethylethylene (TME) has been widely studied and its unique course established[145] as

$$O_2(a^1\Delta_g) + TME \longrightarrow Me_2C(OOH) \cdot CMe{=}CH_2 \qquad (2.74)$$

Three different detection techniques have been used in flow-discharge studies[144-146] to measure $k_{2.74}$. The value obtained by Gleason et al.[145] is an order of magnitude lower than the other two and further evidence that their technique, based on a cobalt oxide probe, may be in error is provided by the fact that a similar measurement of $k_{2.75}$ is also lower by almost two orders of magnitude compared to the others;

$$O_2(a^1\Delta_g) + DMF \longrightarrow Me{\overset{\displaystyle \diagup\!\!\!\diagdown}{\underset{O\!-\!O}{\diagdown\!\!\!\diagup}}}{-}O{-}{\diagdown}Me \qquad (2.75)$$

In view of some doubt regarding the effectiveness of physical quenching of $O_2(a^1\Delta_g)$ by 2,5-dimethylfuran (DMF)[145], only $k_{2.74} = (7.15 \pm 0.88) \times 10^8$ $\exp((-14\,620 \pm 420)/RT)$ 1 mol^{-1} s^{-1} over 298–460 K, derived[146] by use of e.s.r. detection of $O_2(a^1\Delta_g)$, can be regarded as reliable at present.

According to selection rules[147] for singlet oxygen reactions, $O_2(b^1\Sigma_g^+)$, in contrast to $O_2(a^1\Delta_g)$, is expected to be unreactive towards olefins and dienes in concerted addition reactions. In keeping with this, little difference has been found in the rate constants for removal of $O_2(b^1\Sigma_g^+)$ by acetylenes, olefins, alkanes and molecules such as CO_2, NH_3 and H_2O, indicating only physical quenching[143, 148, 149]. However, there is some tentative evidence from flow-discharge studies[144] that $O_2(b^1\Sigma_g^+)$ does react rapidly with C_6–C_8 olefins. Some support comes from a rate constant of c. 5×10^8 1 mol^{-1} s^{-1} for $O_2(b^1\Sigma_g^+)$ removal by TME [143], some 500 times larger than for the *chemical* reaction between $O_2(a^1\Delta_g)$ and TME.

2.10 BIMOLECULAR REACTIONS

Since Sullivan[55] elegantly demonstrated that the H_2–I_2 reaction involved a termolecular iodine-atom reaction

$$I + I + H_2 \longrightarrow 2HI \qquad (2.76)$$

rather than the traditional four-centre mechanism, there has been doubt as to whether any reactions at all proceed by the latter mechanism. Extending the original suggestion of Noyes[150], Porter et al.[6] have made a detailed trajectory analysis to show that the large activation energy for formation of the four-centre complex from $H_2 + I_2$ arises from the momentum-conserving motions of nuclei. Hoffmann[151] and Cusachs et al.[152] have invoked orbital correlation to produce similar but less quantitative conclusions.

Noyes[153] has predicted from theory that reactions between halogen molecules should proceed by a truly bimolecular mechanism. There has been no clear demonstration of this in the gas phase but, in stopped-flow experiments, Goldfinger *et al.*[154] have found first-order dependence on both reactants, independent of nitrogen carrier-gas concentration, for the reactions

$$Cl_2 + HBr \longrightarrow HCl + BrCl \qquad (2.77)$$

$$BrCl + HBr \longrightarrow HCl + Br_2 \qquad (2.78)$$

At 300 K, $k_{2.78} > k_{2.77} = 30 \pm 5 \ 1 \ mol^{-1} \ s^{-1}$, indicating an activation energy of $\leqslant 60 \ kJ \ mol^{-1}$ for typical pre-exponential factors. This would be inconsistent with an atomic mechanism.

The results obtained in shock-tube studies of isotope-exchange reactions by Bauer and his co-workers are more difficult to interpret. Typically, when H_2–D_2 exchange at large dilution in argon was studied[10], approximately second-order behaviour was found overall, but included an almost first-order dependence on argon concentration. In the range 1060–1420 K, established rate constants for H_2 and D_2 dissociation could not account for the rate of HD formation; nor was the activation energy of *c.* 177 kJ mol^{-1} consistent with an atomic or direct bimolecular mechanism, which also would not have involved argon. Bauer interprets these, and similar results, in terms of four-centre complex formation which is favoured by critical vibrational excitation (and hence bond extension) of one of the reactants. Calculations with independent data for vibrational excitation rates by argon approximately matched the experimental results but it is unfortunate that no experiments were performed with other diluents as further tests.

Theoretical analyses of H_2–D_2 exchange have failed to confirm the existence of an easily accessible four-centre transition state. Morokuma *et al.*[155] conclude that translational energy should enhance exchange more than vibrational. Conroy and Malli[156], Wilson and Goddard[157] and Rubinstein and Shavitt[158] have all attempted to find a transition-state geometry with a calculated minimum activation energy as low as that found experimentally, but without success, although the last[158] found some indications that an isosceles trapezoid or kite might suffice. It is perhaps notable that these geometries demand unequal bond-lengths for the reactants. It must be concluded that at present there is no satisfactory alternative mechanism nor any good theoretical justification for the present one.

A number of other bimolecular reactions have been studied recently, namely reaction of CO with N_2O [99], O_2 [159] and NO_2 [160, 161] and the disproportionation of NO [162]. There are however no mechanistic difficulties here since they only involve single-atom transfer.

References

1. Donovan, R. J. and Husain, D. (1970). *Chem. Rev.,* **70**, 489
2. Polanyi, J. C. and Wong, W. H. (1969). *J. Chem. Phys.,* **51**, 1439
3. Mok, M. H. and Polanyi, J. C. (1969). *J. Chem. Phys.,* **51**, 1451
4. Mok, M. H. and Polanyi, J. C. (1970). *J. Chem. Phys.,* **53**, 4588
5. Jaffe, S. B. and Anderson, J. B. (1968). *J. Chem. Phys.,* **49**, 2859
6. Porter, R. N., Thompson, D. L., Sims, L. B. and Raff, L. M. (1970), *J. Amer. Chem. Soc.,* **92**, 3208
7. Anlauf, K. G., Maylotte, D. H., Polanyi, J. C. and Bernstein, R. B. (1969). *J. Chem. Phys.,* **51**, 5716

8. Polanyi, J. C. and Tardy, D. C. (1969). *J. Chem. Phys.*, **51**, 5717
9. Anderson, J. B. (1970). *J. Chem. Phys.*, **52**, 3849
10. Bauer, S. H. and Ossa, E. (1966), *J. Chem. Phys.*, **45**, 434
11. Stedman, D. H., Stefferson, D. and Niki, H. (1970). *Chem. Phys. Lett.*, **7**, 173
12. Wray, K. L., Feldman, E. V. and Lewis, P. F. (1970). *J. Chem. Phys.*, **53**, 4131
13. Braun, W., Bass, A. M., Davis, D. D. and Simmons, J. D. (1969), *Proc. Roy. Soc. (London)*, **A312**, 417
14. Meaburn, G. M. and Perner, D. (1966). *Nature (London)*, **212**, 1042
15. Burns, G. and Browne, R. J. (1970). *J. Chem. Phys.*, **53**, 3318
16. Nelson, L. Y. and Pimentel, G. C. (1967). *J. Chem. Phys.*, **47**, 3671
17. Ip, J. K. K. and Burns, G. (1969). *J. Chem. Phys.*, **51**, 3414
18. Ip, J. K. K. and Burns, G. (1969). *J. Chem. Phys.*, **51**, 3425
19. Boyd, R. K., Burns, G., Lawrence, T. R. and Lippiatt, J. H. (1968). *J. Chem. Phys.* **49**, 3804
20. Blake, J. A., Browne, R. J. and Burns, G. (1970). *J. Chem. Phys.*, **53**, 3320
21. De Graffe, B. A. and Lang, K. J. (1970). *J. Phys. Chem.*, **74**, 4181
22. Wong, K. T. and Armstrong, D. A. (1970). *Can. J. Chem.*, **48**, 2426
23. Burns, G., LeRoy, R. J., Morriss, D. J. and Blake, J. A. (1970). *Proc. Roy. Soc. (London)*, **A316**, 81
24. Troe, J. and Wagner, H. Gg. (1967). *Z. Phys. Chem. N. F.*, **55**, 326
25. Lee, Y. T., LeBreton, P. R., McDonald, J. D. and Herschbach, D. R. (1969). *J. Chem. Phys.*, **51**, 455
26. Blake, J. A., Boyd, R. K., Ip, J. K. K. and Burns, G. (1969). *Astronaut. Acta*, **14**, 487
27. Westenberg, A. A. and de Haas, N. (1968). *J. Chem. Phys.*, **48**, 4405
28. Benson, S. W., Cruikshank, F. R. and Shaw, R. (1969). *Int. J. Chem. Kinet.* **1**, 29
29. Davis, D. D., Braun, W. and Bass, A. M. (1970). *Int. J. Chem. Kinet.*, **2**, 101
30. Snider, N. S. (1970). *J. Chem. Phys.*, **53**, 4116
31. Golden, D. M. and Benson, S. W. (1969). *Chem. Rev.*, **69**, 125
32. King, K. D., Golden, D. M. and Benson, S. W. (1970). *Trans. Faraday Soc.*, **66**, 2794
33. Amphlett, J. C. and Whittle, E. (1970). *Trans. Faraday Soc.*, **66**, 2016
34. Clyne, M. A. A. and Cruse, H. W. (1970). *Trans. Faraday Soc.*, **66**, 2227
35. Bennett, J. E. and Blackmore, D. R. (1968). *Proc. Roy. Soc. (London)*, **A305**, 553
36. Bennett, J. E. and Blackmore, D. R. (1970). *J. Chem. Phys.*, **53**, 4400
37. Azatyan, V. V., Romanovich, L. B. and Fillipov, S. B. (1968). *Kinet. Katal.*, **9**, 1188
38. Eberius, H., Hoyermann, K. and Wagner, H. Gg. (1969). *Ber. Bunsenges. Phys. Chem.*, **73**, 962
39. Larkin, F. S. (1968). *Can. J. Chem.*, **46**, 1005
40. Ham, D. O., Trainor, D. W. and Kaufman, F. (1970). *J. Chem. Phys.*, **53**, 4395
41. Roberts, R. E., Bernstein, R. B. and Curtiss, C. F. (1969). *J. Chem. Phys.*, **50**, 5163
42. Roberts, R. E. and Bernstein, R. B. (1970). *Chem. Phys. Lett.*, **6**, 282
43. Hurle, I. R., Jones, A. and Rosenfeld, J. L. J. (1969). *Proc. Roy. Soc. (London)* **A310**, 253
44. Halstead, C. J. and Jenkins, D. R. (1970). *Comb. and Flame*, **14**, 321
45. Hurle, I. R., Mackey, P. and Rosenfield, J. L. J. (1968). *Ber. Bunsenges Phys. Chem.*, **72**, 991
46. Kaufman, F. (1969). *Ann. Rev. Phys. Chem.*, **20**, 45
47. Westenberg, A. A. and de Haas, N. (1967). *J. Chem. Phys.*, **47**, 1393
48. Quickert, K. A. and LeRoy, D. J. (1970). *J. Chem. Phys.*, **53**, 1325
49. LeRoy, D. J., Ridley, B. A. and Quickert, K. A. (1967). *Discuss. Faraday Soc.*, **44**, 92
50. Tweedale, A. and Laidler, K. J. (1970). *J. Chem. Phys.*, **53**, 2045
51. Geddes, J., Krause, H. F. and Fite, W. L. (1970). *J. Chem. Phys.*, **52**, 3296
52. Wood, G. O. and White, J. M. (1970). *J. Chem. Phys.*, **52**, 2613
53. Fass, R. A. (1970). *J. Phys. Chem.*, **74**, 984
54. Penzhorn, R. D. and Darwent, B. de B. (1968). *J. Phys. Chem.*, **72**, 1639
55. Sullivan, J. H. (1967). *J. Chem. Phys.*, **46**, 73
56. Westenberg, A. A. and de Haas, N. (1968). *J. Chem. Phys.*, **48**, 4405
57. Clyne, M. A. A. and Stedman, D. H. (1966). *Trans. Faraday Soc.*, **62**, 2164
58. Kurylo, M. J. and Timmons, R. B. (1969). *J. Chem. Phys.*, **50**, 5076
59. Kurylo, M. J., Hollinden, G. A. and Timmons, R. B. (1970). *J. Chem. Phys.*, **52**, 1773
60. Walker, R. W. (1968). *J. Chem. Soc. A*, 2391
61. Azatyan, V. V. and Fillipov, S. B. (1969). *Dokl. Akad. Nauk. SSSR*, **184**, 625
62. Baldwin, R. R. and Melvin, A. (1964). *J. Chem. Soc.*, 1785
63. Baldwin, R. R., Hopkins, D. E. and Langford, D. H. (1969). *Trans. Faraday Soc.*, **66**, 189

64. Campbell, J. M., Strausz, O. P. and Gunning, H. E. (1969). *Can. J. Chem.,* **47,** 3759
65. Eyre, J. A., Hikida, T. and Dorfman, L. M. (1970). *J. Chem. Phys.,* **53,** 1281
66. Woolley, G. R. and Cvetanovic, R. J. (1969). *J. Chem. Phys.,* **50,** 4697
67. Mihelcic, D. and Schindler, R. N. (1970). *Ber. Bunsenges Phys. Chem.,* **74,** 1280
68. Kurylo, M. J., Peterson, N. C. and Braun, W. (1970). *J. Chem. Phys.,* **53,** 2776
69. Oref, I. and Rabinowitch, B. S. (1967). *J. Phys. Chem.,* **47,** 5219
70. Barker, J. R., Keil, D. G., Michael, J. V. and Osborne, D. J. (1970). *J. Chem. Phys.,* **52,** 2079
71. Halstead, M. P., Leathard, D. A., Marshall, R. M. and Purnell, J. H. (1970). *Proc. Roy. Soc. (London),* **A316,** 575
72. Braun, W. and Lenzi, M. (1967). *Discuss. Faraday Soc.,* **44,** 252
73. Daby, E. E. and Niki, H. (1969). *J. Chem. Phys.,* **50,** 4705
74. Davies, P. B., Thrush, B. A. and Tuck, A. F. (1970). *Trans. Faraday Soc.,* **66,** 886
75. Basco, N., James, D. G. L. and Suart, R. D. (1970). *Int. J. Chem. Kinet.,* **2,** 215
76. Carr, R. W., Gay, I. D., Glass, G. P. and Niki, H. (1968). *J. Chem. Phys.,* **49,** 846
77. Shui, V. H., Appleton, J. P. and Keck, J. C. (1970). *J. Chem. Phys.,* **53,** 2547
78. Jordan, J. E., Colgate, S. O., Amdur, I. and Mason, E. A. (1970). *J. Chem. Phys.,* **52,** 1143
79. Campbell, I. M. and Thrush, B. A. (1967). *Proc. Roy. Soc. (London),* **A296,** 201
80. Shui, V. H., Appleton, J. P. and Keck, J. C. (1970). *M.I.T. Fluid Mechanics Lab. Publication,* 70–3
81. Baulch, D. L., Drysdale, D. D., Horne, D. G. and Lloyd, A. C. (1969). *High Temperature Reaction Rate Data,* Rept. No. 4, 11 (University of Leeds, England)
82. Westenberg, A. A., Roscoe, J. M. and de Haas, N. (1970). *Chem. Phys. Lett.,* **7,** 597
83. Clark, I. D. and Wayne, R. P. (1970). *Proc. Roy. Soc. (London),* **A316,** 539
84. Luiti, G. (1968). *Atti. Accad. Naz. Lincei, Rend. Cl. Sci. Fiz. Mat. Natur.,* **45,** 358
85. Jacob, A. and Winkler, C. A. (1970). *Can. J. Chem.,* **48,** 1774
86. Slack, M. W. and Fishburne, E. S. (1970). *J. Chem. Phys.,* **52,** 5830
87. Arrington, C. A., Bernadini, O. O. and Kistiakowsky, G. B. (1969). *Proc. Roy. Soc. (London),* **A310,** 161
88. Safrany, D. R. and Jaster, W. (1968). *J. Phys. Chem.,* **72,** 3305
89. Herron, J. T. and Huie, R. E. (1968). *J. Phys. Chem.,* **72,** 2235
90. Basco, N., Nicholas, J. E., Norrish, R. G. W. and Vickers, W. H. J. (1963). *Proc. Roy. Soc. (London),* **A272,** 147
91. Sauer, M. C. (1967). *J. Phys. Chem.,* **71,** 3311
92. Klein, F. S. and Herron, J. T. (1966). *J. Chem. Phys.,* **44,** 3645
93. Kaufman, F. and Kelso, J. R. (1967). *J. Chem. Phys.,* **46,** 4541
94. Slanger, T. G. and Black, G. (1970). *J. Chem. Phys.,* **53,** 3717
95. Hochnanadel, C. J., Ghormley, J. A. and Boyle, J. W. (1968). *J. Chem. Phys.,* **48,** 2416
96. Mulcahy, M. F. R., Steven, J. R., Ward, J. C. and Williams, D. J. (1969). *12th International Combustion Symposium,* 323. (Pittsburgh: Combustion Institute)
97. Slanger, T. G. and Black, G. (1970). *J. Chem. Phys.,* **53,** 3722
98. Donovan, R. J., Husain, D. and Kirsch, L. J. (1970). *Trans. Faraday Soc.,* **66,** 2551
99. Lin, M. C. and Bauer, S. H. (1969). *J. Chem. Phys.,* **50,** 3377
100. Clyne, M. A. A. and Thrush, B. A. (1962). *Proc. Roy. Soc. (London),* **A269,** 404
101. Mulcahy, M. F. R. and Williams, D. J. (1968). *Trans. Faraday Soc.,* **64,** 59
102. Kondratiev, V. N. and Intezarova, E. I. (1969). *Int. J. Chem. Kinet.,* **1,** 105
103. Albers, E. A., Hoyermann, K., Wagner, H. Gg. and Wolfrum, J. (1969). *12th International Combustion Symposium,* 313. (Pittsburgh: Combustion Institute)
104. Kurylo, M. J., Hollinden, G. A., Le Fevre, H. F. and Timmons, R. B. (1969). *J. Chem. Phys.,* **51,** 4497
105. Gehring, M., Hoyermann, K., Wagner, H. Gg. and Wolfrum, J. (1969). *Ber. Bunsenges Phys. Chem.,* **73,** 956
106. Becker, K. H. and Bayes, K. D. (1967). *J. Phys. Chem.,* **71,** 371
107. Foner, S. N. and Hudson, R. L. (1970). *J. Chem. Phys.,* **53,** 4377
108. Clark, T. C., Garnett, S. H. and Kistiakowsky, G. B. (1970). *J. Chem. Phys.,* **52,** 4692
109. Herron, J. T. (1969). *Int. J. Chem. Kinet.,* **1,** 527
110. Herron, J. T. and Huie, R. E. (1969). *J. Phys. Chem.,* **73,** 3327
111. Herron, J. T. and Huie, R. E. (1969). *J. Phys. Chem.,* **73,** 1326
112. Clough, P. N., Schwartz, S. E. and Thrush, B. A. (1970). *Proc. Roy. Soc. (London),* **A317,** 575
113. Creek, D. M., Melliar-Smith, C. M. and Jonathan, N. (1970). *J. Chem. Soc. A.,* 646

114. Williamson, D. G. and Bayes, K. D. (1970). *J. Phys. Chem.*, **73**, 1232
115. Bradley, J. N. and Tse, R. S. (1969). *Trans. Faraday Soc.*, **65**, 2685
116. Westenberg, A. A. and de Haas, N. (1969). *J. Phys. Chem.*, **73**, 1181
117. Hoyermann, K., Wagner, H. Gg. and Wolfrum, J. (1969). *Z. Phys. Chem.*, *N.F.*, **63**, 193
118. James, G. S. and Glass, G. P. (1969). *J. Chem. Phys.*, **50**, 2268
119. Fair, R. W. and Thrush, B. A. (1969). *Trans. Faraday Soc.*, **65**, 1557
120. Breckenridge, W. H. and Taube, H. (1970). *J. Chem. Phys.*, **53**, 1750
121. Donovan, R. J., Kirsch, L. J. and Husain, D. (1969). *Nature (London)*, **222**, 1164
122. Breen, J. E. and Glass, G. P. (1970). *J. Chem. Phys.*, **52**, 1082
123. Dixon-Lewis, G., Wilson, W. E. and Westenberg, A. A. (1966). *J. Chem. Phys.*, **44**, 2877
124. Wilson, W. E. and O'Donovan, J. T. (1967). *J. Chem. Phys.*, **47**, 5455
125. Bérces, T. and Förgeteg, S. (1970). *Trans. Faraday Soc.*, **66**, 640
126. Greiner, N. R. (1969). *J. Chem. Phys.*, **51**, 5049
127. Drysdale, D. D. and Lloyd, A. C. (1970). *Oxidation and Combustion Reviews*, **4**, 93
128. Wong, E. L. and Belles, F. E. (1970). *N.A.S.A. Tech. Note D5707*, N70-20629
129. Kochubei, V. F. and Moin, F. B. (1969). *Kinet. Katal.*, **10**, 1203
130. Cupitt, L. T. and Glass, G. P. (1970). *Trans. Faraday Soc.*, **66**, 3007
131. Gehring, M., Hoyermann, K., Wagner, H. Gg. and Wolfrum, J. (1970). *Z. Naturforsch*, **25a**, 675
132. Greiner, N. R. (1970). *J. Chem. Phys.*, **53**, 1070
133. Baker, R. R., Baldwin, R. R. and Walker, R. W. (1970). *Trans. Faraday Soc.*, **66**, 2812
134. Greiner, N. R. (1970). *J. Chem. Phys.*, **53**, 1284
135. Browne, W. G., Porter, R. P., Verlin, J. D. and Clark, A. H. (1969). *12th International Combustion Symposium*, 1035. (Pittsburgh: Combustion Institute)
136. Clyne, M. A. A. and Coxon, J. A. (1968). *Proc. Roy. Soc. (London)*, **A303**, 207
137. Johnston, H. S., Morris, E. D. and Van den Bogaerde, J. (1969). *J. Amer. Chem. Soc.*, **91**, 7712
138. Clyne, M. A. A. and Cruse, H. W. (1970). *Trans. Faraday Soc.*, **66**, 2214
139. Wayne, R. P. (1969). *Advan. Photochem.*, **7**, 400
140. Clark, I. D., Jones, I. T. N. and Wayne, R. P. (1970). *Proc. Roy. Soc. (London)*, **A317**, 407
141. Izod, T. P. J. and Wayne, R. P. (1969). *Proc. Roy. Soc. (London)*, **A314**, 111
142. Biedenkapp, D. and Bair, E. J. (1970). *J. Chem. Phys.*, **52**, 6119
143. Stuhl, F. and Niki, H. (1970). *Chem. Phys. Lett.*, **7**, 473
144. Herron, J. T. and Huie, R. E. (1969). *J. Chem. Phys.*, **51**, 4164
145. Gleason, W. S., Broadbent, A. D., Whittle, E. and Pitts, J. N. (1970). *J. Amer. Chem. Soc.*, **92**, 2068
146. Hollinden, G. A. and Timmons, R. B. (1970). *J. Amer. Chem. Soc.*, **92**, 4181
147. Kearns, D. R. (1969). *J. Amer. Chem. Soc.*, **91**, 6554
148. Filseth, S. V., Zia, A. and Welge, K. H. (1970). *J. Chem. Phys.*, **52**, 5502
149. Stuhl, F. and Niki, H. (1970). *Chem. Phys. Lett.*, **5**, 573
150. Noyes, R. M. (1968). *J. Chem. Phys.*, **48**, 323
151. Hoffmann, R. (1968). *J. Chem. Phys.*, **49**, 3739
152. Cusachs, L. C., Krueger, M. and McCurdy, C. W. (1968). *J. Chem. Phys.*, **49**, 3740
153. Noyes, R. M. (1966). *J. Amer. Chem. Soc.*, **88**, 4318
154. Goldfinger, P., Noyes, R. M. and Wen, W. Y. (1969). *J. Amer. Chem. Soc.*, **91**, 4003
155. Morokuma, K., Pedersen, L. and Karplus, M. (1967). *J. Amer. Chem. Soc.*, **89**, 5064
156. Conroy, H. and Malli, G. (1969). *J. Amer. Phys.*, **50**, 5049
157. Wilson, C. W. and Goddard, W. A. (1969). *J. Chem. Phys.*, **51**, 716
158. Rubinstein, M. and Shavitt, I. (1969). *J. Chem. Phys.*, **51**, 2014
159. Dean, A. M. and Kistiakowsky, G. B. (1970). *J. Chem. Phys.*, **53**, 830
160. Thomas, J. H. and Woodman, G. R. (1967). *Trans. Faraday Soc.*, **63**, 2728
161. Burcat, A. and Lifshitz, A. (1970). *J. Phys. Chem.*, **74**, 263
162. Camac, M. and Feinberg, R. M. (1967). *11th International Combustion Symposium*, 137 (Pittsburg: Combustion Institute)

3
Reactions of Free Radicals

E. WHITTLE
University College, Cardiff

3.1 INTRODUCTION

The field of free radical reactions is so extensive that limitations of space enforce here a selective approach. Therefore in this Chapter, attention is focused mainly on the quantitative aspects of reactions of radicals containing three or more atoms: most of the reactions occur in the gas phase but some reference to relevant solution work is made. Oxidation is not dealt with save for a brief discussion of reactions of specific radicals with O$_2$ in Section 3.4.3. The production of radicals is also largely neglected. Even with these limitations, recent work is so extensive that this Chapter deals almost exclusively with work published in 1969 and 1970: however, a comprehensive coverage of this period is attempted (apart from Russian journals published in late 1970). Previous work is mentioned where necessary to put the recent work into perspective. Where 'present work' is compared to 'previous work', references to the latter are *not* given since they are available in the 'present work' referred to.

The superscripts * and † refer to electronically and vibrationally excited molecules respectively. All rate constants and A factors are in cm^3 mol^{-1} s^{-1} and activation energies are in kcal mol^{-1} (1 kcal mol^{-1} = 4.18 kJ mol^{-1}).

3.2 COMBINATION REACTIONS OF RADICALS

3.2.1 Introduction

Combination reactions of radicals are currently of interest mainly for two reasons. Consider for example the reaction

$$CH_3 + CH_3 \xrightarrow{k_c} C_2H_6 \qquad (3.1)$$

In a system where CH_3 radicals are present, their concentration can be calculated if the rate of formation of C_2H_6 is measured and if k_c is known. This procedure is widely used to obtain rate constants for radical abstraction reactions (and occasionally addition reactions) where in effect reactions such as that depicted in equation (3.1) are used as references. Hence it is desirable that k_c and the associated Arrhenius parameters A_c and E_c should be known for a wide range of radicals.

Combination reactions are also of interest because they are exothermic and hence produce vibrationally 'hot' molecules which may undergo further reaction, e.g.,

$$CH_3 + CH_2F \rightarrow CH_3CH_2F^\dagger \rightarrow CH_2{=}CH_2 + HF$$

In this Chapter, discussion is largely confined to the measurement of k_c, A_c and E_c since the fate of the hot products is more appropriately described in Chapter 1.1.

3.2.2 Combination of alkyl radicals

It is widely accepted that combination of alkyl radicals

$$R + R \xrightarrow{k_c} R_2 \tag{3.2}$$

occurs with zero activation energy and little steric restriction so that combination occurs at almost every collision, with $\log k_c$ (cm^3 mol^{-1} s^{-1}) =

Table 3.1 Rate constants for radical combination*

Radical pair	$\log k_c$	Temp. (°C)	Ref.	Radical pair	$\log k_c$	Temp. (°C)	Ref.
$CH_3 + CH_3$	13.42	25	1	$CF_2Cl + CF_2Cl$	13.14‡	213	13
	13.39	20, 127	2		13.20‡	268	13
	13.51	133–190	3	$CF_2 + CF_2$	13.45§	25–173	14
	12.96†	827–1127	4	$CF_2 + CFCl$	11.8	25	14
$CH_3 + H$	13.01	20	10	$CH_3CO + CH_3O$	12.7	25	20
$C_2H_5 + H$	13.56	25	11	$(CH_3)_3Si + Si(CH_3)_3$	5.5	—	15
allyl + allyl	12.93	25	8		12.34	25	16
	12.87	640	9		14.25	44–126	17
	12.70	790	9	$NCl_2 + NCl_2$	11.74	−14–100	18, 19
$CF_3 + CF_3$	12.95	25	12				
	12.85	60	12				

*k_c in $cm^3 mol^{-1} s^{-1}$.
†mean value.
‡overall k for production of $C_2F_4Cl_2 + (CF_2Cl + CF_2)$.
§at a mean temperature of 100 °C.

$\log A_c \approx 13.5$ and $E_c = 0$. Experimental proof of this is however scanty. For the key reaction (3.1), most authors use Shepp's mean value of $\log k_c = 13.34$ in the range 125–175 °C obtained in 1956. However, valuable new data have now appeared as follows (numerical results are in Table 3.1).

The flash photolysis of azomethane with excess N_2 was studied[1] and the extinction coefficient of CH_3, ε_{CH_3}, and hence its absolute concentration was measured. The decay of CH_3 with time was bimolecular, was independent of pressure, and led to a value of k_c at 25 °C. Similar work[2] led to values of ε_{CH_3} and k_c at 20 °C in excellent agreement with Reference 1 and k_c was unchanged at 127 °C. Other new values of k_c obtained by Biordi[3] in the range 133–190 °C have a mean value given in Table 3.1. Finally, we have new high-temperature data[4] on k_c at 827–1127 °C from separate pyrolyses of azomethane and $Hg(CH_3)_2$ in a shock tube. All the new values given in Table 3.1 are in excellent agreement, and indicate that $E_c = 0$ and $\log k_c = 13.4 \pm 0.1$.

Reaction (3.1) should become third order at sufficiently low pressures and the conditions where this occurs have received much attention. Casas *et al.*[5] suggest that the change begins at about 10 Torr (of $CH_3COCH(CH_3)_2$). They review thoroughly all published data on the effect of pressure on k_c. Other work[6,7] indicates that reaction (3.1) will in general be truly bimolecular above about 50 Torr, but of course the pressure at which the change occurs will depend on temperature and on the nature of the third body available to deactivate $C_2H_6^\dagger$.

The combination of allyl radicals was also studied by flash photolysis[8], and the value of k_c at 25 °C agrees well with that obtained[9] in the range 640–790 °C (see Table 3.1). This suggests that $E_c = 0$ for allyl combination.

The rate constant for combination of H and CH_3 in Table 3.1 was obtained by a new mass-spectrometric method[10]. The total pressure was made up to 6.7 Torr with helium, so although the value of k_c is close to that for other radicals, it is not clear if this is the true bimolecular rate constant. This criticism does not apply to the k_c value obtained[11] for $H + C_2H_5$ in the presence of 50 Torr He because of the different method used.

3.2.3 Combination of halogenated alkyl radicals

Another important reference reaction is

$$CF_3 + CF_3 \rightarrow C_2F_6 \tag{3.3}$$

It has been accepted that $E_c = 0$, and Ayscough's value at 127 °C (obtained in 1956) of $\log k_c = 13.36$ is widely used. The decay of CF_3 radicals, produced by flash photolysis, was monitored by i.r. absorption[12] and, although the kinetic analysis is more complex than that for reaction (3.1) discussed above, k_c (3.3) was measured at 25 °C and 60 °C. Extrapolation to 127 °C yields a value within a factor of two of Ayscough's result. All these results in the range 25–127 °C lead to E_c (3.3) = 0.3 to 2.5 with '1.0 kcal mol^{-1} being the most probable value'. However, earlier data favours $E_c = 0$.

The rotating sector method was applied to the combination of CF_2Cl radicals[13] to obtain k_c which was independent of temperature, suggesting that $E_c \leqslant 2$ kcal mol^{-1} and is probably zero. The reaction $CF_2 + CF_2 \rightarrow C_2F_4$ was studied by flash photolysis[14] from which $E_c \approx 0.4$ kcal mol^{-1}. The combination reaction has an unusually low steric factor of $\sim 10^{-4}$, and the reasons for this are discussed. The value of k_c for the reaction $CF_2 + CFCl$ was also measured.

3.2.4 Combination of other radicals

Before 1969, there was no data on combination of radicals containing Si. Thynne[15] used data on the pyrolysis of $[(CH_3)_3Si]_2$ to calculate that

$\log k_c = 5.5$ for the reaction

$$(CH_3)_3Si + (CH_3)_3Si \rightarrow (CH_3)_3Si\,Si(CH_3)_3 \qquad (3.4)$$

Subsequently k_c was measured more directly in solution[16] and in the gas phase[17]; the resulting values of $\log k_c = 12.34$ and 14.25, respectively, (the latter being independent of temperature) are much more acceptable and correspond to combination at almost every collision; hence $E_c \approx 0$.

The values of k_c for the combination of NCl_2 radicals[18, 19] and for the reaction $CH_3CO + CH_3O$ [20] are given in Table 3.1. Studies[21] of the reaction between CH_3 and $CH_3CH{=}CHCH_2\cdot$ indicate that, in the latter radical, rotation about the allylic double bond is fast enough to give both *cis* and *trans* combination products, unlike similar reactions in solution. Other radical combinations are discussed in Refs. 22–25.

3.2.5 Determination of ϕ factors

Possible combination reactions of unlike radicals A and B are

$$A + A \rightarrow A_2; \quad B + B \rightarrow B_2; \quad A + B \rightarrow AB$$

The so-called ϕ factor is defined as

$$\phi_{A,B} = \text{Rate (AB)}/[\text{Rate}(A_2)]^{\frac{1}{2}}[\text{Rate (B}_2)]^{\frac{1}{2}} \qquad (3.5)$$

If $E_c = 0$ in all cases and if there are no steric restrictions to combination, then we would expect that $\phi \approx 2$; this has been confirmed for many pairs of radicals. For equation (3.5) to lead to $\phi \approx 2$, R_{AB} etc. should be interpreted as the *total* rate of formation of products from encounters between A and B, and hence it should include both combination and disproportionation products. References to new data on ϕ for the following pairs of radicals are: $CH_3/(CH_3)_2CH$[5, 255]; CH_3/CH_2F[26]; CF_3/CF_2H[27]; CF_2H/CF_2Cl[254]; C_2H_5/CH_3CHF[28]; $CCl_3/CHCl_2$ [29, 134]; CCl_3/CH_3CHCl [30]; CF_3/CF_2Cl[31]. All these values of ϕ were approximately equal to two, and several were independent of temperature suggesting that for the various reactions $E_c = 0$. The only anomalous result[32] is $\phi(C_2H_5, \text{allyl}) = 6.8$, from which it was concluded that, if for all reactions $E_c = 0$, the steric factor for combination of allyl radicals is $\approx 10^{-7}$. However, this disagrees with new data on k_c for allyl radicals (see Section 3.2.2 and Table 3.1).

3.3 DISPROPORTIONATION AND COMBINATION OF RADICALS

Most encounters between radicals can lead to combination or disproportionation, e.g.

$$C_2H_5 + C_2H_5 \underset{k_d}{\overset{k_c}{\lessgtr}} \begin{array}{l} \nearrow C_4H_{10} \\ \searrow C_2H_4 + C_2H_6 \end{array}$$

New data on k_d/k_c ratios for various radicals are given in Table 3.2.

The ratio k_d/k_c is known to increase with increasing complexity of the radical structure, and this is expressed quantitatively in the widely-used equation

$$\log k_d/k_c = a(\Delta S_d^0 - \Delta S_c^0) + b \qquad (3.6)$$

where ΔS^0 are the entropy changes for disproportionation and combination, and a and b are *arbitrary* constants: values of a = 0.131 and b = 5.47 have been proposed. Some recent data[33, 34] fits equation (3.6) quite well. The theoretical significance of the constants a and b has been analysed by Konar[35] as follows.

Possible reactions for radicals P and Q are

$$P + Q \underset{k_{-d}}{\overset{k_d}{\rightleftharpoons}} \text{olefin} + \text{alkane} \qquad P + Q \underset{k_{-c}}{\overset{k_c}{\rightleftharpoons}} PQ$$

It is usually accepted that $E_c - E_d = 0$ (the results in Table 3.2 support this) so that $k_d/k_c = A_d/A_c$. It can then be shown that

$$\log (k_d/k_c) = (\Delta S_d^0 - \Delta S_c^0)/2.3R + \log (A_{-d}/A_{-c}) \qquad (3.7)$$

hence by comparison of equations (3.6) and (3.7), a = 1/2.3R and b =

Table 3.2 Ratios of rate constants k_d k_c for radical disproportionation and combination*

Radical pair	k_d/k_c	Temp. (°C)	Ref.	Radical pair	k_d/k_c	Temp. (°C)	Ref.
Et + Et	0.137	83–132	34	$CH_3CHF + CH_3CHF$	0.21	23	43, 44
i-Pr + i-Pr	0.61	26–207	42	$CH_3CF_2 + CH_3CF_2$	0.55	23	43, 44
n-Bu + n-Bu	0.12	68–140	38	$CF_2Cl + CF_2Cl$	0.17	25	254
t-Bu + t-Bu	2.3	24–127	36		0.09	213	13
i-Pe + i-Pe	0.13	68–140	38		0.2	268	13
allyl + allyl	0.008	164–190	33	$CHCl_2 + Et$	0.07	25	29
Et + n-Bu	0.059	68–140	38	$CCl_3 + CH_3CHCl$	0.25	25	30
n-Bu + Et	0.066	68–140	38	$NO + n-PrO$	0.43	100–150	45
i-Pr + i-Pe	0.087	68–140	38	$NO + CH_3O$	0.13	25	46
i-Pe + i-Pr	0.265	68–140	38	$N_2H_3 + N_2H_3$	~4†	25	47
$CF_3 + CF_2H$	0.08	25	27				
$CF_3 + i-Pr$	2.2	30–115	42				

also

$$C_2H_5 + C_2H_5O \begin{cases} \rightarrow C_2H_5OC_2H_5 & 80\% \\ \rightarrow C_2H_6 + CH_3CHO & 3\% \\ \rightarrow CH_2{=}CH_2 + CH_3CH_2OH & 17\% \end{cases} \quad 34$$

$$C_2H_5 + CH_2{=}CHCH_2 \begin{cases} \rightarrow C_2H_5CH_2CH{=}CH_2 & 88\% \\ \rightarrow C_2H_6 + CH_2{=}C{=}CH_2 & 3\% \\ \rightarrow C_2H_4 + CH_3CH{=}CH_2 & 9\% \end{cases} \quad 115{-}160 \quad 32, 34$$

$$CH_2{=}CHCH_2 + \overset{\cdot}{\bigcirc} \begin{cases} \rightarrow CH_2{=}CH-CH_2- & 40\% \\ \rightarrow CH_2{=}CH-CH_2- & 51\% \\ \rightarrow CH_2{=}CHCH_3 + & 9\% \end{cases} \quad 130{-}200 \quad 48$$

*The first radical of each pair gains H (or Cl in case of CF_2Cl).
†See Ref. 47 for reactions involved.

$\log (A_{-d}/A_{-c})$. Konar applied equation (3.6) to all available data on k_d/k_c and obtained $R = 4.5\,\text{kcal mol}^{-1}$. This failure of equation (3.6) to give the correct value of R may be because $E_c - E_d$ is not always zero, or because b = log (A_{-d}/A_{-c}) may vary with the radicals involved.

A new value of k_d/k_c for Bu^t radicals[36] agrees with those previous values where the radicals were generated from azo compounds. However, different values of 3–4.5 were previously obtained when *other* sources of Bu^t radicals were used. This difference may be caused by excess vibrational energy of the radicals in the latter cases[36]. The best current value of k_d/k_c (Bu^t) is probably that listed in Table 3.2.

The data in Table 3.2 on k_d/k_c for halogenated alkyl radicals indicate that introduction of Cl or F into the radical favours disproportionation.

The BEBO method was applied to radical disproportionation[37] assuming a head-to-tail abstraction, and the calculated values of k_d/k_c agree with observed values to within a factor of two for many radicals. This supports the idea of a loose head-to-tail transition state as proposed by Benson in 1964.

Inel[38] has re-interpreted published work[39] on addition of C_2H_5 radicals to C_2H_4. Various k_d/k_c values, re-calculated by Inel, are given in Table 3.2.

Other work on radical disproportionation (not mentioned in Table 3.2) is described in Refs. 22, 40, 41, 49 and 50.

3.4 REACTIONS OF RADICALS WITH NO, NO_2 AND O_2

3.4.1 Reactions of radicals with NO

Reactions between radicals and NO are fast, hence the efficiency of NO as a radical scavenger. The reaction

$$CH_3 + NO + M(?) \rightarrow CH_3NO + M \qquad (3.8)$$

is of particular interest, and has now been studied by flash photolysis[1]. The second-order rate constant was independent of pressure above about 200 Torr N_2 with $k_{3.8}(\infty) = 2.4 \times 10^{12}\,\text{cm}^3\,\text{mol}^{-1}\,\text{s}^{-1}$ at 25 °C. This is probably the most reliable current value, and is approximately three times greater than previous values. Reaction (3.8) is slower than the combination of CH_3 radicals by a factor of ten.

When NO scavenges radicals, the product RNO may undergo a variety of subsequent reactions. If reaction (3.8) occurs at low concentrations of CH_3 and NO, the CH_3NO disappears by dimerisation[51]. The rate of formation of dimer was measured directly by mass spectrometry, from which $k(\text{dimerisation}) = 2.2 \times 10^4\,\text{cm}^3\,\text{mol}^{-1}\,\text{s}^{-1}$ at 25 °C. This agrees with previous results of Christie and co-workers, but not with others. Under certain conditions, reactions such as that depicted in equation (3.8) may initiate disproportionation of NO to $N_2 + NO_2$ via

$$CH_3NO \xrightarrow{\;NO\;} CH_3(NO)_2 \xrightarrow{\;NO\;} CH_3-\underset{\underset{O}{\overset{|}{N}}}{N}-\underset{\underset{O}{\overset{|}{N}}}{O} \rightarrow CH_3 + N_2 + NO_3$$

$$NO_3 + NO \longrightarrow 2NO_2$$

New confirmation of this mechanism has been provided by Gilbert et al.[52] (see also Refs. 20 and 53).

Radicals may undergo disproportionation with NO as well as combination, and examples are given in Table 3.2 (Refs. 45 and 46). For the reaction

$$C_2H_5 + NO \rightarrow C_2H_4 + HNO \tag{3.9}$$

it has been found[49, 54] that $k_{3.9} = 3.5 \times 10^{10} \exp(-11\,000/RT)\,cm^3\,mol^{-1}\,s^{-1}$; this is more typical of H-abstraction than normal disproportionation. The technique of molecular modulation spectroscopy was used[46] to measure k_d/k_c for $NO + CH_3O$ (see Table 3.2). In addition, it was estimated that for

$$CH_3O + HNO \rightarrow CH_3OH + NO \tag{3.10}$$

$k_{3.10} = 3 \times 10^{13}\,cm^3\,mol^{-1}\,s^{-1}$ at 25 °C.

A complication in using NO as a scavenger in photochemistry is that it may react with non-radical *molecules* in excited electronic states. Thus, in the photolysis of CH_3CHO with NO[55], CH_3CONO is formed not only from $CH_3CO + NO$ but also via

$$^1CH_3CHO^* + NO \rightarrow CH_3CONO + ?$$

Furthermore, $^3CH_3CHO^*$ reacts with NO to give a long-lived intermediate. The reactions of RCO radicals with NO are also discussed in Refs. 20 and 54. Other work on radicals plus NO is described in Refs. 56–59.

3.4.2   Reactions of radicals with NO_2

Diethyl ketone was photolysed in the presence of NO and NO_2, and from this[53] and other work it was estimated that, for the reactions

$$RCO + NO \rightarrow RCONO \tag{3.11}$$
$$RCO + NO_2 \rightarrow R + CO_2 + NO \tag{3.12}$$

$k_{3.12}/k_{3.11} = 10.5$ (R = Et) and 4.9 (R = Me) over the range 25–75 °C.

3.4.3   Reactions of radicals with O_2

Baldwin and co-workers have developed a method of measuring rate constants for reactions of the type

$$R + O_2 \rightarrow products$$

by adding a hydrocarbon RH to slowly reacting mixtures of $H_2 + O_2$, usually at 480 °C. Several reactions are possible, e.g. with C_2H_6[60]

$$C_2H_5 + O_2 \begin{cases} \rightarrow C_2H_4 + HO_2 & (3.13) \\ \rightarrow CH_3CHO + OH & (3.14) \\ \rightarrow C_2H_4O + OH & (3.15) \end{cases}$$

$$\frac{k_{3.13}}{k_{3.14}} = 270$$

$$\frac{k_{3.13}}{k_{3.15}} = 35$$

Absolute values of k were measured for reaction (3.13) and its analogues for several other alkyl radicals[61-63].

The reaction between CH_3 and O_2 was studied[64] in the range 155–185 °C, the main products being CH_2O, CH_3OH and CH_3O_2H. For the reaction

$$CH_3 + O_2 \rightarrow CH_3O_2 \qquad (3.16)$$

$k_{3.16} = 4 \times 10^{10}$ cm^3 mol^{-1} s^{-1}, independent of temperature; this agrees well with previous work. The products of reaction of C_2H_5 and C_2H_5CO with O_2 were determined[65] by photolysis of diethyl ketone with O_2 in the range 25–200 °C.

3.5 REACTIONS INVOLVING ABSTRACTION BY RADICALS

3.5.1 Introduction with short sermon

Most radical abstraction reactions involve H-atom transfer, e.g.

$$R + SH \xrightarrow{k_H} RH + S \qquad (3.17)$$

The main emphasis in current work is on determination of the Arrhenius parameters k_H, A_H and E_H, usually by measuring k_H relative to the combination rate constant k_c for

$$R + R \xrightarrow{k_c} R_2 \qquad (3.18)$$

This procedure is very widely used, but the kinetic validity of the treatment is seldom adequately demonstrated for each system. For a given pair of reactants $R + SH$, about 6–20 runs are usually done to determine $k_H/k_c^{\frac{1}{2}}$ over a range of temperature and hence obtain $E_H - \frac{1}{2}E_c$ and $A_H/A_c^{\frac{1}{2}}$. Usually however, temperature is the *only* variable and it is seldom that, at a given temperature, the ratio $k_H/k_c^{\frac{1}{2}}$ is shown to be independent of, for example, the concentration of RH (as it should be if true rate constants are being measured). It would be better to select four temperatures and do four runs at each temperature varying [RH] over as wide a range as possible. The difficulties caused by failure to do this are referred to later (see Sections 3.5.2.2 and 3.5.4). Some authors have thoroughly checked their kinetics (see for example Refs. 66 and 67) but this is not common.

3.5.2 Abstraction of hydrogen by radicals

3.5.2.1 Results

Many H-abstraction reactions have been studied in the last 2 years and these are summarised in Table 3.3. Lack of space prevents the listing of Arrhenius parameters for most of these reactions but, in view of current interest in reactions of radicals with Si compounds, the parameters for these reactions are given in Tables 3.4 and 3.5. It is impossible to discuss all the new data on abstraction, but it is worth noting that new data on CH_3 + neopentane[68] and

Table 3.3 Studies of the abstraction reaction R+SH → RH+S

R = CH₃	Ref.	R = CF₃	Ref.	R	SH	Ref.
SH	68	SH		CF₂H	H₂, D₂	77
neopentane	6, 114	neopentane	95	CF₂H	hydrocarbons	77
toluene	182	cyclanes	123, 124	CF₂H	CF₂HCOCF₂H	77
cis-but-2-ene	115	(CH₃)₂CHCH(CH₃)₂	181	CF₂H	CF₂HCOCF₃	27
substituted toluenes	116			CF₂Cl	CF₃H	132
bicyclo[2.1.1] hexane		esters	71, 74	CF₂Cl	fluorinated SH	76, 133
		ethers	117	CHCl₂	CHCl₃	134
CH₃I	96	ethylene oxide	125	C₂H₅	H₂, D₂	135
CF₂HCl	132	aromatic aldehydes	126	C₂D₅	D₂	113
esters	71, 74	CD₃COCD₃	127	C₂H₅	CH₃CHO	136
ethers	117	CH₃COC₂H₅	128	C₂H₅	(C₂H₅)₂O	136
biacetyl	7	CH₃COCF₃	127	CH₃CD₂	(CH₃CD₂)₂CO	137
△CHO	78	CF₂HCOCF₃	27	C₂H₅	allylic compounds	32, 34
azomethane	36	tetramethyl urea	129	n-C₃H₇	C₃H₈	138
amides	118, 119	alcohols†	130	c-C₃H₅	c-C₃H₅CHO	78
CH₃NO₂	253	Br₂, Cl₂, BrCl	101	t-C₄H₉		36
phenol	120	HCl, H₂S	75	allyl	esters	33, 48
SF₆	69	NH₃	67, 72, 131	C₆H₅CH₂	toluene	139
DBr	70	SF₆	69	(CH₃)₃CO	alkanes	79
H₂S	121	H₂S	183	(CF₃)₂NO	hydrogen halides	140, 141
HN₃	122			(CF₃)₂NO	other halides	141
O₂	64			SO₂*	alkanes	142–144

†in aqueous solution

on CF_3 and $CH_3 + SF_6$ [69] seem more reasonable than previous results. Also noteworthy is the *direct* measurement of k_D for $CH_3 + DBr$ using the new technique of very low pressure pyrolysis[70].

3.5.2.2 Comparison of the reactivity of CH_3 and CF_3 radicals

These radicals continue to receive more attention than any others. It used to be thought that, for reactions of CH_3 and CF_3 with a given SH,

$$CH_3 + SH \rightarrow CH_4 + S \qquad (3.19)$$
$$CF_3 + SH \rightarrow CF_3H + S \qquad (3.20)$$

the difference $\Delta E = E_{CH_3} - E_{CF_3} \approx 2\text{--}3$ kcal mol^{-1}. The following typical results indicate that this is an over-simplification.

SH	Cyclopropane	C_2H_6	*$HCOOCH_3$	HCl	$SiHCl_3$
E_{CH_3}	13.3	11.9	9.9	4.4	4.3
E_{CF_3}	8.7	8.4	8.7	5.2	6.0

*indicates H abstracted. Data (in kcal mol^{-1}) from Refs. in Table 3.3 or cited therein.

Clearly there is a trend from $\Delta E = +4.6$ for cyclopropane to $\Delta E = -1.7$ kcal mol^{-1} for $SiHCl_3$: in other words, ΔE depends on the type of substrate SH [71]. Now reaction (3.20) is more exothermic than (3.19) by about 2.5 kcal mol^{-1} and the extra driving force provided by this could explain $\Delta E \approx 1\text{--}3$ kcal mol^{-1} (but see Section 3.5.4). Values of ΔE in this range are observed when SH is a non-polar hydrocarbon, and it is significant that, as ΔE decreases across the above series, the polarity of SH increases in the sense $(\delta+)H$—$S(\delta-)$, which would cause increased repulsion of the electrophilic CF_3 radical and thus tend to increase E_{CF_3}. The importance of polar effects has been questioned by LeFevre and Timmons[72] who measured E for $CF_3 + NH_3$ and confirmed previous values. They therefore concluded that for NH_3, $E_{CH_3} - E_{CF_3} = 9.9 - 8.3 = 1.6$ kcal mol^{-1}, which is reasonable for a non-polar molecule but not for NH_3. However, the reaction between CF_3 and NH_3 was subsequently studied in detail[67] over a large temperature range, and it was found that (a) the Arrhenius plot is curved and (b) the value of k_H was *not* independent of the NH_3 pressure. Hence no reliable E exists for $CF_3 + NH_3$.

Other examples of polar effects appear in Refs. 66, 71, 73–76.

3.5.2.3 Reactions of other halogenated radicals

The radicals CHF_2 and CH_2F have been shown[77] to abstract H less readily from hydrocarbons than do CH_3 and CF_3. This is at least in part because the bond dissociation energy $D(C$—$H)$ in CH_2F_2 and CH_3F is less than in either CH_4 or CF_3H, hence the analogues of reactions (3.19) and (3.20) are less exothermic. Studies of abstraction of H by CF_2Cl from a variety of almost fully-fluorinated cyclanes[76] reveal striking variations of A_H and E_H with ring size.

3.5.2.4   Reactions of alkyl and alkoxy radicals

Novel results are as follows. Cyclopropyl radicals were generated[78] by reaction of $CH_3 + c\text{-}C_3H_5CHO$, and E_H for abstraction of H from $c\text{-}C_3H_5$ CHO by cyclopropyl and by $\cdot CH_3$ were almost identical. The compound 1,4-cyclohexadiene is a very efficient scavanger of radicals[33, 36], not by addition to the double bonds, as might be expected, but because of facile H-abstraction at the 3-position. The activation energies for abstraction of primary, secondary and tertiary-H in hydrocarbons by the tert-butoxy radical[79] indicate that it is very reactive – about the same as CF_3 or CH_3O.

3.5.2.5   Reactions of NF₂ and NCl₂ radicals

NF_2 radicals are conveniently generated from N_2F_4 by

$$N_2F_4 \rightleftharpoons 2NF_2 \tag{3.21}$$

Thermodynamic data on this equilibrium are reviewed[80] and the first determination of ΔH_f^0 (HNF_2, g) was made[81]. Other relevant thermodynamic data are quoted in Refs. 82 and 83. The kinetics of the N_2F_4 pyrolysis have been investigated[84].

Reactions of the type

$$NF_2 + RCHO \rightarrow HNF_2 + RCO \tag{3.22}$$

show little change in E_H for a variety of RCHO [85, 86]. This agrees with previous observations on the reactions of a given radical with several aldehydes, and suggests, but does not prove, that $D(RCO—H)$ is fairly constant in aldehydes. Refs. 85 and 86 also include detailed studies of the reactions of the RCO radicals formed in reaction (3.22). Irradiation of N_2F_4 at 253.7 nm in the presence of hydrocarbons[87] initiates the reactions

$$NF_2 + h\nu \rightarrow NF + F \text{ and } F + RH \rightarrow HF + R$$

since light is absorbed, not by the N_2F_4, but by the NF_2 radicals formed by reaction (3.21).

The NCl_2 radical was generated[19] in a flow system, and rate constants were measured for

$$Cl + NCl_2 \rightarrow Cl_2 + NCl \text{ and } O + NCl_2 \rightarrow NO + Cl_2$$

It is suggested that reaction between two NCl_2 radicals may lead, not to combination, but to

$$NCl_2 + NCl_2 \rightarrow N_2Cl_3 + Cl$$

with $N_2 + Cl_2$ as the final products.

3.5.2.6   Reactions of radicals with compounds of silicon and other Group IV elements

Much new work has appeared, and in Table 3.4 Arrhenius parameters for the reaction of both CH_3 and CF_3 radicals with various $R_3Si—H$ compounds are listed. Other data are given in Table 3.5. Several reviews of silicon radical chemistry have appeared[88-90].

The interpretation of the results in Table 3.4 is impeded by the lack of reliable Si—H bond dissociation energies, but the best current values (in

Table 3.4   Abstraction of hydrogen by CH₃ and CF₃ radicals from compounds of silicon and other Group IV elements†

Compound	CH₃			CF₃		
	log A	E	Ref.	log A	E	Ref.
SiH₄	11.8	6.9	145	11.9	4.9	112
	11.8	7.0	92	11.9	5.1	91
SiD₄	12.0	8.2	92	12.0	6.0	112
Si(CH₃)₄	12.6	11.0	107	11.9	7.2	147
	11.5	10.3	145, 146	12.0	7.6	91, 146
	11.6	10.2	147	11.9	7.3	95
‡	11.8	10.4	147			
(CH₃)₃SiH̶*	11.3	7.8	145	12.3	5.6	91
(C̶*H₃)₃SiH				12.3	5.6	91
SiHCl₃	10.8	4.3	148	11.8	6.0	148
C(CH₃)₄	12.3	12.0	68	12.0	8.4	95
Ge(CH₃)₄	11.8	9.6	107	11.7	7.4	95
Sn(CH₃)₄	11.1	8.6	107	11.7	7.3	95
Pb(CH₃)₄	10.2	7.4	107			

*indicates H abstracted.
†A in cm³mol⁻¹s⁻¹, E in kcal mol⁻¹.
‡with CD₃ not CH₃.

Table 3.5   Abstraction of hydrogen by various radicals from compounds of silicon and other Group IV elements† R+SH→RH+S

R	SH	log A	E	Ref.	R	SH	log A	E	Ref.
CH₃	Si₂H₆	12.0	5.6	92	CF₃	CH₃SiF₃	12.0	11.7	66
	Si₂D₆	12.2	7.0	92		(CH₃)₂SiF₂	12.3	10.5	66
	C₆H₅SiD̶*₃	12.0	8.0	92		(CH₃)₃SiF	12.4	9.5	66
	C₆H̶*₅SiD₃	10.0	5.9	92		CH₃SiCl₃	11.3	9.4	73
C₂H₅	Si(CH₃)₄	11.9	11.4	147		(CH₃)₂SiCl₂	11.8	9.2	73
	SiHCl₃	11.5	5.3	148		(CH₃)₃SiCl	12.3	9.1	73

*indicates H abstracted.
†A in cm³mol⁻¹s⁻¹, E in kcal mol⁻¹.

kcal mol⁻¹) for compounds of particular interest, together with the C—H analogues are,

$D(SiH_3—H) \approx 95$ $D(CH_3—H) = 104$
$D((CH_3)_3Si—H) = 81 \pm 2$ $D((CH_3)_3C—H) = 92$

The use of these figures, together with data from Table 3.4, leads to the following generalisations.

(a) For a given compound R₃Si—H, $E_{CH_3} - E_{CF_3} \approx 2$ kcal mol⁻¹ just as in hydrocarbons (see Section 3.5.2.2).

(b) Abstraction from a given compound R₃Si—H by CH₃ or CF₃ is much easier than from the corresponding R_3C—H, as expected in view of the weaker Si—H bonds.

Other results are harder to explain. It is striking that E_H is almost unchanged, from SiH_4 to $(CH_3)_3Si$—$\overset{*}{H}$, whether the attacking radical is CH_3 or CF_3, in spite of the decrease from 95 to 81 kcal mol^{-1} in $D(Si$—H$)$. This may indicate[91] that the difference in $D(Si$—H$)$ is really less than 14 kcal mol^{-1}. Another unexpected result is that for abstraction of D by CH_3, E_D is almost the same[92] for SiD_4 and $C_6H_5SiD_3$. This was taken to indicate little Si—D bond weakening in $C_6H_5SiD_3$, i.e. little resonance stabilisation of the $C_6H_5SiD_2$ radical (in contrast to the benzyl radical). However, this data could also indicate that $D(SiD_3$—D$)$ is less than had been supposed.

A strong polar effect was observed[66] in abstraction of H by CF_3 in the methylfluorosilanes, where E_H increases along the series $(CH_3)_4Si$——— CH_3SiF_3. This type of deactivation by fluorine atoms attached to an atom (in this case Si) *adjacent* to the carbon atom carrying the H to be abstracted has previously been observed in carbon compounds, e.g., CH_3CF_3. However, no such deactivation was observed in the corresponding methylchlorosilane series[73], and this was explained in terms of the inductive electron withdrawal by Cl being largely cancelled by $d\pi \leftarrow p\pi$ electron transfer in the Si—Cl bond.

Novel reactions occur when CF_3 reacts with Group IV tetramethyls. The products include CH_4 and C_2H_6, which indicates that CH_3 radicals are present[93]: the proposed origin is a *displacement* reaction which was studied in most detail for $Sn(CH_3)_4$.

$$CF_3 + Sn(CH_3)_4 \rightleftharpoons CF_3Sn(CH_3)_4 \rightarrow CH_3 + CF_3Sn(CH_3)_3$$

The products from these systems[93, 94, 95] also include $CH_2{=}CF_2$ produced by

$$CF_3 + M(CH_3)_4 \rightarrow CF_3H + (CH_3)_3MCH_2$$

$$CF_3 + (CH_3)_3MCH_2 \rightarrow (CH_3)_3MCH_2CF_3^{\dagger} \rightarrow (CH_3)_3MF + CH_2{=}CF_2$$

This occurs for M = Si, Ge, Sn, but not C.

3.5.2.7 *Reactions of hot radicals*

The low intensity photolysis of CH_3I yields CH_4 produced by abstraction of H by CH_3 from CH_3I, but in the high intensity flash photolysis[96] 'extra' CH_4 is formed, probably by a reaction involving $CH_3^{\dagger}$.

The photolyses of CF_3I (253.7 nm) and CF_3Br (185 nm) are thought[97] to yield hot CF_3 radicals, with those from CF_3I carrying excess translational energy, whereas CF_3 from CF_3Br appears to be vibrationally excited. The γ-radiolysis of CF_3I[98] leads to the formation of CF_4, possibly by

$$CF_3^* + CF_3I \rightarrow CF_4 + CF_2I$$

If correct, this is one of the rare examples of a fluorine atom transfer. The effect of added NO and O_2 has also been discussed[99, 100]. The formation of

CF_4 was also observed in similar experiments[101], but the proposed mode of formation, namely,

$$CF_3 + F \rightarrow CF_4$$

seems improbable in view of the exothermicity of this reaction. The effect of added halogens was also investigated.

3.5.3 Abstraction of halogens by radicals

Information on reactions of the type

$$R + SX \rightarrow RX + S$$

where X = halogen, continues to be sparse, mainly because with alkyl and substituted-alkyl radicals, H-abstraction, rather than halogen abstraction, usually occurs. In contrast, halogen atoms are readily abstracted by silyl radicals, and this is explained by differences in the strengths of the bonds broken and formed, e.g., if we compare the reactions of CH_3 and $SiCl_3$ radicals with say C_2H_5Cl, the changes in enthalpy favour H-abstraction by CH_3 and Cl abstraction by $SiCl_3$, mainly because $D(SiCl_3\text{---}Cl) > D(C_2H_5\text{---}Cl)$ whereas $D(SiCl_3\text{---}H)$ is probably $< D(C_2H_4Cl\text{---}H)$. From studies of the competitive reactions[102, 103],

$$SiCl_3 + CH_3Cl \rightarrow SiCl_4 + CH_3 \tag{3.23}$$
$$SiCl_3 + RX \rightarrow SiCl_3X + R \tag{3.24}$$

$A_{3.23}/A_{3.24}$ and $E_{3.23} - E_{3.24}$ have been obtained for a variety of RX compounds. There is a fair correlation between variations in $E_{3.24}$ and $D(R\text{---}X)$ and the data fit Evans–Polanyi type plots (see Section 3.5.4).

Studies of the photolysis of CH_2Br_2 in the presence of CCl_4 [104] gave $E = 14.3$ kcal mol^{-1} for the reaction

$$CH_2Br + CCl_4 \rightarrow CH_2BrCl + CCl_3$$

but some questionable assumptions are involved. The reaction

$$CCl_3 + CCl_3Br \rightarrow CCl_4 + CCl_2Br \tag{3.25}$$

occurs when CCl_3Br is photolysed[105], and the Arrhenius parameters are $A_{3.25} = 7.1 \times 10^9$ cm^3 mol^{-1} s^{-1}, $E_{3.25} = 10.5$ kcal mol^{-1}. However, the value of $A_{3.25}$ is derived using A_c for CCl_3 combination $(= 10^{13.9}$ cm^3 mol^{-1} s^{-1}) whereas lower values have also been proposed[106].

Halogen transfer has already been mentioned in Section 3.5.2.7, Refs. 98–101.

3.5.4 Quantitative theoretical interpretations of abstraction data

Progress in this field continues to be slow. Most treatments involve semi-empirical attempts to explain variations in activation energies, yet many large variations in A factors occur which have not been satisfactorily explained. For example, in the reactions of CF_3 with $(CH_3)_4M$, (M = C, Si, Ge,

Sn), both A_H and E_H are virtually constant[95], yet in the corresponding reactions involving CH_3 [107] (and including M = Pb) there is a progressive decrease in E_H from M = C to M = Pb, which is off-set by a corresponding decrease in A_H from $10^{12.0}$ for $(CH_3)_4C$ to $10^{10.2}$ for $(CH_3)_4Pb$. Even more spectacular changes in A_H and E_H were observed when CF_2Cl abstracts H from heavily-fluorinated cyclanes[76], yet there were quite small variations in k_H. A slick explanation often given is that there must be 'complications' (seldom identified) in the kinetics, so that self-compensation of A and E occurs. This must happen occasionally, but there are now too many examples of changes in both A and E to brush them aside so easily. This problem emphasises the *very strong need* for the thorough checking of the validity of all rate constants as mentioned in Section 3.5.1.

The commonest semi-empirical approach to discussion of activation energies is to use the Evans–Polanyi (E–P) equation which relates E_H and ΔH^0 for a series of reactions involving the same radical, as in equation (3.17). The equation can be used in either of the equivalent forms

$$E_H = \alpha \Delta H^0 + c \tag{3.26}$$
$$= \alpha[D(S\text{—}H) - \beta] \tag{3.27}$$

where α, β and c are arbitrary constants with $0 \leqslant \alpha \leqslant 1$. Equation (3.27) works well, even for a polar radical like CF_3, provided the substrates S—H are non-polar, e.g. hydrocarbons. It is normal to use equation (3.27) to predict the effect on E_H of a change in $D(S\text{—}H)$, and then ascribe to polar effects any discrepancies between E_H (obs.) and E_H (calc.). The main weakness of the approach is that the constant α is different for each radical, and must be determined using experimental data on E_H and $D(S\text{—}H)$ for SH where it is hoped that polar effects are negligible. Thus α *cannot be predicted* for a given radical, though we know from experience that the more exothermic the reactions of a given radical the less selective will be the radical and the closer α will be to zero. Conversely, unreactive radicals, where reaction (3.17) is endothermic, are very selective and α is close to unity.

The most important development for some time in this approach is the work of Kagiya et al.[108] in which the above intuitive relationship between α and ΔH^0 is placed on a quantitative basis so that α *can be predicted* for any given radical. In the process

$$R + HS \rightarrow R\text{----}H\text{----}S \rightarrow RH + S$$

the authors describe the stretching of the H—S bond to reach the transition state by means of a Morse curve in which the Morse constant 'a' is continuously modified by approach of the radical R; the same applies when the transition state is approached from the right-hand side with stretching of the R—H bond. The following equation is then derived*

$$E_H = D_i[(1-2\delta)D_f + \delta^2 D_i]^2/(D_f - \delta^2 D_i)^2 \tag{3.28}$$

where $D_i = D(H\text{—}S)$, $D_f = D(R\text{—}H)$ and $\delta = \exp(-0.0190\Delta H^0)$.

Equation (3.28) is much more sophisticated than the E–P equation because to predict E_H only $D(S\text{—}H)$ and $D(R\text{—}H)$ need be known. The equation was

*The symbol δ has been used instead of the α used in Ref. 108 to avoid confusion with the α of the E–P equations (3.26) and (3.27).

shown to work well for a wide range of reactions involving attack both by radicals and atoms and with transfer of halogen as well as hydrogen.

In a subsequent paper[109], the authors derived a simplified form of equation (3.28), namely,

$$E = D_i - 0.085D_f(1 + 0.0944D_i) \qquad (3.29)$$

applicable to reactions with ΔH^0 in the range $+20$ to -20 kcal mol^{-1}. Equation (3.29) was applied, for example, to the well-known sodium flame reactions for which the experimental data gives $\alpha = 0.25$ in the E–P equation, whereas equation (3.29) predicts results equivalent to $\alpha = 0.22$. Equation (3.29) can also be expressed in a form equivalent to the Hammett equation.

Finally, other semi-empirical approaches to calculation of A and E are to be found in Refs. 110 and 111 while applications of Johnson's BEBO method are given in Refs. 112 and 113.

3.6   REACTIONS BETWEEN RADICALS AND OLEFINS

3.6.1   Introduction

Apart from studies of polymerisation which are not discussed here, current interest in reactions of radicals with olefins is focused mainly on two questions, (i) how readily does a given radical add to the double bond of a given olefin?, (ii) at which end of the double bond does addition occur? It should be noted that the radical may also abstract as an alternative to addition, but for many olefins addition is faster than abstraction.

3.6.2   Determination of the Arrhenius parameters of addition reactions

This is inherently difficult because the addend from radical plus olefin is itself a radical and may react in a variety of ways. There are two main methods of measuring the rate constant for addition, k_{Add}.

3.6.2.1   The direct method

As an example, suppose CF_3I is photolysed in the presence of ethylene[149]: some of the expected reactions are

$$CF_3I + h\nu \rightarrow CF_3 + I \qquad (3.30)$$
$$CF_3 + CH_2{=}CH_2 \rightarrow CF_3CH_2CH_2 \qquad (3.31)$$
$$CF_3CH_2CH_2 + CF_3I \rightarrow CF_3CH_2CH_2I + CF_3 \qquad (3.32)$$

If all the addend radicals from reaction (3.31) are removed by reaction (3.32), then the rate of formation of $CF_3CH_2CH_2I$ is a direct measure of R_{Add} (R denotes rate) for reaction (3.31). Hence k_{Add} can be determined if the stationary concentration of CF_3 radicals can be calculated. It must be stressed that the method works well here, partly because the I atoms from reaction (3.30) are unreactive and do not add to C_2H_4 to initiate chains, and partly because the iodine atom transfer by reaction (3.32) is facile so there is complete scavenging of the addend radical produced by reaction (3.31). Hence, telomerisation should not occur. These conditions can generally be attained when any RI compound is used as a radical source, but now consider the following more complex system.

Suppose SF_5Cl is photolysed in the presence of C_2H_4 [150] under conditions such that the *overall* addition reaction is exclusively

$$SF_5Cl + CH_2{=}CH_2 \rightarrow SF_5CH_2CH_2Cl \qquad (3.33)$$

i.e. no telomerisation occurs. The initial photolytic step

$$SF_5Cl + hv \rightarrow SF_5 + Cl \qquad (3.34)$$

could be followed by either or both of the reaction sequences,

$$SF_5 + CH_2{=}CH_2 \xrightarrow{\text{slow}} SF_5CH_2CH_2$$
$$SF_5CH_2CH_2 + SF_5Cl \xrightarrow{\text{fast}} SF_5CH_2CH_2Cl + SF_5$$

Sequence (1)

or

$$Cl + CH_2{=}CH_2 \xrightarrow{\text{slow}} ClCH_2CH_2$$
$$ClCH_2CH_2 + SF_5Cl \xrightarrow{\text{fast}} ClCH_2CH_2SF_5 + Cl$$

Sequence (2)

Both sequences give the same product but involve initiation by different addition reactions. If the olefin is unsymmetrical, e.g. $CH_2{=}CF_2$, there is a further complication. If the product were entirely $SF_5CH_2CF_2Cl$, and if only Sequence (1) occurred, then we would conclude that SF_5 adds only to the ${=}CH_2$ end of the olefin; however, if only Sequence (2) occurred, then addition would involve a Cl atom attacking ${=}CF_2$. Thus, if *both* radicals produced in step (3.34) are reactive, a product analysis alone will *not* identify the rate-determining addition step and rigorous kinetic analysis is necessary.

3.6.2.2   *The indirect method*

This involves competition between reactions of the type

$$R + \text{olefin} \rightarrow \text{adduct} \qquad (3.35)$$
$$R + SH \rightarrow RH + S \qquad (3.36)$$

where SH has a readily-abstractable H-atom. By measuring the yield of RH with and without added olefin, the ratio $k_{3.35}/k_{3.36}$ can be obtained. The method is intrinsically less sound than the direct method, but can yield reliable relative Arrhenius parameters, i.e. $A_{3.35}/A_{3.36}$ and $E_{3.35} - E_{3.36}$, if the kinetic analysis is thorough.

3.6.3   Quantitative data on reactions of radicals with olefins

In this Section, we discuss data on addition reactions obtained by kinetic studies. New Arrhenius parameters for addition of alkyl radicals to olefins are listed in Table 3.6, and it is clear that most interest is centred on reactions of halogenated radicals. Reactions of SF_5 radicals are also included. The trends in reactivity seen in Table 3.6 are confirmed by many semi-quantitative results, mainly from Haszeldine and co-workers, in which products were

Table 3.6 Arrhenius parameters[‡] for addition of radicals to olefins
$R + olefin \rightarrow addend$

R	Olefin	log A	E	Ref.	R	Olefin	log A	E	Ref.
CH_3	C_2F_4	11.92	5.70	181	CH_2F	C_2H_4	10.6	4.3	185
CH_3	cis-but-2-ene	10.65	7.00	182	CF_2Br	$CH_2{=}CH_2$	8.3	1.0	186
Et	C_2H_4	11.17	7.56	39		$\overset{*}{C}HF{=}CF_2$	10.0	3.3	186
Pr^i	C_2H_4	11.22	6.93	39		$\overset{*}{C}F_2{=}CHF$	10.3	4.0	186
Bu^n	C_2H_4	10.34	6.70	39					
$Pent^i$	C_2H_4	10.5	6.4	39	CF_2Cl	(six-membered ring, F)	14.3	15.4	133‖
						(five-membered ring, F)	9.7	6.6	133
CF_3	C_2H_4	11.39	2.37	{ 149, 159, 183 }	H—(five-membered ring, F)		11.0	7.9	133
	$\overset{*}{C}H_2{=}CF_2$	10.95	4.37†	149	Cl—(five-membered ring, F)		8.7	4.7	133
	$CH_2{=}\overset{*}{C}F_2$	—	>8.4†	149					
	$CF_2{=}CF_2$	10.5	1.4	181	CCl_3	$\overset{*}{C}H_2{=}CHCN$	10.2	3.3	184
						$\overset{*}{C}H_2{=}CHCF_3$	10.4	4.8	184
C_3F_7	$CH_2{=}CH_2$	(10.5)	(2.0)§	184					
	$\overset{*}{C}H_2{=}CHF$	10.1	3.1	184	SF_5	$CH_2{=}CH_2$	8.6	1.9	150
	$\overset{*}{C}H_2{=}CF_2$	10.2	3.9	184		$\overset{*}{C}H_2{=}CF_2$	8.6	2.0	160
	$\overset{*}{C}HF{=}CH_2$	9.9	5.4	184		$\overset{*}{C}HF{=}CF_2$	9.4	3.4	160
	$\overset{*}{C}HF{=}CF_2$	10.8	4.1	184					
	$\overset{*}{C}F_2{=}CH_2$	9.0	9.2	184					
	$\overset{*}{C}F_2{=}CHF$	10.8	5.8	184					

*point of addition of radical.
†calculated by combining results in Refs. 149, 183.
§values in parentheses estimated and used as standards for remaining data.
‖selection of results from Ref. 133.
‡A in $cm^3 mol^{-1} s^{-1}$, E in $kcal\ mol^{-1}$.
Other data are as follows: competitive results[42] on CF_3 + propylene and $CH_3CH{=}C(CH_3)_2$; reactions of CCl_3 with alkenylsilanes[187].

determined quantitatively but without detailed kinetic studies (see Refs. 151–158). Reactions of CF_3 with C_2H_4 and I_2 were studied competitively[159].

3.6.4 Patterns of reactivity in radical addition to olefins

To indicate the variable reactivity of radicals towards a given olefin, Table 3.7 lists Arrhenius parameters for reactions of radicals with C_2H_4. In contrast, the effect of introduction of substituents into the olefin is shown by the results in Table 3.8. From these results and previous work, the following generalisations can be made:

(a) There is little change in E_{Add} for addition of various alkyl radicals to a given olefin.

(b) Replacement of H in the CH_3 radical by halogen facilitates addition, presumably because the radical becomes more electrophilic. However, the chemistry and vibrational frequencies, good agreement between calculations

Table 3.7 Arrhenius parameters for addition of radicals to ethylene*

Radical	log A	E	Radical	log A	E
CH_3	11.1	6.8	CF_3	11.4	2.4
	11.9	7.9	CH_2F	10.6	4.3
C_2H_5	11.2	6.9	CF_2Br	8.3	1.0
$n\text{-}C_3H_7$	10.4	5.1	CCl_3	9.5	3.2
$i\text{-}C_3H_7$	11.2	6.9	SF_5	8.6	1.9
$n\text{-}C_4H_9$	10.3	6.7			
$i\text{-}C_5H_{11}$	10.5	6.4			

*data in this table are from Table 3.6 or from references cited therein; A in $cm^3 mol^{-1} s^{-1}$, E in kcal mol^{-1}.

Table 3.8 Effect of substituents in olefins on radical addition‡

	log A	E		log A	E
$CF_3 + CH_2{=}CH_2$	11.4	2.4	$C_3F_7 + CH_2{=}CH_2$	(10.5)	(2.0)†
$\overset{*}{C}H_2{=}CF_2$	11.0	4.4	$\overset{*}{C}H_2{=}CF_2$	10.2	3.9
$CH_2{=}\overset{*}{C}F_2$	—	>8.4	$CH_2{=}\overset{*}{C}F_2$	9.0	9.2
$CF_2{=}CF_2$	10.5	1.4	$\overset{*}{C}HF{=}CF_2$	10.8	4.1
			$CHF{=}\overset{*}{C}F_2$	10.8	5.8

*point of addition of radical.
†assumed values in parentheses.
‡A in $cm^3 mol^{-1} s^{-1}$, E in kcal mol^{-1}.

reactivities of CF_2Br and CCl_3 are surprisingly high. SF_5 is also very reactive, and it should be noted that addition of SF_5 becomes reversible at higher temperatures[150, 160, 161].

(c) The introduction of CH_3 groups into the olefin reduces E_{Add} for electrophilic radicals.

(d) Replacement of H by F in the olefin has complex consequences, no doubt because of the conflict between I and M effects in the halogens. Let us consider the result of introduction of F into $H_2\overset{\alpha}{C}{=}\overset{\beta}{C}H_2$. It is seen from Table 3.8 that F on C_β usually makes addition at C_α more difficult, but the effect is fairly small, whereas addition at C_β becomes much more difficult, e.g. in $CH_2{=}CF_2$. However, CF_3 adds more readily to both $CH_2{=}CH_2$ and $CF_2{=}CF_2$ than to either end of $CH_2{=}CF_2$, so clearly asymmetry and therefore polarity in the olefin is important.

3.6.5 Interpretation of patterns of reactivity

At present, there is no satisfactory theoretical explanation of the above variations because so many factors may contribute: these are summarised below.

(a) The exact nature of the addend is not clear since it can be regarded as either a σ- or π-radical, e.g.

$$R + CH_2{=}CH_2 \longrightarrow R{-}CH_2{-}CH_2 \text{ or } \overset{R}{CH_2 {\cdots} CH_2}$$

$$\sigma\text{-radical} \qquad \pi\text{-radical}$$

It is usually assumed that a σ-radical is formed, but this may not always be correct. The point is thoroughly discussed by Stefani et al.[162].

(b) Assuming a σ-addend, the strength of the R—C bond formed (i.e. the stability of the addend radical) is important.

(c) The π-electron density in the double bond has a strong effect, e.g. substituents like CH_3 facilitate addition by electrophilic radicals because of the increased π-electron density. In this connection, the cyclopropyl radical is apparently nucleophilic[162] since, along the series $CH_2{=}CH_2$ to $(CH_3)_2C{=}C(CH_3)_2$, there is an increase in ease of addition of CF_3 but a decrease for cyclopropyl.

(d) The position is complicated even more by the occurrence of changes in A factors as well as activation energies (see Table 3.7); for striking examples, see Ref. 133.

Finally, mention should be made of a purely theoretical calculation[163] of the potential energy surface for addition of CH_3 to $CH_2{=}CH_2$, with the calculated $E_{Add} = 7\text{--}10 \text{ kcal mol}^{-1}$ agreeing quite well with experiment.

3.6.6 Miscellaneous reactions of radicals with olefins

Radicals usually add to olefins rather than abstract H, but a striking exception occurs with 1,4-cyclohexadiene[33, 36], where CH_3 and allyl radicals react exclusively by

$$R + \bighexagon \longrightarrow RH + \bighexagon$$

because allylic stabilisation of the product radical makes H-abstraction easy. An interesting method of measuring relative rates of addition of CH_3S radicals to olefins has been described[164]: it involves scavenging of adduct radicals by O_2.

The CCl radical was generated indirectly by flash photolysis and its rate of decay in the presence of several olefins was measured[165, 166]. Apart from work already mentioned in text or tables, the reactions of olefins with the following have been investigated: H_2S (with C_2H_2)[167], the stable $(CF_3)_2NO$

radical[168, 169], NH_2Cl[170], $SiCl_4$[171], NO[172], CCl_4[173], O_2[174], electronically-excited benzene[175], halogenated ethyl radicals[176] and NF_2[87, 177]. The γ-radio-lysis of C_2F_4 gave information on $CF_3 + C_2F_4$[178].

3.6.7   Reactions of cyclobutadiene

The preparation of cyclobutadiene (CB) has long been a challenge to chemists. A preliminary report[179] of its existence appeared in 1967, and further results from the same laboratory have now appeared[180]. This elegant work substantiates theoretical predictions that CB should behave as a biradical. The previous work indicated that photolysis of tricarbonylcyclobutadiene iron (TCI) generates CB, and this has now been convincingly confirmed by isolation of the expected addition products when alkynes and alkenes are present, e.g.

If TCI is photolysed alone, the products include benzene and acetylene produced by

whereas if O_2 is present an aldehyde is formed

3.7   REACTIONS OF CARBENES

3.7.1   Introduction

The most intensively studied carbene is methylene and, in the past, the results have led to more confusion and apparent contradictions than for any other radical. However, many of these difficulties were resolved when it was realised that CH_2 exists in both singlet and triplet states which have entirely different reactivities. The recent appreciation that $^1CH_2$ can be converted to $^3CH_2$ by collisions will no doubt remove other apparent contradictions (see Section 3.7.3.2).

   Many reactions of carbenes are highly exothermic and lead to vibrationally 'hot' products which may rearrange or decompose: these processes are more appropriately discussed in Chapter 1. Our present discussion will therefore deal mainly with reactions of carbenes: the ultimate fate of the products will receive little attention.

3.7.2 Structures and spin states of carbenes

When Herzberg and co-workers first identified CH_2 by flash photolysis, the spectrum was interpreted as indicating a *linear* triplet ground state, $(^3B_1)$, but calculations[188, 189] suggest that $^3CH_2$ is bent with $H\hat{C}H = 135$–138 degrees, also that the lowest-energy $^1CH_2$, $(^1A_1)$, has $H\hat{C}H = 108$ degrees (see also Ref. 190). Recent e.s.r. work[191, 192] confirms that the ground state is indeed $^3CH_2$, and the experimental angle, observed for the first time[192], is $H\hat{C}H = 136$ degrees.

Unlike CH_2, the ground states of CF_2 and CHF appear to be singlets[14, 193], as is the ground state of SiH_2[194, 195] with $H\hat{S}iH = 92$ degrees[196]. The carbene NCCCN is linear[197]. Thermodynamic properties of some carbenes are discussed in Refs. 166 and 198.

3.7.3 Generation and reactions of methylene, CH_2

3.7.3.1 Generation of CH_2

Standard methods involve photolysis of ketene, K, diazomethane, DAM, and diazirine, also pyrolysis of DAM. The photolysis of K produces both $^1CH_2$ and $^3CH_2$, but the precursor[199] of $^3CH_2$ is probably *not* $^3CH_2CO^*$. This is confirmed by flash photolysis studies of K [200] which show that the initial product is $^1CH_2$, with $^3CH_2$ appearing later (see Section 3.7.3.2). Other work[201] shows that photolysis of $^{14}CH_2CO$ yields $CH_2{}^{14}CO$, possibly via an oxirene structure. It is well known that the relative yields of $^1CH_2$ and $^3CH_2$ from K depend on the wavelength used for photolysis. The variation of the relative yields with wavelength has been studied for both K [199] and DAM [202]. However, the $^3CH_2$ may be produced indirectly via $^1CH_2$ (see Section 3.7.3.2). A promising method of production of CH_2 involves far u.v. photolysis[203, 204] of C_3H_8 which leads to

$$C_3H_8 + h\nu \rightarrow {}^1CH_2 + C_2H_6$$

without production of $^3CH_2$. The reactivity pattern of the $^1CH_2$ was identical with that produced by conventional means, in spite of the high energy quanta used. A similar photolysis[204] of cyclopropane also yielded $^1CH_2 + C_2H_4$. CH_2 was produced[205] by pyrolysis of $(CH_3)_3SiCH_2CH=CH_2$. Finally, we note a complication whereby a reaction between the *source* of CH_2 and an added substrate[206] may simulate a reaction of CH_2 itself, e.g.

$$CH_2N_2 + HCl \rightarrow CH_3Cl + N_2$$

which occurs readily at 25 °C.

3.7.3.2 Intersystem crossing from $^1CH_2$ to $^3CH_2$

The occurrence of the process

$$^1CH_2 + M \rightarrow {}^3CH_2 + M \qquad (3.37)$$

is now well established. The fact that $^1CH_2$ may thus be converted to $^3CH_2$ is probably responsible for much of the apparent variations in the relative yields of $^1CH_2$ and $^3CH_2$ from a given source since the ratio will be pressure-dependent. The flash photolysis work[200] mentioned in Section 3.7.3.1 indicates that the only *primary* process in the photolysis of K is

$$CH_2CO + h\nu \rightarrow {}^1CH_2 + CO$$

followed by reaction (3.37).

Table 3.9 contains absolute and relative rate constants for reaction (3.37) with various M. The reaction is relatively slow compared to other reactions of $^1CH_2$ since it occurs at about 1 in 100–1000 collisions, depending on the

Table 3.9 Absolute* and relative rate constants for the reaction $^1CH_2 + M \rightarrow {}^3CH_2 + M$ (equation (3.37))

Absolute $k_{3.37}$

M	CH_4	He	Ar	N_2	at 25 °C
$k_{3.37}$	9.6×10^{11}	1.8×10^{11}	4.0×10^{11}	5.4×10^{11}	Ref. 200

Relative $k_{3.37}$

(a) relative to $^1CH_2 + CH_2CO \rightarrow C_2H_4 + CO$ (3.37a)

M	He	Ar	Kr	Xe	N_2	N_2O	CF_4	C_2F_6	SF_6
$k_{3.37}/k_{3.37a}$	0.018	0.014	0.033	0.074	0.052	0.10	0.047	0.11	0.045

all at 25 °C, Ref. 251.

(b) relative to $^1CH_2 + C_3H_8 \rightarrow n\text{-}C_4H_{10} + i\text{-}C_4H_{10}$ (3.37b)

M	He	Ar	Xe	N_2	CF_4	all at 25 °C
$k_{3.37}/k_{3.37b}$	0.009	0.025	0.05	0.037	0.047	Ref. 252
		0.024				Ref. 203, 204.

(c) relative to $^1CH_2 + CH_2N_2 \rightarrow C_2H_4 + N_2$ (3.37c)

M = N_2

Temp. (°C)	240	260	383	Ref. 207
$k_{3.37}/k_{3.37c}$	0.0021	0.0024	0.0023	

*absolute rate constants in $cm^3 mol^{-1} s^{-1}$.

third body. Hence it is unlikely that hot $^1CH_2^\ddagger$ is involved. For M=N_2, the value of $k_{3.37}$ is independent of temperature[207, 208]. It has been suggested[209] that the intersystem crossing is really a two-stage process involving

$$^1CH_2 \rightleftharpoons {}^3CH_2^\ddagger \qquad (3.38)$$
$$^3CH_2^\ddagger + M \rightarrow {}^3CH_2 + M \qquad (3.39)$$

Calculations indicate that equilibrium (3.38) is rapidly attained with about a

1:1 mixture of $^1CH_2$ and $^3CH_2^\dagger$ so that the bimolecular process (3.39) is rate-determining. However, other authors disagree[210].

3.7.3.3 Reactions of $^1CH_2$

It is known that $^1CH_2$ inserts into R—H bonds and adds to double bonds with retention of stereochemical configuration. Most of the recent work on CH_2 has concentrated on $^1CH_2$ since simultaneous reactions of $^3CH_2$ can be eliminated by scavenging it with O_2. This assumes that reaction betwen $^1CH_2$ and O_2 is slow compared to its other reactions, but it is now suggested[211] that $k(^1CH_2 + O_2) \approx k$(insertion by $^1CH_2$), hence the minimum of O_2 needed

Table 3.10   Absolute* and relative rate constants for reactions of CH₂

$^1CH_2$

Reaction		k	Temp. (°C)	Ref.
$^1CH_2 + H_2 \rightarrow CH_4^\dagger$		4.2×10^{12}	25	200
$^1CH_2 + CH_4 \rightarrow C_2H_6^\dagger$		1.1×10^{12}	25	200
$^1CH_2 + CH_2CO \rightarrow C_2H_4 + CO$ (a)		$\dfrac{k_a}{k_b} = 8$	25	251
$^1CH_2 + CO \rightarrow CH_2CO$ (b)				

$^1CH_2 \; + \; \diagup\!\!\!\diagdown\!\!\!\diagup$ (a) ↗ pentene† (insertion) (b) ↘ $\triangle^\dagger$ (addition) $\quad \dfrac{k_a}{k_b} = 0.77 \quad$ 25 $\quad$ 218

$^3CH_2$

Reaction		k	Temp. (°C)	Ref.
$^3CH_2 + {}^3CH_2 \rightarrow C_2H_2 + H_2$		3.2×10^{13}	25	200
$^3CH_2 + H_2 \rightarrow CH_3 + H$		$\leqslant 3 \times 10^{10}$	25	200
$^3CH_2 + CH_4 \rightarrow CH_3 + CH_3$		$\leqslant 3 \times 10^{10}$	25	200
$^3CH_2 + CO \rightarrow CH_2CO$ (a)		$\dfrac{k_a}{k_b} \sim 3$	25	225
$^3CH_2 + CH_2Cl_2 \rightarrow CH_3 + CHCl_2$ (b)				

*absolute rate constants in $cm^3 mol^{-1} s^{-1}$.

to remove $^3CH_2$ should be used. Recent data on rate constants for reaction of $^1CH_2$ are listed in Table 3.10. New studies[211] of insertion of $^1CH_2$ into C_3H_8 and n-C_4H_{10} give $k_{sec}/k_{prim} = 1.20$ and $k_{tert}/k_{prim} = 1.38$ per C—H bond, in good agreement with previous work.

Relative rates of insertion into C—H and Si—H bonds have been measured[212, 213], and for $(CH_3)_2SiH_2$ and $(CH_3)_3Si$—H the results are $k_{Si-H}/k_{C-H} = 6.9$ and 7.2 respectively (per bond). Thus, insertion into Si—H is one of the fastest known reactions of $^1CH_2$. Insertion by $^1CH_2$ is also discussed in Ref. 214.

Published work shows that both $^1CH_2$ and $^3CH_2$ react with benzene to give mainly toluene, T, and cycloheptatriene, CH. New work[215] shows that the ratio T/CH is little changed if O_2 is added. Probably

$$^1CH_2 \text{ and } ^3CH_2 + \hexagon \longrightarrow \text{\raisebox{0pt}{structure}} \begin{cases} \longrightarrow T \quad (3.40) \\ \\ \searrow CH \quad (3.41) \end{cases}$$

and it is suggested that reaction (3.40) requires a third body. The ratio of T/CH increases with temperature and with decreasing wavelength used to generate the CH_2 from ketene.

Reactions of $^1CH_2$ with olefins are discussed in Refs. 216–218. The possibility of *abstraction* by $^1CH_2$ is discussed below.

3.7.3.4 Reactions of $^3CH_2$

The reaction

$$Hg(^3P_1) + CH_2CO \rightarrow {}^3CH_2 + CO + Hg(^1S_0)$$

might be expected to provide a clean source of $^3CH_2$, but there is evidence[219] that $^1CH_2$ is also formed even though the process is spin-forbidden; recent work[220] confirms this.

Recent data on rate constants of reactions of $^3CH_2$ are listed in Table 3.10. It is usually assumed that $^3CH_2$ abstracts H from RH, but there have been suggestions that insertion of $^3CH_2$ may also occur. A careful study[221] of the reactions of $^3CH_2$ with neopentane shows that $^3CH_2$ reacts only by abstraction, with about the same reactivity as CH_3. However, there is evidence of a novel reaction involving transfer of a CH_3 group[222]

$$^3CH_2 + (CH_3)_4C \rightarrow CH_3CH_2 + (CH_3)_3C$$

The reaction between $^3CH_2$ and *cis*-but-2-ene is discussed in Ref. 223. Experiments on photolysis of $^{14}CH_2CO$ with CD_4 [224] indicate no reaction of $^3CH_2$ with CD_4, even after 10^9 collisions. If CH_2 and CH_3 are simultaneously generated by co-photolysis of $^{14}CH_2CO + CD_3N{=}NCD_3$ in the presence of CD_4, the $^3CH_2$ is removed by

$$^3(^{14}CH_2) + CD_3 \rightarrow {}^{14}CH_2CD_3^\ddagger \rightarrow CH_2{=}CD_2 + D \qquad (3.42)$$

If the CD_4 is replaced by a reactive RH, reaction (3.42) is suppressed because $^3CH_2$ also abstracts H from RH. The collision efficiences for reaction of $^3CH_2$ with CH_2CO and C_2H_4 are $< 10^{-7}$ and $< 10^{-5}$ respectively.

When K or DAM is photolysed with CH_2Cl_2, abstraction of both H and Cl occurs but insertion does *not*; yet both $^1CH_2$ and $^3CH_2$ are present[225]. This is explained by the reactions

$$^3CH_2 + CH_2Cl_2 \rightarrow CH_3 + CHCl_2 \qquad (3.43)$$
$$^1CH_2 + CH_2Cl_2 \rightarrow CH_2Cl + CH_2Cl \qquad (3.44)$$

Presumably reaction (3.44) is faster than the expected insertion.

3.7.4 Reactions of halogenated carbenes

3.7.4.1 CF$_2$ and CHF

The rate constant for combination of CF$_2$ radicals was discussed in Section 3.2.3.

CF$_2$ and CHF are produced by the reactions of F* with CF$_4$, CF$_3$H, etc[226], and the reactions of the carbenes with hydrogen halides, HX, have been studied[227], e.g.

$$CF_2 + HX \rightarrow CF_2HX$$

For CF$_2$, $k_{HI}/k_{HBr} \approx 70$ and $k_{HBr}/k_{HCl} \approx 50$. CFH inserts faster into HCl than does CF$_2$, (note that the ground states of CF$_2$ and CHF are singlets – see Section 3.7.2).

An important study[228] involves generation of ^{1}CF$_2$ by photolysis of difluorodiazirine in the presence of pairs of olefins. The relative Arrhenius parameters indicate that CF$_2$ is electrophilic and the results are discussed in terms of a polar transition state. For the reaction

$$CF_2 + C_2F_4 \rightarrow c\text{-}C_3F_6 \qquad (3.45)$$

the value of $k_{3.45} = 8.7 \times 10^7 T^{\frac{1}{2}} \exp(-3200/RT)$ cm^3 mol^{-1} s^{-1} was obtained by flash photolysis[14, 165, 229]. The pyrolysis of CF$_3$H is known to produce CF$_2$ + HF and the final products include C$_2$F$_4$ and C$_3$F$_6$. New evidence[230] suggests that the C$_3$F$_6$ is formed via

$$CF_2 + CF_2 \rightarrow CF_2 = CF_2$$
$$CF_2 + C_2F_4 \rightarrow c\text{-}C_3F_6^\ddagger \rightarrow CF_3CF = CF_2$$

and not via c-C$_4$F$_8$ as previously proposed.

3.7.4.2 CCl$_2$, CHCl and CCl

CCl$_2$ is produced by pyrolysis of CHCl$_3$ vapour, and when the gas stream is passed into a solution of olefins in an alkane at low temperature (-75 to $-152\,^\circ$C) addition to the double bond occurs and relative k_{Add} values have been measured for several olefins[231]. The results indicate that CCl$_2$ is electrophilic. This is confirmed by other work[232] which indicates that ^{1}CCl$_2$ is involved. The vapour-phase reaction of CCl$_2$ (from CHCl$_3$ pyrolysis) with pyrole gives substituted pyridines[233].

Tritium-labelled CTCl was used[234] to study the reactions

$$CTCl + CH_2{=}CH_2 \longrightarrow \overset{T\diagdown\,\diagup Cl}{\triangle} \qquad (3.46)$$

$$CTCl + R_3Si{-}H \longrightarrow R_3SiCHTCl \qquad (3.47)$$

and the following values of $k_{3.47}/k_{3.46}$ (per Si—H bond) were obtained: (CH$_3$)$_3$SiH, 6.2; (CH$_3$)$_2$SiH$_2$, 1.0; SiH$_4$ <0.1. This again shows that addition of a carbene to an olefin is not the fastest carbene reaction (see Section

3.7.3.3, Refs. 212 and 213). It also follows that reaction (3.46) cannot occur at every collision. The reactions of CCl with olefins and saturated compounds are discussed in Refs. 165 and 166 and the possibility of insertion by CCl is considered (see also Section 3.6.6).

3.7.5 Reactions of silylenes and germylenes

Relevant reviews are given in Refs. 88–90. It is now established[194, 235–237] that the pyrolysis of Si_2H_6 differs from that of C_2H_6 in that the initial step is

$$Si_2H_6 \rightarrow SiH_2 + SiH_4$$

This is followed by insertion reactions of SiH_2 (probably $^1SiH_2$–see Section 3.7.2) e.g. $SiH_2 + Si_2H_6 \rightarrow Si_3H_8$. In mixed C—Si compounds, SiH_2, like other carbenes, (Section 3.7.4.2) inserts preferentially into the Si—H bond[235]. There is no sign of insertion into the C—Si bond. It is found[194] that SiH_2 inserts seven times faster into Si—H in $(CH_3)_2SiH_2$ than in Si_2H_6, hence SiH_2 appears to be electrophilic: this is confirmed by the absence of insertion into $SiHCl_3$. The above results suggest that SiH_2 is more selective in insertion than is $^1CH_2$.

The far u.v. photolysis of CH_3SiH_3 involves several primary processes[195], two important ones being production of SiH_2 and CH_3SiH which then undergo insertion reactions. The rearrangement of CH_3SiH to CH_2SiH_2 appears to be unfavourable unlike the corresponding process in CH_3CH.

The pyrolysis[194] of Ge_2H_6 involves initiation by

$$Ge_2H_6 \rightarrow GeH_2 + GeH_4$$

Pyrolysis[238] of $(CH_3)_3Si_2H(CH_3)_2$ and $(CH_3)_3Si(CH_3)_2SiH(CH_3)_2$ indicates that $(CH_3)_2Si$ is formed which inserts into Si—H bonds. In contrast, no silylenes are formed on pyrolysis of $(CH_3)_3SiH$ [239] or CH_3SiH_3 [240], nor is GeH_2 formed from CH_3GeH_3 [240]. Pyrolysis of $(CH_3)_2CH_3OSiSiOCH_3$ $(CH_3)_2$ yields the silylene $(CH_3)_2Si$, and the products indicate the first observation[241] of insertion of a silylene into an Si—O bond.

3.7.6 Miscellaneous carbene reactions

The photolysis of C_3O_2 leads to carbene formation

$$C_3O_2 + h\nu \rightarrow CO + CCO$$

and it is proposed[242] that the proportion of ^3CCO increases as the photolysis wavelength decreases below 290 nm. The reaction between CCO and cyclopropene was studied. When C_3O_2 is photolysed[243] with CH_4 at 147 nm, possible reactions are

$$C_3O_2 + h\nu \rightarrow (C + 2CO) \text{ or } (CCO^* + CO)$$

followed by

$$C + CH_4 \rightarrow C_2H_4^* \quad \text{or} \quad CCO^* + CH_4 \rightarrow C_2H_4^* + CO$$

The results can be explained either in terms of reactions of carbon atoms or electronically-excited C_2O^*, but not vibrationally-excited C_2O.

Photolysis of C_3O_2 at 147 nm with added CH_3F leads to the reactions[244]

$$^1C_2O^* + CH_3F \rightarrow CO + C_2H_3F^\dagger \xrightarrow{\ M\ } C_2H_3F$$
$$\searrow HF + C_2H_2$$

Pyrolysis of $F_3SiCF_2CHF_2$ produces CHF_2CF which inserts into Si—D bonds and also, but less readily, into Si—Cl and Si—Br bonds[245]. The relative ease of insertion into various Si—H and Si—Cl bonds has been measured.

The pyrolysis of chlorodiazirines was shown[246] to be a general way of producing chlorocarbenes, e.g.

$(R = alkyl)$

The photolysis

produces a carbene[247]. The yields of products were pressure-dependent, and have been discussed in terms of the relation between spin-state stereochemistry and conservation of orbital symmetry. Pyrolysis of $CH_3C_6H_4CH$ $=N_2$ produces tolyl carbenes[248] which undergo interesting rearrangements. The role of carbenes in the vapour-phase Wolf rearrangement has been studied[249]. A novel reaction leading to elimination of a carbon atom from a carbene has been described[250].

References

1. Basco, N., James, D. G. L. and Suart, R. D. (1970). *Int. J. Chem. Kinetics*, **2**, 215
2. Van Den Bergh, H. E., Callear, A. B. and Norstrom, R. J. (1969). *Chem. Phys. Lett.*, **4**, 102
3. Biordi, J. C. (1969). *Mellon Institute Radiation Research Laboratories Quart. Rep.*, **47**, 12
4. Clark, T. C., Izod, T. P. J., Di Valentin, M. A. and Dove, J. E. (1970). *J. Chem. Phys.*, **53**, 2982
5. Casas, F., Previtali, C., Grotewold, J. and Lissi, E. A. (1970). *J. Chem. Soc. A*, 1001
6. Dunlop, A. N., Kominar, R. J. and Price, S. J. W. (1970). *Can. J. Chem.*, **48**, 1269
7. Hole, K. J. and Mulcahy, M. E. R. (1969). *J. Phys. Chem.*, **73**, 177
8. Van Den Bergh, H. E. and Callear, A. B. (1970). *Trans. Faraday Soc.*, **66**, 2681
9. Golden, D. M., Gac, N. A. and Benson, S. W. (1969). *J. Amer. Chem. Soc.*, **91**, 2136
10. Dodonov, A. F., Lavrovskaya, G. K. and Tal'rose, V. L. (1969). *Kinet. Katal.*, **10**, 477
11. Kurylo, M. J., Peterson, N. C. and Braun, W. (1970). *J. Chem. Phys.*, **53**, 2776
12. Ogawa, T., Carlson, G. A. and Pimentel, G. C. (1970). *J. Phys. Chem.*, **74**, 2090
13. Majer, J. R., Olavesen, C. and Robb, J. C. (1969). *Trans. Faraday Soc.*, **65**, 2988
14. Tyerman, W. J. R. (1969). *Trans. Faraday Soc.*, **65**, 1188
15. Thynne, J. C. J. (1969). *J. Organometal. Chem.*, **17**, 155
16. Frangopol, P. T. and Ingold, K. U. (1970). *J. Organometal. Chem.*, **25**, C9

17. Cadman, P., Tilsley, G. M. and Trotman-Dickenson, A. F. (1970). *Chem. Commun.*, 1721
18. Clark, T. C. and Clynne, M. A. A. (1969). *Trans. Faraday Soc.*, **65**, 2994
19. Clark, T. C. and Clynne, M. A. A. (1970). *Trans. Faraday Soc.*, **66**, 372
20. Avery, H. E., Hayes, D. M. and Phillips, L. (1969). *J. Phys. Chem.*, **73**, 3498
21. Yokoyama, N. and Brinton, R. K. (1969). *Can. J. Chem.*, **47**, 2987; Yokoyama, N. (1970). *Bull. Chem. Soc. Jap.*, **43**, 2975
22. Chakrovorty, K., Pearson, J. M. and Szwarc, M. (1969). *Int. J. Chem. Kinetics*, **1**, 357; Stefani, A. P., Thrower, G. F. and Jordan, C. F. (1969). *J. Phys. Chem.*, **73**, 1257
23. Umanskii, V. M. and Stepukhovich, A. D. (1969). *Zh. Fiz. Khim.*, **43**, 2490
24. Stepukhovich, A. D. and Umanskii, V. M. (1969). *Usp. Khim.*, **38**, 1355
25. Umanskii, V. M. and Stepukhovich, A. D. (1969). *Zh. Fiz. Khim.*, **43**, 2050
26. Kerr, J. A., O'Grady, B. V. and Trotman-Dickenson, A. F. (1969). *J. Chem. Soc. A*, 274
27. Pritchard, G. O. and Perona, M. J. (1969). *Int. J. Chem. Kinetics*, **1**, 413
28. Scott, P. M. and Jennings, K. R. (1969). *J. Phys. Chem.*, **73**, 1513
29. Yu, W. H. S. and Wijnen, M. H. J. (1969). *J. Chem. Phys.*, **52**, 2736
30. Yu, W. H. S. and Wijnen, M. H. J. (1969). *J. Chem. Phys.*, **52**, 4166
31. Majer, J. R., Olavesen, C. and Robb, J. C. (1969). *J. Chem. Soc. A*,, 893
32. James, D. G. L. and Kambanis, S. M. (1969). *Trans. Faraday Soc.*, **65**, 1357
33. James, D. G. L. and Kambanis, S. M. (1969). *Trans. Faraday Soc.*, **65**, 1350
34. James, D. G. L. and Troughton, G. E. (1969). *Trans. Faraday Soc.*, **65**, 763
35. Konar, R. S. (1970). *Int. J. Chem. Kinetics*, **2**, 419
36. James, D. G. L. and Suart, R. D. (1969). *Trans. Faraday Soc.*, **65**, 175
37. Thommarson, R. L. (1970). *J. Phys. Chem.*, **74**, 938
38. Inel, Y. (1970). *J. Phys. Chem.*, **74**, 2581
39. Watkins, K. W. and O'Deen, L. A. (1969). *J. Phys. Chem.*, **73**, 4094
40. Benson, S. W. and Bott, J. (1969). *Int. J. Chem. Kinetics*, **1**, 451
41. Setser, D. W. and Richardson, W. C. (1969). *Can. J. Chem.*, **47**, 2593
42. Cadman, P., Inel, Y. and Trotman-Dickenson, A. F. (1970). *J. Chem. Soc. A*, 1207
43. Scott, P. M. and Jennings, K. R. (1969). *J. Phys. Chem.*, **73**, 1513
44. Scott, P. M. and Jennings, K. R. (1969). *J. Phys. Chem.*, **73**, 1521
45. East, R. L. and Phillips, L. (1970). *J. Chem. Soc. A*, 331
46. McGraw, G. E. and Johnston, H. S. (1969). *Int. J. Chem. Kinetics*, **1**, 89
47. Steif, L. J. (1970). *J. Chem. Phys.*, **52**, 4841
48. James, D. G. L. and Kabanis, S. M. (1969). *Trans. Faraday Soc.*, **65**, 2081
49. Eastmond, G. B. M. and Pratt, G. L. (1970). *J. Chem. Soc. A*, 2333
50. Halstead, C. J. and Jenkins, D. R., (1969). *Trans. Faraday Soc.*, **65**, 3013
51. Thomassy, F. A. and Lampe, F. W. (1970). *J. Phys. Chem.*, **74**, 1188
52. Gilbert, J. R., Lambert, R. M. and Linnett, J. W. (1970). *Trans. Faraday Soc.*, **66**, 2837
53. Christie, M. I. and Voisey, M. I. (1969). *Trans. Faraday Soc.*, **65**, 408
54. Eastmond, G. B. M. and Pratt, G. L. (1970). *J. Chem. Soc. A*, 2329, 2337
55. Christie, M. I. and Edwards, J. M. (1969). *J. Chem. Soc. A*, 1134
56. Hogue, J. W. and Levy, J. B. (1969). *J. Phys. Chem.*, **73**, 2834
57. Reed, S. F. (1970). *J. Org. Chem.*, **35**, 3961
58. Schuchmann, H.-P. and Laidler, K. J. (1970). *Int. J. Chem. Kinetics*, **2**, 349
59. Tattershall, B. W. (1970). *Chem. Commun.*, 1522
60. Baldwin, R. R., Hopkins, D. E. and Walker, R. W. (1970). *Trans. Faraday Soc.*, **66**, 189
61. Baldwin, R. R., Walker, R. W. and Longford, D. H. (1969). *Trans. Faraday Soc.*, **65**, 792
62. Baker, R. R., Baldwin, R. R. and Walker, R. W. (1969). *Chem. Commun.*, 1382
63. Baker, R. R., Baldwin, R. R. and Walker, R. W. (1970). *Trans. Faraday Soc.*, **66**, 2812, 3016
64. Sokolova, N. A., Nikisha, L. V., Polyak, S. S. and Nalbandyan, A. B. (1969). *Dokl. Akad. Nauk. SSSR*, **185**, 850
65. Kallend, A. S. and Pitts, J. N., Jr. (1969). *J. Amer. Chem. Soc.*, **91**, 1269
66. Bell, T. N. and Zucker, U. F. (1970). *J. Phys. Chem.*, **74**, 979
67. Morris, E. R. and Thynne, J. C. J. (1970). *Int. J. Chem. Kinetics*, **2**, 257
68. Kerr, J. A. and Timlin, D. (1969). *J. Chem. Soc. A*, 1241
69. LeFevre, H. E., Kale, J. D. and Timmons, R. B. (1969). *J. Phys. Chem.*, **73**, 1614
70. Gac, N. A., Golden, D. M. and Benson, S. W. (1969). *J. Amer. Chem. Soc.*, **91**, 3091
71. Arthur, N. L. and Gray, P. (1969). *Trans. Faraday Soc.*, **65**, 424
72. LeFevre, H. F. and Timmons, R. B. (1969). *J. Phys. Chem.*, **73**, 3854

73. Bell, T. N. and Zucker, U. F. (1970). *Can. J. Chem.*, **48**, 1209
74. Ferguson, K. C. and Pearson, J. T. (1970). *Trans. Faraday Soc.*, **66**, 910
75. Arthur, N. L. and Gray, P. (1969). *Trans. Faraday Soc.*, **65**, 434
76. Leyland, L. M., Majer, J. R. and Robb, J. C. (1970). *Trans. Faraday Soc.*, **66**, 901
77. Pritchard, G. O. and Perona, M. J. (1969). *Int. J. Chem. Kinetics*, **1**, 509
78. Kerr, J. A., Smith, A. and Trotman-Dickenson, A. F. (1969). *J. Chem. Soc. A*, 1400
79. Brockenshire, J. L., Nechvatal, A. and Tedder, J. M. (1970). *Trans. Faraday Soc.*, **66**, 2029
80. Pankratov, A. V., Zercheninov, A. N., Chesnokov, V. I. and Zhdanova, N. N. (1969). *Zh. Fiz. Khim.*, **43**, 398
81. Reference 80, p. 394
82. Zercheninov, A. N., Chesnikov, V. I. and Pankratov, A. V. (1969). *Zh. Fiz. Khim.*, **43**, 380, 390
83. Pankratov, A. V. (1969). *Zh. Fiz. Khim.*, **43**, 403
84. Cherednikov, V. N., Pereverzev, V. S. and Ryabov, V. P. (1969). *Zh. Neorg. Khim.*, **14**, 873
85. Cadman, P., Dodwell, C., Trotman-Dickenson, A. F. and White, A. J. (1970). *J. Chem. Soc. A*, 2371
86. Cadman, P., Trotman-Dickenson, A. F. and White, A. J. (1970). *J. Chem. Soc. A*, 3189
87. Bumgardner, C. L., Lawton, E. L., McDaniel, K. G. and Carmichael, H. (1970). *J. Amer. Chem. Soc.*, **92**, 1311
88. Atwell, W. H. and Weyenberg, D. R. (1969). *Angew. Chem. Int. Ed. Engl.*, **8**, 469
89. Jackson, R. A. (1969). *Adv. Free Radical Chem.*, **3**, 231
90. Davidson, I. M. T. (1971). *Quart. Rev. Chem. Soc.*, **25**, 111
91. Morris, E. R. and Thynne, J. C. J. (1970). *Trans. Faraday Soc.*, **66**, 183
92. Strausz, O. P., Jakubowski, E., Sandhy, H. S. and Gunning, H. E. (1969). *J. Chem. Phys.*, **51**, 552
93. Bell, T. N. and Platt, A. E. (1970). *Chem. Commun.*, 325
94. Bell, T. N. and Zucker, U. F. (1969). *Can. J. Chem.*, **47**, 1701
95. Bell, T. N. and Platt, A. E. (1970). *Int. J. Chem. Kinetics*, **2**, 299
96. Mains, G. J. and Lewis, D. (1970). *J. Phys. Chem.*, **74**, 1694
97. Fass, R. A. and Willard, J. E. (1970). *J. Chem. Phys.*, **52**, 1874
98. McAlpine, I. and Sutcliffe, H. (1969). *J. Phys. Chem.*, **73**, 3215
99. McAlpine, I. and Sutcliffe, H. (1970). *J. Phys. Chem.*, **74**, 848
100. McAlpine, I. and Sutcliffe, H. (1970). *J. Phys. Chem.*, **74**, 1422
101. Shah, P. G., Stranks, D. R. and Cooper, R. (1970). *Aust. J. Chem.*, **23**, 253
102. Kerr, J. A., Smith, B. J. A., Trotman-Dickenson, A. F. and Young, J. C. (1968). *J. Chem. Soc. A*, 510
103. Cadman, P., Tilsley, G. M. and Trotman-Dickenson, A. F. (1969). *J. Chem. Soc. A*, 1370
104. Hautcloque, S. and Pham, M. (1969). *C. R. Acad. Sci. Ser. C*, **268**, 1575
105. Sidebottom, H. W., Tedder, J. M. and Walton, J. C. (1969). *Trans. Faraday Soc.*, **65**, 755
106. De Maré, G. R. and Huybrechts, G. (1968). *Trans. Faraday Soc.*, **64**, 1311
107. Chaudhry, A. U. and Gowenlock, B. G. (1969). *J. Organometal. Chem.*, **16**, 221
108. Kagiya, T., Sumida, Y., Inoue, T. and Dyachkovskii, F. S. (1969). *Bull. Chem. Soc. Jap.*, **42**, 1812
109. Kagiya, T., Sumida, Y. and Inoue, T. (1969). *Bull. Chem. Soc. Jap.*, **42**, 2422
110. Afanas'ev, I. B. (1969). *Dokl. Akad. Nauk. SSSR*, **187**, 1287
111. Salomon, M. (1970). *Int. J. Chem. Kinetics*, **2**, 175
112. Jacubowski, E., Sandhu, H. S., Gunning, H. E. and Strausz, O. P. (1970). *J. Chem. Phys.*, **52**, 4242
113. Hikida, T. and LeRoy, D. J. (1969). *J. Amer. Chem. Soc.*, **91**, 7675
114. Price, S. J. W. and Richard, J. P. (1970). *Can. J. Chem.*, **48**, 3209
115. Pryor, W. A., Tonellato, U., Fuller, D. L. and Jumonville, S. (1969). *J. Org. Chem.*, **34**, 2018
116. Srinivasan, R. and Sonntag, F. I. (1969). *Can. J. Chem.*, **47**, 1627
117. Arthur, N. L., Gray, P. and Herod, A. A. (1969). *Can. J. Chem.*, **47**, 1347
118. Gray, P. and Leyshon, L. J. (1969). *Trans. Faraday Soc.*, **65**, 780
119. Boden, J. C. and Back, R. A. (1970). *Trans. Faraday Soc.*, **66**, 175
120. Shishkina, L. N. and Berezine, I. V. (1969). *Zh. Fiz. Khim.*, **43**, 912

121. Gray, P., Herod, A. A. and Leyshon, L. J. (1969). *Can. J. Chem.,* **47,** 689
122. Konar, R. S. and Darwent, B. de B. (1970). *Can. J. Chem.,* **48,** 2280
123. Jones, S. H. and Whittle, E. (1970). *Int. J. Chem. Kinetics,* **2,** 479
124. Infelta, P. P. and Schuler, P. H. (1969). *J. Phys. Chem.,* **73,** 2083
125. Jones, S. H. and Whittle, E. (1970). *Can. J. Chem.,* **48,** 3601
126. Majer, J. R., Naman, S-A. M. A. and Robb, J. C. (1969). *Trans. Faraday Soc.,* **65,** 3295
127. Pritchard, G. O. and Perona, M. J. (1970). *Int. J. Chem. Kinetics,* **2,** 281
128. McGee, T. H. and Waring, C. E. (1969). *J. Phys. Chem.,* **73,** 2838
129. Majer, J. R., Naman, S-A. M. A. and Robb, J. C. (1970). *J. Chem. Soc. B,* 93
130. Bullock, G. and Cooper, R. (1970), *Trans. Faraday Soc.,* **66,** 2055
131. Gray, P., Arthur, N. L. and Lloyd, A. C. (1969). *Trans. Faraday Soc.,* **65,** 775
132. Leyland, L. M., Majer, J. R. and Robb, J. C. (1970). *Trans. Faraday Soc.,* **66,** 898
133. Reference 132, p. 904
134. Dickey, L. C. and Firestone, R. F. (1970). *J. Phys. Chem.,* **74,** 4310
135. Baldwin, R. R., Walker, R. W. and Langford, D. H. (1969). *Trans. Faraday Soc.,* **65,** 2116
136. Hocklein, G. and Freeman, G. R. (1970). *J. Amer. Chem. Soc.,* **92,** 6118
137. Barta, C. I. and Gordon, A. S. (1970). *J. Phys. Chem.,* **74,** 2285
138. Berkeley, R. E., Woodall, G. N. C., Strausz, O. P. and Gunning, H. E. (1969). *Can. J. Chem.,* **47,** 3305
139. Jackson, R. A. and O'Neill, D. W. (1969). *Chem. Commun.,* 1210
140. Eméleus, H. J., Spaziante, P. M. and Williamson, S. M. (1969). *Chem. Commun.,* 768
141. Eméleus, H. J., Spaziante, P. M. and Williamson, S. M. (1970). *J. Inorg. Nucl. Chem.,* **32,** 3219
142. Rao, T. N., Collier, S. S. and Calvert, J. G. (1969). *J. Amer. Chem. Soc.,* **91,** 1609
143. Okuda, S., Rao, T. N., Slater, D. H. and Calvert, J. G. (1969). *J. Phys. Chem.,* **73,** 4412
144. Timmons, R. B. (1970). *Photochem. Photobiol.,* **12,** 219
145. Morris, E. R. and Thynne, J. C. J. (1969). *J. Phys. Chem.,* **73,** 3294
146. Morris, E. R. and Thynne, J. C. J. (1969). *J. Organometal. Chem.,* **17,** 3
147. Kerr, J. A., Stephens, A. and Young, J. C. (1969). *Int. J. Chem. Kinetics,* **1,** 339
148. Reference 147, p. 371
149. Braslavsky, S. F., Casas, F. and Cifuentes, O. (1970). *J. Chem. Soc. B,* 1059
150. Sidebottom, H. W., Tedder, J. M. and Walton, J. C. (1969). *Trans. Faraday Soc.,* **65,** 2103
151. Gregory, R., Haszeldine, R. N. and Tipping, A. E. (1969). *J. Chem. Soc. C,* 991
152. Gregory, R., Haszeldine, R. N. and Tipping, A. E. (1970). *J. Chem. Soc. C,* 1750
153. Haszeldine, R. N., Keen, D. W. and Tipping, A. E. (1970). *J. Chem. Soc. C,* 414
154. Haszeldine, R. N., Lythgoe, S. and Robinson, P. J. (1970). *J. Chem. Soc. B,* 1634
155. Banks, R. E., Haszeldine, R. N. and Morton, W. D. (1969). *J. Chem. Soc. C,* 1947
156. Liška, F. and Šimek, S. (1970). *Collect. Czech. Chem. Commun.,* **35,** 1752
157. Mews, R. and Glemser, O. (1969). *Chem. Ber.,* **102,** 4188
158. Freidlina, R. K., Terent'ev, A. B. and Ikonnikov, N. S. (1970). *Dokl. Akad. Nauk SSSR,* **193,** 605
159. Weir, R. A., Infelta, P. P. and Schuler, R. H. (1970). *J. Phys. Chem.,* **74,** 2596
160. Sidebottom, H. W., Tedder, J. M. and Walton, J. C. (1970). *Trans. Faraday Soc.,* **66,** 2038
161. Sidebottom, H. W., Tedder, J. M. and Walton, J. C. (1970). *Chem. Commun.,* 253
162. Stefani, A. P., Chuang, L-Y. Y. and Todd, H. E. (1970). *J. Amer. Chem. Soc.,* **92,** 4168
163. Basilevsky, M. V. and Chlenov, I. E. (1969). *Theor. Chim. Acta,* **15,** 174
164. Graham, D. M. and Soltys, J. F. (1970). *Can. J. Chem.,* **48,** 2173
165. Tyerman, W. J. R. (1969). *J. Chem. Soc. A,* 2483
166. Tyerman, W. J. R. (1969). *Trans. Faraday Soc.,* **65,** 2948
167. Majer, J. R., Morton, J. and Robb, J. C. (1969). *J. Chem. Soc. B,* 301
168. Makarov, S. P., Englin, M. A. and Mel'nikova, A. V. (1969). *Zh. Obshch. Khim.,* **39,** 538
169. Mel'nikova, A. V., Baranaev, M. K., Makarov, S. P. and Englin, M. A. (1970). *Zh. Obshch. Khim.,* **40,** 382
170. Prakash, H. and Sisler, H. H. (1970). *J. Org. Chem.,* **35,** 3111
171. Lengyel, B., Knausz, D., Székely, T. and Telegdi, L. (1970). *Acta. Chim. Acad. Sci. Hung.,* **64,** 155
172. Egger, K. W. (1969). *Int. J. Chem. Kinetics,* **1,** 297; Egger, K. W. and Jola, M. (1970). *Int. J. Chem. Kinetics,* **2,** 265; Egger, K. W. and Jola, M. (1969). *Helv. Chim. Acta,* **52,** 449

173. Énglin, B. A. and Onishenko, T. A. (1969). *Izv. Akad. Nauk SSSR Ser. Khim.*, 1906; Osipov, B. N. and Englin, B. A. (1969). *Izv. Akad. Nauk SSSR Ser. Khim.*, 2430
174. Hagiwara, M., Okamoto, H., Kagiya, T. and Kagiya, T. (1970). *J. Polym. Sci., (A-1)*, **8**, 3295
175. Morikawa, A., Brownstein, S. and Cvetanović, R. J. (1970). *J. Amer. Chem. Soc.*, **92**, 1471
176. Moore, L. O. (1970). *J. Phys. Chem.*, **74**, 3603
177. Bumgardner, C. L. and McDaniel, K. G. (1969). *J. Amer. Chem. Soc.*, **91**, 1032; Fokin, A. V., Studner, Y. N. and Proskin, N. A. (1969). *Zh. Obshch. Khim.*, **39**, 1762; Kosyrev, Y. M., Bagryantsev, V. F., Brusentsova, N. A. and Fokin, A. V. (1969). *Izv. Akad. Nauk SSSR Ser. Khim.*, 2597; Fokin, A. V., Zimin, V. I., Studnev, Y. N. and Korotkov, V. F. (1970). *Zh. Org. Khim.*, **6**, 880
178. Cooper, R. and Roy, C. R. (1970). *Aust. J. Chem.*, **23**, 939
179. Tyerman, W. J. R., Kato, M., Kebarle, P., Masamune, S., Strausz, O. P. and Gunning, H. E. (1967). *Chem. Commun.*, 497
180. Font, J., Barton, S. C. and Strausz, O. P. (1970). *Chem. Commun.*, 499
181. Sangster, J. M. and Thynne, J. C. J. (1969). *Int. J. Chem. Kinetics*, **1**, 571
182. Yokoyama, N. and Brinton, R. K. (1969). *Can. J. Chem.*, **47**, 2987
183. Sangster, J. M. and Thynne, J. C. J. (1969). *J. Phys. Chem.*, **73**, 2746
184. Gibb, J., Peters, M. J., Tedder, J. M., Walton, J. C. and Winton, K. D. R. (1970). *Chem. Commun.*, 978
185. Sangster, J. M. and Thynne, J. C. J. (1969). *Trans. Faraday Soc.*, **65**, 2110
186. Tedder, J. M. and Walton, J. C. (1970). *Trans. Faraday Soc.*, **66**, 1135
187. Sakurai, H., Hosomi, A. and Kumada, M. (1969). *J. Org. Chem.*, **34**, 1764
188. Bender, C. F. and Schaefer, H. F. (1970). *J. Amer. Chem. Soc.*, **92**, 4984
189. Harrison, J. F. and Allen, L. C. (1969). *J. Amer. Chem. Soc.*, **91**, 807
190. Goldberg, M. C. and Riter, J. R. (1969). *J. Chem. Phys.*, **50**, 547
191. Bernheim, R. A., Bernard, H. W., Wang, P. S., Wood, L. S. and Skell, P. S. (1970). *J. Chem. Phys.*, **53**, 1280
192. Wasserman, E., Yager, W. A. and Kuck, V. J. (1970). *Chem. Phys. Lett.*, **7**, 409
193. Jacox, M. E. and Milligan, D. E. (1969). *J. Chem. Phys.*, **50**, 3252
194. Estacio, P., Sefcik, M. D., Chan, E. K. and Ring, M. A. (1970). *Inorg. Chem.*, **9**, 1068
195. Obi, K., Clement, A., Gunning, H. E. and Strausz, O. P. (1969). *J. Amer. Chem. Soc.*, **91**, 1622
196. Milligan, D. E. and Jacox, M. E. (1970). *J. Chem. Phys.*, **52**, 2594
197. Smith, W. H. and Leroi, G. E. (1969). *Spectrochim. Acta, Part A*, **25**, 1917
198. Farber, M., Frisch, M. A. and Chung Ko, H. (1969). *Trans. Faraday Soc.*, **65**, 3202
199. Grossman, M., Semeluk, G. P. and Unger, I. (1969). *Can. J. Chem.*, **47**, 3079
200. Braun, W., Bass, A. M. and Pilling, M. (1970). *J. Chem. Phys.*, **52**, 5131
201. Russell, R. L. and Rowland, F. S. (1970). *J. Amer. Chem. Soc.*, **92**, 7508
202. Taylor, G. W. and Simons, J. W. (1970). *Can. J. Chem.*, **48**, 1016
203. Koob, R. D. (1969). *J. Phys. Chem.*, **73**, 3168
204. Dhingra, A. K. and Koob, R. D. (1970). *J. Phys. Chem.*, **74**, 4490
205. Sakurai, H. and Hosomi, A. (1970). *Chem. Commun.*, 767
206. Baer, T. and Bauer, S. H. (1970). *J. Amer. Chem. Soc.*, **92**, 4769
207. Shteinman, A. A. (1970). *Zh. Fiz. Khim.*, **44**, 1389
208. Gabunia, M. B. and Shteinman, A. A. (1970). *Zh. Fiz. Khim.*, **44**, 1682
209. Chang, T. Y. and Basch, H. (1970). *Chem. Phys. Lett.*, **5**, 147
210. Eder, T. W. and Carr, R. W. (1970). *J. Chem. Phys.*, **53**, 2258
211. Johnson, R. L., Hase, W. L. and Simons, J. W. (1970). *J. Chem. Phys.*, **52**, 3911
212. Hase, W. L., Brieland, W. G. and Simons, J. W. (1969). *J. Phys. Chem.*, **73**, 4401
213. Hase, W. L. and Simons, J. W. (1970). *J. Chem. Phys.*, **52**, 4004
214. Kerr, J. A., O'Grady, B. V. and Trotman-Dickenson, A. F. (1969). *J. Chem. Soc. A*, 275
215. Hwang, D-Y. and Ho, S. Y. (1970). *J. Chin. Chem. Soc. Taipe.*, **17**, 26
216. Rynbrandt, J. D. and Rabinovitch, B. S. (1970). *J. Phys. Chem.*, **74**, 1679
217. Simons, J. W. and Taylor, G. W. (1969). *J. Phys. Chem.*, **73**, 1274
218. Taylor, G. W. and Simons, J. W. (1970). *J. Phys. Chem.*, **74**, 464
219. Montague, D. C. and Rowland, F. S. (1968). *J. Phys. Chem.*, **72**, 3705
220. Frey, H. M. and Walsh, R. (1969). *Chem. Commun.*, 158
221. Frey, H. M, and Walsh, R. (1970). *J. Chem. Soc. A*, 2115
222. Frey, H. M. and Walsh, R. (1969). *Chem. Commun.*, 159
223. Eder, T. W. and Carr, R. W. (1969). *J. Phys. Chem.*, **73**, 2074

224. Lee, P. S. T., Russell, R. L. and Rowland, F. S. (1970). *Chem. Commun.*, 18
225. Clark, W. D., Setser, D. W. and Siefert, E. E. (1970). *J. Phys. Chem.*, **74,** 1670
226. Tang, Y-N., Smail, T. and Rowland, F. S. (1969). *J. Amer. Chem. Soc.*, **91,** 2130
227. Smail, T. and Rowland, F. S. (1970). *J. Phys. Chem.*, **74,** 1866
228. Mitsch, R. A. and Rodgers, A. S. (1969). *Int. J. Chem. Kinetics*, **1,** 439
229. Tyerman, W. J. R. (1969). *Trans. Faraday Soc.*, **65,** 163
230. Politanski, S. F. (1969). *Kinet. Katal.*, **10,** 500
231. Skell, P. S. and Cholod, M. S. (1969). *J. Amer. Chem. Soc.*, **91,** 6035, 7131; *ibid.*, (1970), **92,** 3522
232. Sadler, I. H. (1969). *J. Chem. Soc. B*, 1024
233. Baker, F. S., Busby, R. E., Igbal, M., Parrick, J. and Shaw, C. J. G. (1969). *Chem. Ind. (London)*, 1344
234. Tang, Y. N., Daniel, S. H. and Wong, N-B. (1970). *J. Phys. Chem.*, **74,** 3148
235. Bowrey, M. and Purnell, H. (1970). *J. Amer. Chem. Soc.*, **92,** 2594
236. Davidson, I. M. T. (1970). *J. Organometal. Chem.*, **24,** 97
237. Tebben, E. M. and Ring, M. A. (1969). *Inorg. Chem.*, **8,** 1787
238. Sakurai, H., Hosomi, A. and Kumada, M. (1969). *Chem. Commun.*, 4
239. Davidson, I. M. T. and Lambert, C. A. (1969). *Chem. Commun.*, 1276
240. Kohanek, J. J., Estacio, P. and Ring, M. A. (1969). *Inorg. Chem.*, **8,** 2516
241. Atwell, W. H., Mahone, L. G., Hayes, S. F. and Uhlmann, J. G. (1969). *J. Organometal. Chem.*, **18,** 69
242. Peterson, R. F., Baker, R. J. K. and Wolfgang, R. L. (1969). *Tetrahedron Lett.*, 4749
243. Stief, L. J. and DeCarlo, V. J. (1969). *J. Amer. Chem. Soc.*, **91,** 839
244. Tschuikow-Roux, E. and Kodama, S. (1969). *J. Chem. Phys.*, **50,** 5297
245. Haszeldine, R. N., Tipping, A. E. and Watts, R. O'B. (1969). *Chem. Commun.*, 1364
246. Frey, H. M. and Liu, M. T. H. (1970). *J. Chem. Soc. A*, 1916
247. Guarino, A. and Wolf, A. P. (1969). *Tetrahedron Lett.*, 655
248. Baron, W. J., Jones, M. and Gaspar, P. P. (1970). *J. Amer. Chem. Soc.*, **92,** 4739
249. Thornton, D. E., Gosavi, R. K. and Strausz, O. P. (1970). *J. Amer. Chem. Soc.*, **92,** 1768
250. Shevlin, P. B. and Wolf, A. P. (1970). *Tetrahedron Lett.*, 3987
251. Cox, R. A. and Preston, K. F. (1969). *Can. J. Chem.*, **47,** 3345
252. Eder, T. W., Carr, R. W. and Koenst, J. W. (1969). *Chem. Phys. Lett.*, **3,** 520
253. Fedorova, T. V., Ballod, A. P. and Shtern, V. Y. (1970). *Kinet. Katal.*, **11,** 1052
254. Pritchard, G. O. and Perona, M. J. (1969). *J. Phys. Chem.*, **73,** 2944
255. Scala, A. A. (1970). *J. Phys. Chem.*, **74,** 2639

4
Experimental Studies of Hot Atom Reactions

F. S. ROWLAND
University of California

4.1 INTRODUCTION

For most of the last century of chemical kinetics, the principal source of the energy needed to cause chemical change has come from the thermal motion of the molecules themselves. As a consequence, the theories concerned with the quantitative evaluation of the rates of chemical reactions have been strongly influenced by the necessity for explanation of the evermore abundant facts about thermal reactions. These accumulated experiments have certainly demonstrated that the energy requirement for reaction, coupled with the collision of two molecular species possessing this energy in sum, is of overwhelming importance in the determination of the order of magnitude of thermal reaction rate constants, while other parameters are of relatively minor importance. Thus, a major part of the research in chemical kinetics has been directed toward knowledge of the two parameters, A and E_a, of the Arrhenius expression for the rate of a chemical reaction, $k = A \exp(-E_a/RT)$. The correlation of these Arrhenius activation energies with the fraction of molecular collisions possessing more than a specified minimum energy in a Maxwell–Boltzmann distribution of energetic molecules satisfactorily rationalises to reasonable accuracy the quantitative aspects of such reactions.

Further understanding of the intricacies of thermal chemical reaction often depends upon precise comparison of the relative rates of reaction of molecules differing by one controllable parameter (e.g. the substituent effects of physical organic chemistry), by a satisfactory theoretical understanding of the magnitude of the measured activation energies, and by similar elaborations of the original Arrhenius concept.

An alternative method for the study of chemical kinetics has also been potentially available, but has not been widely used until approximately the past 10–15 years; chemical reactions are now being examined in systems involving *other* distributions of energy than the Maxwell–Boltzmann, i.e. those in which the distribution in energies of the reacting molecular species are *not* those characteristic of thermal equilibrium. Among these 'non-Boltzmann' systems, those in which certain atoms possess substantial excess kinetic energy have played a prominent role, and are often described under the colloquial title 'hot atom chemistry'. Despite the fuzzy aspects of this nomenclature – the atoms are *not* in high-temperature thermal equilibrium – the terminology is now sufficiently widely used not to be very misleading. Many examples also can be given of 'hot' radical reactions in which the probabilities of reaction in collision are enhanced by non-thermal energy distributions in a particular group of radicals – as for instance, in the chemical activation systems involving the addition of thermal H atoms to olefins in a highly exothermic process. Similarly, 'hot' molecules have been created by various processes: the chemical activation of radical combination and of singlet methylene insertion, photo-excitation, etc. Neither energetic radicals nor molecules are discussed in this Chapter, except insofar as they are formed as products of energetic atom reactions.

A further introductory comment is worthwhile here: *chemical* change has usually been narrowly limited either explicitly or by inference to rearrangement or re-distribution of the valence electrons which account for molecular bonding among groups of atoms. The other articles in this Review have accepted a similar limitation. However, the rearrangement of the constituents of the atomic nucleus – radioactive decay, as well as other *nuclear* chemical reactions – are also an important aspect of the whole problem of the rates of alteration of the components of the material world. The naturally occurring thermonuclear reactions – especially in astronomical bodies – are also strongly temperature-dependent as a consequence of the over-riding importance of the high energy tail of the Maxwellian distribution of the kinetic energies of nuclei. The temperature scale for these nuclear reactions, has however, been transported from the chemist's familiar 10^2–10^3 K to the 10^7–10^8 K characteristic of the stellar interiors.

While the natural nuclear processes often involve Maxwellian distributions of kinetic energies, the *laboratory* study of the rates of nuclear reactions has long since been carried out with highly non-Boltzmann systems – with controlled beams of particles of precisely known energy. The ordinary terminology of nuclear reactions therefore involves the concept of the *cross-section* for a given process – effectively, the probability of removal from a particle beam by a particular reaction (*partial* cross-section) or by the sum of all possible removal processes (*total* cross-section). The usual chemical kinetic expression of the rate constant for a chemical reaction can then, by analogy, be described as the integration of the appropriate reaction cross-sections over a Maxwellian distribution of molecular energies.

One of the ultimate goals of hot atom chemistry, and of non-Boltzmann chemistry in general, lies in the description of chemical reactions in terms of the respective cross-sections for each of the various processes possible at all energies. The present situation is far short of this distant goal, and usually now involves the averaging of cross-sections over a different broad energy distribution (often only indistinctly known) of atomic or molecular kinetic energies. However, in the process of such investigations many new reaction modes have been discovered, and have been described with both qualitative and semi-quantitative treatments of simple models for reaction cross-sections. Most of the information about microscopic differential cross-sections remains to be discovered.

In this Chapter, we do not attempt a comprehensive review of all of the current facets of hot atom chemistry. After qualitative identification of the hot products in several systems, some general questions invite consideration for further understanding of these reactions. What are the mechanisms of these reactions?; the stereochemistry?; the time-scale?; the energy ranges involved?; the controlling factors in determining the course of reaction? The succeeding Sections provide some information relative to these questions.

4.2   SOURCES OF HOT ATOMS

4.2.1   Nuclear reactions

While atoms with high translational energies are formed in many chemical systems, there are four principal experimental routes to information about such chemical processes. The most frequently used sources to date have been the exceptionally energetic atoms formed in nuclear reactions, as illustrated for tritium atoms ($_1^3H$) in equation (4.1).

$$_0^1n + _2^3He \rightarrow _1^1H + _1^3H \tag{4.1}$$

In this nuclear reaction, a small fraction of the mass of the reactants is converted into kinetic energy of the two products, furnishing monoenergetic tritium atoms with 192 000 eV of kinetic energy. The important characteristics of all nuclear reactions used in hot atom experiments are (a) high kinetic energy, and (b) the atom of interest is radioactive with a convenient half-life. The radioactivity is important because it permits subsequent tracer detection of the labelled products of the chemical reactions of these atoms. Tables of the nuclear properties of frequently used isotopes have been given elsewhere[1,2].

4.2.2   Photolysis

A second valuable source of hot atoms is the short wavelength photolysis of an appropriate molecule, as illustrated by the formation of 2.9 eV H atoms in the dissociation of HBr (bond energy 3.75 eV) with 1849 Å light (6.70 eV) from a mercury resonance lamp.

Other molecules that have been successfully photolysed are HCl, HI, H_2O, H_2S, and their various isotopic counterparts, i.e., DBr, TBr, DI, etc[3-10]. When molecules with three or more atoms are used, the partitioning of the excess energy into the rotational and vibrational motion of polyatomic fragments must also be satisfactorily measured.

4.2.3 Chemical accelerators

The third primary laboratory source of energetic atoms is the chemical accelerator — a class of devices intended to produce controlled beams of energetic atoms (or molecules) in the experimentally difficult region around 2–50 eV in which many of the reactions of current chemical interest are found[11, 12].

4.2.4 Computer simulation

The fourth important experimental technique for investigating hot atom reactions is the computer simulation of the collisions between energetic atoms and substrate molecules[13-15].

4.2.5 Advantages and limitations

There is not sufficient space here to allow any detailed consideration of the advantages and limitations of each technique. Briefly, however, the nuclear reaction method provides atoms of kinetic energy much higher (100–10^6 eV) than any possible energies for bond-forming reactions for these atoms. Since in almost every system some of these atoms *fail* to form a bond while kinetically hot, and become thermalised, the nuclear procedures insure complete sampling of the entire range of energies over which reactions can occur. Thus, *all* processes should be observable in the nuclear systems.

Since reactions can be initiated by atoms with essentially *any* chemically significant energy, the observed reaction yields in the nuclear systems are an average taken over a wide range of energies — yields from a distribution that is clearly non-Boltzmann, but nonetheless an integral and not a differential measurement.

When atoms can be injected into a system with a controlled energy within the chemical reaction range (as with atomic beams and with photochemically produced hot atoms), a closer approach can be made to a differential measurement of reaction yields versus energy. However, unless the possibility of reaction is carefully limited to the first collision, the hot atom may lose energy in a series of non-reactive encounters and the observed reaction yields again become integrals over a broad distribution of energies. None of the systems yet in use for hot atom reactions contains this single collision provision (i.e. crossed beams of the two reactants), and one can only, therefore, measure indirectly differential yields or cross-sections versus kinetic energy[12].

Precision control of the initial atomic energies permits the determination of energy thresholds for particular reactions simply by noting the appearance of new reaction products as the initial kinetic energy of the atoms is increased. This determination of thresholds is one of the most important experimental advantages of these two methods, and is discussed in detail later. At the present time, photochemical formation of hot atoms offers precision control of initial energies but only for energies up to about 3 or 4 eV. The beam machines provide atoms with less precision in initial energy, but with initial energies covering the entire chemically-significant energy range.

4.3   REACTIONS INITIATED BY HOT ATOMS

4.3.1   Known and unknown products

In principle, hot atom chemistry can be studied for any of the more than 100 known elements, and studies have been made with more than 30 elements[16, 17]. However, the number of elements for which appreciable understanding has been gained is scarcely more than a half-dozen: hydrogen, carbon, the halogens, perhaps phosphorus. One important reason for this relative scarcity of information about hot atom processes is the corresponding dearth of experimental knowledge of *thermal* atom reactions for the great majority of elements. Indeed, the investigation of the reactions of atomic carbon has been pioneered through the extensive studies of ^{14}C and ^{11}C formed in nuclear reactions[1, 2]. More recently, the study of fluorine atom chemistry is frequently much easier with ^{18}F atoms from (γ, n) or $(n, 2n)$ reactions than by more standard methods for the formation of thermal fluorine atoms[18-20].

Experimental studies can then be crudely divided into (a) those in which the hot atom processes lead to the formation of a known product, often by a previously unknown route, and (b) those in which some of the products are actually previously unknown chemical species.

While the formation of previously unknown chemical species is a scientifically valuable aspect of hot atom chemistry, little information of *kinetic* interest has been developed from such experiments, and our attention will be concentrated upon reaction systems in which the products are well-known, readily identifiable chemical entities.

4.3.2   Monovalent versus polyvalent atoms

A second general classification of elements can be based upon the number of bonds that must be formed by the energetic atom in attaining the identified product. Elements reacting as monovalent species (e.g. H, F, Cl) need involve only one reaction step, and their reactions are usually simpler to understand than those of polyvalent species such as carbon, silicon, or phosphorus[1, 2]. This Review has been arbitrarily devoted to a more extensive discussion of the atomic reactions of hydrogen and of the halogens, thereby omitting some very interesting chemistry of ^{11}C, ^{32}P, ^{31}Si, etc.

4.3.3   Radio gas chromatography

Progress in the identification of the main paths of hot chemical reaction has been strongly dependent upon the development of suitable analytical techniques, especially that of radio gas chromatography. Other techniques have also been applied, but the quantitative data used in testing various hypotheses have very largely been obtained through gas proportional or scintillation counting of the separated labelled molecules contained in the effluent stream from a gas chromatograph[21-23].

4.4   REACTIONS OF ENERGETIC MONOVALENT ATOMS

4.4.1   Reactions of energetic tritium atoms

The basic hot reaction processes identified for energetic tritium atoms have been established through experiments with many substrate molecules, and can be classified into three general categories[1, 2, 16, 17, 24]:

(a) abstraction of hydrogen atoms, as in equation (4.2);

$$T^* + RH \rightarrow HT + R \qquad (4.2)$$

(b) substitution for another atom or radical, as in equation (4.3);

$$T^* + RX \rightarrow RT^* + X \qquad (4.3)$$
$$(X = H, D, F, Cl, Br, I, NH_2, OH, CH_3, R, etc.)$$

(c) addition to a π-bond system, such as olefins, acetylenes, or aromatic molecules;

$$T^* + \quad \begin{matrix} \diagdown \\ / \end{matrix} C = C \begin{matrix} \diagup \\ \diagdown \end{matrix} \quad \rightarrow \quad \begin{matrix} \diagdown \\ \end{matrix} CT—C \begin{matrix} \diagup \\ \diagdown \end{matrix} \cdot \quad * \qquad (4.4)$$

The addition of thermal hydrogen atoms to a π-system is itself quite exothermic; when reaction is initiated by an atom with excess kinetic energy, the resulting radical is even more highly vibrationally excited, symbolised by the asterisk in equation (4.4). The substitution product, RT, formed in equation (4.3), has been found experimentally to carry high vibrational energy in most, if not all, cases and is also shown as an excited species marked with an asterisk (Section 4.7).

The most important additional reaction modes in providing a diversity of products in hot atom systems are the secondary isomerisation or decomposition of the highly excited products formed in the initial primary substitution or addition reactions.

The yields of some other reactions which probably occur are difficult to measure because of the instability of the product, or because of the lability of the newly formed tritium bond. Such possible reactions are typified by the abstraction of Cl with the formation of TCl, and the substitution of T for H atoms in systems such as H_2O or NH_3, with the formation of HTO or NH_2T. In each of these cases, the tritium label on the product is readily lost through isotopic exchange with H atoms in similar bonding positions, leaving a non-radioactive molecule undetected by the usual experimental procedures. While the yields of such labile species have been measured in some experiments, and could readily be measured in most, the labile tritium yields normally represent a sum of quite unrelated hot and thermal reactions (e.g., reaction of thermalised T atoms with a scavenger such as O_2 or Br_2) and are of little quantitative value in mechanistic discussions.

4.4.2 Reactions of fluorine and other halogen atoms

A survey of energetic ^{18}F reactions leads to a very similar summary of the common reactions, as in equations (4.5) to (4.7),

$$^{18}F + RH \rightarrow H^{18}F + R \tag{4.5}$$
$$^{18}F + RX \rightarrow R^{18}F + X \tag{4.6}$$
$$(X = H, D, F, Cl, Br, I, CH_3, R, etc.)$$
$$^{18}F + CH \equiv CH \rightarrow CH^{18}F \dot{=} CH \tag{4.7}$$
(and for other π-bonded molecules)

and the same pattern is also found with energetic ^{38}Cl (or ^{39}Cl) atoms. The study of translationally hot bromine or iodine atoms activated by nuclear recoil is complicated by the frequent presence of Auger cascades among the nuclear processes[16, 17]. These cascades form highly charged radioactive ions, and their subsequent reactions reflect the factors involved in ionic reactions, and not those characteristic of 'hot' atoms. Qualitatively, however, the same basic reaction types appear also to occur for all of the energetic halogen atoms.

4.4.3 Thermal and hot reactions: criteria

Among these general classes of hot reaction are several which have also been accepted as characteristic thermal reactions, as with the abstraction of H from alkanes by either tritium or fluorine atoms. In these cases, there is no clear-cut borderline between hot and thermal. The (as yet, unmeasured) differential cross-sections for abstraction presumably extend continuously from the several eV region (or higher) down to energies low enough to permit observable reaction with a Maxwellian distribution of atoms (see Section 4.6.2).

Among the criteria used to establish the 'hot' nature of such reactions are[1, 2]:

(a) observation of reactions unknown in the corresponding thermal systems, the ideal scavenger being a molecule with high reactivity towards thermal species and no reactivity at all towards hot species;

(b) insensitivity to the concentration of scavenger present in the system;

(c) insensitivity to the temperature of the reacting substrate molecules;

(d) progressive suppression of the hot reactions by the inclusion of increasing amounts of a non-reactive moderator molecule (e.g. Ar, Ne, etc.), which removes the excess kinetic energy of the atom before chemical reaction can occur.

Naturally, reactions which are classified as both hot and thermal processes may turn out not to be completely insensitive to variables such as scavenger concentration. For example, the abstraction of H atoms from weak C—H bonds (or from Si—H bonds) is progressively suppressed by increasing concentrations of O_2, while the substitution of T for H atoms in most molecules is quite insensitive to O_2 concentrations at the scavenger level. Furthermore, while it is quite reasonable to find that reactions initiated by 7 eV atoms are not much influenced by an additional 0.1 eV in a substrate molecule (as would result from a 100 °C change in substrate temperature for butene), the extra energy is not so negligible for reactions initiated by 1 eV atoms.

Renewed searches for temperature effects in hot atom reactions have shown that changes of 100–200 °C have no appreciable effect on the yields for the substitution of energetic T for H atoms in cyclobutane[25] or methane[26]. On the other hand, the decomposition of excited butyl radicals formed by H atom addition to 1-butene or 2-butene is measurably affected by the temperature change from 24 °C to 125 °C [26, 27]. Even in this case, however, the temperature effect is essentially upon the secondary chemical reaction; the initial addition reaction of the energetic T atom is apparently not influenced by the extra energy in the substrate molecule.

4.5 STEREOCHEMISTRY OF THE SUBSTITUTION PROCESS AT ASYMMETRIC CARBON POSITIONS

4.5.1 Substitution of T atoms for H atoms at asymmetric carbon positions

Experimental investigations of the stereochemistry of the hot substitution process have been an area of primary interest since the discovery that the substitution reactions occurred with good yield in the gas phase[1, 2]. Probably the greatest stereochemical interest has centred around the retention–inversion problem during the replacement by an energetic monovalent recoil atom of one of the four substituents of an asymmetric carbon atom. The initial isolation of glucose-t from T* reactions with glucose — coupled with the intramolecular determination that tritium atoms were distributed among all C—H positions — immediately indicated that substitution *with retention of configuration* was a high yield process, while the absence of galactose-t indicated that the inversion process was not an important process. Earlier experiments with crystalline L(+)-alanine and with gaseous sec-butyl alcohol (both D and L) showed heavy preferences for retention of configuration with little or no positive evidence for the existence of the inversion reaction. The accumulated experimental difficulties involved in the resolution of the optically active forms from a racemic mixture, followed by degradation to determine intramolecular tritium location, prevented highly accurate measurements with either of these systems.

A series of experiments have recently been carried out — again using molecular targets with more than one asymmetric centre — in particular, using the *dl* and *meso* forms of 2,3-dihalobutanes and of $(CHFCl)_2$. In these instances, the *dl* and *meso* forms are readily separable by gas chromatographic methods and the overall efficiency of the separation process permits shorter irradiation times (and less radiation damage) in the preparation of samples.

The experiments involving tritium substitution for H in $(CHFCl)_2$ again show that retention of configuration is the predominant mode of substitution; with much higher accuracy, they also show no evidence $(<1\%)$ for the inversion mechanism in the gas phase, as illustrated in Figure 4.1 [28]. At this

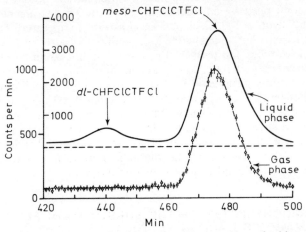

Figure 4.1   Radio gas chromatographic separation of tritium-labelled *dl*- and *meso*-CHFClCTFCl from reactions of energetic tritium atoms with *meso*-$(CHFCl)_2$; — liquid phase; ϙ gas phase (From Palino, F. G. and Rowland, F. S.[28] by courtesy of the American Chemical Society)

time, there is no positive evidence for the inversion of configuration of any molecule during the substitution of energetic T atoms for H atoms in the gas phase. A tabular summary of all of the retention/inversion experiments with recoil tritium atoms is given in Reference 28.

These experimental investigations have been limited to molecules containing asymmetric carbon atoms, and therefore to substitution of T for H atoms at positions to which the other three substituents have all been heavy groups. The question of whether substitution with inversion can occur for molecules with several light substituents, in particular CH_4, can only be dealt with by inference from these experiments. However, in trajectory calculations, the processes of inversion and retention *can* be distinguished, at least in principle, for molecules such as CH_4. This stereochemical problem is discussed again in the Section connected with the most recent calculations of energetic tritium atom trajectories with CH_4 (Section 4.10.2).

4.5.2   Liquid phase experiments with tritium

In the liquid phase experiments, some of the inversion product is usually observed, as shown in Figure 4.1, and two different hypotheses are possible:

(a) the inversion product can be formed by the recombination of an atom with a radical or of two radicals, following the isomerisation of the radical around the asymmetric carbon atom position;

(b) the process of substitution of T for H atoms with inversion is invariably accompanied by such substantial amounts of residual excitation energy that the product molecule always decomposes in gas-phase experiments. In the case of $(CHFCl)_2$, the complete decomposition of an excited CHFClCTFCl product at the gas pressures used in the experiment would require a minimum excitation energy of $\gtrsim 6$ eV. Such an excitation energy is approximately that found for the substitution of T for CH_3 groups, so the hypothesis is not a highly improbable one.

4.5.3 Substitution near threshold: the substitution of T for X atoms

The interesting complementary question of retention–inversion for photo-chemically formed, relatively less energetic, hot tritium atoms has not yet been experimentally soluble[28]. The rather high threshold energy for T for H atom substitutions at such positions and the photochemical complications involved in the photolysis of the TBr with chloroalkanes (i.e., CHFClCHFCl) which are also subject to photolytic dissociation, make these experiments quite complicated.

No successful measurements of the stereochemistry of reactions involving the substitution of T for X atoms have been reported except for X = H. Since the energy deposited on a product molecule is often very much higher for reactions involving the substitution of T for Cl atoms or reactions involving the substitution of T atoms for CH_3 groups (Section 4.7.4), very little primary substitution product is expected to survive under typical gas-phase irradiation conditions, and the stereochemical measurement becomes much more difficult in terms of product yields.

4.5.4 Substitution by halogen atoms

The stereochemistry of the gas-phase substitution by energetic chlorine and fluorine atoms at saturated asymmetric carbon atom positions has also been investigated, utilising 2,3-dichlorobutane[29] and $(CHFCl)_2$ [30] respectively as substrate molecules. In each case, the predominant mode of substitution is retention; in each, there is no positive evidence that the inversion process gives measurable yields in the gas phase.

In halogen systems, condensed phase studies have shown appreciable amounts of the labelled inversion products. Since these amounts vary quite widely with the particular conditions of irradiation, there is no simple characteristic ratio of retention/inversion products for the condensed phases. The occurrence of substantial yields of inversion products from cage recombination processes has been postulated. (Good evidence for such cage processes exists for ^{18}F reactions with CH_3F measured over a wide range of molecular densities[31]). Similarly, the occurrence of substitution with inversion as a primary step has also been postulated in particular cases; while substantial yields of inversion product are incontrovertibly present, the mechanism of formation in the condensed phases is still not clear.

An interesting new set of experiments demonstrates that the amount of inversion product that is observed in the liquid phase can be varied substantially by changing the nature of the solvent, and the hypothesis has been put forth that the controlling factor in these inversions is the percentage contribution of the various conformers of the substrate molecules[32]. More tests of this hypothesis will certainly be required, and the question of retention/inversion in the liquid phase will continue to be actively investigated.

4.6   THRESHOLDS FOR ATOMIC REACTIONS

4.6.1   Threshold for reaction of D with H_2

The threshold energy for an atomic reaction can be defined as the minimum kinetic energy of the atom required for reaction with a stationary molecule. The cross-section for this reaction can be expected to increase at energies above this threshold, and then at still higher energies to level off, to decline, and eventually to become zero again. As the kinetic energy is increased through these stages, thresholds for additional reaction channels will be passed, and additional reaction products will be observed. As lower energy processes decrease in cross-section, they are usually supplanted by different reactions requiring more energy—for example, abstraction and substitution can be replaced by chemi-ionisation.

The threshold energy for the reaction of deuterium atoms with hydrogen, as in equation (4.8),

$$D + H_2 \rightarrow HD + H \qquad (4.8)$$

has been the subject of a very detailed study by Kuppermann et al.[33–36]. The reaction yield has been measured as a function of the initial kinetic energy of the deuterium atom, as formed in the u.v. photolysis of DI or DBr. Deuterium atoms which collide with H_2 and which undergo only non-reactive scattering to an energy below the threshold can eventually react with DX to form D_2; the relative yields of HD and D_2 serve to measure the 'hot' and thermal reaction components. The actual experimental technique requires the presence of macroscopic quantities of the DX source, with the consequent additional possibility of hot reaction of D with DX to form D_2. The hot yield from reaction with H_2 is determined by extrapolation of the HD/D_2 yield ratio to zero concentration of DX.

The fraction of deuterium atoms undergoing reaction (4.8) falls from 0.66 at a laboratory energy of 2.86 eV to near zero at a laboratory energy in the vicinity of 0.48 eV. If the H_2 were actually stationary, then this would indicate a near zero cross-section for a relative centre-of-mass energy of 0.24 eV, since the D and H_2 molecules are of almost identical mass. However, because of the thermal motion of the H_2 molecules, the relative energy of the initial collisions between D and H_2 is no longer monoenergetic, but is spread quite substantially to both higher and lower energies. Collision of the energetic atom with a molecule moving towards it thus can have substantially greater kinetic energy than possible with the energy of the atom alone, and reactions can be observed in a room temperature gas at energies lower than in a hypothetical gas with no molecular motions. When corrections are made for this spread in energies, the threshold for reaction (4.8) is estimated to be 0.33 ± 0.02 eV [33].

4.6.2 Threshold for abstraction from alkanes by H atoms

Similar measurements for the abstraction of D from the secondary positions of $n\text{-}C_4D_{10}$ have led to an estimate of 0.35 ± 0.02 for the threshold of reaction (4.9) [6, 37, 38].

$$H + n\text{-}C_4D_{10} \rightarrow HD + sec\text{-}C_4D_9 \qquad (4.9)$$

The abstraction of D from a primary position of the same molecule has been estimated by differential comparison of HD yields from $n\text{-}C_4D_{10}$ and $CD_3CH_2CH_2CD_3$ to have a somewhat higher threshold of 0.45–0.55 eV.

The total yield of reaction (4.9) for a specific initial laboratory kinetic energy of the H atoms is effectively an integral over a very non-Boltzmann distribution of H atom collisions with C_4D_{10}. As the initial H atom energy is slowly varied, this integral yield is also varied, providing the possibility of an evaluation of the differential yield. The actual distribution versus H atom kinetic energy for collisions with C_4D_{10} depends upon the energy losses

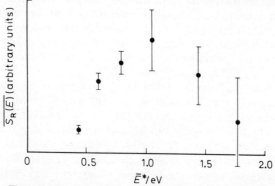

Figure 4.2 Cross-section versus laboratory energy for reaction $H + n\text{-}C_4D_{10} \rightarrow HD + sec\text{-}C_4D_9$
(From Gann, Ollison and Dubrin[6], by courtesy of the American Institute of Physics)

of the H atom in non-reactive collisions in the system, a parameter for which good measurements are not available. These energy losses have, however, been estimated through measurement of the effect of dilution with xenon upon the yield of reaction (4.9), together with independent experimental evaluations of the H–Xe potential[6]. The calculated cross-section versus energy for this reaction, using an r^{-7} potential for H–Xe, is shown in Figure 4.2. The unfolding of the integral yield into the differential cross-section is not very sensitively dependent upon the precise form of the H–Xe potential, and a very similar shape is also found for an r^{-4} potential.

The cross-section illustrated here rises to a maximum at an energy of about 1 eV and appears to be falling toward zero at a quite low initial kinetic energy. The evaluation at higher energies becomes progressively more difficult as the differential increment to the observed total yield becomes fractionally of less importance, and no measurements were made for initial laboratory H atom energies greater than 2.05 eV. Obviously no information is contained in these data about the possible existence of additional maxima at higher energy for abstraction reactions (see Section 4.10.2).

Both of these threshold measurements for abstraction involve the lowest energy reaction path available with the alkane substrates, and therefore are also the reactions observed thermally. The threshold energy and activation energy for a given reaction should have similar but not necessarily identical numerical values, while representing quite different fundamental concepts. The activation energy corresponding to a given threshold energy is quite dependent upon the functional form of the cross-section immediately above the threshold. The relationships between threshold energies and activation energies have been discussed in detail, with examples both from neutral atom and ion–molecule reactions[39, 40].

4.6.3 Thresholds for substitution reactions

With higher initial hydrogen atom kinetic energies, substitution reactions have also been observed, with a slight modification in the reaction scheme. When HX is used both as hot H atom source for reactions with a deuterated alkane and as thermal scavenger in the system, each initial abstraction reaction leads to the formation of one molecule of HD, and, after reaction of the residual sec-C_4D_9 radical with HX, one molecule of C_4D_9H. An initial substitution reaction by a hot H atom forms C_4D_9H directly, and one molecule of HD after reaction of the displaced D atom with HX. The HD + C_4D_9H combination is therefore identical whether the initial hot reaction is abstraction or substitution, and the initial step cannot be distinguished. However, when the thermal scavenger is not identical to the hot hydrogen atom source, the initial steps are easily separated. With TBr as the source and Br_2 the scavenger, the hot abstraction with CH_4 leads to the formation of HT (and CH_3Br), while the substitution reaction leads to CH_3T (and HBr)[7]. (In this system, *total* hot yields are no longer easily measurable, for the thermalised T atoms react with Br_2 to re-form TBr, and there is no molecular record of the number of thermalised atoms.)

Variation of the initial kinetic energy of T atoms from u.v. photolysis of TBr has shown that the threshold energy for substitution into CD_4 is approximately 1.5 eV[41]. In these measurements, lacking values for the absolute hot yields, the thresholds are determined by extrapolation of yield ratios, e.g., CD_3T/DT, towards zero.

4.6.4 Thresholds in beam experiments

Threshold energies can also be determined with atomic beams of well-controlled initial energies. The integral yield for substitution of energetic tritium into solid cyclohexane has been measured over initial energies between 1 and 200 eV [11]. In the initial apparatus, the tritium was accelerated as T^+, and solid targets were used to minimise vapour pressure problems. The T^+ ions were neutralised in the target, and then reacted as energetic T atoms. After correction for a small residual yield of T^+ reactions, the threshold for substitution into cyclohexane was estimated as 1.5 ± 0.5 eV, while that for formation of hexene was estimated as 4.5 ± 0.5 eV.

The integral yields rise rapidly with increasing energy between 2 eV and about 20 eV, and very slowly thereafter. Differentiation of such integral yield curves cannot be done with precision, particularly in the known presence of unwanted ionic processes leading to the same products. Nevertheless, it is qualitatively clear that appreciable cross-sections for substitution can be found in the 3–15 eV range; whether the cross-sections remain significant above 15 eV is a moot point awaiting more precise experiments. In a later version of this apparatus [42], the T^+ ions are neutralised in the gas phase by charge exchange, and impinge on the solid target as T atoms. No experimental results have yet been published with this improved technique.

The initial experimental results have now been published for controlled energy beams (5–100 eV) of T_2 (and of T_2^+) crossed with a gaseous sheath of n-butane [12] and of 1-butene [43], permitting direct measurement of differential yields versus energy. No such machine has been operated yet with beams of neutral atoms, but such a development will certainly occur in the near future.

4.7 EXCITATION ENERGIES OF HOT REACTION PRODUCTS

4.7.1 Energetics of hot atom reactions

Measurements of the threshold energies for hot reactions and of the integral hot yields versus energy provide an estimate of the energy of the hot atom during its initial approach to a successful encounter. However, these experiments give no information about the ultimate disposal of this excess energy during the course of reaction. For atom-for-atom substitution reactions, additional information can in principle be obtained from measurement of the kinetic energy of the replaced atom. It has been reported that a hot deuterium atom is released when the reaction $H + HD \rightarrow H_2 + D$ is initiated by a hot H atom [44].

4.7.2 Internal energy of molecules formed in hot reactions

An extensive series of measurements has been made of the vibrational energy of the product molecules formed by hot substitution reactions. The prototypical experiment involved measurement of the yields versus pressure of c-C_4H_7T formed by recoil tritium substitution into c-C_4H_8, and of its secondary decomposition product, $CH_2{=}CHT$ [45]. The increased yields of c-C_4H_7T with increasing pressure are assumed to arise from successful collisional stabilisation, and lead readily to estimates of the lifetimes of the excited molecules if not stabilised. With the further assumption of random distribution of internal energy in accord with RRKM (Rice–Ramsperger–Kassel–Marcus) theory, these rates of decomposition can be converted into excitation energies. The experimental behaviour of many systems suggests that the usual result is a broad distribution of excitation energies, with a substantial fraction of the product molecules capable of secondary isomerisation or decomposition. No measurements have been made of the kinetic energies of the replaced atoms, and therefore there is no direct indication as to whether an excitation energy of 5 eV corresponds to a 10 eV T atom knocking out a 5 eV H atom or a 5.5 eV T atom giving rise to a 0.5 eV H atom.

Measurements in the condensed phases usually show some decomposition persisting under these high collision density conditions, indicating that very high excitation energies are found in a small fraction of the successful reactions. For example, the decomposition of c-C_4H_7T proceeds in 5% yield in solid cyclobutane at 77 K — in contrast to 50% yield at about 100 torr (and 25 °C)[25]. Very high excitation energies are also indicated by the observations of highly endothermic reactions (e.g. $CF^{18}F$ from $CF_3^{18}F$ formed by $^{18}F/F$ in CF_4)[47], and of tertiary reaction sequences: $T^* + CH_3CHF_2 \rightarrow C_2H_3TF_2^* + H$; $C_2H_3TF_2^* \rightarrow C_2H_2TF^* + HF$; $C_2H_2TF^* \rightarrow CH{\equiv}CT + HF$[48].

4.7.3 Median excitation energies for T atom for H atom substitutions

Measurements of the pressure dependence of the yields of primary versus secondary products have now been carried out with a number of systems, as summarised in Table 4.1, and several broad general conclusions can be stated. First, all studies of small molecules show decomposition of the primary T for H substitution product, indicating that residual excitation energies of several eV or more are a general phenomenon in most systems. Second, the absence of any CH_2TNC from $T^* + CH_3NC$ in the gas phase indicates that (in this system, at least) the T for H substitution reaction normally occurs with residual excitation energy greater than 2 eV[46, 50]. Finally, the calculated median excitation energies for reactions involving substitution of T for H atoms with different substrate molecules are not identical, suggesting that the residual excitation energy may be a function of the chemical environment of the individual C—H bond involved. This conclusion is tentative, for it depends critically upon the assumptions that (a) the particular mode of introduction of the excitation energy (hot substitution) has no unusual effects upon the rates of decomposition of the molecules involved, i.e. that the RRKM theory *is* applicable to these molecules, and (b) that RRKM calculations of excitation energies for different molecules can be closely compared without serious error.

With small molecules, the 'trapping' of a tritium atom by the residual radical and the simultaneous loss of a slow-moving H atom leaves the nascent molecule with a higher fraction of its excitation in rotation than in other chemical activation experiments. Computer simulations of the reactions of energetic tritium atoms with CH_3NC suggest that this excess rotational energy plays a very important role in determining the course of reac-

Table 4.1 Estimates of median energy deposition for energetic atom reactions

Reference	Reaction	Substrate	Primary product	Secondary product	Act. energy (eV)	Estimated median energy (eV)
45	T/H	$c\text{-}C_4H_8$	$c\text{-}C_4H_7T$	C_2H_3T	2.7	5 ± 1
49		CH_4	CH_3T	CH_2T	4.5	3.5 ± 1
50		CH_3NC	CH_2TNC	CH_2TNC	1.6	>4
58		CH_3CF_3	CH_2TCF_3	$CHT{=}CF_2$	3.0	4.7 ± 1
51		spiro-pentane	$CHT{-}C(CH_2)(CH_2)(CH_2)$ (spiro structure)	$CHT{=}C{=}CH_2$ or $CHT{=}CH_2$	2.5	~5
25	T/D	$c\text{-}C_4D_8$	$c\text{-}C_4D_7T$	C_2D_3T	2.7	5 ± 1
52	T/CH$_3$	(cyclic CH_2 structure)	(cyclic CH_2 structure)		2.7	5 ± 1
53	T + olefin	$CH_3CHCH_2CHCH_3$	CH_3CHCH_2CHT	$CH_2{=}CHT$	~2.7	$6{-}8$
53		1-butene	$CH_3CH_2CHCH_2T$	$CH_2TCH{=}CH_2$	1.4	$\leqslant0.5$
53		cis-2-butene	$CH_3CHTCHCH_3$	$CH_3CH{=}CHT$	1.4	$\leqslant0.5$
47	$^{18}F/F$	CF_4	$CF_3{}^{18}F$	$CF^{18}F$	~9.6	$\geqslant10$
60		CH_3CF_3	$CH_3CF_2{}^{18}F$	$CH_2{=}CH^{18}F$	3.0	8.3 ± 1.5
54		$c\text{-}C_4F_8$	$c\text{-}C_4F_7{}^{18}F$	several	3.2	10 ± 2
54		$c\text{-}C_3F_6$	$c\text{-}C_3F_5{}^{18}F$	$CF_2{=}CF^{18}F$	1.7	~10
60	$^{18}F/H$	CH_3CF_3	$CH_2{}^{18}FCF_3$	$CH^{18}F{=}CF_2$	~3.0	6.3 ± 1.0

tion, and that CH_2TNC^* just formed by substitution of a H atom by a T atom may be a 'non-RRKM' molecule[46, 55].

The ratio of $CH_2{=}C{=}CHT/CH_2{=}CHT$ from the decomposition of spiro-pentane-*t* formed by T^* reaction with spiro-pentane has been measured to be 1.0 ± 0.1 at 700 torr, as expected for excitation energy distributed over the entire molecule[51]. No clear-cut evidence has been found in a hot atom system for the non-randomised decomposition behaviour found with hexafluorobicyclopropyl-d_2 as formed by a methylene addition process[56].

4.7.4 Median excitation energies for the substitution of CH_3 groups and D atoms, respectively, by T atoms

The replacement of a CH_3 group in a substituted cyclobutane is accompanied by substantially higher excitation energies (median: 6–8 eV) than found for the substitution of H atoms by T atoms[52]. The present experiments are unable to distinguish whether this results from greater average energy of the tritium atom prior to methyl replacement, or lesser kinetic energy of CH_3 than H when replaced.

At a given gas pressure, a higher fraction of $c\text{-}C_4D_7T^*$ molecules are collisionally stabilised than are $c\text{-}C_4H_7T^*$. However, after correction for the isotope effect in decomposition rates for a given excitation energy, the median energy for T for D substitution in $c\text{-}C_4D_8$ is estimated to be within 0.2 eV of the median excitation energy for T for H substitution in $c\text{-}C_4H_8$ [46].

4.7.5 Median excitation energies for substitution reactions initiated by halogen atoms

The qualitative observation of secondary decomposition products following $^{18}F/H$ reactions with cyclanes[57], the finding of $CF^{18}F$ and $CH^{18}F$ from ^{18}F pius fluoromethanes[47], and numerous decomposition products in perfluoroalkane systems[54] all are sufficient indication that secondary reactions are very important in hot fluorine atom systems. A detailed quantitative study of the replacement of F by ^{18}F in CH_3CF_3 shows extensive decomposition to $CH_2{=}CF^{18}F + HF$ (or the alternate isotopic path to $H^{18}F$) [58–60]. Comparison with an RRKM calculation for this molecule indicates that the median excitation energy is in the range of 8 eV per molecule. While this comparison involves extrapolation of RRKM calculations to energies considerably higher than the 3–5 eV range in which most RRKM calculations are calibrated, the semi-quantitative estimate of residual excitation energies in the 10 eV or higher range is certainly correct. The observation of $CF^{18}F$ from $CF_3{}^{18}F$ requires a minimum excitation energy of 182 kcal mol^{-1} (loss of F_2) and a probable higher minimum of 220 kcal mol^{-1}, corresponding to the loss of two F atoms instead of F_2 [47, 54]. This estimate is basically a thermochemical estimate of 10 eV or greater excitation energy after $^{18}F/F$ substitution in CF_4, in excellent agreement with the RRKM-based value for CH_3CF_3.

The same CH_3CF_3 system also furnishes evidence for the decomposition of $CH_2{}^{18}FCF_3$ to $CH^{18}F{=}CF_2 + HF$, and therefore for high excitation energies for the $^{18}F/H$ reaction – estimated median, 6.3 ± 1 eV [60].

Similar observations have also been reported for several systems involving the substitution of energetic ^{38}Cl and ^{39}Cl atoms in various chlorocarbons[61], and for $^{80}Br^m$ reactions with both cyclopropane ($^{80}Br^m/H$) and cyclopropyl bromide ($^{80}Br^m/Br$) [62].

4.7.6   Excitation energies following hot addition

The excitation energies of the radicals formed by addition of recoil tritium atoms to olefins have also been evaluated through pressure dependence studies. Most addition reactions occur at tritium kinetic energies less than 0.5 eV (and total excitation energies <2.2 eV) [53].

In each of the systems described in this Section, the monitored secondary reaction of the product molecule is the same reaction found by thermal excitation of the molecule or radical. As the excitation energy increases, however, additional modes of secondary decomposition appear, reflecting the non-Boltzmann distribution of vibrational energies in these product molecules. For example, the addition of T atoms to 2-butene with the formation of excited $CH_3CHT\dot{C}HCH_3^*$ leads normally to exothermic decomposition to $CH_3 + CH_3CH{=}CHT$; with kinetically hot T* atoms, however, $CH_3CHTCH{=}CH_2$, corresponding to slightly endothermic H atom loss, is also found.

4.8   SUBSTITUENT INFLUENCES ON PRIMARY REACTION YIELDS

4.8.1   Effects of secondary decomposition on measured primary reaction yields

The qualitative observation that secondary decomposition processes can seriously deplete the surviving residual yields of primary reactions places some stringent limitations upon primary yield measurements: the secondary reaction products must also be known and measured for accurate estimates of original primary yield. Since this caveat applies equally to energetic tritium, ^{18}F, ^{38}Cl, ^{39}Cl and probably $^{80}Br^m$, little reliance can be placed upon inferences drawn from the surviving yields found in gas-phase experiments for any of these atoms near one atmosphere in pressure.

4.8.2   Substituent effects on reactions involving T for H replacement

The original primary yields for T for H substitution reactions with several alkanes and haloalkanes are illustrated in Figure 4.3. These measurements were made in both gas and liquid phases, with appropriate measurement and correction for secondary decomposition in all cases[63]. The correlation

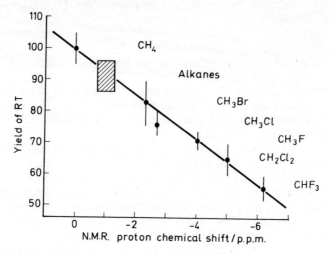

Figure 4.3 Yield of RT substitution product per C—H bond correlated with n.m.r. proton chemical shift
(From Tang, Y.-N.[63] by courtesy of the American Chemical Society)

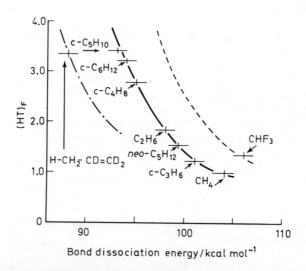

Figure 4.4 Yield of HT from $T + RH$ in excess $c\text{-}C_4F_6$ versus bond dissociation energy (hypothetical lines for varying excitation energies of residual radicals)

of primary yield with the proton n.m.r. shift in the substrate molecule suggests that the effect of halogen substituents on the T for H atom reaction is an electron density effect. Steric effects of halogen substituents are therefore negligible upon the yields of the T for H atom reaction. A negligible effect has also been found for alkyl steric obstruction of T/H substitution by alkyl substituents[52, 64].

4.9 BOND ENERGY EFFECTS IN ABSTRACTION REACTIONS

4.9.1 Abstraction by energetic tritium atoms

The yields for the abstraction of hydrogen from C—H bonds by energetic tritium atoms have been shown to correlate well with the known bond dissociation energies of alkanes and cyclanes, as shown in Figure 4.4[65, 66]. Deviations from this correlation with molecules such as CHF_3 [67] and $CH_3CD=CD_2$ [68] have been attributed, respectively, to the slight and large spatial alterations existing between the molecular distances and angles and those of the residual CF_3 and $CH_2=CD=CD_2$ radicals. It has been suggested that the time-scale for the abstraction process is too rapid (~ 2–5×10^{-14} s) for complete adjustment of the radicals to their equilibrium configuration prior to removal of the H atom.

4.9.2 Estimates of N—H and Si—D bond dissociation energies

Measurement of the per bond HT yields from various amines again indicates that the abstraction process proceeds in progressively higher yield as the N—H bonds weaken, as in $NH_3 < CD_3NH_2 < (CD_3)_2NH$ [69]. Similar measurements of DT yields from methylsilanes deuterated in all silicon–hydrogen positions show only slight variations for primary, secondary, and tertiary Si—D bonds, consistent with differences in bond dissociation energies of only 1 or 2 kcal mol^{-1} among them[70].

4.9.3 Abstraction of H by ^{18}F atoms

Quite recent measurements of the yields of abstraction from alkanes by energetic and by moderated ^{18}F atoms indicate that these reaction yields decrease per bond in the series $C_2H_6 > CH_4 > CH_3CF_3$ [71, 72]. This trend is clearly parallel to the bond dissociation energy trend established for atomic bromine reactions with the same molecules. The measurements are as yet too fragmentary to permit detailed comparison of thermal and hot ^{18}F atom reactions.

4.9.4 High energy mode of hydrogen abstraction by energetic tritium atoms: stripping?

The lowest activation energy reaction for tritium atoms with alkanes is the abstraction of hydrogen atoms, and reaction by this mechanism is also found in good yield for initially very energetic tritium atoms. Whether this mechanism is sufficient to account for essentially all of the observed HT in recoil tritium systems has been the subject of some disagreement[7, 73–76], and an additional high energy mode of formation of HT has been proposed, involving the 'stripping' of an H atom from the molecule during the passage of the tritium atom. The Author's opinion is that the experimental evidence for stripping in these systems is unconvincing, but the question is still an open one.

4.10   TRAJECTORY CALCULATIONS OF HOT ATOM REACTIONS

4.10.1   Trajectories for $T+H_2$ reactions

Calculations of the classical trajectories of atoms approaching and interacting with molecules have been widely applied in the past 15 years, limited chiefly by the accuracy of available potential energy surfaces for the interacting atoms and molecules, and by the currently available computational capabilities. Trajectories of chemical reactions initiated by energetic atoms have so far been calculated for only a few rather simple systems, but the field is rapidly developing and is certain to bring both more accurate calculations for the simple systems, and expansion to the more complicated many-atom systems. The first hot atom trajectory calculation involved the system $T^* + H_2$ [13] (and its isotopic counterparts, $T+D_2$ and $T+HD$, which were carried out on the same potential surface). Comparisons of these calculations with experimentally measured ratios of reaction yields (HT/DT from mixtures of H_2 and D_2) have been moderately successful. There is, however, substantial disagreement between the HT/DT ratio measured in argon-moderated H_2/D_2 mixtures and the values calculated from the cross-sections found from the trajectory calculations[77, 78]. Experimental measurements have also been made for the HT/DT ratio from HD, with agreement that the value is less than unity (0.7–0.9) for hot tritium atoms from both nuclear recoil[79] and from the photolysis of TBr [80].

4.10.2   Trajectories for $T+CH_4$ reactions

Two calculations have now appeared for the system $T^* + CH_4$, one involving the approximation that CH_3 is a centre of potential of mass 15, i.e., CH_4 is treated essentially as a diatomic molecule with atoms of mass 1 (or 2) and 15 (or 31) [14]. Both the abstraction and substitution reactions were observed in amounts reasonably consistent with experimental yields. Detailed information was obtained concerning impact parameters, angular scattering of products, effect of mass changes, energy distribution, etc. The second calculation involves the more complex system of six independently moving atoms[15]. This was made more feasible for computation by certain limitations accepted for the potential surface, the most important probably being that the back-side trio of H atoms were not permitted to undergo an inversion step. A comparison of the cross-sections calculated for the abstraction of H by these two methods is shown in Figure 4.5. The high energy mode for the formation of HT found in the three-particle calculation has disappeared from the six-particle calculation; possibly it was an artifact of the lesser interference offered to the passage of a tritium atom by a single point of mass 15 versus a CH_3 with structural size.

The estimates of T for H atom substitution probabilities arising from these two calculations invite comparison with the laboratory stereochemical experiments concerning retention and inversion at asymmetric carbon atoms. In the 'diatomic' methane calculation, most substitution processes

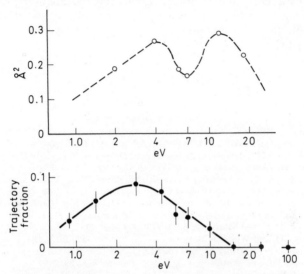

Figure 4.5 Calculated cross-sections versus energy for $T + CH_4 \rightarrow HT + CH_3$ reaction. Upper Figure from Kuntz, P. J. *et al.*[14], lower Figure from data given in Bunker, D. L. and Pattengill, M. D.[15], by courtesy of the American Institute of Physics

corresponded to an inversion process with the replaced H atom leaving from the opposite side of the structureless CH_3 group to that of the original T atom approach. However, the very lack of structure for CH_3 makes interpretations of stereochemical significance quite uncertain.

Substitution with inversion was not permitted in the six-atom $T–CH_4$ trajectory calculations by the particular approximations used for the potential energy surface. Here, any kind of T for H atom substitution was found to be quite scarce for each of several potential surface modifications — much less substitution was found than necessary for agreement with laboratory experiment. A substantial fraction of these trajectories, however, were terminated by accumulated calculational errors (designated as errors of the second kind) with the T atom making repeated collisions with several H atoms in the molecule. A reasonable hypothesis is that these trajectories might be transformed into substitutions with inversion on a less restrictive potential energy surface.

The laboratory experiments for the foreseeable future are limited to those discussed in Section 4.5.1, in which the asymmetric carbon atom has three heavy substituents plus one H atom. In these cases, retention of configuration is the only substitution process which is observed. The most likely exceptions to such an observational rule, of course, would be precisely those molecules with several easily movable H substituents, with CH_4 as the most favourable case of all for the inversion process. The relative yields of T for H atom reactions in CH_4 versus other molecules do not indicate that substitution into CH_4 occurs with unusually high yield, which would seem to imply that substitution into CH_4 occurs by the same mechanism found with several heavy substituents, and is almost exclusively a retention process[28]. This hypothesis is unlikely to be tested experimentally in the next few years; the possible inversion hypothesis for computed trajectories will, however, soon be tested.

References

1. Wolfgang, R. (1965). *Prog. Reaction Kinetics*, **3**, 97
2. Stocklin, G. (1969). *Chemie heisser Atome*. (Weinheim: Verlag Chemie)
3. Martin, R. M. and Willard, J. E. (1964). *J. Chem. Phys.*, **40**, 3007
4. Betts, J. A. (1971). *Reaction of Monoenergetic Deuterium Atoms with Hydrogen and Monoenergetic Hydrogen Atoms with Per-deutero Methane (Ph. D. Thesis*, California Institute of Technology; available from University Microfilms, Ann Arbor, Michigan).
5. Shvedchikov, A. P. (1967). *Russian Chem. Rev.* (Engl. Trans.). **36**, 202
6. Gann, R. G., Ollison, W. M. and Dubrin, J. (1971). *J. Chem. Phys.*, **54**, 2304
7. Chou, C. C. and Rowland, F. S. (1969). *J. Chem. Phys.*, **50**, 5133
8. Compton, L. E. and Martin, R. M. (1970). *J. Chem. Phys.*, **52**, 1613
9. Wood, G. O. and White, J. M. (1970). *J. Chem. Phys.*, **52**, 2613
10. Vermeil, C. (1970). *Israel J. Chem.*, **8**, 147
11. Menzinger, M. and Wolfgang, R. (1969). *J. Chem. Phys.*, **50**, 2991
12. Beatty, J. W. and Wexler, S. (1971). *J. Phys. Chem.*, **75**, 2417
13. Karplus, M., Porter, R. and Sharma, R. (1966). *J. Chem. Phys.*, **45**, 3871
14. Kuntz, P. J., Nemeth, E. M., Polanyi, J. C. and Wong, W. H. (1970). *J. Chem. Phys.*, **52**, 4654
15. Bunker, D. L. and Pattengill, M. D. (1970). *J. Chem. Phys.*, **53**, 3041
16. *Chemical Effects of Nuclear Transformations* (1961). 2 Volumes. (Vienna: International Atomic Energy Agency)
17. *Chemical Effects of Nuclear Transformations* (1965). 2 Volumes. (Vienna: International Atomic Energy Agency)
18. Todd, J. F. J., Colebourne, N. and Wolfgang, R. (1967). *J. Phys. Chem.*, **71**, 2875
19. Smail, T., Miller, G. E. and Rowland, F. S. (1970). *J. Phys. Chem.*, **74**, 4080
20. Krohn, K. A., Parks, N. J. and Root, J. W. (1971). *J. Chem. Phys.*, in press
21. Wolfgang, R. and Rowland, F. S. (1958). *Anal. Chem.*, **30**, 903
22. Lee, J. K., Lee, E. K. C., Musgrave, B., Tang, Y.-N., Root, J. W. and Rowland, F. S. (1962). *Anal. Chem.*, **34**, 741
23. Stocklin, G., Cacace, F. and Wolf, A. P. (1963). *Z. Anal. Chem.*, **194**, 406
24. Rowland, F. S. (1970). *Molecular Beams and Reaction Kinetics*, Schlier, Ch., (Ed.), 108. (New York: Academic Press)
25. Hosaka, A. and Rowland, F. S. (1971). *J. Phys. Chem.*, **75**, 3781
26. Kushner, R. and Rowland, F. S. (1972). *J. Phys. Chem.*, in press
27. Kushner, R. and Rowland, F. S. (1969). *J. Amer. Chem. Soc.*, **91**, 1539
28. Palino, G. F. and Rowland, F. S. (1971). *J. Phys. Chem.*, **75**, 1299
29. Wai, C. M. and Rowland, F. S. (1970). *J. Phys. Chem.*, **74**, 434
30. Palino, G. F. and Rowland, F. S. (1971). *Radiochim. Acta*, **15**, 57
31. Richardson, A. E. and Wolfgang, R. (1970). *J. Amer. Chem. Soc.*, **92**, 3480
32. Vasaros. L., Machulla, H. J. and Stocklin, G. (1972). *J. Phys. Chem.*, in press
33. White, J. M. and Kuppermann, A. (1966). *J. Chem. Phys.*, **44**, 4352

34. Kuppermann, A. (1967). *Nobel Symposium 5*, Claesson, S., (Ed.), p. 131. (New York: Interscience)
35. White, J. M., Davis, D. R., Betts, J. A. and Kuppermann, A. (1971). Manuscript contained in Reference 4
36. Davis, D. R., Betts, J. A., White, J. M. and Kuppermann, A. (1971). Manuscript contained in Reference 4
37. Gann, R. G., Ollison, W. M. and Dubrin, J. (1970). *J. Amer. Chem. Soc.*, **92**, 450
38. Rebick, C. and Dubrin, J. (1970). *J. Chem. Phys.*, **53**, 2079
39. Menzinger, M. and Wolfgang, R. (1969). *Angew. Chem.*, **8**, 438
40. LeRoy, R. L. (1969). *J. Phys. Chem.*, **73**, 4338
41. Chou, C. C. and Rowland, F. S. (1969). *J. Chem. Phys.*, **50**, 2763
42. Wolfgang, R. (1970). Private communication
43. Beatty, J. W., Pobo, L. G. and Wexler, S. (1971). *J. Phys. Chem.*, **75**, 2407
44. White, J. M. (1969). *Chem. Phys. Lett.*, **4**, 441
45. Lee, E. K. C. and Rowland, F. S. (1963). *J. Amer. Chem. Soc.*, **85**, 897
46. Bunker, D. (1972). Private communication
47. Tang, Y.-N., Smail, T. and Rowland, F. S. (1969). *J. Amer. Chem. Soc.*, **91**, 2130
48. Smith, W. S. and Tang, Y.-N. (1971). Presented at 161st Meeting of the American Chemical Society, Los Angeles, March, 1971
49. Tang, Y.-N. and Rowland, F. S. (1968). *J. Phys. Chem.*, **72**, 707
50. Ting, C. T. and Rowland, F. S. (1970). *J. Phys. Chem.*, **74**, 4080
51. Tang, Y.-N. and Su, Y.-Y. (1970). Presented at 160th Meeting of the American Chemical Society, Chicago, Sept. 1970
52. Ting, C. T. and Rowland, F. S. (1970). *J. Phys. Chem.*, **74**, 445
53. Kushner, R. and Rowland, F. S. (1971). *J. Phys. Chem.*, in press
54. McKnight, C. F. and Root, J. W. (1969). *J. Phys. Chem.*, **73**, 4430
55. Harris, H. H. and Bunker, D. L. (1971). *Chem. Phys. Lett.*, **11**, 433
56. Rynbrandt, J. D. and Rabinovitch, B. S. (1971). *J. Phys. Chem.*, **75**, 2171
57. Tang, Y.-N. and Rowland, F. S. (1967). *J. Phys. Chem.*, **71**, 4576
58. McKnight, C. F., Parks, N. J. and Root, J. W. (1970). *J. Phys. Chem.*, **74**, 217
59. Krohn, K. A., Parks, N. J. and Root, J. W. (1971). *J. Chem. Phys.*, **55**, 5771
60. Krohn, K. A., Parks, N. J. and Root, J. W. (1971). *J. Chem. Phys.*, **55**, 5785
61. Tang, Y.-N., Smith, W. S., Williams, J. L., Lowery, K. and Rowland, F. S. (1971). *J. Phys. Chem.*, **75**, 440
62. Wai, C. M. and Jennings, R. L. (1971). *J. Phys. Chem.*, **75**, 2698
63. Tang, Y.-N., Lee, E. K. C., Tachikawa, E. and Rowland, F. S. (1971). *J. Phys. Chem.*, **75**, 1290
64. Smail, T. and Rowland, F. S. (1970). *J. Phys. Chem.*, **74**, 456
65. Root, J. W., Breckenridge, W. and Rowland, F. S. (1965). *J. Chem. Phys.*, **43**, 3694
66. Tachikawa, E. and Rowland, F. S. (1968). *J. Amer. Chem. Soc.*, **90**, 4767
67. Tachikawa, E. and Rowland, F. S. (1969). *J. Amer. Chem. Soc.*, **91**, 559
68. Tachikawa, E., Tang, Y.-N. and Rowland, F. S. (1968). *J. Amer. Chem. Soc.*, **90**, 3584
69. Tominaga, T. and Rowland, F. S. (1968). *J. Phys. Chem.*, **72**, 1399
70. Hosaka, A. and Rowland, F. S. (1971). Unpublished results
71. Williams, R. L. and Rowland, F. S. (1971). *J. Phys. Chem.*, **75**, 2709
72. Merrill, J., Parks, N. L. and Root, J. W. (1971). Private communication
73. Wolfgang, R. (1969). *Accounts Chem. Res.*, **2**, 248
74. Chou, C. C., Smail, T. and Rowland, F. S. (1969). *J. Amer. Chem. Soc.*, **91**, 3104
75. Baker, R. T. K. and Wolfgang, R. (1968). *J. Amer. Chem. Soc.*, **90**, 4473
76. Baker, R. T. K., Silbert, M. and Wolfgang, R. (1970). *J. Chem. Phys.*, **52**, 1120
77. Seewald, D., Gersh, M. and Wolfgang, R. (1966). *J. Chem. Phys.*, **45**, 3870
78. Porter, R. N. and Kunt, S. (1970). *J. Chem. Phys.*, **52**, 3240
79. Seewald, D. and Wolfgang, R. (1967). *J. Chem. Phys.*, **46**, 1207
80. Chou, C. C. and Rowland, F. S. (1967). *J. Chem. Phys.*, **46**, 812

5
Chemiluminescent Reactions

TUCKER CARRINGTON
York University, Ontario

and

J. C. POLANYI
University of Toronto, Ontario

5.1 INTRODUCTION

Chemiluminescence is the emission of radiation from the products of chemical reactions. The observation of such emission implies that the reaction products are in a state of thermal disequilibrium with respect to their surroundings. It is customary to take a more restrictive view of chemiluminescence and to apply the term only to emission from reaction products which are excited in some non-thermal fashion. If this proviso is not made, then it becomes impossible to distinguish the case where emission is a direct consequence of the liberation of energy by a chemical reaction from the more common case where emission arises simultaneously from direct and from indirect causes (i.e. simply from heating).

To take an example, flames at atmospheric pressure are strongly luminescent but, in the absence of disequilibrium among the forms of motion responsible for the luminescence, it is impossible even to surmise what may be the direct contribution of chemiluminescence to the observed emission. At low pressures flames exhibit symptoms of non-equilibrium, indicative of the preferential excitation of certain degrees of freedom in the products of chemical reactions. This disequilibrium consists in the emergence of significantly different temperatures characteristic of the distribution of the emitting species over electronic states, vibrational states and rotational states; $T_e \neq T_V \neq T_R$. One can usefully describe such a system as being *chemiluminescent*. For chemical reactions at very low pressures non-Boltzmann distributions are observed in several degrees of freedom. Such a high degree of disequilibrium is only achieved at pressures below those required to sustain flames, since flames are dependent on chain reactions and these do not take place at very low pressures.

Under these circumstances it ceases to be possible to describe the distribution over electronic, vibrational or rotational states by characteristic temperatures. The only method of characterising the energy distribution is by cataloguing the vibrational populations in the various vibrational levels (quantum numbers, v) of each electron state, and also the rotational populations in the various levels (J) of each of these v-states. This large matrix of populations contains a great deal of information concerning the tendency of the chemical reactions involved in the particular reactive system to channel their exothermicity into specific types of motion in the newly formed reaction products. This information, in turn, can be used as a clue to the nature of the forces that govern the atomic and molecular rearrangements in chemical reactions. This latter information is then embodied in some sort of a model of the chemical reaction; this model is usually based on certain hypotheses concerning the form of the potential energy hypersurface (which records the potential energy of the system of nuclei and electrons as a function of their nuclear configuration). The reaction is described as taking place 'across a potential energy surface' which possesses certain characteristic features.

When the reaction releases sufficient energy that electronically excited states of the products are accessible, it becomes necessary, in order to arrive at an understanding of the product energy distribution, to consider the forces that are operative as the system proceeds across more than one potential-energy hypersurface, as well as to consider the likelihood of transfer to and fro between such surfaces. We shall postpone the consideration of this more difficult case and shall discuss, in the first place, those reactions that lead (or are believed to lead) to products in their ground electronic state.

It should be recognised that even in this simpler case the possibility cannot be excluded that the existence of paths leading to electronically excited products could affect the energy distribution of the products formed in their ground electronic state. By use of the semi-classical terminology already employed, it is possible to distinguish two ways in which the existence of intersecting energy surfaces could affect the outcome on the lower surface. The system could transfer to the upper surface and then revert to the lower surface *en route* to the formation of products; the outcome would then be affected by the nature of the forces operating on both surfaces. Alternatively, certain paths across the lower surface could to some extent be 'blocked', so far as the formation of electronic ground-state products is concerned, since these paths lead through configurations in which the upper surface closely approaches the lower one and consequently the anticipated ground-state products are seldom formed.

5.2 VIBRATIONAL-ROTATIONAL EXCITATION

5.2.1 Theoretical

The theoretical problem is to compute the extent of vibrational, rotational (and hence translational) excitation in the products of exothermic reaction. In the absence of electronic excitation in the products, the reaction can be regarded, for simplicity, as taking place over a single potential energy hypersurface. For the simple exchange reaction, $A + BC \rightarrow AB + C$, this potential function is $V(r_{AB}, r_{BC}, r_{AC})$. The outcome of reactive collisions in this potential field can be obtained by attempting the solution of an intricate problem in quantum mechanical scattering[1-10]. Fortunately, classical mechanics provides for the present an entirely adequate basis for calculation of product energy distributions[4, 6, 11-13].

The classical equations of motion are, of course, insoluble in closed form even for the simple three-particle system $A + BC$. However, numerical integration of the equations of motion on a computer (as pioneered by Wall, Hiller and Mazur[14-16]) is practical even for many-atom systems. The largest system treated in this fashion until now has been a six-atom system[17].

The major problem in predicting product vibrational and rotational excitation arises from the paucity of our knowledge of the potential energy hypersurfaces. The variational solution of the Schröedinger equation has been achieved to chemical accuracy (i.e. to plus or minus a few kilocalories per mole) for two systems; H_3^+ and H_3. A complete potential energy hypersurface has been computed only for the simpler of these two systems, H_3^+ [18], and has been used for classical trajectory studies[19]. Further work on these lines, for systems of greater chemical interest, is anticipated, but progress will be slow since the solution of the Schröedinger equation for many-electron systems to this level of accuracy is still enormously expensive in computer time.

Meanwhile a semi-empirical theoretical approach is being explored. Potential energy hypersurfaces of contrasting types have been constructed, in order to ascertain which features play the dominant role in determining the outcome of reactive encounters. This type of work has been reviewed recently[20]. In considering chemiluminescent processes we are particularly concerned with exothermic reactions. Following a proposal made 30 years ago[21] particular attention has been paid in investigations of exothermic potential energy surfaces to the location along the reaction path of the down-hill slope[22-29].

Potential energy surfaces which have the down-hill slope situated predominantly along the coordinate of approach, r_{AB}, are commonly referred to as 'attractive', and those with the major part of the down-hill slope along r_{BC} are termed 'repulsive'[25]. It had been predicted[21] that the attractive surfaces would favour vibrational excitation in the newly formed chemical bond (i.e. AB in the reaction $A + BC \rightarrow AB + C$), whereas the repulsive type of interaction would lead to product translation. The contemporary investigations, with high-speed computers, have broadly speaking confirmed this prediction. They have, however, revealed a substantially more intricate pattern of behaviour. Some of the *tentative* conclusions from recent computations are listed below. The list is a selective one in which the items appear to be of interest in connection with some of the experimental results which will be reviewed in Sections 5.1.2.1 to 5.1.2.5.

5.2.1.1 *Attractive and repulsive surfaces*

For a given mass combination the efficiency of conversion of the reaction exothermicity into vibration in the newly formed bond, increases as the surface becomes increasingly attractive[22-27]. A very crude index of the extent of vibrational excitation in the new bond can be obtained by drawing a rectangular path across the potential energy surface appropriate to collinear reaction and calling the energy release along r_{AB} the attractive energy release, $A_\perp$ and the remainder the repulsive energy release, $R_\perp$ [27].

5.2.1.2 Curvature of the minimum path and mixed energy release

A more refined criterion of surface character takes account of the curvature of the minimum path. Energy released on the curved portion of the reaction path has been termed 'mixed energy release' (M) [27]. The term 'mixed' was coined to indicate that in this type of energy release both r_{AB} and r_{BC} are changing. A criterion of surface character which defines A_M, M_M and R_M in terms of the minimum path across the collinear surface, is an improvement over one that only considers the energy release along a rectangular path (see above). However, A_M, M_M and R_M are still a poor index of the tendency to form vibrationally excited product, since the reactive trajectories diverge significantly from the minimum path[27].

5.2.1.3 Mass combination and mixed energy release

The combination of masses involved in the reactive encounter has a profound effect on the outcome. A criterion of reaction dynamics which takes account of the curvature of the reaction path and the effect of mass combination is one which is based on apportionment of a single collinear trajectory between attractive, mixed and repulsive energy release: A_T, M_T and R_T. The sum $(A_T + M_T)$ is an approximate index of the tendency of a particular potential energy surface to convert reaction energy into vibration in the new bond, for a given mass combination[27].

5.2.1.4 The light-atom anomaly

The mass combinations with $m_A \gg m_C$ (especially if m_B is large) have a very large M_T, and those with $m_A \ll m_C$ (especially if m_B is large) have a very small M_T (the m values refer to the masses of the atomic species). The mass combination $L + HH$ (where L denotes a light atom and H a heavy atom; e.g. $H + X_2$, where X_2 is a halogen) has, consequently, the smallest possible M_T. On repulsive surfaces, for which $A_\perp$, and hence also A_T, is small, this mass combination has a small $(A_T + M_T)$, and correspondingly low vibrational excitation. The effect is called the 'light-atom anomaly'[*][27-30]. All

---

* Ref. 22b reported a light-atom anomaly, but on a highly attractive surface, where Ref. 27 showed none. Ref. 23 using a new type of attractive surface showed no light-atom anomaly, in accord with Ref. 27. Recently Bunker and co-workers (Ref. 30) have again looked for a light-atom anomaly, but have found no evidence of one; it seems probable that a more repulsive surface than theirs is required in order to show this effect, which is an important effect on repulsive LEPS (London–Eyring–Polanyi–Sato) surfaces.

other mass combinations channel vibrational excitation into the new bond with greater efficiency on repulsive surfaces $(A_T + M_T)$ is sizeable, owing to the fact that M_T is sizeable. The actual mechanism by which 'mixed energy release' channels B·C repulsion efficiently into product vibration can be pictured in a number of ways. A very simple representation is shown in

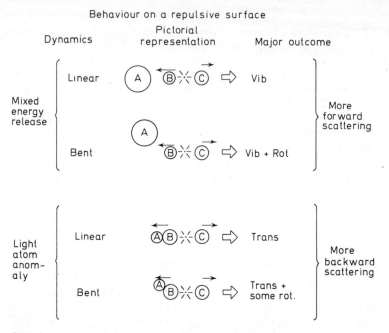

Figure 5.1 A simple pictorial representation of two types of energy release that have been observed on repulsive potential energy surfaces. For the majority of mass combinations mixed energy release yields internal excitation; for the minority with mass combination $m_A \ll m_B$ and m_C the light-atom anomaly results in substantial product translational energy

(From Polyani[20], by courtesy of the American Chemical Society)

Figure 5.1. When the attacking atom is heavy (and consequently slow) the repulsion is seen to be released while the new bond is still extended. This has the result that B recoils from C to give internal excitation in AB, instead of AB recoiling as a whole to give translation.

5.2.1.5 *Product rotational excitation*

Trajectory studies on many types of potential energy surface have shown that reagent orbital angular momentum, L, has a positive correlation with product rotational angular momentum, J (e.g. Ref. 23). A high degree of rotational excitation in reaction products may therefore be indicative of a large $L(|L| = \mu v_{rel.} b$, where μ is the reduced mass of the reagents, $v_{rel.}$ is their relative speed of approach and b is the impact parameter for the reactive collision). Rotational excitation can also arise from the energy released by the chemical reaction. If the energy surface is repulsive, then mixed energy release in bent configurations (see Figure 5.1) will lead to enhanced rotational excitation in the products, at the expense of vibrational excitation; there will, in this case, be an inverse correlation between vibration and rotation in the products.

5.2.1.6 Product angular distribution

A parameter which can only have subtle consequences for the nature of the chemiluminescence, but is nonetheless of considerable interest in terms of the molecular mechanics of the chemiluminescent reaction, is the angular distribution of the newly formed reaction products. In atomic reactions the new molecule, AB, is said to be 'forward' scattered if it is ejected along the continuation of the direction from which the atomic reagent approached. For a given mass combination, forward scattering is in general more probable if the surface is attractive[32, 33]. Certain types of secondary encounter on an attractive surface (particularly 'clutching' secondary encounters, such as are involved in 'migration', see (f) below) can produce very sharp forward scattering[31]. On a repulsive energy surface the light-atom anomaly produces the sharpest backward scattering of AB, owing to the high translational energy in the products which causes AB to recoil sharply backwards[33, 34]. Mixed energy release, by contrast, channels the repulsive energy into internal excitation of AB, hence in many encounters AB moves away from the reaction zone approximately along the continuation of the direction of approach of A; i.e. in a more-or-less forward direction[33, 35].

5.2.1.7 Direct and indirect encounters

Both direct and indirect reactive encounters are observed on potential energy surfaces that are entirely free from potential energy hollows (and would not therefore be suspected of producing *long-lived* intermediates). A direct encounter is one in which the reagents approach to produce a strong interaction and then the products separate continuously after this first strong interaction[36]. In an indirect (also termed 'complex' or 'snarled') encounter the products begin to separate and then come together again in one or more secondary encounters. Secondary encounters are most common on attractive surfaces since there is the least tendency for the products to repel one another and separate cleanly. Secondary encounters are less common on repulsive surfaces. They do occur on repulsive surfaces in cases where the mass combination produces mixed energy release, i.e. internal excitation, since the repulsion can not in this case invariably separate the products before vibration or rotation results in a secondary collision.

Secondary encounters are of two types; 'clutching' and 'clouting', i.e. the products may be pulled together or pushed apart. Migration, in which an attacking species interacts strongly with one end of the molecule under attack and then proceeds to react with the other end, constitutes a type of clutching secondary encounter[31, 32, 37].

5.2.1.8   Correlation between percentage attractive energy release and barrier height

In related families of exchange reactions there is evidence of a strong correlation between decreasing barrier height and 'earlier' barrier location[38-40]. This shift in barrier location in turn correlates strongly with increased percentage of attractive energy release[40]. Consequently one would expect to see a systematic increase in the fractional conversion of the exothermicity into vibration and a greater tendency for forward scattering, in members of a related family of reactions that have successively lower activation barriers. This effect should be most marked for the case of a light attacking atom ($m_A \ll m_C$), which gives low vibrational excitation and sharp backward scattering on a repulsive surface (see Section 5.2.1.4).

5.2.1.9   Favoured reagent energy distributions for exothermic and endothermic reactions

An important outcome of studies of vibrational and rotational (and hence translational) energy distribution among products of exothermic reactions has been an increase in our knowledge of the role of these various degrees of freedom in promoting endothermic reaction. Trajectory studies have shed light on this same question. Studies have been made of contrasting thermoneutral potential energy surfaces (a) with the crest of the energy barrier displaced slightly into the entry valley of the collinear surface (i.e. along r_{AB}) and (b) with the crest of the barrier displaced slightly into the exit valley (along r_{BC}). Surfaces (a) and (b) can be summarised as having 'early' and 'late' barriers, respectively. Translational energy is conducive to reaction across surfaces with early barriers whilst vibrational energy is most effective with late barriers[41, 42]. These contrasting requirements for reagent energy are understood easily, since in the first case the resistance is met as A approaches BC (taking $A + BC$ as an example), whereas in the second case the resistance arises as B is separated from C; the approach of A to BC constitutes relative translation, the separation of B from C is involved in BC vibration.

Inspection of potential energy surfaces for substantially exothermic reactions, ranging from attractive to repulsive, shows that, in either case, in the exothermic direction the barrier is of the 'early' type[42]. Reagent translation is therefore most effective in bringing about reaction. In the endothermic direction the barrier is of the late variety. (It is, moreover, a very high barrier, so that considerable importance attaches to the optimal choice of the reagent energy distribution). For substantially endothermic reactions the degree of freedom most conducive to reaction will be reagent vibration.

5.2.1.10   Vibrational excitation in 'new' and 'old' bonds, of reaction products

Trajectory calculations on four-atom exothermic exchange reactions, $A + BCD \rightarrow AB + CD$ indicate that the 'new' bond AB will in general take up a greater percentage of the energy release as vibration, than will the 'old' bond CD[43]. This is understandable since the energy can be released to varying degrees in the form of attraction as A approaches B, yielding up to $\approx 100\%$ conversion of the exothermicity into AB vibration. By contrast, CD, an already existent bond, is less susceptible to strong perturbation. The CD equilibrium bond distance, for example, is unlikely, except in rare cases, to differ very greatly from BCD to CD.

5.2.1.11 Effect of downhill slope

In general, the slope of the downhill region of the potential energy surface is not as important a characteristic as its location; attractive energy release differs more markedly from repulsive, than more forceful repulsive release differs from less forceful[23]. (The slope of the potential is the force.) Nonetheless, in cases where vibrational excitation is inefficient, the repulsive force does become a significant parameter. One example is the mass combination $L + HH$ on a repulsive surface[44]. Another example is vibrational excitation in the old bond in a reaction $A + BCD \rightarrow AB + CD$ [45]. In both cases the effect of increasing the repulsive force, without altering the percentage of the total energy released as repulsion, is to increase the efficiency of vibrational excitation. The explanation is evident from a consideration of the dynamics in the limit of infinite force, i.e. impulsive release of the repulsive energy[25, 26, 46-49]; in this case the repulsion produces the maximum possible differential force on the adjacent bond (since it applies force at first only to the nearest atom); this is most favourable for vibrational excitation.

5.2.1.12 Vibrational and rotational excitation in the products of unimolecular dissociation

It is now well understood (particularly as a consequence of the work of Rabinovitch and associates[50]) how one can form molecules with internal energy in excess of that required for unimolecular dissociation. Some of the earliest trajectory calculations were concerned with the dynamics of unimolecular dissociation in triatomic molecules; $ABC^\dagger \rightarrow AB + C$ [51, 52]. Under normal circumstances it appeared that excess energy would be inefficiently channelled into vibrational excitation in AB [52].

In the example just cited, the reaction path sloped up hill all the way from reagent $(ABC^\dagger)$ to product. In other cases the surface may at first slope up hill and then down again due to the formation of new bonds (cf. Section 5.2.2.6). Similar questions may then be asked regarding the location (in the hyper-space of the energy surface) of the downward-sloping region, as were asked for other reactions in the preceding sections[53]. It is worth considering this energy (total $E_c - \Delta H_0^0$, where E_c is the classical height of the energy barrier en route to dissociation and $-\Delta H_0^0$ is the enthalpy change in going from reagent to products) separately from any energy the dissociating molecule may have in excess of E_c. The excess energy is likely to be more-or-less randomly distributed among the products. The energy release, $E_c - \Delta H_0^0$, by contrast, can be channelled into specific degrees of freedom in the products. If it is liberated 'attractively' as new bonds relax to their normal separations, the more efficient vibrational excitation will occur. If it is liberated 'repulsively', product vibrational excitation will be less. If the repulsion is released in such a manner as to apply a torque to one or both of the fragments of dissociation, rotational excitation will result.

5.2.2 Experimental

Chemiluminescence is evidence of excitation in products of chemical reaction. It is most often studied for the information that it can give concerning the *initial* excitation of product, newly formed as a result of reaction. Chemiluminescence is only one method that might be used in order to study what may be termed the 'detailed rate constant' into specified vibrational, rotational and (hence) translational states of the product; $k(v, J, T)$ (where v and J are the product vibrational and rotational quantum numbers, and T is the relative translational energy of the product). The range of methods that are and that will soon be available, is indicated in a recent, but sketchy, survey[54]. Two more detailed reviews have been published[55, 56].

The two principal methods in this field other than the chemiluminescence approach are the molecular beam method and absorption spectroscopy. The beam method is an extremely powerful one in which the relative translational energy, T in the reaction products and their scattering angles (see Section 5.2.1.6 above) are measured. The absorption spectroscopy approach is only beginning to be exploited as a quantitative tool for obtaining $k(v)$ [57, 58]. The procedure is to record u.v. absorption spectra during, and immediately after, a brief photolytic flash; this method yielded qualitative information early in the history of this subject[59].

Progress in the area of quantitative measurement employing chemiluminescence has been sufficient to warrant this additional review. Quite recently the chemiluminescence method has yielded rate constants $k(v, J, T)$ at a level of detail not yet matched by other approaches.

Several variants of the chemiluminescence method are summarised briefly below. In each method infrared chemiluminescence due to a vibrational transition is observed. The methods differ in the manner in which they attempt to extract information concerning the initial energy distribution in the reaction products from what is, to a greater or lesser extent, a relaxed distribution. With the exception of the arrested relaxation method [(c)] only the initial *vibrational* distribution, reflecting $k(v)$, is sought. (Note that the symbols k_v and $k(v)$ are used interchangeably). Rotational relaxation is so rapid under the conditions of methods (a), (b), (d) and (e) that correction for this relaxation is not feasible. Method (c) reduces rotational relaxation to the point where only a minor correction is required in order to extract the initial rotational distribution.

A typical pattern of vibrational relaxation is shown in Figure 5.2. In the example, relaxation is by radiation only. The mean radiative lifetime for the relaxing molecule, HCl, from $v = 7$ to states $v \leqslant 6$ is 6.1 ms. From Figure 5.2a it is evident that appreciable relaxation has occurred in 10 ms. Observation along the line-of-flow (Figure 5.2b) gives a mean distribution which appears significantly less relaxed.

The following techniques have been used to obtain initial distributions.
(a) *Flow.* [For $k(v)$] – The reagents are swept by fast laminar flow (10^{-2}–10

torr total pressure) past a single observation window located as close as possible to the point of mixing. This method was used in the early work on infrared chemiluminescence, in conjunction with calculations of the extent of relaxation[60, 61]. It is a hazardous method. The best chance of success is, of course, if the flow is so rapid that the correction is small[53]. A variant measures the relaxation by formation of the same product molecule in a

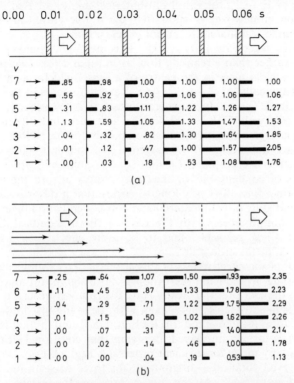

Figure 5.2 Relative vibrational distributions (indicated by the lengths of the horizontal black lines) at successive times, with continuous formation of HCl in level $v = 7$ (only) starting at $t = 0.00$. Subsequent distributions are shown at intervals of 10 ms. Relaxation is by radiation only. In (a) the emitting gas is observed instantaneously at the stated times, and in (b) the emitting gas is observed from $t = 0.00$ up to the stated time (i.e. the recorded population is an integral from $t = 0$ to some $t > 0$). (From Polanyi[26], by courtesy of the Optical Society of America)

known initial distribution (using a reaction studied by some other method) in the same, or similar, environment[62, 63]. Inevitably the environment is somewhat altered if another set of reagents is introduced.

(b) *Measured relaxation.* [For $k(v)$] – Several observation windows are placed at intervals along the line of flow, in a system resembling that described in (a)[34, 62, 64]. Provided that the observation windows are situated at distances corresponding to times during which relaxation is moderate, a simple graphical extrapolation back to zero time will yield fairly dependable values for the relative $k(v)$ value (see for example Figure 5 of Ref. 65). In the case of the reaction $H + Cl_2 \rightarrow HCl + Cl$ a very detailed numerical analysis has been made of the combined effects of reaction, diffusion, flow, radiation and collisional deactivation[65]. A detailed model of this sort permits, in effect, a more intelligent extrapolation to $t = 0$. The consequences of various limiting assumptions regarding the nature of the relaxation process, can be tested. In the case of $H + Cl_2$, the result of numerical analyses based on a number of relaxation models was to show that the simple graphical extrapolation procedure under-estimated the extent of relaxation.

(c) *Arrested relaxation.* $[k(v, J, T)]$ – This method[34, 66] makes use of molecular flow, rather than streaming flow as in (a) and (b). Two uncollimated beams of reagents meet in the centre of a vessel which has a background pressure of typically $10^{-4} - 10^{-5}$ Torr. Reaction occurs at the intersection of these 'beams'. The products are transferred, with few secondary collisions, either to a cold wall (77 K, or 20–40 K) which surrounds the reaction zone, or through a large orifice to the diffusion pump (Figure 1b Ref. 34). The cold walls may act (depending on the reagents) as a conventional cryo-pump, or they may act as a cryo-pump for non-condensables by trapping them in a large excess of condensable, or they may simply adsorb the vibrationally excited product for sufficiently long to ensure that it desorbs only as vibrational ground state ($v = 0$) material and consequently no longer contributes to the chemiluminescence. (In the case of the free radical $OH^\dagger$, room-temperature silica gel was found to be more effective as a deactivating surface than a metal surface at 77 K[76].)

The proof of the effectiveness of this combination of molecular flow plus rapid 'pumping' in reducing relaxation to an insignificant amount is to be found (i) in the fact that the steady-state distribution observed under these conditions is in satisfactory agreement with the initial distribution obtained by extrapolation using method (b) (see Figure 5.3), and (ii) that the *rotational* distribution is largely unrelaxed, implying that collisional deactivation of vibrators (a very much less efficient process) is negligible.

The observed rotational distribution for three background pressures is shown in Figure 5.4[67]. At the lowest pressure the extent of rotational relaxation is minor. It can be quite accurately corrected for, with a model which assumes that the probability of rotational deactivation is proportional to $\exp(-\Delta E_J/kT)$ (where ΔE_J is the energy difference between levels J and $J-1$)[67]. The steady-state distribution then yields $k(v, J)$.

The total energy available for distribution among the products, $E'_{tot.}$, is calculable using the formula[66, 68–70]

$$E'_{tot.} = -\Delta H_0^0 + E_a + RT + \tfrac{3}{2}RT,$$

where $-\Delta H_0^0$ is the zero point exothermicity, $(E_a + RT)$ is the mean reagent translational energy[71], and $\tfrac{3}{2}RT$ the mean reagent rotational energy. (An uncertainty of $\pm 2RT$ in this expression would be equivalent to $\approx \pm 1.2$ kcal, which represents typically an uncertainty of ± 2–5% in $E'_{tot.}$ less than the uncertainty in the measurement of the populations $N_{v,J}$). If $E'_{tot.}$ and $k(v, J)$ are known it is possible to obtain the translational energy distributions[70] by difference, and hence obtain $k(v, J, T)$. This function can most

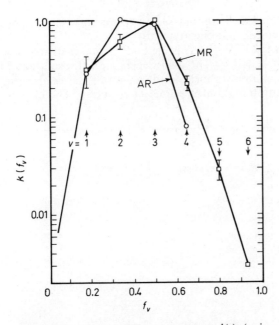

Figure 5.3 The detailed rate constant $k(v)$ (v is indicated across the centre of the Figure) and $k(f_v)$ (where f_v is the fraction of the 48.6 kcal mol^{-1} available energy going into vibration) for the reaction $H + Cl_2 \rightarrow HCl + Cl$ as obtained from the method of measured relaxation (MR) and the method of arrested relation (AR). (From Pacey and Polanyi[65], by courtesy of the Optical Society of America)

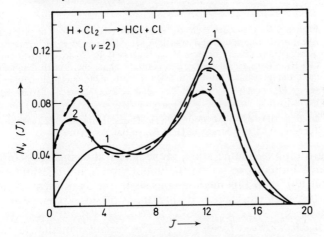

Figure 5.4 The solid lines show rotational distributions (for $v = 2$, formed in the reaction $H + Cl_2 \rightarrow HCl + Cl$) observed at successively lower pressures ($3 \rightarrow 1$). The heavy broken lines show relaxed distributions at pressures 2 and 3, as computed from the least-relaxed distribution at pressure 1. (From Polanyi and Woodall[67], by courtesy of the American Institute of Physics)

easily be recorded as a contour plot; two examples are given in Figure 5.5 and 5.6 (one typical, and one atypical).

(d) *Chemical laser.* $[k(v)]$ — This method[72, 73] makes use of the gain equation for a chemical laser[26, 77]. The gain depends on the rotational temperature (assumed Boltzmann). In method (d) the rotational temperature is allowed to rise following initiation of pulsed laser action. The changing gain with

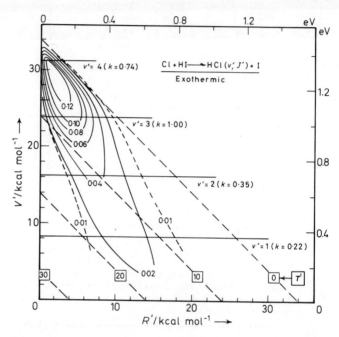

Figure 5.5 Contours indicate, on a relative scale, values of product vibrational, rotational and translational energies (V, R, T; ignoring quantisation) which correspond to equal $k(V, R, T)$ for the reaction $Cl + HI$. Translational energy T is constant along the broken diagonal lines, and increases from right to left. The limits of the shaded domain are set by $E'_{tot} = 33.9$ kcal mol^{-1}. The value of k beside each v level is k_v, normalised to unity for the most-populated level; $k_v = \sum_J k(v, J)$ for a given v. (Based on Maylotte, Ph.D. Thesis, University of Toronto, Toronto, 1968, by courtesy of D. H. Maylotte)

increasing rotational temperature, immediately after initiation of the laser pulse, allows limits to be set on the initial vibrational distribution, i.e. on $k(v)$. A difficulty with this method consists in the assessment of the extent to which the laser emission is itself responsible for a change in the gain with time.

(e) *Equal gain (chemical laser).* $[k(v)]$ — This variant of method (a) represents an improvement in accuracy. By external heating the rotational temperature of the gas mixture is adjusted until a laser pulse is obtained which corresponds to the simultaneous initiation of two vibration-rotation transitions, with equal gain. The ratio N_v/N_{v-1} is then obtained for the first transition to reach laser threshold. A correction for collisional vibrational deactivation is made by extrapolation to zero pressure, in which the limit, $N_v/N_{v-1} = k(v)/k(v-1)$.

Table 5.1 summarises results obtained by methods (a) to (e). The table is subdivided as indicated below. Outstanding features will be listed, briefly, for each category (see also Ref. 20).

Table 5.1 Initial vibrational and rotational excitation as obtained from infrared chemiluminescence experiments. The quantities $\bar{f}_{V'}$, $\bar{f}_{R'}$ and $\bar{f}_{T'}$ are the mean fractions of the total available energy, $E'_{tot.}$, that are channelled into vibration or rotation in a newly formed molecule (e.g. $AB^\dagger$ in $A + BC \rightarrow AB^\dagger + C$) and into translational energy in the products. Where $\bar{f}_{R'}$ is given, the full matrix, $k(v', J')$, for the detailed rate constant is known, i.e. triangular plots of $k(V', R', T')$ or $k(V, R, \varepsilon')$ are available. Data which is marked 'Prelim.' represents preliminary results

Section	Reaction	Method[a]	Vibrational distribution[b]	$E'_{tot.}$ (kcal mol^{-1})	$\bar{f}_{V'}$	$\bar{f}_{R'}$	$\bar{f}_{T'}$	Ref.
5.2.2.1	$A + BC \rightarrow AB^\dagger + C$							
	(a) $H + Cl_2 \rightarrow HCl + Cl$	A.R.	$k_1 = 0.28, k_2 = 1.00,$ $k_3 = 0.92, k_4 = 0.08,$ $k_5 \ll 0.08$	48.6	0.39	0.07	0.54	34, 68, 70, 89
		M.R.	$k_1 = 0.31, k_2 = 0.61,$ $k_3 = 1.00, k_4 = 0.22,$ $k_5 = 0.03, k_6 = 0.003$	48.6	0.42			34, 65
	(b) $D + Cl_2 \rightarrow DCl + Cl$	A.R.	$k_1 \sim 0.10, k_2 = 0.37,$ $k_3 = 1.00, k_4 = 0.88,$ $k_5 = 0.27, k_6 = 0.05$	49.6	0.40	0.10	0.50	70, 89
	(c) $H + Br_2 \rightarrow HBr + Br$	A.R.	$k_1 \ll 0.1, k_2 = 0.19,$ $k_3 = 0.89, k_4 = 1.00,$ $k_5 = 0.23$	43.8	0.56	0.05	0.39	34, 89
	(d) $H + F_2 \rightarrow HF + F$	M.R. (Prelim)	$k_1 = 0.2, k_2 = 0.3,$ $k_3 = 0.5, k_4 = 0.65,$ $k_5 = 1.00, k_6 = 0.95$	106	~ 0.5			(91), 92
	(e) $F + H_2 \rightarrow HF + H$	A.R.	$k_1 = 0.31, k_2 = 1.00,$ $k_3 = 0.48$	34.7	0.67	0.07	0.26	69, 70
		E.G.	$k_1 = 0.18, k_2 = 1.00$	34.7				72, 74
	(f) $F + D_2 \rightarrow DF + D$	A.R.	$(k_1 \approx 0.3)^c, k_2 < 0.7(\approx 0.5)^c$ $k_3 = 1.00, k_4 = 0.72$	34.4	0.69	0.06	0.25	70
		E.G.	$k_2 = 0.63, k_3 = 1.00$	34.7				74
	(g) $F + HCl \rightarrow HF + Cl$	M.R. (Prelim)	$k_1 \sim 0.7, k_2 = 1.00,$ $k_3 \sim 0.15$	~ 35				64
	(h) $Cl + HI \rightarrow HCl + I$	A.R.	$k_1 = 0.18, k_2 = 0.32,$ $k_3 = 1.00, k_4 = 0.74$	33.9	0.71	0.13	0.16	35, 88
	(i) $Cl + DI \rightarrow DCl + I$	A.R.	$k_1 \approx 0.08, k_2 = 0.12,$ $k_3 = 0.33, k_4 = 0.68,$ $k_5 = 1.00, k_6 = 0.05$	34.0	0.71	0.14	0.15	35, 88
	(j) $Cl + HBr \rightarrow HCl + Br$	A.R. (Prelim)	Efficient vib. excitation	~ 17				34, 88
	(k) $Br + HI \rightarrow HBr + I$	A.R. (Prelim)	Efficient vib. excitation	~ 18				34, 88
	(l) $O + CS \rightarrow CO + S$	Flow (Prelim)	Efficient vib. excitation	$\sim 77 \pm 5$	> 0.5			94
5.2.2.2	$A + BCD \text{ (or } BCDE) \rightarrow AB^\dagger + CD^{(\dagger)} \text{ (or } CDE^\dagger)$							
	(a) $H + O_3 \rightarrow OH + O_2$	A.R. (Prelim)	$k_6 < 0.4, k_7 \sim 0.4, k_8 \sim 0.8,$ $k_9 = 1.00$	~ 79	Large			76

Section	Reaction	Method[a]	Vibrational distribution[b]	$E'_{tot.}$ (kcal mol^{-1})	$\bar{f}_{V'}$	$\bar{f}_{R'}$	$\bar{f}_{T'}$	Ref.
	(b) $H + F_2O \rightarrow HF + FO$	Flow (Prelim)	$k_1 = 1.1, k_2 = 1.4,$ $k_3 = 1.00, k_4 = 0.71,$ $k_5 = 0.45, k_6 = 0.22,$ $k_7 = 0.011, k_8 = 0.012$	~ 93	$\geqslant 0.32$			90
	(c) $H + Cl_2O \rightarrow HCl + ClO$	Flow (Prelim)	$k_1 \sim 0.3, k_2 \sim 0.9,$ $k_3 = 1.00, k_4 \sim 0.4, k_5 \sim 0.1$	~ 70	> 0.30			63
	(d) $H + Cl_2S \rightarrow HCl + SCl$	M.R.	$k_1 = 1.0, k_2 = 1.6,$ $k_3 = 1.00, k_4 = 0.4,$ $k_5 \approx 0.1$	$\sim 48^e$	$> 0.35^{f, g}$			62
		A.R.	$k_1 = 0.53, k_2 = 0.72,$ $k_3 = 1.00, k_4 = 0.83,$ $k_5 = 0.25$	$\sim 48^e$	0.43^f	0.19^d	0.38	66
	(e) $H + Cl_2S_2 \rightarrow HCl + S_2Cl$	M.R.	$k_2 = 0.80, k_3 = 1.00,$ $k_4 = 0.92, k_5 = 0.40$	~ 51	> 0.36			62
	(f) $N + NO_2 \rightarrow NO + NO$	Flow	Vib. excitation up to $v' \geqslant 5$	~ 80				85
5.2.2.3	$A + BCD \rightarrow ABC^{\dagger} + D$							
	(a) $N + NO_2 \rightarrow N_2O + O$	Flow	$\sim 0.70 \, (v_1 = 0, v_2 \geqslant 1,$ $v_3 = 1)^h$ $\sim 0.20 \, (v_1 = 1, v_2 \geqslant 2,$ $v_3 = 0)$ $\sim 0.03 \, (v_1 \geqslant 1, v_3 \geqslant 1)$	~ 44	$\sim 0.35^h$			85
5.2.2.4	$AB + CDE \rightarrow ABC^{\dagger} + DE$							
	(b) $NO + O_3 \rightarrow NO_2 + O_2$	Flow	Inefficient vib. excitation	~ 50				86
5.2.2.5	$A + Polyatomic \rightarrow AB^{\dagger} + \dots$							
	(a) $F + CH_4 \rightarrow HF + CH_3$	E.G.	$k_1 \approx 0.18, k_2 = 1.00$	34.5				74
	(b) $F + CHX_3 \rightarrow HF + X_3$ (X = F, Cl)	C.L.	Less efficient than $F + CH_4$	~ 44 (X = Cl)				87
5.2.2.6	$Polyatomic^{\dagger} \rightarrow AB^{\dagger} + \dots$							
	(a) $CH_3CF_3 \rightarrow$ $CH_2CF_2 + HF$	C.L.	$k_0 = 1.00, k_1 = 1.1–1.7$	~ 72				73
		Flow[i]	$k_1 = 1.00, k_2 = 0.4,$ $k_3 = 0.1, k_4 = 0.03$	~ 72	0.13^j			53
	(b) $H_2C = CCl_2 \rightarrow HCl + \dots$ (1,1-dichloroethylene)	E.G.	$k_2 = 1.15, k_3 = 1.00$	~ 120				75
	(c) $cis\text{-}ClHC = CHCl \rightarrow$ $HCl + \dots$ (cis-1,2-dichloroethylene)	E.G.	$k_1 = 1.2, k_2 = 1.00$					75
	(d) $trans\text{-}ClHC = CHCl \rightarrow$ $HCl + \dots$ (trans-1,2-dichloroethylene)	E.G.	$k_1 = 1.4, k_2 = 1.00$					75
	(e) $CH_3NF_2 \rightarrow 2HF^{\dagger} + HCN$		$k_0 = 1.00, k_1 = 0.33–0.37$	$\leqslant 50$				93

a   Flow $\equiv$ Fast flow without measurement of relaxation; M.R. $\equiv$ Measured relaxation; A.R. $\equiv$ Arrested relaxation; C.L. $\equiv$ Chemical laser; E.G. $\equiv$ Equal-gain chemical laser method.

b k_1 means $k(v = 1)$, etc., similarly for $k_2, k_3 \dots$ These values are relative to the value listed as 1.00.

c   Numbers in parenthesis assume continued parallelism of $k(f_v)$ with those for $F + H_2$.

d   This is the mean over a multiply peaked rotational distribution; see text.

e   Heydtmann and Polanyi[66] give experimental evidence that suggests $E'_{tot.} \approx 43$ kcal mol^{-1} may be a better value; 48 kcal mol^{-1} has been used to conform with Johnson, Perona and Setser[62].

f   Using $k_0 = 0.3$ extrapolated from $k_1, k_2 \dots k_5$.

g   The authors gave > 0.37 using $k_0 = 0.0$ and $E'_{tot.} = 48$.

h v_1 and v_3 are stretching vibrations (which take up 0.15 of $E'_{tot.}$); v_2 is a bending vibration (0.20 of $E'_{tot.}$). Note that the authors used $E'_{tot.} = -\Delta H_0^0 = 41.7$ kcal mol^{-1} without allowance for reagent energy; we have based our fractions on $E'_{tot.} = 44$ kcal mol^{-1}.

i   Relaxation (radiational) was corrected for by computer analysis of the build-up and loss of HF† in the 2.5×10^{-3} s viewing time.

j   Taking the mean value of $k_1/k_0 = 0.75$ from the C.L. results.

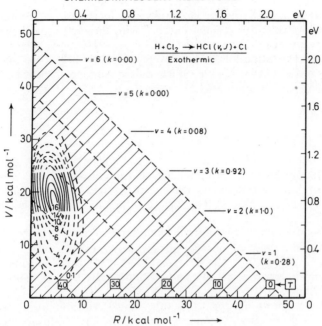

Figure 5.6 Plot of $k(V, R, T)$ for the reaction $H + Cl_2$ (see caption to Figure 5.7). $E'_{tot.} = 48.6$ kcal mol^{-1}. (Based on Anlauf, Ph.D. Thesis, University of Toronto, Toronto, 1968, by courtesy of K. G. Anlauf)

5.2.2.1 A+BC → AB†+C

The efficiency of conversion of the available energy, $E'_{tot.}$, into vibration (symbolised†) tends to be lower for reactions involving hydrogen as the attacking atom; this is thought to be evidence of the light-atom anomaly on a repulsive surface (see Section 5.2.1.4, and cf. Figures 5.5 and 5.6)[24, 35, 78, 79].

The $H + Br_2$ reaction is more efficient in converting $E'_{tot.}$ into vibration than is $H + Cl_2$ [34]; since the energy barrier is lower for $H + Br_2$, this is in accord with the expectation that lower barriers in related families of reaction are 'earlier', and hence correspond to less repulsive energy release (see Section 5.1.1.8).

For $D + Cl_2$, $F + D_2$ and $Cl + DI$ the values of $k(v)$ are markedly different than for the corresponding reactions involving H ($H + Cl_2$, $F + H_2$ and $Cl + HI$). The fractional conversion of $E'_{tot.}$ into vibration is, however, virtually unchanged from the H to the D case. These observations are in accord with the results of classical trajectory calculations[35, 78-81], and lend support to the view that such calculations provide an adequate approximation to the real (quantum mechanical) phenomenon.

The $H + Cl_2$ case (Figure 5.6) shows little increase in rotational excitation in lower v levels; this seems to accord with the light-atom behaviour on a repulsive surface (Figure 5.1 and Ref. 79). $Cl + HI$ shows a large increase in rotational excitation for lower v levels, which is likely to be due at least in part to mixed energy release on a repulsive surface, in bent configurations (Figure 5.1, and Section 5.2.1.5).

5.2.2.2   A+BCD (or BCDE) → AB†+CD† (or CDE†)

The reaction $H + O_3$ is extremely efficient in converting $E'_{tot.}$ into vibration in the new bond and shows no evidence of the light-atom anomaly (characteristic of repulsive surfaces). The surface is therefore likely to be an attractive one (see Section 5.2.1.1).

For one example of infrared chemiluminescence ($H + NOCl \rightarrow HCl + NO$[82, 83]) and three of absorption spectroscopy ($O + NO_2 \rightarrow O_2 + NO$ [84], $O + CS_2 \rightarrow CO + CS$ [45], $O + ClO_2 \rightarrow O_2 + ClO$ [58]) there is evidence that the new bond is much more efficiently vibrationally excited than the old one (see Section 5.2.1.10).

Multiple rotational peaks within single vibrational levels, possibly indicative of migratory secondary encounters, have been observed for $H + Cl_2S \rightarrow HCl + SCl$ [66]. It remains to be established, however, whether these multiple rotational peaks originate in the primary reaction.

5.2.2.3   A+BCD → ABC†+D

This type of four-atom reaction, in which the new molecule is triatomic, is particularly difficult to study due to rapid relaxation. The pioneering results reported in Table 5.1 [85] suggest that a further study using the arrested relaxation approach would be of interest.

5.2.2.4   AB+CDE → ABC†+DE

The several vibrational frequencies of the new molecule raise the same problems as in Section 5.2.2.3. In addition, it is found that a significant fraction of the product ABC is formed in an electronically excited state for $NO + O_3$ [86].

5.2.2.5   A+ Polyatomic → AB†+ . . .

The most interesting question here is what fraction of $E'_{tot.}$ becomes internal excitation in the polyatomic fragment, with its many available degrees of freedom. It has been argued that for $F + CH_4$ little energy flows into CH_3 [74], but the question is not yet settled.

5.2.2.6 Polyatomic $^\dagger \to AB^\dagger + \ldots$

In the case of $CH_3CF_3^\dagger$ dissociation[53] it was clear the 30 kcal 'excess' energy (see Section 5.1.1.12) could not account for the vibrational excitation in the HF^+, but that specific channelling of the additional 42 kcal $(E_c - \Delta H_0^0)$ released in traversing the downhill slope on the energy surface was required even for the rather modest vibrational excitation observed. A predominantly repulsive energy surface was proposed for this case[53]. In the case of the dichloroethylenes a particularly notable feature is the stereospecificity of vibrational excitation in the new bond[75].

5.2.2.7 More complex systems

The reactions included in Table 5.1 were identified fairly clearly as the elementary processes which gave rise to infrared chemiluminescence. Studies been made of infrared chemiluminescence from more complicated systems, for which the elementary step cannot be so easily identified. For example chemiluminescence has been observed from atomic oxygen reaction with ethylene[95], and atomic oxygen plus acetylene, carbon suboxide (C_3O_2) and cyanoacetylene, HC_3N [96-98]. The $O + C_2H_2$ case has been studied particularly well in its qualitative details, with a single-window flow method. At low flows of acetylene it appeared that the reactions $O + C_2H_2 \to CO + CH_2$ $(-\Delta H \approx 48$ kcal mol$^{-1})$ and $O + CH_2 \to CO + 2H$ $(-\Delta H \approx 72$ kcal mol$^{-1})$ predominated, and gave rise to $CO^\dagger$ in $v \leqslant 14$. At high acetylene flows, and also in the presence of added hydrogen, a further secondary reaction appeared likely; $O + CH \to CO + H$ $(-\Delta H \approx 176$ kcal mol$^{-1})$. This $A + BC$ reaction yielded $CO^\dagger$ in a vibrationally excited state with an extraordinarily high quantum number, $v \leqslant 33$ (corresponding to $\leqslant 163$ kcal mol^{-1} of vibrational energy). The CO emission in $v \leqslant 8$ recorded from the reaction atomic oxygen plus carbon suboxide[98] was attributed to $O + C_3O_2 \to 3CO$ $(-\Delta H \approx 115$ kcal mol$^{-1})$. The maximum vibrational excitation, $v = 8$, would be thermodynamically accessible even if the reaction partitioned the product energy equally in vibrational excitation of all three CO molecules.

5.3 ELECTRONIC EXCITATION

We have chosen to review a limited area in order to discuss reactions of special interest in detail. For each of several types of reactions, we will discuss a particular example, and mention only briefly work on other examples of the same type. Because of the great importance for chemiluminescence of potential surfaces and their intersections, we will stress cases where some knowledge of these surfaces is available. This effectively limits discussion to triatomic systems. Electronic chemiluminescence is closely related to other kinetic processes, such as reactions or quenching of electronically excited states, and photodissociation. For a given triatomic system we will include data from these sources when they can be combined with observations of chemiluminescence to give more information about potential surfaces involved. Most of the work cited has appeared after earlier reviews[55, 99, 164, 211], which serve as a starting point.

5.3.1 Potential surfaces and their intersections

The concept of potential energy surfaces, based on the Born–Oppenheimer separation of electronic and nuclear motion, is the fundamental approximation in most treatments of the vibrations of polyatomic molecules[100] and slow collisions of heavy particles[101]. Many processes producing electronic chemiluminescence involve an interaction or intersection of two potential surfaces[102]. In this Section we discuss briefly the concept and definition of a potential surface, the geometry of surface intersections and the question of transitions from one surface to another.

5.3.1.1 Concept and definition of a single surface

A potential surface can be defined theoretically or experimentally.

(a) *Theoretical definitions of a potential surface*[103, 104, 113, 119] — The Hamiltonian operator, H, for a molecule may be expressed as a sum of two parts: $T(R)$, involving derivatives with respect to coordinates specifying the configuration of the nuclei, and $H_{el.}(r, R)$, the electronic Hamiltonian in which R appears only as a parameter, and r refers to electron coordinates. We can express H as a matrix in any electronic basis, and these basis states will in general depend on R parametically. The diagonal element H_{ii} in this basis gives the energy of the ith electronic basis state. The energy H_{ii} depends on the nuclear coordinates R and has the significance of a potential energy function. In this general sense there is a set of potential functions corresponding to every choice of basis set. The adiabatic basis, which diagonalises $H_{el.}$, gives the exact adiabatic potential functions. These are the functions sought after in most calculations[105, 106]. The adiabatic basis is conceptually simple and uniquely specified, but it is by no means the only basis which is useful. In many cases a diabatic basis is physically more reasonable and computationally more convenient. Diabatic potential curves pass smoothly through an intersection, whereas adiabatic curves avoid the intersection by a change of slope which may be quite abrupt when the interaction is weak.

(b) *Experimental 'definition' of a potential surface* — From this point of view, the potential surface is simply that function of R, which when used in the appropriate theoretical formulation, predicts the observed results of one (or preferably many) experiments. These experiments may be elastic scattering[107], inelastic scattering[108], reactive scattering, photodissociation[109, 110], predissociation[111], and of course infrared and electronic spectroscopy of bound states[100, 102].

5.3.1.2 *Intersections and interactions of potential surfaces*

In a reaction of the type:

$$A + BC \rightarrow AB^* + C \qquad (5.1)$$

(where the asterisk indicates electronic excitation) reactants and products can correlate with a single adiabatic state of ABC only if A and C are chemically different. In this case the reaction can occur on a single adiabatic surface and intersections with another surface are likely to be unimportant since they occur in a space of fewer than the full number of dimensions specifying the configuration of the nuclei. In other words, if a reaction can occur on a single surface, it is likely to do so. There are however exceptions to this rule[210].

On the other hand, the reaction:

$$A + BA \rightarrow AB^* + A \qquad (5.2)$$

cannot occur on a single adiabatic surface. The reactants approach on a surface correlating with A + AB, but they must separate on a different surface correlating with A + AB*. A transition between adiabatic surfaces must occur at or near an intersection. The rate of processes like (5.2) therefore depends critically on the geometry of these intersections. This is true whether or not the over-all process in (5.2) occurs by atom exchange.

Three types of intersection between exact adiabatic surfaces can be distinguished[102]. An orbitally degenerate state (e.g. Π of $C_{\infty v}$ or E of C_{3v}) will split into two states of different symmetry species when the molecule is distorted to a lower symmetry. These states therefore intersect in the symmetrical conformation. The ground and first excited states of H_3 have a conical intersection of this type at equilateral conformations[112, 113]. Another example is the glancing intersection of A′ and A″ states in the linear conformation in which they form the two components of a Π-state.

A second type of intersection occurs when states of different species, like A_1 and B_2 of C_{2v}, intersect 'accidentally', but repel each other in neighbouring C_s conformations, since they both become A′ in that point group. In this type of intersection, the two surfaces have the same species, e.g. NO_2 and HNO (discussed later) and $NaCl_2$ [209]. This type of intersection in a linear molecule is discussed by Herzberg[114].

A third type of intersection occurs for example between the two lowest states of H_3^+. The ground surface, in regions corresponding to H_2 with H^+ at infinity, intersects the excited surface for H_2^+ with H at infinity. This intersection is avoided at finite distances, but the surfaces approach sufficiently closely to permit transitions from one to the other, producing charge exchange[115-117, 196]. An entirely analogous chemiluminescence reaction is[118]

$$H(2^2P) + H_2(X^1\Sigma_g^+) \rightarrow H(1^2S) + H_2(B^1\Sigma_u^+) \qquad (5.3)$$

in which the excited species play the roles of the ions just discussed. In (5.3) the reactant $H_2(X)$ must be vibrationally excited[118].

5.3.1.3 Radiationless transitions between electronic states

Much of the confusion surrounding this subject is due to incomplete understanding of the several basis sets which may be used, and of their physical
significance. The Hamiltonian operator may be expressed as a matrix in
any electronic basis set, which may be chosen with the hope that physical
appropriateness will lead to computational convenience. The magnitude
and physical significance of the off-diagonal elements of H, which induce
transitions between basis states, may be quite different in different bases.
In the adiabatic basis, transitions are induced by terms derived from the
momentum of the nuclei. Alternatively, one can define a diabatic basis in
which transitions between basis states are induced by parts of the electronic
Hamiltonian not diagonalised in the basis sets, and effects of nuclear motion
do not appear in the off-diagonal terms. This matter has been clarified in
several recent papers[103, 104, 113, 119, 120, 213, 214], devoted largely to diatomic molecules. They will be useful also for understanding the behaviour of polyatomic
molecules[113]. The coupling terms connecting adiabatic states have been
derived for charge exchange in the H_3^+ case[115]. These results have been used
in a 'surface hopping' model[116, 117] where a classical trajectory on one surface
can jump to another to complete the charge exchange reaction. Though this
particular case is not chemiluminescence, the method is equally applicable
to cases where one of the products is electronically excited.

5.3.2 Radiative recombination of atoms

Several types of recombination are possible, principally distinguished by
the presence or absence of a third body, and by the interaction, or lack of it,
of a second electronic state in the recombination step.

5.3.2.1 Two-body recombination on a single curve

The only well documented case is [121, 212]

$$He(2^1S) + He \rightarrow He_2(A^1\Sigma_u^+) \rightarrow 2He + h\nu \ (600 \ \text{Å bands}) \qquad (5.4)$$

where emission occurs from continuum levels above the A state potential
well. The reverse of this process has also been discussed[122].

5.3.2.2 Two-body recombination with diabatic curve crossing

Here the transition is from a continuum translational state of one potential
to a bound vibrational state of another, which can radiate to a lower bound
or repulsive state. It is often more useful to consider this process as resonance
scattering, described by the Breit–Wigner cross-section. The compound
state can decay via predissociation or spontaneous radiation, and the width
of the resonance is the sum of these widths.

One of the best studied cases of this type is[123-126]

$$N(^4S) + O(^3P) \leftrightarrows NO(a^4\Pi) \leftrightarrows NO(C^2\Pi) \rightarrow NO(X^2\Sigma) + h\nu(\delta \text{ bands}) \quad (5.5)$$

This is a particularly favourable example for several reasons. The perturbing state, $a^4\Pi$, is fairly well known, the radiative transition $C \rightarrow X$ is strongly allowed, and the lifetime of the predissociating levels of the C-state is sufficiently short so that third body effects are minor even up to pressures of a few torr. The resonances (rotational levels in $v = 0$ of the C-state) are sufficiently narrow that there is no overlapping.

Recent work on reaction (5.5) has concerned the position of the dissociation limit relative to the lowest rotational level in the C-state. Studies of the reverse of reaction (5.5), the pre-dissociation observed in fluorescence studies[123, 124], indicate that several of the lowest levels of the C-state lie below the ground state dissociation limit, i.e. they have negative energy with respect to that natural zero. In the complete absence of collision effects, molecules excited to these levels cannot dissociate, and in the two-body radiative recombination these levels are not populated. In the fluorescence experiments, collisions can excite C-state molecules from negative to positive levels, which can dissociate. This quenching of the fluorescence has been used to estimate the number of negative energy levels[123]. In the radiative recombination, collisions can populate the negative energy levels, so that lines from these are no longer missing in the emission spectrum. This may be the explanation for the fact that spectra at very high resolution[127] show full intensity in the lowest lines.

Other reactions of the type discussed in Section 5.3.2.2 are the radiative recombination (inverse predissociation) of $O + H$ into the $A^2\Sigma^+$ state, for which there is a recent theoretical calculation[128], and $S + S$ into $B^3\Sigma_u^- v'$ $\geqslant 10$ [129-131]. Recombination of ground state oxygen atoms into the $B(^3\Sigma_u^-)$ state, emitting the Schumann–Runge bands, may also be a two-body process[197, 198].

5.3.2.3 Three-body recombination on a single potential curve

The general reaction:

$$X + Y + M \rightarrow XY^* + M \quad (5.6)$$

can occur wherever there is a bound electronic excited state XY^* which correlates with ground state $X + Y$. The recombination is basically like the 'ordinary' kind, such as $H + H + M \rightarrow H_2 + M$ which involves no electronic excitation. However, if light emission is to be observed, the state XY^* must have a more or less allowed transition to the ground state, and must emit before it is electronically quenched by M, X or Y. These quenching processes can produce a dependence of intensity on concentration (equation (5.7))

$$I = k_7[X]^l[Y]^m[M]^n \quad (5.7)$$

which is quite different from the simple form $l = m = n = 1$ expected for simple recombination. In addition to electronic quenching, XY^* may be vibrationally relaxed before it emits, thus shifting the emission to a different wavelength range.

(a) *Recombination into the* $A^3\Sigma_u^+$ *state of* O_2 — Emission of the Herzberg bands in gas flowing from a discharge through O_2 is due to a recombination of type (5.6)[132],

$$O(^3P) + O(^3P) + O_2 \rightarrow O_2(A^3\Sigma_u^+) + O_2 \qquad (5.8)$$

O_2 is an efficient quencher, so that in the rate expression (5.7), $n = 0$ when $1 \leqslant P_{O_2} \leqslant 10$ torr. The resulting second-order rate coefficient, k_7, is a composite of the radiative transition probability per second (A), quenching rate coefficient (k_Q) and (k_8). For recombination emitting the Herzberg bands[132], $k_7 \doteq k_8 A / k_Q = 2.5 \times 10^{-21} \text{ cm}^3 \text{ s}^{-1}$.

Another example of type 5.3.2.3 is the recombination of nitrogen atoms into the $A(^3\Sigma_u^+)$ state, included in the next section. This appears to make a major contribution to the total atom recombination rate[133].

5.3.2.4   Three-body recombination with curve crossing

This most complicated case can occur in two ways. The overall process is again $X + Y + M \rightarrow XY^* + M$, followed by radiation from XY^*. The important difference with respect to the case of Section 5.3.2.3 is that here the excited state XY^* does not correlate with the ground state atoms $X + Y$. As $X + Y$ collide on potential curve A correlating with those atoms in their ground states, the third body M may induce a transition to a bound vibrational level of the electronically excited state B correlating with $X + Y^*$. If the state A has bound vibrational levels, the same result can be achieved in two steps. The colliding $X + Y$ are first stabilised by M into a bound level of A, which then undergoes a second collision inducing the transition $A \rightarrow B$. In either case, a major problem is to establish the identity of the precursor state A, since in general there will be several states correlating with ground state atoms, and some of them may not be known spectroscopically. Furthermore, the third body M and the atoms themselves may cause electronic or vibrational deactivation of A or B. These processes will determine the way the light yield of the recombination depends on [X], [Y] and [M], and how it is distributed over wavelength.

(a) *Recombination of nitrogen atoms* — All the intricacies just mentioned are abundantly present in this case[134, 135]. All we can do here is to present some of the basic problems which have been emphasised recently and to restrict attention to processes leading to emission from $N_2(B^3\Pi_g)$. A review of the literature produces a certain humility, 'One cannot be dogmatic about the full details of this reaction mechanism ...'[136].

The first problem is the identity of the precursor state. The candidates are $A^3\Sigma_u^+$, which is well known spectroscopically, and the weakly bound $^5\Sigma_g^+$, known through its predissociation of the $B(^3\Pi_g)$ and $a(^1\Pi_g)$ states. This predissociation is the strongest argument for the involvement of the $^5\Sigma$ state and implies that recombination into levels of the B-state above the $N + N$ energy limit ($v' = 12$, $N > 32$) does not require a third body. Lower levels of the B-state may be populated by vibrational relaxation of those initially excited levels[137]. The alternative route to B, $v < 12$ in the $^5\Sigma$-mechanism, requires a collision-induced transition to B which simultaneously changes the vibrational quantum by several units[138].

The after-glow emission is strongly quenched by N_2 [136, 138, 141] at pressure above a few tenths of a torr unit[138], corresponding to a quenching cross-section* for the B-state of $Q \approx 10$ Å^2. Independent quenching measurements on the B-state produced by photodissociation of N_2O in the vacuum ultra-violet region gave a cross-section of the same order of magnitude[140]. This rapid quenching means that the total rate of recombination into excited states is considerably greater than the observed emission rate[133], and may be too large to be accounted for by recombination into $^5\Sigma$ alone[139]. The $A^3\Sigma_u^+$ state provides another recombination path, and a natural way to populate levels near $v = 6$ in the B-state. Lower levels of the B-state may be populated through the mediation of the $B'^3\Sigma_u^-$ or $^3\Delta_u$ states, though these do not correlate with ground state atoms and hence must have their own precursors.

Gases such as Ar or He which quench the afterglow more weakly than N_2 lead to emission with a relaxed vibrational distribution, probably due to relaxation in the precursor state[141]. In addition to these collision effects, there is the possibility of quenching by N atoms themselves. The proportionality of intensity to $[N]^2$ suggests that any such effect is small. Nitrogen is known to be an efficient quencher of low vibrational levels of the A-state[142], and this has been used as an argument against the A-state as a precursor[137]. It is possible however, that higher vibrational levels are quenched less efficiently[136]. Some evidence has recently been obtained for quenching of the B-state by N [141], and the conversion of B to the $a'^1\Sigma_u^-$ state in a collision with N has been observed, leading to vacuum ultraviolet emission[195].

Like most recombination reactions, the afterglow has a small negative temperature dependence[205].

It may seem that the nitrogen afterglow is interesting simply because it has interested so many people for so many years. The fact that it involves, in unusually vivid form, the complexities that must influence most diatomic chemiluminescence systems is a better reason for its notoriety. Its many admirers have the obligation to show that understanding of the nitrogen afterglow can lead to a deeper understanding of recombination chemiluminescence in general.

(b) *Recombination in other systems* — The recombination of halogen atoms shows most of the same features as the nitrogen atom recombination. In Cl_2, most of the observed emission is from the $0_u^+ (^3\Pi)$ state[143-145], which correlates with ground state $Cl(^2P_{\frac{3}{2}}) +$ excited $Cl(^2P_{\frac{1}{2}})$. The dependence of the emission intensity on pressure and composition indicates that the 0_u^+ state (populated up to $v' = 14$) is vibrationally relaxed by Cl_2, and strongly quenched electronically by Cl [143, 144]. As an added complication, the efficiency of this quenching by Cl appears to depend on the Cl_2 concentration, perhaps because different vibrational levels in the 0_u^+ state are quenched at different rates.

---

*The thermal average cross-section is $Q = k/\bar{v}$, where k is in units of cm^3 s^{-1}, $\bar{v} = (8 \, kT/\pi\mu)^{\frac{1}{2}}$ is the average relative speed and μ is the reduced mass.

The radiative recombination of Cl atoms has also been observed in shock tubes[146], in equilibrium at temperatures near 1650 K. At 5000 Å the 0_u^+ ($^3\Pi$) and 1_u($^1\Pi$) states contribute approximately equally to the observed emission. Below 4800 Å the 1_u($^1\Pi$) state predominates.

Because of the efficient and rather complex quenching phenomena, there is no reliable estimate of absolute radiative recombination rates in the halogens. The radiative path probably contributes at least a few per cent of the total recombination rate.

The recombination $N+O$ to give $NO(B^2\Pi)$ and β-band emission is another process of type 5.2.2.4, which show a complex dependence on [N] and [O] [199].

5.3.3   Recombination of an atom with a diatomic molecule

In contrast to the systems treated in Section 5.2.2, the recombination of an atom with a diatomic molecule must be treated as a triatomic system, and discussed in terms of one, or several interacting potential surfaces in a four-dimensional space. The general reaction is:

$$X + YZ \rightarrow XYZ^* \rightarrow XYZ + h\nu. \qquad (5.9)$$

As written, it must be supplemented immediately with two questions: does the state from which emission is observed correlate with the ground state reactants and is a third body M involved? The path from observation of chemiluminescence to an understanding of the interactions of the potential surfaces involved is sufficiently ill-conditioned to maintain controversies on these matters for years. Progress is most rapid when the chemiluminescence results can be supplemented with other information about the same surfaces derived from theoretical calculations, absorption and emission spectra, photodissociation, fluorescence and the study of other reactions occurring on the same surfaces. There are a few triatomic systems of kinetic interest for which this approach shows some promise, for example NO_2, HNO, O_3. Probably the most famous reaction of type (5.9) is the recombination of O with NO.

5.3.3.1   Recombination of O with NO

We will take this reaction to be the archetype of its class and examine it in some detail, trying to bring together as much theoretical and experimental information as can be brought to bear on the central problem which is the interaction of O and NO on the potential surfaces of NO_2 and the energy transfer processes which accompany this interaction.

(a) *Electronic states, potential surfaces and spectrum of* NO_2 – There have been a number of recent calculations[147-152]. There are five states lying within a few electron volts of the ground state as indicated in Figure 5.7. The linear $^2B_1$-state is probably responsible for the long wavelength part of the visible absorption, while the $^2B_2$-state is the upper state for the absorption in the region 3500–6200 Å. In addition to these allowed transitions there may be weak absorption in the forbidden transition leading to the $^4A_2$-state. The notorious complexity of the visible spectrum is a result of interactions among these states and with the ground state.

The vertical excitation energies are of primary importance in understanding the absorption spectrum. For the chemiluminescence, however, much more

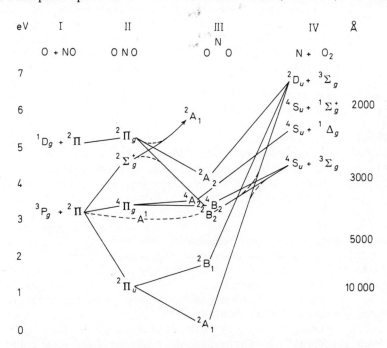

Figure 5.7 Correlation diagram for NO_2, taken in part from Ref. 152. I is the $O + NO$ limit. II is linear ONO with bond lengths equal to their equilibrium value in the ground state. III is the equilibrium conformation of the ground state. IV is the $N + O_2$ limit. Solid lines between I and II refer to linear approach of O to NO. Between II, III and IV, they represent correlations in C_{2v} symmetry. Dotted lines are correlations in C_s symmetry and have only qualitative meaning. The energy of the linear states is also rather uncertain

extensive regions of the potential surfaces are important, particularly the correlation with fragments $O + NO$ (see Figure 5.7). The predissociation of NO_2, which occurs below 3980 Å in the absorption spectrum may be thought of as due to an interaction between the $^2B_2$-state and a linear excited $^2A_1$-state which correlates with ground states of $O + NO$. This view of the predissociation is strengthened by studies of the photodissociation recoil velocity and angular distributions[153, 154]. The angular distribution indicates that the transition moment lies in the plane of the molecule parallel to the line joining the oxygen atoms, hence that the upper state has $^2B_2$-symmetry. Additional evidence that this is the upper state has recently been obtained in an analysis of the fluorescence spectrum[155].

(b) *Lifetime, quenching and energy transfer in the fluorescence of NO_2 –*

Fluorescence experiments provide another source of information about excited states of NO_2. The radiative lifetime of the visible fluorescence has been measured using pulse techniques[156, 157], and the phase-shift method[158, 159], and the range of values is 40–90 μs. The upper state, nominally $^2B_2$, is strongly mixed with the $^4A_2$-state by spin–orbit coupling, and with high vibrational levels of the ground $^2A_1$-state by vibronic coupling. Note that the $^2B_1$-state cannot mix vibronically with ground state since these states have different species in the C_s-point group. This mixing with states not optically connected to the ground state probably explains the fact that the radiative lifetime is considerably longer than the value calculated from the integrated absorption coefficient[159]. Recently much shorter lifetimes, ≈ 1 μs, have been measured[160] in a region of the spectrum in which the relatively unperturbed $^2B_1$-state[161] is excited. The concept of a single electronic lifetime is of limited value even for diatomic molecules and its use is even more questionable for polyatomics.

Electronic quenching and vibrational energy transfer are observed in NO_2 fluorescence and the results should be useful in interpreting the chemiluminescence. Three recent fluorescence studies agree that collision induced vibrational energy transfer in the emitting state is fast. Rate coefficients have been reported giving cross sections of the order of 50 Å², derived from step-ladder models[158, 159] with a step size of roughly 1000 cm^{-1}. In experiments with laser excitation[162], with bandwidth only slightly larger than the Doppler width, fluorescence from the initially excited level can be distinguished from emission from other levels populated by collisions. The cross-section for this vibrational energy transfer is approximately 40 Å². Electronic quenching of the total fluorescence corresponds to a cross-section of c. 3 Å². Just what is meant by vibrational energy transfer in electronically excited NO_2 is somewhat unclear because of the strong spontaneous vibronic interaction which apparently exists in the absence of collisions. The distinction between vibrational and electronic energy transfer may not be entirely meaningful[159].

(c) *Thermal emission* – When flowing NO_2 is heated in a furnace to temperatures in the range 970–1340 K, thermal emission is observed[163] at wavelengths longer than 3768 Å. It appears to be a continuum with a weak diffuse band structure superimposed. This structure is quite similar to that observed in the chemiluminescence emission and in the absorption spectrum. One the basis of rather crude models, radiative lifetimes and upper state energies have been derived as a function of wavelength of the thermal emission. This is the triatomic analogue of experiments on thermal emission of Cl_2 mentioned earlier[146].

(d) *Chemiluminescence* – The preceding discussion attempts to collect information which can be applied to understanding the chemiluminescence reaction:

$$O + NO + M \rightarrow NO_2^* + M \qquad (5.10)$$

We give here a very brief summary of the major features established by many workers and reviewed elsewhere[55, 125, 164, 200]. More recent results are then discussed.

The emission spectrum begins at wavelengths slightly below the dissociation limit, 3980 Å, and extends to at least 1.4 μm, with a maximum near 6200 Å. There is some diffuse band structure, especially near the short wavelength limit, which correlates rather well with structure in the observed absorption spectrum. For the most part, the emission appears to be continuous, although this is probably due to incomplete resolution of a very dense discrete structure. The intensity is proportional to the concentration product $[N][O]$ under all conditions. The over-all intensity and its distribution over wavelength depend only slightly on the identity of the carrier gas M, and are independent of pressure of a given carrier in the torr range. In this second-order region the rate coefficient for light emission is $k_{em.} = 6.5 \times 10^{-17}$ cm^3 s^{-1}, corresponding to an efficiency of $c.$ 10^{-6} per collision.

Recent work on the chemiluminescence has centred on the dependence of the intensity and the spectral distribution on M and the rationalisation of the results with the fluorescence experiments. A careful analysis of the relationship between chemiluminescence and fluorescence is clearly the most fruitful way to proceed. In both cases the $^2B_2$ and $^2B_1$ states are involved, but the $^2B_1$ state is probably more important in chemiluminescence than in fluorescence excited in the blue or violet[152]. Actually, the upper 'state' is a dense system of vibronic levels derived from the several low-lying excited electronic states, as well as the ground state. The extreme complexity of the observed spectrum, the dramatically large isotope effect[161] and the anomalously long radiative lifetime are all consequences of this vibronic interaction. Internal conversion and intersystem crossing in molecules of this type have been the object of much recent theoretical work[165]. Clearly the preparation of the excited state is somewhat different in fluorescence, as compared to chemiluminescence, and the quantum mechanical aspects of state preparation should be borne in mind[165]. It is unlikely, however, that these quantum mechanical differences play a significant role in experiments to date and we will not consider them further.

In a typical fluorescence experiment, all the excited molecules start out with a reasonably well-defined energy depending on the exciting wavelength and bandwidth. In chemiluminescence, on the other hand, a third body is involved in the preparation of the initial state and it may leave the excited NO_2 molecule in an excited state well below the dissociation limit. This initial energy spread produced by the third body corresponds only roughly to a broad exciting bandwidth in the fluorescence experiment. Recent work has done much to show that the chemiluminescence and fluorescence results are closely related and consistent.

The dependence of the chemiluminescence on inert gas pressure $[M]$ has been a matter of controversy until recently. The fluorescence experiments show strong quenching at pressures above 0.1 torr. The chemiluminescence intensity is independent of $[M]$ in this pressure range. These two facts are consistent if the recombination requires a third body. The intensity should then be proportional to M at sufficiently low pressure. For M = Ar the half-quenching pressure is $(k_Q\tau)^{-1} \approx 30$ m torr and recent work[166-168], has demonstrated that the chemiluminescence shows a transition to third-order pressure-dependence in this region. For other third bodies the transition between second- and third-order chemiluminescence occurs at pressures roughly consistent with fluorescence-quenching half pressures.

Two-body radiative recombination is possible in an $O + NO$ collision, since it can occur as the reverse of the observed predissociation[161]. At sufficiently low pressures, third-body processes will be negligible and two-body recombination must predominate. It has recently been observed at pressures below a few m torr[167]. Two-body recombination should lead to emission corresponding to an upper state energy up to a few kT above the dissociation limit. Emission at wavelengths down to 3833 Å with an energy 1300 cm^{-1} above the predissociation limit at 3979 Å, has been observed, in one investigation[163], but not otherwise[168]. The spectrum of the $O + NO$ chemiluminescence shifts slightly to the blue at low pressures, consistent with the fact that fluorescence emission at shorter wavelengths is quenched more strongly than that further to the red[158]. This effect of O_2 and Ar on the intensity distribution in the chemiluminescence has been used to estimate quenching constants in fair agreement with fluorescence values[168].

Since the chemiluminescence was last reviewed[55, 125, 164], new work has confirmed the importance of NO dimers and clusters when the chemiluminescence is observed in nozzle expansions[169, 170]. Absolute intensity measurements imply that nearly every collision of O with an NO cluster produces NO_2^* which can emit light[170]. This is consistent with the estimate that nearly every 3-body collision in the ordinary chemiluminescence produces an excited molecule.

In Figure 5.7, it should be noted the potential surfaces of NO_2 are relevant to a number of different processes, namely quenching of $O(^1D)$ by NO [191], quenching of $O_2(^1\Delta)$ by N [192, 193], the excitation processes $N + O + O \rightarrow NO + O(^1S)$ [204] and the reaction of ground states $N + O_2 \rightarrow NO + O$ [193], in addition to chemiluminescence and predissociation.

After a survey of the $O + NO$ chemiluminescence, one realizes that the problem has become more complex with every investigation of it. As each new level of depth or detail is reached, finer difficulties arise from the resolution of coarser ones. The reason for continuing to attack this particular reaction (and for devoting so much attention to it in this review) is that it serves as a model for illustrating concepts and developing techniques which are important in many other chemiluminescent processes. Some of these will be mentioned briefly now.

5.3.3.2 *Radiative recombination in other systems*

There are several other triatomic systems for which radiative recombination can be discussed along the lines used for $O + NO$. The requirements are information about energies and correlations of excited states, an interpretation of the absorption spectrum and photodissociation, measurements of the radiative lifetime and quenching, and, of course, careful measurements of the chemiluminescence itself over a wide range of wavelength and pressure.

(a) *Recombination of* $O + SO$ — The situation here is quite similar to that for $O + NO$, but the spectrum is better known[102, 201]. Radiative-lifetime[55, 171, 172] and quenching measurements[173, 174] are available and perturbations in excited states are again important[175, 207]. The dependence on third body is probably similar to that for $O + NO$ and there is a small negative temperature coefficient[176, 177]. There has been a suggestion that cascade emission of two photons may occur in the chemiluminescence[178]. In a flow system at pressures down to 10 mTorr, radiative recombination at the wall has been observed[206].

(b) *Recombination* $H + NO$ — For this system, the addition to work already reviewed[55, 125, 164, 179] concerns potential surfaces and their intersections. Emission is observed from the $^1A''$ state, and shows a predissociation at the $H(^2S) + NO(^2\Pi)$ energy limit. This predissociation is also seen in absorption. The fact that it is weak indicates that direct recombination into the $^1A''$ state may be improbable. Recent calculations[180, 181] show that the $^3A''$ state is unlikely to serve as a precursor, since it is repulsive at large distances.

(c) *Recombination of* $O + CO$ — There has been little work on this reaction since it was last reviewed[55, 164, 182]. The crucial feature here is that the ground state, $CO_2(^1\Sigma_g^+)$, does not correlate with $O(^3P) + CO(^1\Sigma^+)$. Recombination must begin on the excited $^3B_2$ surface, whether or not radiation is ultimately produced[202]. In the discussion of the recombination the radiative and non-radiative processes are closely intertwined[183, 184]. This case is a particularly good illustration of the importance of considering these processes together.

(d) *Recombination of* $Ba + Cl_2$ — Here is a well established two-body radiative recombination studied in a crossed-beam apparatus[185]. The major emitter is $BaCl_2^*$, though some $BaCl^*$ is formed also. However the most probable reaction path gives $BaCl^\dagger$. Little is known about the excited states of $BaCl_2$, but the reaction presumably proceeds by an electron jump mechanism.

5.3.4   Atom transfer reactions

The general type is equation (5.10) in which either product may be excited.

$$A + BC \rightarrow \begin{cases} AB^* + C & (5.10.1) \\ AB + C^* & (5.10.2) \end{cases}$$

If A and C are the same chemical species, both reactions must go through a C_{2v} conformation, and neither can occur adiabatically. There is in this case an alternate route to the same products, involving an inelastic collision which does not involve atom transfer. This process cannot proceed adiabatically either. If A, B, and C are all different, these reactions may or may not occur adiabatically, depending on the energies and symmetry species involved.

5.3.4.1   *Excitation of the diatomic product*

There has been little recent work on reactions of type (5.10.1) in which the reactants are in their ground states. One example is[185] equation (5.11)

$$Ba + Cl_2 \rightarrow BaCl^* + Cl \qquad (5.11)$$

although the principal product is $BaCl^\dagger$, and some $BaCl_2^*$ is formed, as mentioned previously. The reaction

$$C + CCl \rightarrow C_2^* + Cl \tag{5.12}$$

has been proposed[186] as the origin of some of the many band systems of C_2 observed in the reaction $K + CCl_4$ in diffusion flames.

There has been recent work on reactions of type (5.10.1) in which one of the reactants is electronically excited[187], for example

$$O(^1D) + O_2(^3\Sigma_g^-) \rightarrow O(^3P) + O_2(^1\Sigma_g^+) \tag{5.13}$$

Here it must be admitted however that it is not known whether the reaction involves atom exchange. Whichever mechanism is involved, it accounts for most of the rapid quenching of $O(^1D)$ by O_2 [188–190]. A similar reaction in the nitrogen afterglow is

$$N(^4S_u) + N_2(B^3\Pi_g) \rightleftarrows N_2(a'^1\Sigma_u^-) + N(^4S_u) \tag{5.14}$$

which appears to be sufficiently fast that the populations of the a′ and B states are comparable[195]. It is not clear what states of N_3 are involved in this reaction, but it cannot occur on a single adiabatic surface.

5.3.4.2   *Excitation of the atomic product*

Reactions of the type (5.10.2) are quite rare. An example is

$$Cl(^2P_u) + Na_2(^1\Sigma^+) \rightarrow NaCl(^1\Sigma^+) + Na(^2P_u) \tag{5.15}$$

which has been studied recently in a crossed beam experiment[208]. The cross-section is in the range 10–$100\,Å^2$ for (5.15) and also for the corresponding reaction with K_2. The reactants correlate with three states of the triatomic intermediate, $2\,^2A' + {}^2A''$. One of the $^2A'$ states correlates adiabatically with ground state products, but the other two correlate with $NaCl + Na^*$. Configuration interaction with an ionic state is important in lowering the energy of the surface on which the reaction proceeds. In other words, this is another electron jump mechanism.

Acknowledgements

We thank L. A. Burnell, M. Krauss, R. K. Preston and W. R. Thorson for manuscripts prior to publication, the Chemical Kinetics Information Center, U.S. National Bureau of Standards, for a long list of pertinent references, J. L. Schreiber for help in the literature search; and the National Research Council of Canada and NASA (grant NGR 52-026-028) for financial support.

References

1. Levine, R. D. (1969). *Quantum Mechanics of Molecular Rate Processes* (Oxford: Clarendon Press)
2. Mortensen, E. M. and Pitzer, K. S. (1962). *Chem. Soc., Specl. Publ.*, **16**, 57
3. Mortensen, E. M. (1968). *J. Chem. Phys.*, **48**, 4029

4. Karplus, M. and Tang, K. T. (1967). *Discus. Faraday Soc.*, **44**, 56
5. Tang, K. T., Kleinman, B. and Karplus, M. (1969). *J. Chem. Phys.*, **50**, 1119
6. Rankin, C. C. and Light, J. C. (1969). *J. Chem. Phys.*, **51**, 1701
7. Diestler, D. J. and McKoy, V. (1968). *J. Chem. Phys.*, **48**, 2951
8. Diestler, D. J. (1969). *J. Chem. Phys.*, **50**, 4746
9. Truhlar, D. G. and Kuppermann, A. (1970). *J. Chem. Phys.*, **52**, 3841
10. Gelb, A. and Suplinskas, R. J. (1970). *J. Chem. Phys.*, **53**, 2249
11. Miller, G. and Light, J. C. (1971). *J. Chem. Phys.*, **54**, 1635, 1643
12. McCullough, E. A., Jr. and Wyatt, R. E. (1969). *J. Chem. Phys.*, **51**, 1253; McCullough, E. A., Jr. and Wyatt, R. E. (1971). *J. Chem. Phys.*, **54**, 3578, 3592
13. Miller, W. H. (1970). *J. Chem. Phys.*, **53**, 1949, 3578
14. Wall, F. T., Hiller, L. A., Jr. and Mazur, J. (1958). *J. Chem. Phys.*, **29**, 255; (1961). ibid., **35**, 1284
15. Bunker, D. L. (1966). *Gas Kinetics* (Oxford: Pergamon Press)
16. Laidler, K. J. (1969). *Theories of Chemical Reaction Rates,* 153 (New York: McGraw-Hill Book Co.)
17. Bunker, D. L. and Pattengill, M. D. (1970). *J. Chem. Phys.*, **53**, 3041
18. Csizmadia, I. G., Kari, R. E., Polanyi, J. C., Roach, A. C. and Robb, M. A. (1970). *J. Chem. Phys.*, **52**, 6205
19. Csizmadia, I. G., Polanyi, J. C., Roach, A. C. and Wong, W. H. (1969). *Can. J. Chem.*, **47**, 4097
20. Polanyi, J. C. (1972). *Accounts Chem. Res.*, **5**, April
21. Evans, M. G. and Polanyi, M. (1939). *Trans. Faraday Soc.*, **35**, 178
22. (a) Blais, N. C. and Bunker, D. L. (1962). *J. Chem. Phys.*, **37**, 2713; (b) same authors (1963). ibid., **39**, 315
23. Bunker, D. L. and Blais, N. C. (1964). *J. Chem. Phys.*, **41**, 2377
24. Polanyi, J. C. (1962). *Transfert d'Energie dans les Gaz,* 177, 526 (New York: Interscience); Polanyi, J. C. and Rosner, S. D. (1963). *J. Chem. Phys.*, **38**, 1028
25. Polanyi, J. C. (1963). *J. Quant. Spectrosc. Radiat. Transfer,* **3**, 471
26. Polanyi, J. C. (1965). *J. Appl. Optics., Suppl.,* **2**, 109
27. Kuntz, P. J., Nemeth, E. M., Polanyi, J. C., Rosner, S. D. and Young, C. E. (1966). *J. Chem. Phys.*, **44**, 1168
28. Karplus, M. and Raff, L. M. (1964). *J. Chem. Phys.*, **41**, 1267
29. Raff, L. M. and Karplus, M. (1966). *J. Chem. Phys.*, **44**, 1212
30. Bunker, D. L. and Parr, C. A. (1970). *J. Chem. Phys.*, **52**, 5700
31. Kuntz, P. J., Nemeth, E. M. and Polanyi, J. C. (1969). *J. Chem. Phys.*, **50**, 4607
32. Kuntz, P. J., Mok, M. H. and Polanyi, J. C. (1969). *J. Chem. Phys.*, **50**, 4623
33. Polanyi, J. C. and Schreiber, J., to be published
34. Anlauf, K. G., Kuntz, P. J., Maylotte, D. H., Pacey, P. D. and Polanyi, J. C. (1967). *Discus. Faraday Soc.*, **44**, 183
35. Anlauf, K. G., Polanyi, J. C., Wong, W. H. and Woodall, K. B. (1968). *J. Chem. Phys.*, **49**, 5189
36. Polanyi, J. C. (1967). *Discus. Faraday Soc.*, **44**, 293
37. Kuntz, P. J., Mok, M. H., Nemeth, E. M. and Polanyi, J. C. (1967). *Discus. Faraday Soc.*, **44**, 229
38. Hammond, G. S. (1955). *J. Amer. Chem. Soc.*, **77**, 334
39. Polanyi, J. C. (1959). *J. Chem. Phys.*, **31**, 1338
40. Mok, M. H. and Polanyi, J. C. (1969). *J. Chem. Phys.*, **51**, 1451
41. Polanyi, J. C. and Wong, W. H. (1969). *J. Chem. Phys.*, **51**, 1439
42. Mok, M. H. and Polanyi, J. C. (1970). *J. Chem. Phys.*, **53**, 4588
43. Raff, L. M. (1966). *J. Chem. Phys.*, **44**, 1202
44. Laidler, K. J. and Polanyi, J. C. (1965). *Progress in Reaction Kinetics,* vol. 1, ed. by G. Porter (Oxford: Pergamon Press)
45. Smith, I. W. M. (1967). *Discus. Faraday Soc.*, **44**, 194
46. Suess, H. (1940). *Z. Physik. Chem.*, **B45**, 312
47. Karl, G., Kruus, P. and Polanyi, J. C. (1967). *J. Chem. Phys.*, **46**, 224
48. Holdy, E. K., Klotz, C. L. and Wilson, K. R. (1970). *J. Chem. Phys.*, **52**, 4588
49. Parrish, D. D. and Herm, R. R. (1970). *J. Chem. Phys.*, **53**, 2431
50. Rabinovitch, B. S. and Setser, D. W. (1964). *Advan. Photochem.*, **3**, 1
51. Bunker, D. L. (1962). *J. Chem. Phys.*, **37**, 393
52. Bunker, D. L. (1964). *J. Chem. Phys.*, **40**, 1946

53. Clough, P. N., Polanyi, J. C. and Taguchi, R. T. (1970). *Can. J. Chem.*, **48**, 2919
54. Polanyi, J. C. (1971). *J. Appl. Optics*, **10**, 1717
55. Carrington, T. and Garvin, D. (1969). *Comprehensive Chemical Kinetics*, Vol. 3, 107, ed. by Bamford, C. H. and Tipper, C. F. H. (Amsterdam: Elsevier)
56. Smith, I. W. M., to be published. *The Role of the Excited State in Chemical Physics*, ed by McGowan, J. W. (New York: Wiley-Interscience)
57. Smith, I. W. M. (1967). *Discus. Faraday Soc.*, **44**, 194
58. Basco, N. and Dogra, S. K. (1971). *Proc. Roy. Soc. (London)*, **A233**, 29
59. Lipscomb, F. J., Norrish, R. G. W. and Thrush, B. A. (1956). *Proc. Roy. Soc. (London)*, **A233**, 455
60. Cashion, J. K. and Polanyi, J. C. (1960). *Proc. Roy. Soc. (London)*, **A258**, 529, 564, 570
61. Garvin, D., Broida, H. P. and Kostkowski, H. J. (1960). *J. Chem. Phys.*, **32**, 880
62. Johnson, R. L., Perona, M. J. and Setser, D. W. (1970). *J. Chem. Phys.*, **52**, 6372
63. Perona, M. J., Setser, D. W. and Johnson, R. J. (1970). *J. Chem. Phys.*, **52**, 6384
64. Jonathan, N., Melliar-Smith, C. M. and Slater, D. H. (1970). *Chem. Phys. Lett.*, **7**, 257
65. Pacey, P. D. and Polanyi, J. C. (1971). *J. Appl. Optics*, **10**, 1725
66. Heydtmann, H. and Polanyi, J. C. (1971). *J. Appl. Optics*, **10**, 1738
67. Polanyi, J. C. and Woodall, K. B. (1972). *J. Chem. Phys.*, **56**, 1563
68. Anlauf, K. G., Maylotte, D. H., Polanyi, J. C. and Bernstein, R. B. (1969). *J. Chem. Phys.*, **51**, 5716
69. Polanyi, J. C. and Tardy, D. C. (1969). *J. Chem. Phys.*, **51**, 5717
70. Anlauf, K. G., Charters, P. E., Horne, D. S., Macdonald, R. G., Maylotte, D. H., Polanyi, J. C., Skrlac, W. J., Tardy, D. C. and Woodall, K. B. (1970). *J. Chem. Phys.*, **53**, 4091
71. Menzinger, M. and Wolfgang, R. (1969). *Angew. Chem.*, **8**, 438
72. Kompa, K. L., Parker, J. H. and Pimentel, G. C. (1968). *J. Chem. Phys.*, **49**, 4527
73. Berry, M. J. and Pimentel, G. C. (1968). *J. Chem. Phys.*, **49**, 5190
74. Parker, J. H. and Pimentel, G. C. (1969). *J. Chem. Phys.*, **51**, 91
75. Berry, M. J. and Pimentel, G. C. (1970). *J. Chem. Phys.*, **53**, 3453
76. Charters, P. E., Macdonald, R. G. and Polanyi, J. C. (1971). *J. Appl. Optics.*, **10**, 1747
77. Patel, C. K. N. (1964). *Phys. Rev.*, **136**, A1187
78. Parr, C. A., Polanyi, J. C. and Wong, W. H., to be published
79. Parr, C. A., Polanyi, J. C., Wong, W. H. and Tardy, D. C., to be published
80. Jaffe, R. L. and Anderson, J. B. (1971). *J. Chem. Phys.*, **54**, 2224
81. Muckerman, J. T. (1971). *J. Chem. Phys.*, **54**, 1155
82. Polanyi, J. C. and Cashion, J. K. (1961). *J. Chem. Phys.*, **35**, 600
83. Charters, P. E., Khare, B. N. and Polanyi, J. C. (1962). *Discus. Faraday Soc.*, **33**, 107
84. Basco, N. and Norrish, R. G. W. (1961). *Nature (London)*, **189**, 455; (1962). *Proc. Roy. Soc. (London)*, **A268**, 291
85. Clough, P. N. and Thrush, B. A. (1969). *Proc. Roy. Soc. (London)*, **A309**, 419
86. Clough, P. N. and Thrush, B. A. (1969). *Trans. Faraday Soc.*, **65**, 23
87. Lin, M. C. and Green, W. H. (1970). *J. Chem. Phys.*, **53**, 3383
88. Maylotte, D. H., Polanyi, J. C. and Woodall, K. B. (1972). *J. Chem. Phys.*, to be published
89. Anlauf, K. G., Polanyi, J. C., Horne, D. S., Macdonald, R. G., Polanyi, J. C. and Woodall, K. B. (1972). *J. Chem. Phys.*, to be published
90. Perona, M. J. (1971). *J. Chem. Phys.*, **54**, 4024
91. Basov, N. G., Kulakov, L. V., Markin, E. P., Nikitin, A. I. and Oraevskii, A. N. (1969). *JETP Lett.*, **9**, 375
92. Jonathan, N., Melliar-Smith, C. M. and Slater, D. H. (1970). *J. Chem. Phys.*, **53**, 4396
93. Padrick, T. D. and Pimentel, G. C. (1971). *J. Chem. Phys.*, **54**, 720
94. Hancock, G. and Smith, I. W. M. (1969). *Chem. Phys. Lett*, **3**, 573
95. Clough, P. N. and Thrush, B. A. (1968). *Chem. Commun.*, **13**, 1351
96. Schwartz, S. E. and Thrush, B. A. (1969). *J. Mol. Spectrosc.*, **32**, 343
97. Creek, D. M., Melliar-Smith, C. M. and Jonathan, N. (1970). *J. Chem. Soc. A*, 646
98. Clough, P. N., Schwartz, S. E. and Thrush, B. A. (1970). *Proc. Roy. Soc. (London)*, **A317**, 575
99. Gilmore, F. R., Bauer, E. and McGowan, J. W. (1969). *J. Quant. Spec. Rad. Transf.*, **9**, 157
100. Herzberg, G. (1945). *Molecular Spectra and Molecular Structure*, Vol. II (London: Van Nostrand)

101. Light, J. C. (1971). *Advan. Chem. Phys.,* **19,** (London: Interscience)
102. Herzberg, G. (1966). *Molecular Spectra and Molecular Structure,* Vol. III (London: Van Nostrand)
103. Smith, F. T. (1969). *Phys. Rev.,* **179,** 111
104. Thorson, W. R., Delos, J. B. and Boorstein, S. A. (1971). *Phys. Rev.,* **A4,** 1052
105. Krauss, M. (1967). *U.S. Nat. Bur. of Standards, Technical Note 438*
106. Krauss, M. (1970). *Ann. Rev. Phys. Chem.,* **21,** 39 (Palo Alto, California: Annual Reviews, Inc.)
107. Buck, U. (1971). *J. Chem. Phys.,* **54,** 1923
108. Breig, E. L. (1969). *J. Chem. Phys.,* **51,** 4539
109. Holdy, K. E., Klotz, L. C. and Wilson, K. R. (1970). *J. Chem. Phys.,* **52,** 4588
110. Shapiro, M. and Levine, R. D. (1970). *Chem. Phys. Lett.,* **5,** 499
111. Child, M. S. (1970). *J. Mol. Spectrosc.,* **23,** 487
112. Porter, R. N., Stevens, R. M. and Karplus, M. (1968). *J. Chem. Phys.,* **49,** 5163
113. Nikitin, E. E. (1968). *Chemische Elementarprozesse,* ed. by Hartmann, H. (New York: Springer Verlag)
114. See Ref. 102, pp. 443-444.
115. Preston, R. K. and Tully, J. C. (1971). *J. Chem. Phys.,* **54,** 4297
116. Tully, J. C. and Preston, R. K. (1971). *J. Chem. Phys.,* **55,** 562
117. Krenos, J. *et al.* (1971). *Chem. Phys. Lett.,* **10,** 17
118. Chow, K-W. and Smith, A. L. (1970). *Chem. Phys. Lett.,* **7,** 127
119. Lefebvre, R. (1971). *Chem. Phys. Lett.,* **8,** 306
120. Sharf, B. and Sibley, R. (1971). *Chem. Phys. Lett.,* **9,** 125
121. Chow, K-W. and Smith, A. L. (1971). *J. Chem. Phys.,* **54,** 1556
122. Sando, K. M. and Dalgarno, A. (1971). *Mol. Phys.,* **20,** 103
123. Callear, A. B. and Pilling, M. J. (1970). *Trans. Faraday Soc.,* **66,** 1618
124. Callear, A. B. and Pilling, M. J. (1970). *Trans. Faraday Soc.,* **66,** 1886
125. Heicklen, J. and Cohen, N. (1968). *Advan. Photochem.,* **5,** 157 (London: Interscience)
126. Gross, R. W. F. and Cohen, N. (1968). *J. Chem. Phys.,* **48,** 2582
127. Akerman, F. and Miescher, E. (1969). *J. Mol. Spectrosc.,* **31,** 400
128. Juliene, P. S., Krauss, M. and Donn, B. (1971). *Astrophys. J.,* **170,** 65
129. Fair, R. W. and Thrush, B. A. (1969). *Trans. Faraday Soc.,* **65,** 1208
130. Bott, J. F. and Jacobs, T. A. (1970). *J. Chem. Phys.,* **52,** 3545
131. Tewarson, A. and Palmer, H. B. (1970). *Thirteenth Symposium on Combustion,* 99
132. McNeal, R. J. and Durana, S. C. (1969). *J. Chem. Phys.,* **51,** 2955
133. Shui, V. H., Appleton, J. P. and Keck, J. C. (1970). *J. Chem. Phys.,* **53,** 2547
134. Wright, A. N. and Winkler, C. A. (1968). *Active Nitrogen* (New York: Academic Press)
135. Anketell, J. and Nicholls, R. W. (1970). *Rep. Prog. Phys.,* **33,** 197
136. Jonathan, N. and Petty, R. (1969). *J. Chem. Phys.,* **50,** 3804
137. Benson, S. W. (1968). *J. Chem. Phys.,* **48,** 1765
138. Brennen, W. and Shane, E. C. (1971). *J. Phys. Chem.,* **75,** 1552
139. Campbell, I. M. and Thrush, B. A. (1967). *Proc. Roy. Soc. (London),* **A296,** 201
140. Young, R. A., Black, G. and Slanger, T. G. (1969). *J. Chem. Phys.,* **50,** 303
141. Brown, R. L. (1970). *J. Chem. Phys.,* **52,** 4604
142. Young, R. A. and St. John, G. A. (1968). *J. Chem. Phys.,* **48,** 895
143. Browne, R. J. and Ogryzlo, E. A. (1970). *J. Chem. Phys.,* **52,** 5774
144. Clyne, M. A. A. and Stedman, D. H. (1968). *Trans. Faraday Soc.,* **64,** 1816
145. Clyne, M. A. A. and Stedman, D. H. (1968). *Trans. Faraday Soc.,* **64,** 2698
146. Palmer, H. P. and Carabetta, R. A. (1968). *J. Chem. Phys.,* **49,** 2466
147. Fink, W. H. (1968). *J. Chem. Phys.,* **49,** 5054
148. Fink, W. H. (1971). *J. Chem. Phys.,* **54,** 2911
149. Del Bene, J. E. (1971). *J. Chem. Phys.,* **54,** 3487
150. Burnelle, L., May, A. M. and Gangi, R. A. (1968). *J. Chem. Phys.,* **49,** 561
151. Gangi, R. A. and Burnelle, L. (1971). *J. Chem. Phys.,* **55,** 843
152. Gangi, R. A. and Burnelle, L. (1971). *J. Chem. Phys.,* **55,** 851
153. Busch, G. E., Mahoney, R. T. and Wilson, K. R. (1972). *J. Chem. Phys.,* in press
154. Diesen, R. W., Wahr, J. C. and Adler, S. E. (1969). *J. Chem. Phys.,* **50,** 3635
155. Abe, K., Myers, F. and McCubbin, T. K. and Polo, S. R. (1971). *Symposium on Molecular Structure and Spectroscopy,* Columbus, Ohio

156. Sackett, P. B. and Yardley, J. T. (1970). *Chem. Phys. Lett.*, **6**, 323
157. Sakurai, K. and Capelle, G. (1960). *J. Chem. Phys.*, **53**, 3764
158. Keyser, L. F., Levine, S. Z. and Kaufman, F. (1971). *J. Chem. Phys.*, **54**, 355
159. Schwartz, S. E. and Johnston, A. S. (1969). *J. Chem. Phys.*, **51**, 1286
160. Sackett, P. B. and Yardley, J. T. (1971). *Symposium on Molecular Structure and Spectroscopy*, Columbus, Ohio
161. Douglas, A. E. and Huber, K. P. (1965). *Can. J. Phys.*, **43**, 74
162. Sakurai, K. and Broida, H. P. (1969). *J. Chem. Phys.*, **50**, 2404
163. Paulsen, D. E., Sheridan, W. F. and Huffman, R. E. (1970). *J. Chem. Phys.*, **53**, 647
164. Thrush, B. A. (1968). *Ann. Rev. Phys. Chem.*, **19**, 371 (Palo Alto, California: Annual Reviews Inc.)
165. Jortner, J., Rice, S. A. and Hochstrasser, R. M. (1969). *Advan. Photochem.*, **7** (London: Interscience)
166. Becker, K. H., Groth, W. and Joo, F. (1968). *Ber. Bunsenges.*, **72**, 157
167. Becker, K. H., Groth, W. and Thran, D. (1970). *Chem. Phys. Lett.*, **6**, 583
168. McKenzie, A. and Thrush, B. A. (1968). *Chem. Phys. Lett.*, **1**, 681
169. Vanpee, M. and Kineyko, W. R. (1970). *J. Chem. Phys.*, **52**, 1619
170. Golomb, D. and Good, R. E. (1968). *J. Chem. Phys.*, **49**, 4167
171. Collier, S. S. *et al.* (1970). *J. Amer. Chem. Soc.*, **92**, 217
172. Smith, W. H. (1969). *J. Chem. Phys.*, **51**, 3410
173. Metee, H. D. (1968). *J. Chem. Phys.*, **49**, 1784
174. Strickler, S. J. and Howell, D. B. (1968). *J. Chem. Phys.*, **49**, 1947
175. Gardner, P. J. (1969). *Chem. Phys. Lett.*, **4**, 167
176. Fletcher, S. R. and Levitt, P. B. (1969). *Trans. Faraday Soc.*, **65**, 1544
177. Cohen, N. and Gross, R. W. G. (1969). *J. Chem. Phys.*, **50**, 3119
178. Reeves, R. R. and Emerson, J. A. (1970). *J. Chem. Phys.*, **52**, 2161
179. Herzberg, G. (1968). *Advan. Photochem.*, **5**, 1 (London: Interscience)
180. Krauss, M. (1969). *J. Res. Nat. Bureau Stds.*, **73A**, 191
181. Salotto, A. W. and Burnelle, L. (1970). *J. Chem. Phys.*, **52**, 2936
182. Baulch, D. L., Drysdale, D. D. and Lloyd, A. C. (1968). *High Temperature Reactions Rate Data, No. 1*, The University, Leeds 2, England
183. Clark, T. C., Garnett, S. H. and Kistiakowsky, G. B. (1970). *J. Chem. Phys.*, **52**, 4692
184. Lin, M. C. and Bauer, S. H. (1969). *J. Chem. Phys.*, **50**, 3377
185. Jonah, C. D. and Zare, R. N. (1971). *Chem. Phys. Lett.*, **9**, 65
186. Tewarson, A., Naegeli, D. W. and Palmer, H. B. (1969). *Twelfth Symposium on Combustion*, 415
187. Donovan, R. J. and Husain, D. (1970). *Chem. Rev.*, **70**, 489
188. McCullough, D. W. and McGarth, W. D. (1971). *Chem. Phys. Lett.*, **8**, 353
189. Izod, T. P. J. and Wayne, R. P. (1969). *Chem. Phys. Lett.*, **4**, 208
190. Snelling, D. R. and Gauthier, M. (1971). *Chem. Phys. Lett.*, **9**, 254
191. Young, R. A., Black, G. and Slanger, T. G. (1968). *J. Chem. Phys.*, **49**, 4758
192. Westenberg, A. A., Roscoe, J. M. and deHaas, N. (1970). *Chem. Phys. Lett.*, **7**, 597
193. Clark, I. D. and Wayne, R. P. (1970). *Proc. Roy. Soc. (London)*, **A316**, 539
194. Clark, I. D. (1970). *Chem. Phys. Lett.*, **5**, 317
195. Campbell, I. M. and Thrush, B. A. (1969). *Trans. Faraday Soc.*, **65**, 32
196. Conroy, H. (1969). *J. Chem. Phys.*, **51**, 3979
197. Sharma, R. D. and Wray, K. L. (1971). *J. Chem. Phys.*, **54**, 4578
198. Myers, B. F. and Bartle, E. R. (1968). *J. Chem. Phys.*, **48**, 3935
199. Campbell, I. M., Neal, S. B., Goldie, M. F. and Thrush, B. A. (1971). *Chem. Phys. Lett.*, **8**, 612
200. Baulch, D. L., Drysdale, D. D. and Horne, D. G. (1970). *High Temperature Reaction Rate Data No. 5*, 24. The University, Leeds 2, England
201. Maria, H. J., Larsen, D., McCarvillu, M. E. and McGlynn, S. P. (1970). *Accounts Chem. Res.*, **3**, 356
202. Rabalais, J. W., *et al.* (1971). *Chem. Rev.*, **71**, 73
203. Wayne, R. P. (1969). *Advan. in Photochem.*, **7**, 311 (London: Interscience)
204. Young, R. A. and Black, G. (1966). *J. Chem. Phys.*, **44**, 3741
205. Gross, R. W. F. (1968). *J. Chem. Phys.*, **48**, 1302
206. McKenzie, A. and Thrush, B. A. (1968). *Proc. Roy. Soc. (London)*, **A308**, 133
207. Rao, T. N. and Calvert, J. G. (1970). *J. Phys. Chem.*, **74**, 681

208. Struve, W. S., Kitigawa, T. and Herschbach, D. R. (1971). *J. Chem. Phys.*, **54**, 2759
209. Lee, Y. T., Gordon, R. J. and Herschbach, D. R. (1971). *J. Chem. Phys.*, **54**, 2410
210. Tully, J. C., Herman, Z. and Wolfgang, R. (1971). *J. Chem. Phys.*, **54**, 1730
211. Vasil'ev, R. F. (1970). *Russ. Chem. Rev.*, **39** (6), 529
212. Sando, K. M. (1971). *Molec. Phys.*, **21**, 439
213. Nikitin, E. E. (1970). *Adv. Quantum Chem.*, **5**, 135
214. O'Malley, T. F. (1971). *Adv. Atomic Molec. Phys.*, **7**

6
Molecular Beam Reactions

J. L. KINSEY
Massachusetts Institute of Technology

6.1   INTRODUCTION

Once the technical limitations which had so long held back the application of molecular beams to studies of reactions were overcome, the field grew quickly. The differential surface ionisation detector, whose discovery in 1955 by Taylor and Datz initiated these investigations, continues to play a special role. However, within the past few years the development of new detection schemes and improvements in source designs have opened entirely new areas. Two main thrusts have become apparent in the evolution of this field, one directed towards diversification of the types of reactions studied and the other towards ever greater refinement of the measurements made on a few simple systems.

Volume 10 of *Advances in Chemical Physics* (1966)[1], was devoted entirely to the subject of molecular beams and contains several articles of direct relevance to this review. The article by Herschbach[2] on reactive scattering is especially recommended as background to the present work, which will concentrate on progress within the past 5 years. Part of the work in that period has already been covered in reviews[3-6] subsequent to 1966, but because of the thoroughness of that paper I have taken it as a starting point. An attempt has been made to include references to all published works on experimental studies of neutral reactions appearing since reference 2 and to all theoretical papers directly related to the interpretation or analysis of such experiments. Other theoretical works are included only if they seemed to have special bearing on beam reactions. I have decided to avoid unpublished works except preprints. The related experimental fields of elastic and inelastic scattering are not covered unless there is particular relevance to reactive scattering, and reactions of beams with surfaces will not be discussed.

6.1.1   Updated list of reactions studied with molecular beams

Table I of reference 2 contains a list of all the reactions that had successfully been studied in molecular beams up to that time. Table 6.1 of this paper brings that list up to date (July 1971).

6.2 CROSS-SECTIONS FOR REACTIVE SYSTEMS

The primary property measured in a molecular beam experiment, depending upon the kind of detection employed, is proportional either to the number of particles striking a given area per second (flux detector) or to the instantaneous number of particles contained in a given volume (density detector). Under the ideal conditions of (i) perfectly monoenergetic incident beams and (ii) perfectly rectangular density distributions in each beam, the contribution to the signal due to the detected particles (no subscript) scattered into solid angle $d^2\Omega$ of the detector at laboratory (LAB) angles Θ and Φ and in the range of laboratory velocities between v and $v + dv$ is[85]:

$$d^3S = n_1 n_2 \,\Delta V t(v)\, v_r \frac{[d^3\sigma(v_r; \Theta, \Phi v)]}{d^2\Omega dv} d^2\Omega dv \qquad (6.1)$$

where n_1 and n_2 are the number densities of the reactants within the volume of intersection ΔV of the two beams; $t(v)$ is the detector efficiency, v_r is the relative speed of the reactants, and $d^3\sigma/d^2\Omega dv$, is the *laboratory (doubly) differential cross-section* for reactive or non-reactive scattering*. Additional subscripts on σ would be needed if the dependence on internal states of the reactants or products should also be indicated. The detector efficiency function, $t(v)$, is a constant for a flux detector and proportional to v for a density detector. (The notation is that of reference 85.) In a real experiment, of course, averaging over the velocity distributions in the incident beam must be taken into account[85].

If the solid angle subtended by the detector is sufficiently small and if the resolution in v is sufficiently fine that $d^2\sigma/d^2\Omega dv$ does not vary appreciably over the range encompassed by a measurement, then the cross-section can be obtained simply by dividing the measured signal by $n_1 n_2 \Delta V \bar{t}(v) \,\Delta^2\Omega\Delta v$ where $\Delta^2\Omega$ is the finite angular size of the detector and $\bar{t}$ is the average of t over the range in v of Δv. For less sensitive measurements the observed signal is the integral of d^3S over the range of solid angle and v. For example, without velocity analysis, integration over the *full* range of v is required. The resulting quantity (after $n_1 n_2 \,\Delta v$ is divided out) is often called the *laboratory differential cross-section* $\langle d^2\sigma/d^2\Omega \rangle$:

$$\langle d^2\sigma/d^2\Omega \rangle = \int_0^\infty [d^3\sigma/d^2\Omega dv]\, dv \qquad (6.2)$$

In order to avoid confusion with $d^3\sigma/d^2\Omega dv$, the term *angular distribution* will occasionally be used in discussing *reactive* scattering.

---

*The use of cross-sections as the fundamental quantity is borrowed from nuclear and x-ray scattering, where it is usually a very good approximation to neglect the motion of the target. In that case $v_1 = v_r$ and $v_r n_1 = \Phi_1$, the flux density of the beam, so that the cross-section appears naturally as the ratio between the observed scattering and the product of the incident flux with the target density. In molecular scattering, the motion of both particles is usually appreciable, so that the cross-section is a derived quantity. Since it always appears in the combination $\kappa(v_r) = v_r\sigma(v_r)$ one could as well use κ, which has the units of a bimolecular rate coefficient. However, the convention of using cross-sections is by now firmly enough established that it will be maintained here.

Table 6.1

Reaction	$\sigma_R^{\ddagger}$	References and comments§
K + HBr	35 ± 9 (7)	7 (VS, NR/OPT)
K + DBr	26 ± 8 (7)	52 (EL/DEFL); 65, 32 (VS/VA), 66 (model)
K + HCl	0.15 (59)	58 (evidence for activation energy), 59
K + HCl† ($v = 1$)	20 (59)	59 (vibrational excited state, chem. laser)
K + TBr		53 (T atom detected)
K + Br$_2$	200–260 (14)	56; 57 (product KBr† used in subseq. reaction); 13 (VA); 14; 52 (EL/DEFL); 25; 73 (VS/VA); 36 (NR/OPT); *16*; *34*; *79–82*
K + I$_2$	127 (24)	56; 14; 36 (NR/OPT); 29 (NR small angles only); 33 (VS/VA); 24; *16*
K + Cl$_2$	260–300 (37)	37; 36 (NR/OPT)
K + ICl → KCl† + I (mainly)	≳ 150 (57)	47
K + IBr	≳ 150 (47)	47
K + MeI	35 (48)	18 (VS/orient); 8 (NR/OPT); 25; 48
K + CF$_3$I		19, 75 (orient.)
K + EtI, Pr^{n}I, Pr^{i}I, Bu^{n}I, Bu^{i}I, Bu^{s}I, n-C$_5$H$_{11}$I, n-C$_7$H$_{15}$I		48
K + Me$_3$CBr		36 (NR/OPT)
K + (CN$_2$)C=C(CN)$_2$ → ?}		
K + BrCN	120 (38)	38
K + ICN	210 (38)	38
K + SCl$_2$	230 (38)	38
K + CBr$_4$	250 (74)	74
K + SnCl$_4$		8 (NR/OPT)
K + SF$_6$		8 (NR/OPT)
K + NO$_2$ → ?		44 (MAG/DEFL) (products in doubt)
K + RbCl → KCl + Rb		55 (CPLX)
K + SO$_2$, CO$_2$, NO ⇌ KOR*		41–43 (VS, CPLX, no reac.); 77 (VS/VA non-reac., CO$_2$ only)
Cs + HBr		52 (EL/DEFL); 36 (NR/OPT)
Cs + TBr		53 (T atom detected)
Cs + Br$_2$	≳ 150 (14)	56; 14; 52 (EL/DEFL), *16*
Cs + I$_2$	195 (24)	14; 20, 24 (non-beam σ_R determ.)
Cs + Cl$_2$	210–310 (37)	37
Cs + ICl, IBr	≳ 150 (47)	47
Cs + MeI		52 (EL/DEFL), 25
Cs + BrCN	250 (38)	38
Cs + ICN	260 (38)	38
Cs + NOCl	130 (38)	38
Cs + SCl$_2$	340 (38)	38
Cs + CCl$_4$	150 (74)	52 (EL/DEFL); 74
Cs + SnCl$_4$	100–400 (74)	74
Cs + CHCl$_3$	40 (74)	74
Cs + SiCl$_4$ → no reaction	≳ 15 (74)	74
Cs + SF$_6$		30 (Stark spectrum, V′ dist.)
Cs + TlCl, TlI		26 (OSC–CPLX)
Cs + MeNO$_2$ → CsNO$_2$ (singlet) + Me	c. 100 (44)	44 (MAG/DEFL)

Table 6.1 – *continued*

Reaction	$\sigma_R^\ddagger$	References and comments§
$Cs + RbCl \rightarrow CsCl + Rb$		55 (CPLX)
$Cs + SO_2, CO_2 \rightleftarrows CSOR*$		41–43 (VS, CPLX)
$Cs + NO_2 \rightarrow ?$		41–43 (VS, CPLX)
$Cs + RI$		*83* (model)
$Rb + Br_2$	$\gtrsim 150$ (14)	14; 39 (rotational state distribution), *16*
$Rb + I_2$	$\gtrsim 150$ (14)	14; *16*
$Rb + Cl_2$	260–310 (37)	37
$Rb + ICl, IBr$	$\gtrsim 150$ (47)	47
$Rb + MeI$		10–12 (orient., VS)
$Rb + NOCl$	100 (38)	38
$Rb + SCl_2$	250 (38)	38
$Rb + PCl_3$	130 (74)	74
$Rb + PBr_3$	250 (74)	74
$Rb + CCl_4$	100 (74)	74
$Rb + BBr_3$	<2 (74)	74 (no reaction)
$Rb + CsCl, KCl$		55 (CPLX, VA)
$Rb + NO_2 \rightarrow ?$		44
$Rb + Me_3NO_2 \rightarrow RbNO_2$ (singlet) $+ Me$	*c.* 100 (44)	44 (MAG/DEFL)
$Na + Br_2$	100 (15)	15
$Na + I_2$	97 (24)	24 (non-beam σ_R determination)
$Na + ICl$	90 (15)	15
$Na + MeI$	5 (15)	15
$Na + NO_2 \rightarrow NaO \; (^2\Pi?) + NO$		44 (MAG/DEFL)
$Na + MeNO_2 \rightarrow NaNO$ (singlet) $+ Me$	*c.* 100 (44)	44 (MAG/DEFL)
$Na + KBr^\dagger \rightarrow K* + NaBr$		57 (chemiluminescence)
$Li + KBr$		*54* (optical model)
$Li + Br_2$	115–146 (62)	61–62 (MAG/DEFL)
$Li + Cl_2$	86–87 (62)	61–62 (MAG/DEFL)
$Li + ICl \rightarrow LiCl + I*(?)$	123–130 (62)	61, 62 (MAG/DEFL)
$Li + SnCl_4$	147–165 (62)	61, 62 (MAG/DEFL)
$Li + PCl_3$	43–55 (62)	61, 62 (MAG/DEFL, CPLX ?)
$Li + NO_2 \rightarrow LiO(^2\Pi) + NO$	15–16 (63)	63 (MAG/DEFL)
$Li + MeNO_2 \rightarrow LiNO_2 + Me$	55–58 (63)	63 (MAG/DEFL)
$Li + SF_6$	17–18 (63)	63 (MAG/DEFL)
$Li + CCl_4$	37–43 (63)	63 (MAG/DEFL)
$Li + MeI$	27 (63)	63 (MAG/DEFL)
$Li + KF \rightarrow LiF + K$		49 (OSC/CPLX)
$Li + KBr \rightarrow LiBr + K$		49 (OSC/CPLX)
$CsCl + KI$		55 (CPLX)
$Na_2 + Cl \rightarrow NaCl + Na* + hv$	10–100 (70)	70 (chemiluminescence)
$K_2 + Cl \rightarrow KCl + K* + hv$	10–100 (70)	70 (chemiluminescence)
$K_2, Rb_2, Cs_2 + H \rightarrow ?$		78
$Cl + Br_2 \rightarrow BrCl + Br$	5–20 (50)	9; 50, 51; 21, 17 (VA)
$Br + I_2$	5–20 (50)	50
$Cl + IBr, BrI$	1–10 (51)	51
$Cl + I_2, Br_2$	1–10 (51)	51
$Br + ClI$	0.2–2 (51)	51
$Ba + NO_2, N_2O \rightarrow BaO* \; (?)$ $+ hv$		60 (chemiluminescence)
$Ca + NO_2, N_2O \rightarrow CaO* \; (?)$ $+ hv$		60 (chemiluminescence)

Table 6.1—*continued*

Reaction	$\sigma_R{}^{\ddagger}$	References and comments§
$Ba + Cl_2 \Big\langle \begin{array}{l} BaCl_2^* + h\nu \\ BaCl^* + Cl + h\nu \end{array}$		46 (chemiluminescence)
$Sr + Cl_2 \Big\langle \begin{array}{l} SrCl_2^* + h\nu \\ SrCl^* + Cl + h\nu \end{array}$		46 (chemiluminescence)
$D + H_2$		22; 31
$H + D_2$		27, 28
$H, D + Cl_2$		40; *64; 67; 84*
$H, D + Br_2$		23; 40
$F + D_2$		68 (VS)
$HI + DI \rightarrow$ no reaction	<0.04 (45)	45 (beam-gas experiment)
$Xe(Fast) + CsI, CsBr,$ $RbI \rightarrow M^+ + X^-$	>10 (72)	72
$H + CCl_4$	*c.* 10 (143)	143 (beam-gas experiment)

*Indicates electronic excitation.
†Indicates vibrational excitation.
‡The best estimates of the total reactive cross-section, σ_R, are given. The numbers in parentheses indicate the reference from which the values were taken
§Abbreviations: VS = velocity selection of one of the primary beams; VA = velocity analysis of products; VS/VA = combined velocity selection and analysis; NR/OPT = optical analysis of non-reactive scattering only; EL/DEF = electric deflection analysis; MAG/DEFL = magnetic deflection discrimination (but not rotational state analysis); CPLX = long-lived complex. References to theoretical papers are italicised.

6.2.1 Transformation to centre-of-mass coordinates

The information contained in the laboratory measurements becomes interpretable in molecular-dynamic terms only when expressed as an equivalent quantity in a coordinate system moving with the centre of mass (CM). There have been a number of treatments of the inverse CM → LAB transformation problem. The most recent and thorough of these is reference 85, which gives a consolidation of previous work with important extensions and clarification of the relationships. References to earlier work can be found there.

The CM differential cross-section $d^3\sigma/d^2\omega\, dw$ (where ω is the CM solid angle and w the CM velocity of the product) is related to the LAB cross-section by a Jacobian, $J(CM \rightarrow LAB)$:

$$\frac{d^3\sigma}{d^2\Omega\, dv} = J(CM \rightarrow LAB)\frac{d^3\sigma}{d^2\omega\, dv} \tag{6.3}$$

where the Jacobian given by

$$J(CM \rightarrow LAB) = \frac{d^3\omega\, dw}{d^2\Omega\, dv} = v^2/w^2 \tag{6.4}$$

Hence, both the forward and reverse transformation are quite straight-forward for the velocity-analysed case since the CM velocity, w, and the CM scattering angles, θ and ϕ, are readily computed from v, Θ, Φ, and the velocity of the centre of mass in the laboratory coordinate system. Inversion of LAB data into CM cross-sections becomes more complicated when the averages over initial conditions are taken into account. In that case, the LAB results are integral transforms of the desired CM quantities[85], and various schemes have been used to invert the transforms. These usually describe $d^3\sigma/d^2\omega\, dw$ as flexible functions of the variables θ and w and then adjust parameters in these functions to achieve a best fit (see reference 33 and references given there).

When there is no velocity analysis even the ideal case presents problems because the transformations are no longer point-to-point in three dimensions: The CM scattering angles and the size of the CM solid angle seen by the detector vary with the (unmeasured) velocity v. Transformation of the variable of integration in equation (6.2) from v to the CM velocity w gives

$$d^2\sigma/d^2\Omega = \int dw \left[\frac{d^3\sigma(\theta, \phi, w)}{d^2\omega dw} \right] \left(\frac{d^2\omega}{d^2\Omega} \right) \tag{6.5}$$

In this equation the CM scattering angles, and the (two-dimensional) Jacobian $(d^2\omega/d^2\Omega)$ are to be considered as functions of w, Θ, Φ. These may be double-valued functions, in which case the integral is understood to be over both branches. The ratio of CM to LAB solid angles is[85]

$$(d^2\omega/d^2\Omega) = v^2/w^2 |\cos \delta| \tag{6.6}$$

where δ is the angle between v and w.

Hence:

$$d^2\sigma/d^2\Omega = \int dw \left[\frac{d^3\sigma}{d^2\omega\, dw} \right] \frac{v^2}{w^2 |\cos \delta|} \tag{6.7}$$

This equation is also the proper one to use for the *elastic scattering* cross-section because the w-dependence of $d^3\sigma/d^2\omega\, dw$ is a δ-function in that case.

Under favourable kinematic conditions, an approximate inversion of $\langle d^2\sigma/d^2\Omega \rangle$ into CM quantities can be achieved, because the transformation between scattering angles in the two coordinate systems may be relatively insensitive to the recoil velocity. If two regions of LAB angles correspond to the same region of CM angles, for example, a sometimes satisfactory procedure is to assume a single δ-function for the w spectrum and to establish the best value for this 'nominal' w by requiring the two redundant branches to give the same result in the CM system. Although this is risky, in some cases where there were subsequent velocity-analysis measurements the results showed the conclusions were not qualitatively incorrect. Before the details of CM $\rightarrow$ LAB transformations came to be fully appreciated, the 'nominal' w values were used in a much cruder way, ignoring the Jacobian factor. Although there may be favourable cases where the Jacobian varies only slightly over the range of measurement, Entemann[25] has pointed out that its neglect can in general lead to seriously inaccurate conclusions.

6.2.2 Non-reactive scattering of reactants: optical model

The differential cross-section $\langle d^2\sigma/d^2\omega \rangle$ of the *unreacted* incident particles in a reactive system can give valuable information, not obtainable from any other measurements at present, on the impact-parameter dependence of the reaction probability. These cross-sections are always observed to fall off towards large scattering angles conspicuously more steeply than for elastic systems of about the same size. A measure of the reaction probability can be obtained by attributing this decrease to depletion of the reactants by the reaction. In the early uses of these ideas[86], the reaction probability (or *opacity function*) $p(b,E)$ was taken as

$$p(b,E) = 1 - \frac{I(E,\theta)}{I_0(E,\theta)} \qquad (6.8)$$

where b is the impact parameter, E the energy, θ the CM scattering angle and $I(E,\theta)$ is used instead of $\langle d^2\sigma/d^2\omega \rangle$ for the *observed* differential cross-section. $I_0(E,\theta)$ is a *reference* cross-section, originally obtained either by visual extrapolation of the smooth small-angle observed cross-section into wider angles or by fitting a potential to the small-angle rainbow structure and then computing the wide-angle cross-section for this potential. The $b \leftrightarrow \theta$ correspondence, taken to be the classical one, can also be used to give p as a function of the *distance of closest approach*, y, as well as the potential energy at that distance. Plots of data using the product $E\theta$ as the independent variable are found to provide a useful scaling for comparing data taken at different energies[87, 88]. This representation is also an extremely useful aid in determining the distances of closest approach at threshold[86].

   This qualitative interpretation has now found more formal expression in terms of the 'optical model'[86, 89-95] in which a *complex* potential is introduced to provide complex phase shifts. The imaginary part of the phase shifts accounts for the non-unitarity of the *elastic part* of the scattering matrix, i.e. the loss due to reaction or inelastic processes. This gives an *elastic* scattering amplitude of the form

$$f(\theta) = (2ik)^{-1} \sum_l (2l+1)\, p_l(\cos\theta)\{\exp(2_i\eta_l)[1-p_l]^{1/2} - 1\} \qquad (6.9)$$

Here η_l is the *real part* of the lth phase shift and p_l, the reaction probability for the lth partial wave, is obtained from the imaginary part δ_l of the lth phase shift by $p_l = 1 - \exp(-4\delta_l)$. Nyeland and Ross[92] have shown that, at least in regions where the $\theta \leftrightarrow b$ correspondence is classically one to one, p can reliably be obtained from equation (6.8), but with I_0 now being interpreted as the cross-section calculated from equation (6.9) setting all $p_l = 0$. In principle, the solution should be obtained iteratively, since the real part of the phase shift also depends on the imaginary part of the potential. Bernstein and Levine[93] have tested this procedure using η values calculated for a Lennard–Jones 12–6 potential together with various opacity functions of the form $p_l = p^0\{1 + \exp[l - l_c]/d\}^{-1}$. As long as the parameters p^0, l_c and d were chosen so that the change of p_l was not too abrupt, the inversion procedure in the single-valued $\theta \leftrightarrow b$ region recovered the input values with acceptable accuracy. Abrupt variation in p_l introduces quantum oscillations into the predicted cross-sections[97] that have not as yet been observed.

When an appreciable reaction probability is manifested in an angular region for which the classical $\theta \leftrightarrow b$ correspondence is not unique, no simple way of calculating p_1 has been developed. Instead a set of η_1 and p_1' must be found that reproduce the observations when placed in equation (6.9). The η values are usually taken from an assumed real potential without further fitting, and p_l is fitted to a convenient functional form[7, 8, 36]. The analysis has recently been modified by Harris and Wilson[76] for use with highly asymmetric reactants by including anisotropy in the reference potential and computing the η values from the 'sudden approximation'.

There are several problems associated with the optical model; the most serious is the strong reliance on the reference potential or its equivalents. Although the *language* of the newer developments in terms of the optical model avoids unaesthetic dependence on such non-operational concepts as 'what the cross-section would have been if the reaction could be turned off', there remain serious questions as to uniqueness and convergence of various fitting models. Moreover, *practice* still relies on η values taken from a 'would have been' potential. In its current form, the optical model also suffers in not separating inelastic scattering and reactive scattering. Rosenfeld and Ross[89] have indicated that treating all non-reactive scattering as elastic may not be seriously in error if the inelasticity is small. However, the estimate of the total reactive cross-section for $K + HCl$ obtained from the non-reactive scattering is much larger than is observed directly[58], presumably because of inelastic processes.

The inclusion of anisotropy in the reference potential moves partially towards treating other kinds of non-reactive scattering; however, it is not clear that inelastic processes can properly be accounted for in the general case without at least solving an already formidable inelastic-scattering reference problem as a first step. Nonetheless, at its worst the optical model analysis amounts to sophisticated curve fitting, and in this respect it is probably no more objectionable than similar procedures in many other areas.

6.2.3 Total reactive cross-sections

The total rate of production of products for a given relative speed of the reactants is given by $\sigma_R(v_r)$, the *total reactive cross-section*.

$$\sigma_R(v_r) = \int d^2\Omega \frac{d^2\sigma}{d^2\Omega}$$

$$= \int dw_3 \int d^2\omega \left[\frac{d^3\sigma}{d^2\omega\, dw_3} \right] \tag{6.10}$$

Of all the quantities measured in beam experiments, σ_R is most closely related to conventional kinetic data. If σ_R is known for all relative speeds (and for all initial *internal* states), the thermal rate coefficient $k(T)$ can be obtained from

$$k(T) = \int_0^\infty v_r \bar{\sigma}_r(v_r; T) f_B(v_r; T) \, dv_r \tag{6.11}$$

where $\bar{\sigma}_R (v_r; T)$ is the thermal *average* of the cross-sections over the internal states at temperature T; $f_B(v_r, T)$ is the normalised Maxwellian velocity distribution.

Because of the spatial dispersion of products, σ_R is not measured directly in molecular beam experiments but must be inferred from the *angular distributions*. These suffer from large uncertainties in the absolute magnitudes because of difficulties in establishing n_1, n_2, ΔV, and t in equation (6.1). Hence, σ_R is among the least reliably determined quantities in beams. Nonetheless, reasonable estimates can be made by careful measurement, and at worst, it should be possible to *order* reactions qualitatively as to the size of σ_R.

Both the elastic and the reactive scattering data can be used to estimate σ_R. Once the opacity function by $p(b)$ is determined as discussed in Section 6.2.2, σ_R can be evaluated by

$$\sigma_R = \int_0^\infty 2\pi b \, p(b) \, db \tag{6.12}$$

If the reaction is one for which the narrow-angle 'rainbow' structure is not significantly affected by reaction (or anisotropy, etc.), the optical analysis is believed to be quite reliable, and this method probably gives the best estimate of σ_R obtainable from beam measurements. In general, the σ_R obtained from equation (6.12) must be taken with the same reservations as the $p(b)$ functions, especially until some reliable independent data are available as a check on some of the systems.

From the reactive scattering, σ_R can be obtained, in principle, by either of the integrations indicated in equation (6.10) if reliable absolute values are available for $\langle d^2\sigma/d^2\Omega \rangle$ or $[d^3\sigma/d^3\omega \, dw]$ over the whole range of variables. Integrating the laboratory cross-sections over all angles has been used up to now only to set rough lower limits on σ_R, since most measurements have not included a sufficient range of laboratory angles for the complete integration.

The centre-of-mass cross-sections have the advantage (for non-oriented beams) of depending only on the single CM scattering angle θ. This allows a reliable evaluation of the integral as long as the significant ranges of θ are spanned, even though laboratory data may be taken only in a single plane. However, the cross-sections so obtained will have any additional uncertainties introduced in the LAB $\rightarrow$ CM transformation. Several schemes have been proposed[14] for 'calibrating' the absolute magnitudes of the differential cross-sections by finding the value of the constant, κ, in

$$\sigma_R \text{ (absolute)} = \kappa \sigma_R \text{ (relative)} \tag{6.13}$$

These are: (a) assume $\kappa = I_{el}$ (absolute)$/I_{el}$ (relative) at some small angle and obtain the absolute elastic (non-reactive) cross-section by computation from a van der Waals interaction potential with a force constant estimated by the Slater–Kirkwood approximation, (b) take $\kappa = \sigma_{TOTAL}(\text{abs})/\sigma_{TOTAL}(\text{rel})$ where σ_{TOTAL} includes both reactive and non-reactive scattering; $\sigma_{TOTAL}(\text{rel})$ is to be obtained by integration of the reactive and non-reactive cross-sections over all angles and σ_{TOTAL}(abs) to be taken from the Landau–Lifschitz approximation for the van der Waals potential mentioned above, (c) the same as (b), except taking σ_{TOTAL}(rel) from the attentuation of the parent beam. Both the latter methods require careful attention to correction for the finite angular resolution. Most of the σ_R values given in Table 6.1 were obtained by one of the three latter methods. All these methods give values in the same neighbourhood for a given reaction and, where the comparison is possible, in reasonable agreement with the results obtained by the non-beam method described below.

Recently, extremely accurate values of σ_R for the reactions of alkali metals with I_2 have been obtained by a non-beam method[20, 24]. In these determinations, a vapour-phase mixture of I_2, the alkali halide, and a substantial excess of xenon are irradiated with a pulse of u.v. light. At the wavelengths used, the only significant effect of the light is to dissociate the alkali halide into ground-state atoms. These are produced uniformly throughout the cell, and after the flash they begin to react with the known concentration of I_2 molecules. The rate of disappearance of the free alkali metal atoms is followed by the decay in the absorption of a resonance line. This gives $\gamma = n\langle v\sigma(v)\rangle$ where n is the number density of I_2 in the cell, and the brackets indicate an average over the distribution in relative velocities v. This distribution is assumed to be Maxwellian at the temperature of the cell because of the moderating effect of the xenon, and an average cross-section $\bar{\sigma}_R$ is then extracted from

$$\bar{\sigma}_R = \gamma/n\bar{v} \tag{6.14}$$

where $\bar{v}$ is the *average* relative velocity at the temperature of the experiment. This equation assumes that σ_R is independent of v. A slight energy dependence, such as that suggested by the data of reference 33, however, would not have much effect on the values obtained. The absolute accuracy of the cross-sections obtained is estimated as 15%, but the *relative* accuracy within the alkali series is c. 5%. The same method has been used[98] to determine cross-sections for collisions of Tl ($6^2P_{\frac{1}{2}}$ and $6^2P_{\frac{3}{2}}$) with I_2. The ground-state atoms ($^2P_{\frac{3}{2}}$) react ($Tl + I_2 \rightarrow TlI + I$), but the principal process removing the metastable atoms seems to be inelastic collisions to give the ground-state atoms. Brus[99] has also applied this technique to measure lifetimes of Na(3^2P) and Tl(7^2S), quenched by I_2, as a function of incident kinetic energy.

6.2.4   Product angular distributions

Since no information of any kind was available on product angular distributions before molecular beam studies were possible, it is not surprising that the subject received meagre attention in the development of the concepts of reaction kinetics. As we have seen in Section 6.2.2, an interpretive scheme based on the optical model has been developed for the *reactant* angular distributions. Although it has some problems, this model allows a plausible interpretation of the observations in terms of simple quantities. The situation for product angular distributions is far less gratifying, where no comparable unifying principle has been developed. Indeed, in classical trajectory studies of reactive scattering, for example, the product angular distribution remains the most troublesome property to correlate with features of the potential surface. However, despite the lack of a *quantitative* theory, there are significant qualitative questions that can be resolved from the product angular distributions: Is the reaction process a direct one or does it go through a long-lived complex? Are the impact parameters giving rise to reactions predominantly large or small?

Roughly speaking, these qualitative correlations rely on the approximate separation of the product scattering angle, θ, for a given set of initial conditions into the sum of contributions, θ_R, as the reactants approach, θ_P as they separate, and θ_C from the motion of the intermediate complex. The distinction between *direct* and *complex* processes can then be drawn according to whether θ_C is small or large, i.e. whether the lifetime of the intermediate is short or long in comparison to its rotational period. If the lifetime is sufficiently long on average to include several rotational periods, the fragments of the disintegration would be expected to be uniformly distributed insofar as allowed by the conservation laws; in any event they should be symmetrically distributed about $\theta = \pi/2$. Hence the simple absence or presence of such symmetry can be used to distinguish the two limiting cases.

A quantum-mechanical rationalisation of the separability of θ into $\theta_R + \theta_C + \theta_P$ follows (at least roughly) from a parametrisation of the S-matrix that Levine has shown[100, 101] to be valid under quite general conditions:

$$S_{\alpha\beta} = \exp(i\delta_\alpha)S'_{\alpha\beta}\exp(i\delta_\beta)$$

where δ_α and δ_β are, respectively, phase-shifts for 'distortion potentials' in the entrance and exit channels, and $S'_{\alpha\beta}$ is an 'interaction region' contribution. A stationary phase treatment along lines similar to those employed in elastic scattering would presumably yield the product scattering angle as the desired sum from the above form of the S-matrix.

The reasons for expecting forward–backward symmetry in long-lived collisions can be obtained quantum mechanically through use of the *collision lifetime matrix* $Q = ihS\dfrac{dS^+}{dE}$ (S is the scattering matrix)[102]. When

the elements of Q are sufficiently large, averaging the differential cross-section over a small range of energy would lead to the cancellation of all cross-terms involving different S-matrix elements in a partial-wave expansion of the scattering amplitude. The remaining terms, all *squares* of functions that are either even or odd under reflection through $\theta = \pi/2$, have the predicted forward–backward symmetry. Actually, for the cross-terms to vanish in an energy average it is not sufficient simply that Q should have large elements.

It is necessary that all the S-matrix elements undergo large changes *relative to each other*, i.e. that the *differences* between the eigenvalues of Q be large. Otherwise it would be possible for the whole S-matrix to change rapidly with energy, but in a way that preserves the relationships between large numbers of elements. Hence it is in principle possible to have long lifetimes without the cross-section exhibiting this symmetry, but it seems likely that generally large elements of Q would also imply the stronger condition for many-channel cases.

6.2.4.1 Direct processes

For a *direct* process, the important contributions to θ, θ_R and θ_P, can be thought of each as one-half an elastic scattering angle. If the distortion potentials in the entrance and exit resemble ordinary elastic potentials, one is then led to the association of *small* scattering angles with *large* impact parameters and vice versa (as in elastic scattering). Hence, a direct reaction with a total reactive cross-section much larger than a hard-sphere value would necessarily involve large impact parameters, and would therefore be expected to produce *forward-peaked* products. This expectation is well borne out by a large number of observations on systems discussed below in Section 6.4.5.

Cross-sections *smaller than* or comparable to hard-sphere values do not necessarily carry the reverse implication, however. To be sure, cases are known where a small reactive cross-section seemingly owes to the participation of only small impact parameters in the reaction, and for these reactions the products are primarily scattered backwards. These are the 'rebound' reactions discussed in reference 2. However, small cross-sections can also arise in principle from predominantly large impact parameters but with small probabilities. The latter situation would favour forward-peaked results. Possible examples of the second behaviour are to be found in the interhalogen exchange reactions (see Section 6.4.3) and in the reaction of Li with NO_2 [63], although it is also possible that these results could be interpreted instead by adjustment of the distortion potentials.

6.2.4.2 Collision complexes

Against the expectations of many people at the time, the earliest molecular beam observations of product distributions clearly did not support the existence of any intermediate complex of appreciable lifetime. On the contrary, the processes studied all appeared to involve reactions occurring in times comparable to a single molecular vibrational period. However, in 1967, Miller, Saffron and Herschbach[55] reported observing angular distributions that were clearly symmetric about $\theta = \pi/2$ for several alkali metal–alkali halide exchange reactions. They presented an analysis in terms of a statistical-complex model that gave a quite good fit to their angular distributions. At the same time Ham, Kinsey and Klein[41–43], reported evidence for sticky collisions in non-reactive systems of alkali atoms and SO_2, NO_2, CO_2 and NO. The demonstration of the existence of long-lived intermediates is not quite so straightforward in this case, since there will always be a forward-peaked 'non-complex' component in the *reactant* angular distribution from large-impact-parameter elastic collisions. However, when the initial and final orbital angular momenta are large compared to the rotational angular momenta, as was the case here, the wide-angle differential scattering cross-section should vary approximately as the cosecant of the scattering angle (except near $\theta = 0$ or $\theta = \pi$) for the long lifetime case, thus rising steeply past $\theta = \pi/2$ as opposed to the continued gentle fall-off expected for impulsive collisions. The non-reactively scattered reactants in the experiments of Miller et al. also exhibited this behaviour. Shortly afterwards, examples were found that seemed to show complexes whose lifetimes are comparable to the rotation period[26] (osculating complex). Since then other systems exhibiting both the behaviour of long-lived complexes[30, 63] and osculating complexes[49, 50] have been observed. It is still unclear exactly what features of the interaction surfaces are required to produce complex rather than direct reactions[103].

6.3 EXPERIMENTAL TECHNIQUES

6.3.1 'Supermachines'

The technical development of the past few years with the greatest potential impact on the field of reactive scattering is the advent of the 'supermachine', i.e., molecular beam apparatus with most or all of the following features: (a) universal detection by electron-impact ionisation and mass analysis of ions; (b) nozzle sources of parent beams (see Section 6.3.2); (c) high pumping speeds and numerous stages of differential pumping; (d) beam modulation coupled either with phase-sensitive detection or gated scaling; (e) the capability of angular scans over a reasonably wide range. Many of these features have already appeared in the early studies of the hydrogen exchange reactions[22, 27, 28]. Two supermachines specifically aimed at reactive scattering capabilities have been described in the literature[104, 105]. Results of investigations using these and others for reactive studies have appeared[9, 16, 21, 31, 40, 50, 51, 68, 72]. Several other supermachines are known to be in various stages of design, construction, or use.

The most critical element in a supermachine is the detector. The best efficiencies achieved for ionisation and detection are only a few tenths of a per cent. In most experiments this will mean that statistical fluctuations in the measurements will be a significant source of noise (often the limiting source)[104]. When this is the case, a figure of merit, τ, can be taken as the ratio of the square of the signal-to-noise ratio R to the measuring interval t [104]:

$$\tau = R^2 t^{-1} = \varepsilon s^2/(s+2\beta)$$

where s and β are, respectively, the concentration of molecules contributing to the signal and background, and ε is the counting rate produced by a unit concentration. For the fractional improvement in τ for fractional changes in s, β, and ε, we find

$$\tau_\varepsilon \equiv (\partial \ln \tau/\partial \ln \varepsilon)_{s,\beta} = 1 \tag{6.15}$$

$$\tau_s \equiv (\partial \ln \tau/\partial \ln s)_{\varepsilon,\beta} = 1 + \frac{2\beta}{s+2\beta} \tag{6.16}$$

$$\tau_\beta \equiv -(\partial \ln \tau/\partial \ln \beta)_{\varepsilon,s} = \frac{2\beta}{s+2\beta} \tag{6.17}$$

so that $\tau_\beta \leqslant \tau_\varepsilon \leqslant \tau_s$; i.e. from any given operating point the greatest improvement is obtained by increasing the concentration of signal molecules, followed by increasing the detection efficiency, followed by reducing the concentration of background molecules. However, the capability of achieving significant change in the different variables might well predicate a different order of priority.

For mass analysis in supermachines, quadrupole mass spectrometers offer the advantages of easily adjustable transmission and resolution, and they are used in most of the machines described in the literature. The electron-impact ionisers used have been mostly of a type utilising space-charge focusing of the positive ions to aid efficient extraction. While this improves the detection efficiency significantly, it has the disadvantage that the efficiency for any given ionic species may be sensitive to the total concentration of positive ions produced, since they partially neutralise the space charge. Thus the ion current can be a non-linear function of the product intensity, or worse, there may be sensitivity to angle-dependent background in the detector of ions at masses rejected by the mass spectrometer.

Beam modulation has long been used as an aid in distinguishing signal from background. A novel twist to this technique is used in the apparatus described in reference 105. These authors modulate *both* beams at different frequencies and detect at the beat frequency. A fixed phase relation is maintained between the two modulated beams by running the two choppers (with unequal numbers of open spaces) off a single shaft.

6.3.2 Nozzle beams

Most techniques, such as velocity selection or state selection, for reducing the spread in initial conditions of the parent beams suffer from being *analytic* rather than *synthetic*, i.e. they remove molecules that do not have the desired properties rather than enhancing the number that do, with attendant intensity reductions. One notable exception to this is the use of supersonic nozzle beams for achieving narrow velocity distributions. In a free hydrodynamic expansion, the Maxwellian velocity distribution collapses into a narrower distribution peaked at a higher value than before by the addition of a flow velocity[106]. There is also a modest increase in the ratio of the on-line integrated beam intensity to the total rate of flow from the course. (Enormous enhancement of beam intensity/flow-rate can be achieved in systems with separately pumped 'skimmer' regions if the flow-rate is taken to be that through the skimmer)[106].

The realisation that many of the advantages of nozzle sources can be obtained with ordinary sources simply by pushing the source pressure beyond the effusive limit, with no necessity for carefully designed nozzles or skimmers, has led to widespread use of 'Laval slits'[2] and other simple nozzle arrangements. Most of the applications in reactive scattering have allowed the free expansion to proceed through the transition to molecular flow, so that a collimator rather than a skimmer defines the beam.

The nozzle-source velocity distribution $f_N(v)$ is usually described in terms of a flow velocity α_N and a transverse temperature T_N: $f_N(v) = Av^2 \exp[-2m(v-\alpha_N)^2/kT_N]$. T_N and α_N are determined by the 'Mach number' M

of the expansion by

$$T_N = T(1 + \frac{\gamma - 1}{2}M^2)^{-1} \tag{6.18}$$

(where T is the source temperature and γ the specific heat ratio C_p/C_v), and

$$\alpha_N^2 = \frac{\gamma}{\gamma - 1} \frac{2k(T - T_N)}{m} \tag{6.19}$$

The Mach number is zero at the throat of the nozzle and increases as long as continuum expansion is maintained, stabilising at a terminal value M_T when molecular flow takes over. At present M_T cannot be predicted with certainty[106, 107]. For polyatomic gases part of the internal energy is also transformed into flow kinetic energy. The values of α_N and T_N then depend on the extent of rotational and vibrational relaxation, i.e. the relaxation determines the 'correct' γ to use in equations (6.18) and (6.19). Usually rotation is fully relaxed and vibration only slightly relaxed.

Transverse temperatures as low as $\frac{1}{20}$ the source temperature are commonly achieved in practice, so that additional mechanical velocity selection is usually not necessary. As the internal degrees of freedom relax, their distributions also collapse, but without very careful measurements it is difficult to determine the new distributions. The increase in the mean velocity of the beam in the expansion can be amplified by 'seeding', i.e. expanding a mixture in which the species of interest is dilute in a lighter carrier. The flow velocity becomes roughly that of molecules of the average molecular weight. Jaffe and Anderson[45] have used seeded beams to reach translational energies as high as 215 kcal mol^{-1} in a study of the hydrogen iodide reaction (see Section 6.4.4). Presumably seeding could also be employed in reverse, slowing the molecules to be used with an excess of heavier inert carrier.

6.3.2.1 Dimer beams

In an expansion *chemical* relaxation can also occur, resulting in the formation of dimers or larger aggregates[107]. Although this possibility is often a disadvantage, since requirements of beam purity may preclude reaching too high a terminal Mach number, it can also be put to use, as was recently shown by an alkali-dimer nozzle source developed by Gordon, Lee and Herschbach[107]. This source has now been used in two studies of the reactions of alkali dimers[70, 78]. A further interesting possibility offered by such recombination sources is that of producing vibrationally excited beams if the dimers are formed in states of high vibrational quantum number and do not fully relax in the continuum flow part of the jet[107].

6.3.3 Combined velocity selection and velocity analysis

The combination of velocity selection of one of the parent beams with velocity analysis of the reaction products has been developed to an art by the group at the University of Wisconsin as they have progressed through a series of studies on the reactions K + HBr and DBr[108, 65, 32], K + Br$_2$[109, 73] and

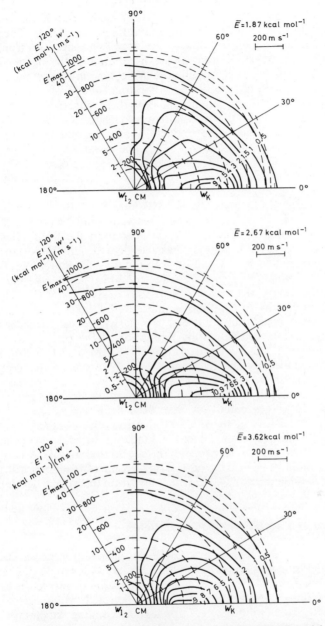

Figure 6.1 KI polar CM contour maps. Best KI CM differential cross-sections $d^3\sigma(\theta, w)/d^2\omega\,dw$ obtained from the laboratory contour maps at *three* values of $\bar{E}$ by an iterative least-squares fitting procedure. All three are normalized to 10 in the peak region. The vectors $\mathbf{W}_{I_2}$ and $\mathbf{W}_K$ are the nominal CM values; the dashed energy circles E'_{max} are the thermodynamic limits for *nominal* collision conditions at each $\bar{E}$ (From Gillen, K. T., *et al.*[33], by permission of the American Institute of Physics)

$K + I_2$ [33]. Compounding the advantages of both techniques permits a LAB $\rightarrow$ CM inversion of the data with a minimum of ambiguity. The product analysis removes guesswork about the recoil velocity spectrum, while reactant selection not only sharpens the distribution in the *magnitude* of the initial relative velocity (and hence in the initial kinetic energy) but also (and equally importantly) it reduces the laboratory spread of CM coordinate systems (origins as well as orientations). With these refinements, investigation of a number of properties that otherwise would be inaccessible may be undertaken: (a) The dependence of the product CM angular- and speed-distribution on the initial relative kinetic energy $\bar{E}$. (b) The variation (within a restricted range) of the total reactive cross-section with the incident relative speed. (c) The extent of dependence of the recoil velocity spectrum of the products on their scattering angle.

The pathological kinematics in $K + HBr$ and $K + DBr$ (see Section 6.4.1) *require* both velocity selection of the reactants and velocity analysis of products to resolve even qualitative behaviour in the CM system when KBr is detected. Selection/analysis has also proved quite important for the other systems studied, $K + Br_2$ and $K + I_2$, but for different reasons. Over the range of initial kinetic energy $\bar{E}$ from 1.9 to 3.6 kcal mol^{-1}, the *shape* of the doubly differential CM cross-section $[d^3\sigma/d^2\omega\ dw]$ for $K + I_2$ was found to be quite insensitive to $\bar{E}$, as shown in Figure 6.1, taken from reference 33. ($K + Br_2$ gives similar results, but with less clarity because the techniques of data interpretation were less fully developed at the time of that study). Also, though significant coupling between the CM θ and E' distributions could be concluded for the KI from $K + I_2$, the extent of this coupling is surprisingly small. Both the above qualitative facts are predicted by the 'stripping models', but without the precision made possible with the selection/analysis they could not have been definitely verified. Moreover, the *quantitative details* of the E' and θ distributions, obtainable with confidence only from measurements of this kind, are the acid test of the ability of various models for accurate prediction (see Section 6.4.5.1). Accumulation of a larger body of such data on a larger group of reactions will provide extremely useful and stringent guidelines for future theoretical developments.

6.3.4   Use of electric and magnetic fields

Beams of neutral molecules can be deflected in inhomogeneous electric or magnetic fields, a property which has been used as a simple discriminator of polar from non-polar or diamagnetic from magnetic molecules as an aid to unambiguous identification of species detected with surface ionisation[61-63, 110]. More sophisticated applications, taking advantage of the rotational-state dependence of the deflection pattern for diatomics, were discussed in reference 2. These studies have now been extended to include $K + HBr$, $Cs + HBr$, $K + Br_2$, $Cs + Br_2$, $Cs + MeI$ and $Cs + CCl_2$ [52], all of which show *small* values of $\bar{E}_R$, the average rotational *energy*.*

---

*According to reference 39, the quantitative results given in reference 52 have since been revised in the Ph.D. Thesis of C. Maltz, Harvard University, 1969.

6.3.4.1 Focusing analysis and the rotational-state distribution

States whose *effective* moments are *negative* are strongly focussed in fields of appropriate geometry[111]. For diatomic or linear molecules in a state whose vibrational, rotational, and space-quantisation quantum numbers are V, J, M the effective moment μ_{eff} is given in terms of the moment of inertia, I, the fixed dipole moment of the molecule, μ, and the field strength, ε, by:

$$\mu_{eff} = \frac{2I\mu^2\varepsilon}{h} \frac{3M^2 - J(J+1)}{J(J+1)(2J-1)(2J+3)} \qquad (6.20)$$

States for which $3M^2 \geqslant J(J+1)$ will be focused in an electrostatic quadrupole field of the proper strength. A new era in the refinement of measurements on reaction products was launched by the study of Grice *et al.*[39] using the field strength dependence of the focused intensity to elucidate the full rotational-state distribution of the product RbBr from $Rb + Br_2$. The observed intensity pattern was congruent to within experimental error with a universal Maxwell–Boltzmann curve, with $T_R/\mu^2 = 8.2$ KD^{-2}, irrespective of the scattering angle or the velocity of the RbBr. Taking the dipole moment of the vibrationally excited RbBr to be 17.5 D as predicted by a polarised-ion model for RbBr[112, 113] yields $T_R = 2550 \pm 300$ K ($\bar{E}_R = 5.0 \pm 0.6$ kcal mol^{-1}). Since the RbBr velocity distribution was also measured in this work, the average vibrational energy (E_V c. 37 kcal mol^{-1}) could be determined by difference.

6.3.4.2 Electric resonance and the vibrational-state distribution

Freund *et al.*[30] have reached another state of elaboration by linking a cross-beam scattering apparatus with an electric resonance spectrometer and observing the radio-frequency Stark spectrum of the CsF product from $Cs + SF_6$. In the spectrometer section of their apparatus, with the usual Rabi A–C–B fields, molecules in definite J, M states are selected by focusing in the static quadrupolar A-field. These are then deflected by the dipolar B-field onto the detector *only* if a $\Delta M = \pm 1$ transition is induced in the C-field region (see Figure 6.2). Since the resonant frequencies for these transitions differ slightly with the vibrational quantum number, a single vibrational state can be made to 'flop-in' at a time. The relative heights of the peaks appearing as the resonance is swept through the spectrum give the relative populations of the various vibrational states for the particular J, M state being focused.

Relative vibrational state populations for $v = 0$ to $v = 4$ or 5 were obtained in this way for a number of rotational states, all with J values between 1 and 4. All the observations were consistent with a Boltzmann distribution in vibrational state, with no significant differences being found for the different rotational states examined. The vibrational temperature found was $T_V = 1120 \pm 90$ K (99% confidence level) when the SF$_6$ source was operated at 230 K, and $T_V = 1270 \pm 140$ K (also 99% level) when it was at 600 K. The Boltzmann-like variation of the vibrational distribution is suggestive of a strongly coupled intermediate complex. This is supported by additional findings of a number of unpublished investigations, quoted by Freund *et al.*, which show: (a) a product angular distribution with forward–backward symmetry, (b) average translational and rotational temperatures for CsF roughly the same as T_V, $T_t \approx 1200$ K, $T_R \approx 1050$. (This illustrates, however, that the rotational states investigated in this study are

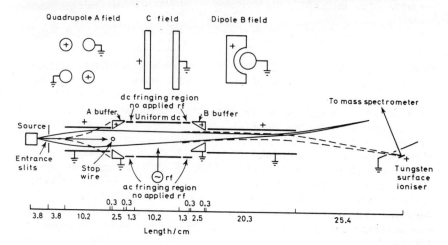

Figure 6.2 Schematic view of electric-resonance spectrometer, showing field configuration used in the work of reference 30. Lower scale gives distances (in centimetres) between successive elements along the beam path (From Freund *et al.*[30], by permission of the American Institute of Physics)

far from typical ones. A typical J would be more like 50 for $T_R \approx 1000$.) The appearance of most of the reaction exothermicity in internal excitation of SF_5 is also said to agree well with an unpublished study of statistical complexes using unimolecular decomposition theory. Preliminary results on $Li + SF_6$, showing $T_V \approx 1800 \pm 200$ K, were also reported.

Techniques similar to these could be used to select initial states rather than to analyse the final ones. Waech and Bernstein[114] have produced measurable intensities of beams in definite V, J, M states using the focussing-resonance techniques, although as yet no collision studies employing this method have been published. Waech *et al.*[115] have also shown that a ten-pole field can focus states of diatomics with *positive* effective moments. However, there is focusing in only one direction transverse to the beam so that a line image rather than a point image is produced, giving poorer intensity than can be obtained with the states of negative effective moments.

6.3.4.3 *Oriented molecules and reactive asymmetry*

Since diatomic (and linear) molecules have *second-order* Stark effects, the focusing characteristics are independent of the sign of M. Symmetric top molecules, on the other hand, by virtue of a twofold degeneracy in the K

quantum number (projection of rotational angular momentum on the molecular symmetry axis), have *first-order* Stark effects. As a consequence, focusing in a six-pole electrostatic field followed by adiabatic passage to a homogeneous field can be used to produce a beam that is *oriented* in laboratory coordinates[116-118]. Because of the high density of rotational states for most symmetric tops, individual J, M, K states are not distinguished. Instead, certain ranges of cos $\theta = MK/J(J+1)$ are transmitted and others eliminated. Jones and Brooks[119] have also shown that asymmetric rotors can be oriented in favourable cases.

In 1966, two groups[10, 18] reported almost simultaneously on the use of oriented MeI beams to demonstrate the difference in reactivity of the two 'ends' of the molecule with alkali metals. Brooks and Jones[18] used potassium, and Beuhler, Bernstein and Kramer[18] used rubidium as the alkali. Both sets of data clearly and directly showed that attack at the I end of MeI is favoured over attack at the Me end. Later, Beuhler and Bernstein improved their experimental method by velocity selecting the MeI beam and aligning the homogeneous field to increase the orientation of the MeI in the centre-of-mass coordinate system[11] and gave extensive quantitative analysis of the data[12].

Brooks has also looked at the reaction of K with oriented CF_3I [19]. Under the assumption that the I atom is the positive end of the dipole (which seems extremely likely in view of the large electronegativity of F), the reported data at a scattering angle of c. 60 degrees in the LAB coordinate system show that reaction favours attack at the CF_3 end of the molecule, despite strong support from other evidence for assuming that the product is KI. Since the publication of these findings, however, Brooks has found another peak in the laboratory angular distribution at wider angles which shows favourable reaction at the I end[110], suggesting the following pretty interpretation: reaction at the CF_3 end proceeds from large impact parameters to give mostly forward scattering in the centre-of-mass coordinate system, while reaction at the I-end involves more nearly head-on collisions and hence largely backwards scattering of the product (presumed to be KI in both cases).

6.3.5 Light

6.3.5.1 Chemiluminescence

The richest details available on the distribution of the total energy of reaction products among the available modes are those from infrared chemiluminescence studies in dilute gases (Chapter 5, Volume 9). Although no i.r. chemiluminescence has been observed yet in molecular beam reaction, *visible* chemiluminescence has been seen in a few cases and promises to be an extremely useful technique for future work. The first example, reported by Moulton and Herschbach[57], was seen in a 'triple beam' experiment. A beam of $KBr^†$, formed by collimation of products of the reaction between beams of K and Br_2, was crossed with a wide beam of Na atoms. The exchange reaction $Na + KBr^† \rightarrow NaBr + K^*$ was then observed via the light emitted by the electronically excited K atom, confirming this two-step process as one route for producing the observed emission of alkali atom resonance radiation in dilute flame experiments. The $K + ClI$ reaction also resulted in a product that gave rise to K^* radiation when crossed with a Na beam, clearly indicating an appreciable yield of $KCl^\dagger$, since formation of $KI^\dagger$ sufficiently excited internally to give light emission in the second reaction is not allowed energetically.

Recently, Struve et al.[70] have seen D-line emission from the reaction of Cl atoms with Na_2 and K_2 dimers: $M_2 + Cl \rightarrow MCl + M^*$ ($\sigma_R \approx 10$–100 Å^2). The emission was also observed when the Cl-atom beam was replaced with a Cl_2 beam, but with the intensity down by about a factor of 500 for Na and about a factor of 30 for K. The sodium emission is presumed to arise either from a small Na_3 component in the beam or from two-collision processes, but an origin from Cl_2 collisions with vibrationally excited K_2 cannot be ruled out for the potassium light.

Ottinger and Zare[60] found intense visible light from reactions of Ba and Ca with NO_2 and N_2O in crossed beams. Only in the case of the $Ba + NO_2$ reaction could the product (BaO) be definitely assigned, but the metal oxides BaO and CaO were thought likely for at least three of the four reactions. The BaO spectrum from $Ba + NO_2$ sets a new lower limit (5.74 eV) for the ground-state dissociation energy of this molecule. More recently, Jonah and Zare[46] have seen chemiluminescence in the formation of barium and strontium dichlorides by radiative association (the inverse of molecular pre-dissociation) in crossed beams: $Ba + Cl_2 \rightarrow BaCl_2^* \rightarrow BaCl_2 + h\nu$. Some chemiluminescence was also attributed to $Ba + Cl_2 \rightarrow BaCl^* \rightarrow BaCl + h\nu$.

6.3.5.2   Optical pumping

The inverse of chemiluminescence, optical pumping of the reactants to favourable states, gives another method for preparing non-equilibrium initial distributions. Like nozzle beams, it is synthetic rather than analytic, so that the intensity can in principle be quite high. Odiorne, Brooks, and Kasper[59] have just finished a study in which a hydrogen chloride chemical laser was used to pump a significant fraction of a HCl beam from the $v = 0$ to the $v = 1$ state, which is not significantly populated at equilibrium at room temperature. The chemical laser derives from flash photolysis of a flowing mixture of HI and Cl_2. It is typical in such systems for a *group* of rotational lines in the vibrational band to 'lase' sequentially following the pulse, so that many of the thermally populated rotational states in the beam become equilibrated with the excited vibrational state during irradiation by the laser pulse. Although the flashes of the initiating lamp last only c. 20 μs (and each laser pulse a much shorter time), the effective duty time is determined by the radiative lifetime of the excited vibrational state, which is of the order of several ms for $HCl^\dagger$ ($v = 1$). In the work of Odiorne et al. the laser beam crosses the collimated molecular beam a short distance before the scattering centre. They estimated that roughly 25% of the beam was excited to $v = 1$ in their experiments. A discussion of the results of this work is to be found in Section 6.4.3.

6.3.6 Miscellaneous technical developments

To enumerate all the recent developments in techniques that have obvious or possible applications to reactive scattering studies would go far beyond the scope of this paper. In this section, I intend only to draw attention to a few of the new techniques whose promise seems to me especially great.

(a) *Internal energy detector* — Gillen and Bernstein[121] observed that the efficiency of a surface ionisation detector in the 'non-detecting' mode for ionising KI is strongly dependent on the internal excitation energy of the KI molecule, and suggest the possible use of this behaviour as a way of measuring average internal energies of reaction products.

(b) *Free radical beam source* — Kalos and Grosser[122] have developed an intense source for beams of alkyl radicals that will surely find many useful applications in beam studies of radical reactions.

(c) *Radioactive beams* — Grover et al.[123] have indicated possibilities for using beams of molecules containing short-lived radioactive nuclides in reactive studies.

(d) *Time-of-flight velocity analysis* — Time-of-flight (TOF) analysis has been widely used in studying the properties of nozzle beams[106] and is beginning to appear in reactive scattering studies. McDonald[124], in an unpublished report, describes an elaborate system using a small digital computer in connection with the TOF measurement. Hirschy and Aldridge[125] describe an interesting method for obtaining the TOF spectrum as the transform of the signal generated by a 'snaggle-toothed' chopper with many teeth and gaps in a pattern representing a pseudo-random maximum-length sequence of binary numbers.

6.4 DISCUSSION OF SPECIFIC SYSTEMS

6.4.1 K + HBr and related reactions

It is ironic that the K + HBr reaction, the wellspring for all the alkali reaction studies, and itself one of the most thoroughly investigated processes[7, 32, 52, 53, 65], remains somewhat enigmatic. The non-reactive scattering[7, 32] is not at issue. With its well-resolved rainbow structure, it affords one of the best examples for optical analysis to obtain the reaction probability as a function of impact parameter. HBr and DBr show qualitatively quite similar differential scattering with a typical elastic appearance at small angles and more rapid fall-off at wider angles. HBr appears to be the more reactive, with a σ_R estimated to be about 40% larger than for DBr[7, 32]. An isotope effect of this magnitude is quite plausible for substitution of D for H.

The reactively scattered products, however, raise some interesting questions that have not been resolved. The three possible isotopic species, HBr, DBr and TBr, have all been investigated but not in the same laboratory nor with the same techniques. With the idea of using the extremely one-sided kinematics to advantage, Martin and Kinsey[53] developed a method for detecting atomic tritium based on selective adsorption of the atoms by molybdenum trioxide films, followed with counting of the weak tritium emission from the surfaces of the films. The LAB angular resolution in these experiments was crude but adequate to give a qualitative CM angular distribution of the product, since the velocity of the tritium atom is an order of magnitude greater than the CM velocity. Most of the product was found at angles corresponding to *backwards* scattering of KBr with respect to the direction of the incident K ($\theta \approx 180$ degrees). Although the magnitude of the recoil velocity of the products was not measured directly, it was felt that the observations were only consistent with a recoil kinetic energy roughly the same as the incident kinetic energy or slightly smaller. The results ·with Cs + TBr were similar.

The HBr and DBr reactions with K were studied by Gillen, Riley and Bernstein[32, 65] using velocity selection and analysis. Their primary K beam was velocity selected with a resolution of 14.4 % (FWHM), and the reactively scattered KBr was velocity analysed (resolution 4.7 %). The resulting laboratory flux maps[32], obtained from hundreds of data points, were subjected to a very careful computer analysis[65] to extract distributions in angles *and* in recoil velocities of the products in the CM coordinate system. The recoil energy distribution was found to peak at a value near the incident translational energy. The angular distributions were presented as 'best fits' to the data, along with limits on either side giving 'reasonable fits'. The best K + HBr angular distribution they give varies linearly, with $d^2\sigma/d^2\omega$ in the *backwards* ($\theta = 180$ degrees) direction being 1.5 that in the forward direction. The 'reasonable fit' range permits generous variation around this curve, but in any event is decidedly backwards. K + DBr, however, shows an almost isotropic distribution, with a slight preference for *forward* scattering of the product, in opposition to the change that would have been predicted on the basis of the smaller reactive cross-section for DBr. The TBr results, showing strongly backwards peaking, thus do not fall into a smooth trend with the HBr, DBr results which move towards more forward peaking as the mass increases.

How is the apparent discrepancy to be understood? It is possible that one of the experimental results is incorrect. The inferences from the TBr experiments are unambiguous, but the method used has not been verified in independent experiments and systematic error cannot be ruled out. These data could also be thought to show some isotropic component of the product distribution in addition to the backwards spike. However, it does not appear that there could be enough to make these data compatible with the HBr, DBr results. Martin and Goldbaum are now engaged in experiments at the University of California (Riverside) to check and extend these results[126]. Their preliminary results appear to confirm the previous findings for the Cs reactions, and consequently it seems likely that the K experiments, when finished, will also reproduce the earlier data.

In contrast, the HBr and DBr reactions employed extremely well tested and widely used techniques that limit any plausible sources of inaccuracy to the interpretation process. Computer inversions of LAB measurements into CM data are always subject to doubts as to uniqueness. This is a particularly important consideration in the present case because of the unfavourable kinematics. However, the limiting curves given by Gillen *et al.* address this problem and would seem to rule out the possibility of any ambiguities requiring serious revision of the qualitative trend presented, unless there is either strong coupling between the angular- and recoil-distributions or the angular divergence of the cross-beam has an unexpectedly large influence on the LAB → CM transformation.

Alternatively, and this seems the most likely interpretation at present, we must conclude that the product angular distributions do not follow a consistent trend in the series HBr, DBr, TBr. This may reflect the importance of quantum affects in these reactions. A large fraction of the energy released in the reaction ($\Delta D°$) comes from differences between the large zero-point vibrational energy of the reactants and that of the products. In view of the extremely large effect of vibrational excitation on the K + HCl reaction[58, 59] (see Section 6.4.3), it would not be surprising to find that the substantial differences in zero-point energies in this series also produce major effects. Why these quantum effects would cause the observed non-uniform tendency is not clear, however.

6.4.2 The hydrogen exchange reaction

The reaction $H + H_2 \rightarrow H_2 + H$ is of enormous interest for obvious theoretical reasons. Two isotopic variants of this reaction have been studied in crossed beams: $H + D_2 \rightarrow HD + D$ [27, 28] and $D + H_2 \rightarrow HD + H$ [22, 31]. All the reported experiments have been basically similar. The HD product was detected using electron impact ionisation and mass analysis with a quadrupole mass spectrometer. The atomic beams (H or D) derived from thermal dissociation sources operated in the neighbourhood of 3000 K without

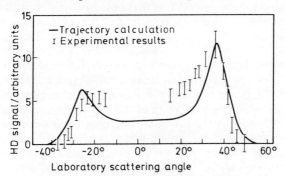

Figure 6.3 In-plane angular distribution of HD. The solid line is the theoretical result from the trajectory study of reference 127, and the data points are those of reference 31. The calculated results are normalised to agree with the experimental value at 36 degrees (From Brumer, P. and Karplus, M.[127], by permission of the American Institute of Physics)

velocity selection. These beams were modulated for lock-in detection of the product by chopping at frequencies in the kHz range and then crossed by DC thermal (77 K or 100 K) molecular beams (H_2 or D_2). Cryopumping at 4 K and differential pumping of the detector were employed.

The reported laboratory angular distribution in the $H + D_2$ study[27, 28], was flat to within the experimental precision in the range c. 15–50 degrees relative to the H beam. The absolute laboratory differential cross-section for reaction, estimated by comparison with the elastic cross-section, agreed satisfactorily with expectations.

The more accurate measurements on the $D + H_2$ reaction extend over negative, as well as positive, angles[22, 31], and show a doubly peaked laboratory distribution. Some information about the velocity spectrum of the HD product was also obtained from the phase of the HD signal. These velocity data and the angular distributions were well fit by assuming a laboratory angular distribution proportional to $(1 - E_0/E) \cos^2 1.35 (\pi - \theta)$ for E (the relative translational energy) greater than E_0 (taken to be 0.33 eV) and in the range where $\cos 1.35 (\pi - \theta)$ is positive[31]. (The scattering angle θ is zero when the product is scattered forward with respect to the incoming D beam in the CM coordinate system.) Brumer and Karplus[127] have recently communicated a trajectory calculation for this reaction, using a previously described surface[128]. Figure 6.3, reproduced from their paper, shows the quite good agreement between the computed and observed results.

6.4.3   Interhalogen reactions

Several reactions of the interhalogen series have now been investigated in crossed beams using universal detection. One of these reactions, $Cl + Br_2 \rightarrow ClBr + Br$, has been independently studied by groups at Freiburg[9], Harvard[50, 51], and Los Alamos[17, 21]. The three results are consistent and show qualitative behaviour resembling that observed in the alkali–halogen reactions: the product angular distribution is strongly forward-peaked, and most of the product energy appears as internal excitation ($\Delta D°$ is 6.7 kcal mol^{-1} for this reaction). Both these features are especially clear in the velocity-analysed data of the Los Alamos experiments[17, 21]. However, the total reactive cross-section, instead of being larger than a hard-sphere value as might have been expected from the product angular distribution (see Section 6.2.4.1), is distinctly smaller than the hard-sphere estimate. σ_R/σ_{el} is set at about 0.015–0.03 by the Los Alamos group[17]. It is not clear whether this is to be attributed to the participation mainly of large impact parameter collisions with small reaction probabilities or to small impact parameter with a surface that appears 'attractive' even for nearly head-on collisions of the reactants. Some light could be shed on this question by an investigation at good resolution of the non-reactive scattering of the reactants (see Section 6.2.2) to ascertain, for example, whether a relatively unperturbed rainbow structure can be seen. However, none of the reported studies used velocity selection of the primary beams and so the rainbow structure could not be resolved.

Another interesting facet of the $Cl + Br_2$ reaction revealed in the velocity analysis experiments is the apparently strong coupling between the angular- and recoil velocity distributions. Some coupling has been seen in the $K + I_2$ reaction, but there it is such a small effect that it only becomes apparent in experiments with velocity selection/analysis[33]. The LAB $\rightarrow$ CM transformation of the data in the $Cl + Br_2$ study was somewhat rudimentary, and it is possible that a more refined analysis of the data would reduce the apparent angle–energy coupling. Even in that event, however, it appears that there is stronger coupling in this case than for any known examples at the present.

The reactions $Br + I_2$, $Cl + IBr$, $Cl + BrI$, $Cl + I_2$ and $Br + ClI$ have also been studied by Lee et al.[51] pursuing evidence for preferred geometry of the reaction intermediates. The reactions $Cl + IBr$ and $Cl + BrI$ have about equal cross-section, and both exhibit forward-peaking (although it is more pronounced for $Cl + BrI$ than $Cl + IBr$). The greater spread of products towards wide angles in $Cl + IBr$ was felt to indicate a slightly longer-living complex and therefore probably more 'attraction', in keeping with the expectation of finding a lower energy for the intermediate with the least electronegative atom in the middle. This idea is further strengthened by the similarities between $Cl + I_2$ and $Cl + IBr$ and between $Cl + Br_2$ and $Cl + BrI$ as well as by the contrast of all these with $Br + ClI$ which has a backward-peaked product distribution and a reactive cross-section smaller by a factor of about 5.

6.4.4   Activation energies: HI + DI and K + HCl

The variation of the macroscopic rates of reactions with temperature has formed the basis for a major part of chemical kinetics. On the microscopic level, the analogous endeavour is investigation of the energy dependence of reactive cross-sections. Since the cross-sections in principle depend not only on the *total* energy but also upon its distribution among the various possible degrees of freedom, there are many directions from which the question of energy dependence can be approached. Most of the reactions that have been studied in beams have been chosen because they have reasonably large cross-sections at modest energies in order that sufficient reaction take place to be observable under beam conditions. Hence, they have been predominantly processes with small or negligible threshold energies, and they have often been appreciably exoergic. To the extent that the range of variation of the reactant energies can be neglected in comparison to the reaction exoergicity in these systems, the reactions may be considered as taking place at a single total energy. Simple application of the principle of microscopic reversibility[129–131] to data on the distributions of product energies among the different possible modes then allows the determination of the dependence of the rates of the reverse (endoergic) reactions on these distributions. The use of microscopic reversibility for a number of kinds of measurements of product distributions has recently been summarised[131].

The data obtainable from the product distributions by microscopic reversibility indicate the favourable dispositions of a *fixed* total energy for bringing about the reverse reactions. This information, although interesting in itself, has no bearing on the macroscopic *activation energy* for these reactions. The latter is determined instead by the energy dependence of the total reactive cross-section *averaged* over all the accessible states at a given total energy. Ordinarily, there has been a strong tendency to regard this as due to the dependence on the translational kinetic energy, probably because it then becomes easy to construct appealing collisional models. However, the non-statistical product distributions that have been observed in many of the beam results and even more dramatically in chemiluminescence studies (see Chapter 5, Volume 9) strongly suggest that it is not uncommon for internal energy to play the predominant role.

Striking direct evidence of the importance of vibrational excitation has been obtained in two recent beam experiments. In the first of these, Jaffe and Anderson[45, 132] have looked at one of the classics of chemical kinetics,

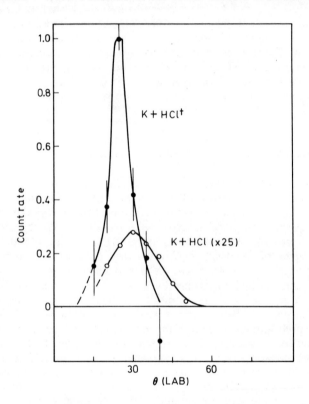

Figure 6.4 Count rate for KCl from HCl † and that for KCl from HCl versus laboratory scattering angle from K beam. Units are arbitrary, but correction has been made for dilution of the pulse of excited molecules by ground state molecules (From Odiorne, T.J. *et al.*[59], by permission of the American Institute of Physics)

the hydrogen iodide reaction. The specific reaction studied, HI + DI, has a well-established experimental activation energy, E_a, of 44 kcal mol^{-1}, although there is some question as to whether the elementary reaction gives HD + I$_2$ or HD + I + I as products. The molecular beam investigation was undertaken to resolve the question of whether the activation barrier can be overcome exclusively with translational energy. To that end a seeded nozzle beam (see Section 6.3.2) of HI, with He or H_2 as the carrier, was fired into a reaction chamber containing about 10^{-4} torr of DI. The contents of the reaction chamber were then probed with a quadrupole mass spectrometer for any HD product formed. The HI laboratory velocities, directly measured by a time-of-flight method, gave relative kinetic energies ranging between 20 and 109 kcal mol^{-1}. Within the sensitivity of the detection scheme, no reaction was observed, even at the highest energies attained (well above E_a). From the measured beam intensity and the calibrated limits of sensitivity, it was concluded that the reactive cross-section was smaller than 0.04 Å^2 at any of the energies investigated. Since this is insufficient to account for the observed thermal rate of the reaction, it is clear that at least some internal excitation of the reactants beyond the thermally populated region is essential to make the HI + DI reaction proceed. This raises interesting questions: Is it necessary that *all* the excitation be internal? What is the relative importance of rotation or vibration? Is it necessary for both reactant molecules to be internally excited?

Another source of direct evidence for the importance of internal excitation is found in the work of Odiorne, Brooks and Kasper[59] on the reaction K + HCl† ($v = 1$). An earlier investigation[58] of the same reaction with the *ground vibrational state* of HCl had revealed a small but measurable amount of product KCl formed; careful comparison with the HBr reaction yielded an estimate of 0.15 ± 0.03 Å^2 for σ_R. A qualitative explanation of this small value, based principally on consideration of angular momentum conservation, led to the postulation of an energy barrier of c. 1.5 kcal mol^{-1} above the slight endoergicity of the reaction.

The vibrationally excited HCl used in the reaction was produced by optical pumping of the HCl beam with the output of a hydrogen chloride chemical laser (see Section 6.3.5.2). The result, illustrated in Figure 6.4, is a dramatic enhancement of the observed KCl product. The authors estimate that the reactive cross-sections for HCl† ($v = 1$) is 100 times that for the ground-state molecule for roughly the same distributions of rotational states in the two instances. Excitation of one quantum of vibration in HCl amounts to an increase of c. 8.5 kcal mol^{-1} in the total energy. It would be extremely interesting to know whether an increase by the same amount in the relative kinetic energy would bring about a similar enhancement in the cross-section.

6.4.5 Stripping reactions

The set of so-called 'stripping reactions' have been more widely investigated than any others. The typical examples are reactions of alkali metals with halogens, but the group now includes reactions of at least one of the alkalis with all the following compounds: Br_2, BrCN, CBr_4, Cl_2, IBr, ICl, ICN, I_2, NOCl, PBr_3, PCl_3, SCl_2, $SnCl_4$ (see Table 6.1 for complete references to each system). The common features, whose similarities with nuclear stripping reactions give rise to the terminology, are: (a) The total reactive cross-sections are quite large ($\sigma_R > 100$ Å^2). (b) The product angular distributions are forward-peaked. (c) The non-reactive scattering falls off very rapidly, showing no small-angle rainbow structure. (d) Only a small fraction of the total energy of the products appears as translational kinetic energy.

Table 6.2 shows estimates of the fraction of products scattered into the forward hemisphere for several typical stripping processes and, for comparison, for one typical 'rebound' reaction[2] $Na + MeI$. Although the

Table 6.2

Reaction	Fraction of products scattered forward	Reference
K, Rb, Cs + Cl_2	0.78*	37
Li + Cl_2	0.69*	62
K, Rb, Cs + Br_2	0.72*	14
Na + Br_2	0.67*	15
Li + Br_2	0.63*	62
K, Rb, Cs + I_2	0.63*	14
K + I_2	0.60	33
Cs + ICl	0.71*	47
Rb + ICl	0.67*	47
K + ICl	0.67*	47
Na + ICl	0.57*	15
Li + ICl	0.63*	62
Na + MeI	0.29*	15

*Obtained by integrations of Legendre expansion of $\langle d^2\sigma/d^2\omega \rangle$ from references. Subject to some uncertainty because of approximate LAB → CM inversion used.

products are definitely scattered forward in these reactions, the preponderance of forward scattering is not as overwhelming as seems to be generally believed.

6.4.5.1 $K + I_2$ with velocity selection and analysis

It is not known just how closely the various stripping processes mirror each other's detailed behaviour. However, in the discussion of models, the reaction $K + I_2$ can be taken as the touchstone because of the wealth of information provided by the velocity selection/analysis experiments of Gillen et al.[33]. Qualitatively, their findings can be summarised as follows:

(a) *Product angular distribution* – The data encompass only part of the full range of CM angles to $\theta \approx 90$ degrees for the runs at average initial translational energy $\bar{E} = 1.87$ and 3.62 kcal, and to $\theta \approx 120$ degrees for $\bar{E} = 2.67$ kcal. Within this range $d^2\sigma/d^2\omega$ clearly peaks at $\theta = 0$, although there is some indication that it may be starting to rise again past $\theta = 90$ degrees in the $\bar{E} = 2.67$ run. Integration of $d^2\sigma/d^2\omega$ over the forward and backward hemispheres (using extrapolated data based on reference 14 for $\theta \geqslant 120$ degrees) indicates that 60% of the products appear in the forward hemisphere.

(b) *Recoil energy distribution* – The distribution in product translational energy E' *peaks* at $E' \approx 1.5$ kcal at all incident energies, falls rather sharply to $E' \approx 5$ kcal, then decreases more gradually to zero at the maximum E'. The *average* $\bar{E}'$, obtained by a rough integration using numbers from Figure 19 of reference 33, is $\bar{E}' \approx 10$ kcal. The recoil *velocity* distributions indicate a bimodal structure that is not obvious in the E' representation of the data. The second maximum corresponds (very roughly) to $E' \approx 15\text{--}20$ kcal. Some speculation as to possible origins of this effect can be found in reference 33.

(c) *Coupling between the CM velocity-angle distribution* — Although such coupling can definitely be discerned in the data, it is quite modest, and there appears to be no simple trend. Careful examination of Figure 18 of reference 33 permits the weak generalisation that as θ increases from zero the amount of products with *large* E' at first decreases, then the trend reverses beyond $\theta \approx 30$ degrees.

(d) *Dependence on initial kinetic energy* — Careful relative normalisation of the data at the three energies studied allows the $\bar{E}$ dependence of σ_R to be established. If it is assumed that $\sigma_R \propto \bar{E}^{-a}$, the 'best' value is $a = 0.2_5 \pm 0.1$, but other candidates such as $a = \frac{1}{3}$ or $a = 0$ cannot be excluded. Over the range of $\bar{E}$ of these experiments the *shape* of $[d^3\sigma/d^2\omega \, dw]$ was quite insensitive to $\bar{E}$. If it is *assumed* to be independent of $\bar{E}$ then less detailed data at many more values of $\bar{E}$ can be used to estimate the dependence of σ_R. These data favour a constant σ_R in the energy region spanned.

6.4.5.2. The spectator stripping model

The spectator model[2, 56, 133], attempts to account for the observed behaviour of the stripping reactions with the extremely simple assumption that the unreacting fragment C in $A + BC \rightarrow AB + C$ acts as a fully disinterested spectator of the reaction, continuing with the same velocity after the reaction as just prior to it. In the simplest form of the model, the motion of C relative to the centre-of-mass of BC (i.e. the rotation and vibration of BC) before the reaction is further ignored[56]. This predicts *all* the products to be scattered *exactly* forward in the CM system and also gives a *unique* recoil translational energy E' in terms of the incident kinetic energy $E : E' = M_A M_C E / M_{AB} M_{BC}$. If the internal motion of reactant BC is taken into account the sharpness of both the θ and the E' distributions is weakened. Since the reactants are oriented randomly before reaction the θ-distribution still centres at $\theta = 0$. The centre of the E' distribution is shifted slightly higher to include a contribution from the internal kinetic energy.

Qualitatively, the spectator model's predictions agree well with the observations of forward-peaking of products and mainly small values of E'. Both forms of the model fail seriously, however, for quantitative predictions. The discrepancy is most clear-cut for $K + I_2$, whose actual angular distribution is only moderately forward-peaked, whereas the spectator model expectation is an extremely sharp distribution. Also, the actual E' distribution is very much broader and has appreciable values well outside the cut-off of the spectator model[25]. For qualitative summary of the behaviour of the large body of data now available on stripping reactions the following rough rules of thumb probably do as well as the models: $E' \approx E$, and $c.$ $\frac{2}{3}$ of the products scatter forward. Quantitative predictions apparently will require more detailed (and more difficult) theoretical treatment.

6.4.5.3 Electron-jump mechanism (harpoon model)

Some years ago an electron-transfer model[2, 134, 135] was proposed by Magee[136] to account for the surprisingly large reactive cross-sections Polanyi's flame experiments indicated for sodium–halogen reactions[137]. This model notes that the two potential surfaces going asymptotically to $M + X_2$ and $M^+ + X_2^-$ would be expected in a first approximation to cross in the neighbourhood of configurations (for the X–X distance near its most probable value) at a M–X_2 distance of $c. r_c = e^2/[\text{I.P.} - \text{E.A.}^{(v)}]$, with I.P. = the alkali ionisation potential and E.A.$^{(v)}$ = the X_2 vertical electron affinity. The 'non-crossing rule' revises this to predict instead that within a small range of $r(M—X_2)$ the charge distribution of the lowest adiabatic state abruptly rearranges to $M^+ + X_2^-$ and the force becomes suddenly strongly attractive (coulombic). Thus, in the parlance, the alkali metal extends its range of force by throwing out a 'harpoon' (the electron) and reeling in its 'whale' (X_2^-) with the coulomb attraction to closer contact and reaction.

In its most primitive form, the electron-jump model is not pushed beyond this essentially two-body picture, and the reactive cross-section is estimated[2, 134] as πr_c^2. Molecular orbital theory or various semiempirical rules can be employed to approximate the X_2^- potential curve for obtaining E.A.$^{(v)}$ if it is not known experimentally[2, 38]. Although many qualitative trends are correctly ordered by this simple method, quantitative failures were noted right away in that the electron affinities needed to give the estimated values of σ_R appeared to be too large[2, 63]. Modification of the simple model to include an increase in the attractive force for $r(M—XY) > r_c$ because of interaction of the ionic and covalent surfaces has been suggested[24, 62, 138] as a possible remedy (see references 24 and 138 for discussions of unpublished theses developing this idea), and some support for such behaviour can be found in the results on non-reactive scattering[36] for these systems. However, the real coup-de-grace for either version of the simple harpoon model is contained in the highly accurate non-beam determinations[20, 24] of σ_R for $M + I_2$ (see Section 6.2.3). No single value of E.A.$^{(v)}$ for I_2 will account for the trend in the series M + Na, K, Rb and Cs either in the crude or modified picture[24]. It is not surprising that an essentially two-body theory fails to account quantitatively for three-body (or many-body) phenomena. For example, the total cross-sections seem to be in the order $\sigma_R(I_2) > \sigma_R(Br_2) > \sigma_R(Cl_2)$ for any given M as predicted by the model, yet the degree of forward-peaking indicates the opposite trend[37] (see Table 6.2).

More elaborate use of the electron-jump model is made in three trajectory studies with potential surfaces having the characteristic abrupt change in the reactant region[16, 34, 79–82]. Each study was able to produce qualitatively satisfactory E' and θ distributions, and they are uniform in concluding that forward-peaking of the products depends essentially on special exit-channel interactions. Unfortunately, however, they do not agree as to the nature of these interactions. Godfrey and Karplus[34] found that weak product interactions gave forward scattering, whereas Blais[16] obtained the same result by offsetting initial repulsion with long-range ion-induced attractive forces. In the work of Kuntz et al.[79-82], the possibility of charge migration in the XY^- was shown to be possibly quite important, with secondary encounters between the products having an essential effect on the product angular distribution. Their surface alone of the three produced the tendency, noted in reference 13, for small scattering angles to be associated with high translational energy of the products for $K + Br_2$. However, it is not clear that this correlation is borne out experimentally by the velocity selection/analysis experiments on the same reaction[73] and it is certainly not conspicuous in the more detailed results[33] for $K + I_2$. Hence, it is still unclear what features must be incorporated into the potential surfaces to produce the characteristics of the stripping reactions. Some form of an electron-transfer model seems likely to be in order, but the additional properties are not yet definitely established.

6.4.6 Lithium atom reactions

Parrish and Herm[61-63] have extended alkali atom reactions to include the reactions of many of the molecules with Li. The high ionisation potential of Li resulted in such a poor efficiency of surface ionisation that the usual differential surface ionisation discrimination of products from reactants could not be used. Instead, magnetic deflection in a Stern–Gerlach field was used to sweep the paramagnetic reactants away from the detector wire. A group of dihalogens, which showed typical stripping characteristics with the other alkalis, gave similar results in most respects with Li except that the fall-off of $d^2\sigma/d^2\omega$ from $\theta = 0$ was somewhat less pronounced, and in Br_2 the peak actually has moved slightly away from $\theta = 0$. The integrated predominance of forward scattering, however, is not very different than for the other alkalis (see Table 6.2). Two polyhalides, $SnCl_4$ and PCl_3, gave products peaked *sideways*, although their total reactive cross-section and recoil energy distributions are otherwise in accord with the stripping picture. (Even for these the integrated forward-scattered fractions, 0.60 for $SnCl_4$ and 0.55 for PCl_3, are not strikingly out of the pattern.) It is possible that all these systems are indicating collisions of longer duration, perhaps because of kinematic effects; the $SnCl_4$ and PCl_3 data are not far from showing the symmetry characteristic of long-lived complexes (see Section 6.2.4.2).

Further studies of reactions with smaller reactive cross-sections augment these studies. Of these, the reaction $Li + NO_2$ is particularly interesting. The large electron affinity that NO_2 is believed to have would predict an electron-jump radius comparable to, or larger than, that for the halogens, but the exoergicity of the overall reaction is small. These features suggest a potential surface having restricted entrance and exit channels and a deep well in the intermediate region, factors that seem likely to favour the formation of a long-lived complex. Instead, sharper forward-peaking of the products is observed than for the $Li + X_2$ reactions, contradicting not only the complex hypotheses but also the correlation usually found between large total cross-sections and forward-scattered products. Their $Li + SF_6$ data do indicate complex formation. Kwei et al.[49] have just reported finding evidence for short-lived complexes in the exchange reactions of Li with KBr and KF.

6.4.7 Miscellaneous reactions

6.4.7.1 $F + D_2$

The reactions $F + H_2$ and $F + D_2$ have been studied in i.r. chemiluminescence[139] and in chemical lasers[140], and the relative rates for product formation in the various energetically accessible rotational and vibrational states are now known. These data fall short of *complete* characterisation of the product states only due to the lack of information about the product angular distribution and its dependence on the internal states. Schaefer et al.[68] have removed this shortcoming with a cross-beam study of $F + D_2$ using a supermachine similar in design[141] to that described in reference 104. Their data, shown in Figure 6.5, were fitted with separate θ and E' distributions for the forward and backward scattering for each vibrational state. Initial guesses for the E' distributions were taken from the chemiluminescence data. Increasing vibrational excitation was found to be accompanied by broader CM angular distributions. Different ranges of impact parameters appear to contribute in the formation of different vibrational states.

6.4.7.2 $H(D) + Cl_2$, Br_2

Another set of reactions for which molecular beam studies now complement i.r. chemiluminescence experiments is that of hydrogen and deuterium atoms with halogen molecules, as recently reported by Grosser and Haberland[40]. These experiments were done in an out-of-plane geometry to reduce the principal source of noise in the detector. Since both the initial and final relative velocities are large compared to the velocity of the centre-of-mass, $\theta \approx \Theta$ and the LAB $\rightarrow$ CM Jacobian should be slowly varying. The data, extending out to $\Theta \approx 120$ degrees are quite asymmetric about 90 degrees with almost all the diatomic product appearing in the *backwards* direction, clearly indicating a direct mechanism for the reaction. The angular distributions for the H and D reactions are indistinguishable. Since there is a large reduced-mass difference for the reactants and only a small one for the products this suggests that the angular distributions are probably determined mainly by the exit-channel interactions. This conclusion is supported by the trajectory studies and i.r. chemiluminescence experiments of Anlauf et al.[142]. A previous molecular beam study[23] of $D + Br_2$ had come to different conclusions about the angular distributions of products, but the present results are said to be confirmed by unpublished work of the Harvard group.

6.4.7.3 Collision-induced dissociation

The recent paper of Tully et al.[72] on collision-induced dissociation of CsI, CsBr and RbI to ion pairs by high energy Xe atoms represents the addition of a new type of process to the molecular beam field. The fast Xe beam was

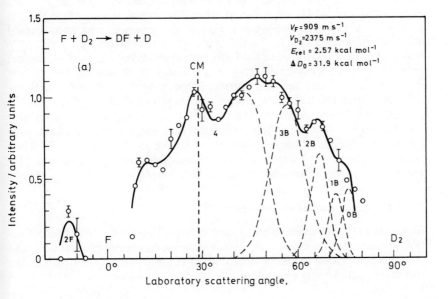

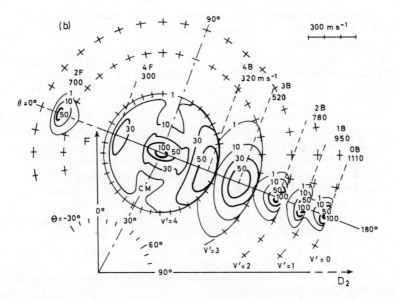

Figure 6.5 (a) Laboratory angular distribution of DF intensity (number density): The open circles are experimental points; the solid line is the calculated total intensity; the dashed curves indicate intensity of DF in particular vibrational states scattered forward (F) or backward (B) in CM frame.
(b) Cartesian contour plot of [DF CM flux]/[DF CM velocity]² in velocity space. The dashed circles denote the largest DF velocity for each vibrational state allowed by energy conservation (From Schafer, T. P. *et al.*[68], by permission of the American Institute of Physics)

produced by the seeded nozzle beam technique (see Section 6.3.2) with H_2 as the carrier (99% H_2 – 1% Xe) at laboratory energies up to 11.0 eV (254 kcal mol^{-1}). The observed ion-pair thresholds agree well with the accepted dissociation energies. Near the threshold, the cross-sections for CsBr were found to be quite sensitive to the CsBr temperature, indicating that internal energy is more effective than translational energy for producing dissociation in that region. At translational energies higher than c. 1.5 eV over the threshold, the dependence on CsBr temperature is less dramatic.

Parks and Wexler have made a similar study on collisions of TlBr with fast Xe and Kr atoms[161]. Their seeded-nozzle system produced atoms with laboratory energies in the ranges 8–15 eV for Xe and 6–10 eV for Kr. They report *absolute* total cross-sections for formation of the positive ions Tl^+, $TlXe^+$, and Tl_2Br^+ (from dimers in the TlBr beam).

6.4.7.4 Beam-gas experiments

Besides the HI + DI experiments discussed in Section 6.4.4, another investigation of reactions of a beam with a stationary gas has been reported by Seidel et al.[69]. The reaction studied was a beam of H atoms with CCl_4 gas. A cross-section of c. 10 Å^2 was reported.

6.5 OTHER RELEVANT STUDIES

In this section I intend merely to draw attention without discussion to a number of works that are of some relevance to the reactive scattering field but did not fit into any of the previous sections.

Among these are experimental papers on: negative surface ionisation[143], the virtues of the 'out-of-plane' configuration for elastic/inelastic scattering studies[144], a high resolution small-angle apparatus[145], and high-energy scattering from reactive systems[146].

Theoretical papers with some relevance are legion. A highly subjective (and somewhat haphazard) group of those particularly relevant to the topic at hand includes treatments of the following: classical trajectory studies (apart from those previously mentioned in other sections)[67, 147, 148]; exact quantum calculations on collinear systems in stationary[64, 84, 149–151] and time-dependent[152, 153] formulations and on a planar system[154]; several approximate or model treatments[155–159]; and a provocative summary[160].

At the time of this writing the programme for the Seventh International Conference on the Physics of Electronic and Atomic Collisions (1971, Amsterdam) has appeared, but the abstracts of papers are not available. There are a number of titles of papers on the programme that suggest these abstracts will be an excellent place for interested readers to take up the trail of the future work in this area.

Acknowledgement

Financial support from the National Science Foundation is gratefully acknowledged.

References

1. Ross, J., Editor (1966). *Advan. Chem. Phys.,* **10,** Molecular Beams. (New York: Interscience Publishers)
2. Herschbach, D. R. (1966). *Advan. Chem. Phys.,* **10,** 319
3. Blythe, A. R., Fluendy, M. A. D. and Lawley, K. P. (1966). *Quart. Rev. Chem. Soc.,* **20,** 465
4. Toennies, J. P. (1968). *Chemische Elementarprozesse* (ed. by Hartmann, J., Heidberg, J., Heydtmann, H. and Kohlmeier, G. H.), 157. (Berlin: Springer-Verlag)
5. Toennies, J. P. (1968). *Ber. Bunsenges. Phys. Chem.,* **72,** 927
6. Steinfeld, J. I. and Kinsey, J. L. (1970). *Progr. Reaction Kinetics,* **5,** 1
7. Airey, J. R., Greene, E. F., Kodera, K., Reck, G. P. and Ross, J. (1967). *J. Chem. Phys.,* **46,** 3287
8. Airey, J. R., Greene, E. F., Reck, G. P. and Ross, J. (1967). *J. Chem. Phys.,* **46,** 3295
9. Beck, D., Engelke, F. and Loesch, J. H. (1968). *Ber. Bunsenges Phys. Chem.,* **72,** 1105
10. Beuhler, R. J., Jr., Bernstein, R. B. and Kramer, K. H. (1966). *J. Amer. Chem. Soc.,* **88,** 5331
11. Beuhler, R. J., Jr. and Bernstein, R. B. (1968). *Chem. Phys. Lett.,* **2,** 166
12. Beuhler, R. J., Jr. and Bernstein, R. B. (1969). *J. Chem. Phys.,* **51,** 5305
13. Birely, J. H. and Herschbach, D. R. (1966). *J. Chem. Phys.,* **44,** 1690
14. Birely, J. H., Herm, R. R., Wilson, K. R. and Herschbach, D. H. (1967). *J. Chem. Phys.,* **47,** 993
15. Birely, J. H., Entemann, E. A., Herm, R. R. and Wilson, K. R. (1969). *J. Chem. Phys.,* **51,** 5461
16. Blais, N. C. (1968). *J. Chem. Phys.,* **49,** 9
17. Blais, N. C. and Cross, J. B. (1970). *J. Chem. Phys.,* **52,** 3580
18. Brooks, P. R. and Jones, E. M. (1966). *J. Chem. Phys.,* **45,** 3449
19. Brooks, P. R. (1969). *J. Chem. Phys.,* **50,** 5031
20. Brodhead, D. C., Davidovits, D. and Edelstein, S. A. (1969). *J. Chem. Phys.,* **51,** 3601
21. Cross, J. B. and Blais, N. C. (1969). *J. Chem. Phys.,* **50,** 4108
22. Datz, S. and Taylor, E. H. (1963). *J. Chem. Phys.,* **39,** 1896
23. Datz, S. and Schmidt, T. W. (1967). *Proceedings of the Fifth International Conference on the Physics of Electronic and Atomic Collisions,* 247. (Leningrad: Publishing House Nauka)
24. Edelstein, S. A. and Davidovits, P. (1971). *J. Chem. Phys.,* **55,** 5164
25. Entemann, E. A. and Herschbach, D. R. (1967). *Discuss. Faraday Soc.,* **44,** 289
26. Fisk, G. A., McDonald, J. D. and Herschbach, D. R. (1967). *Discuss. Faraday Soc.,* **44,** 228
27. Fite, W. L. and Brackman, R. T. (1964). *Atomic Collision Processes,* 955. (Amsterdam: North Holland Publ. Co.)
28. Fite, W. L. and Brackman, R. T. (1965). *J. Chem. Phys.,* **42,** 4057
29. Fluendy, M. A. D., Horne, D. S., Lawley, K. P. and Morris, A. W. (1970). *Molec. Phys.,* **19,** 659
30. Freund, S. M., Fisk, G. A., Herschbach, D. R. and Klemperer, W. (1971). *J. Chem. Phys.,* **54,** 2510
31. Geddes, T., Krause, H. F. and Fite, W. L. (1970). *J. Chem. Phys.,* **52,** 3296
32. Gillen, K. T., Riley, C. and Bernstein, R. B. (1969). *J. Chem. Phys.,* **50,** 4019
33. Gillen, K. T., Rulis, A. M. and Bernstein, R. B. (1971). *J. Chem. Phys.,* **54,** 2831
34. Godfrey, M. and Karplus, M. (1968). *J. Chem. Phys.,* **49,** 3602
35. Greene, E. F. and Ross, J. (1968). *Science,* **159,** 587
36. Greene, E. F., Hoffman, L. F., Lee, M. W. and Ross, J. (1969). *J. Chem. Phys.,* **50,** 3450
37. Grice, R. and Empedocles, P. B. (1968). *J. Chem. Phys.,* **48,** 5352
38. Grice, R., Cosandey, M. R. and Herschbach, D. H. (1968). *Ber. Bunsenges Phys. Chem.,* **72,** 975
39. Grice, R., Mosch, J. E., Safron, S. A. and Toennies, J. P. (1970). *J. Chem. Phys.,* **53,** 3376
40. Grosser, J. and Haberland, H. (1970). *Chem. Phys. Lett.,* **7,** 442
41. Ham, D. O., Kinsey, J. L. and Klein, F. S. (1967). *Discuss. Faraday Soc.,* **44,** 174
42. Ham, D. O. and Kinsey, J. L. (1968). *J. Chem. Phys.,* **48,** 939
43. Ham, D. O. and Kinsey, J. L. (1970). *J. Chem. Phys.,* **53,** 285
44. Herm, R. R. and Herschbach, D. R. (1970). *J. Chem. Phys.,* **52,** 5783

45. Jaffee, S. B. and Anderson, J. B. (1969). *J. Chem. Phys.*, **51**, 1057
46. Jonah, C. D. and Zare, R. N. (1970). *Chem. Phys. Lett.*, **9**, 65
47. Kwei, G. H. and Herschbach, D. R. (1969). *J. Chem. Phys.*, **51**, 1742
48. Kwei, G. H., Norris, J. A. and Herschbach, D. R. (1970). *J. Chem. Phys.*, **52**, 1317
49. Kwei, G. H., Lees, A. B. and Silver, J. A. (1971). *J. Chem. Phys.*, **55**, 456
50. Lee, Y. T., McDonald, J. D., LeBreton, P. R. and Herschbach, D. R. (1968). *J. Chem. Phys.*, **49**, 2447
51. Lee, Y. T., LeBreton, P. R., McDonald, J. D. and Herschbach, D. R. (1969). *J. Chem. Phys.*, **51**, 455
52. Maltz, C. and Herschbach, D. R. (1967). *Discuss. Faraday Soc.*, **44**, 176
53. Martin, L. R. and Kinsey, J. L. (1967). *J. Chem. Phys.*, **46**, 4834
54. Marriott, R. and Micha, D. A. (1969). *Phys. Rev.*, **180**, 120
55. Miller, W. B., Safron, S. A. and Herschbach, D. R. (1967). *Discuss. Faraday Soc.*, **44**, 108; ibid, **44**, 292
56. Minturn, R. E., Datz, S. and Becker, R. L. (1966). *J. Chem. Phys.*, **44**, 1149
57. Moulton, M. C. and Herschbach, D. R. (1966). *J. Chem. Phys.*, **44**, 3010
58. Odiorne, T. J. and Brooks, P. R. (1969). *J. Chem. Phys.*, **51**, 4676
59. Odiorne, T. J., Brooks, P. R. and Kasper, J. V. V. (1971). *J. Chem. Phys.*, **55**, 1980
60. Ottinger, Ch. and Zare, R. N. (1970). *Chem. Phys. Lett.*, **5**, 243
61. Parrish, D. D. and Herm, R. R. (1968). *J. Chem. Phys.*, **49**, 5544
62. Parrish, D. D. and Herm, R. R. (1969). *J. Chem. Phys.*, **51**, 5467
63. Parrish, D. D. and Herm, R. R. (1971). *J. Chem. Phys.*, **54**, 2518
64. Rankin, C. C. and Light, J. C. (1969). *J. Chem. Phys.*, **51**, 1701
65. Riley, C., Gillen, K. T. and Bernstein, R. B. (1967). *J. Chem. Phys.*, **47**, 3672
66. Roach, A. C. (1969). *Chem. Phys. Lett.*, **6**, 389
67. Russell, D. and Light, J. C. (1969). *J. Chem. Phys.*, **51**, 1720
68. Schafer, T. P., Siska, P. E., Parson, J. M., Tully, F. P., Wong, Y. C. and Lee, Y. T. (1970). *J. Chem. Phys.*, **53**, 3385
69. Seidel, W., Martin, H. and Mietzner, F. G. (1965). *Z. Phys. Chem.*, **47**, 348
70. Struve, W. S., Kitagawa, T. and Herschbach, D. R. (1971). *J. Chem. Phys.*, **54**, 2759
71. Truhlar, D. G. (1971). *J. Chem. Phys.*, **54**, 2635
72. Tully, F. P., Lee, Y. T. and Berry, R. S. (1971). *Chem. Phys. Lett.*, **9**, 80
73. Warnock, T. T., Bernstein, R. B. and Grosser, A. E. (1967). *J. Chem. Phys.*, **46**, 1685
74. Wilson, K. R. and Herschbach, D. R. (1968). *J. Chem. Phys.*, **49**, 2676
75. Brooks, P. R. (1971). Private communication
76. Harris, R. M. and Wilson, J. F. (1971). *J. Chem. Phys.*, **54**, 2088
77. Beck, D. and Förster, H. (1970). *Z. Phys.*, **240**, 136
78. Lee, Y. T., Gordon, R. J. and Herschbach, D. R. (1971). *J. Chem. Phys.*, **54**, 2410
79. Kuntz, P. J., Mok, M. H., Nemeth, E. M. and Polanyi, J. C. (1967). *Discuss. Faraday Soc.*, **44**, 229
80. Polanyi, J. C. (1967). *Trans. Roy. Soc. (Canada)*, **5**, 105
81. Kuntz, P. J., Nemeth, E. M. and Polanyi, J. C. (1969). *J. Chem. Phys.*, **50**, 4607
82. Kuntz, P. J., Mok, M. H. and Polanyi, J. C. (1969). *J. Chem. Phys.*, **50**, 4623
83. Parrish, D. D. and Herm, R. R. (1970). *J. Chem. Phys.*, **53**, 2431
84. Miller, G. and Light, J. C. (1971). *J. Chem. Phys.*, **54**, 1643
85. Warnock, T. T. and Bernstein, R. B. (1968). *J. Chem. Phys.*, **49**, 1878; ibid., **51**, 4682
86. Greene, E. F., Moursund, A. L. and Ross, J. (1966). *Advan. Chem. Phys.*, **10**, 135
87. Everhart, E. (1963). *Phys. Rev.*, **132**, 2083
88. Smith, F. T., Marchi, R. P. and Dedrick, K. G. (1966). *Phys. Rev.*, **150**, 79
89. Rosenfeld, J. L. J. and Ross, J. (1966). *J. Chem. Phys.*, **44**, 188
90. Eu, B. C. and Ross, J. (1967). *Discuss. Faraday Soc.*, **44**, 39
91. Sun, H. Y. and Ross, J. (1967). *J. Chem. Phys.*, **45**, 3306
92. Nyeland, C. and Ross, J. (1968). *J. Chem. Phys.*, **49**, 843
93. Bernstein, R. B. and Levine, R. D. (1968). *J. Chem. Phys.*, **49**, 3872
94. Levine, R. D. (1969). *Quantum Mechanics of Molecular Rate Processes*. (Oxford: Clarendon Press)
95. Levine, R. D. and Bernstein, R. B. (1969). *Israel J. Chem.*, **7**, 315
96. Eu, B. C. (1970). *J. Chem. Phys.*, **52**, 3021
97. Hundhausen, E. and Pauly, H. (1965). *Z. Phys.*, **187**, 305
98. Gedeon, A., Edelstein, S. A. and Davidovits, P. (1971). *J. Chem. Phys.*, **55**, 5171

99. Brus, L. E. (1970). *J. Chem. Phys.*, **52**, 1716
100. Levine, R. D. and Bernstein, R. B. (1970). *J. Chem. Phys.*, **53**, 686
101. Reference 94, Section 3.5.6
102. Smith, F. T. (1960). *Phys. Rev.*, **118**, 349
103. Kinsey, J. L. (1971). *Chem. Phys. Lett.*, **8**, 349
104. Lee, Y. T., McDonald, J. D., LeBretton, P. R. and Herschbach, D. R. (1969). *Rev. Sci. Instr.*, **40**, 1402
105. Bickes, R. W. and Bernstein, R. B. (1970). *Rev. Sci. Instr.*, **41**, 759
106. Anderson, J. B., Andres, R. P. and Fenn, J. B. (1966). *Advan. Chem. Phys.*, **10**, 275
107. Gordon, R. J., Lee, Y. T. and Herschbach, D. R. (1971). *J. Chem. Phys.*, **54**, 2393
108. Grosser, A. E., Blythe, A. R. and Bernstein, R. B. (1965). *J. Chem. Phys.*, **42**, 1268
109. Grosser, A. E. and Bernstein, R. B. (1965). *J. Chem. Phys.*, **43**, 1140
110. Gordon, R. J., Herm, R. R. and Herschbach, D. R. (1968). *J. Chem. Phys.*, **49**, 2684
111. Ramsey, N. F. (1956). *Molecular Beams.* (Oxford: Clarendon Press)
112. Maltz, C. (1969). *Chem. Phys. Lett.*, **3**, 707
113. Grice, R. (1970). *Mol. Phys.*, **18**, 545
114. Waech, T. G. and Bernstein, R. B. (1968). *Chem. Phys. Lett.*, **2**, 477
115. Waech, T. G., Kramer, K. H. and Bernstein, R. B. (1968). *J. Chem. Phys.*, **48**, 3978
116. Bennewitz, H. G., Paul, W. and Schlier, Ch. (1955). *Z. Phys.*, **141**, 6
117. Kramer, K. H. and Bernstein, R. B. (1965). *J. Chem. Phys.*, **42**, 767
118. Brooks, P. R., Jones, E. M. and Smith, K. (1969). *J. Chem. Phys.*, **51**, 3073
119. Jones, E. M. and Brooks, P. R. (1970). *J. Chem. Phys.*, **53**, 55
120. Brooks, P. R. (1971). Private communication
121. Gillen, K. T. and Bernstein, R. B. (1970). *Chem. Phys. Lett.*, **5**, 275
122. Kalos, F. and Grosser, A. E. (1969). *Rev. Sci. Instr.*, **40**, 804
123. Grover, J. R., Kiely, F. M., Lebowitz, E. and Baker, E. (1971). *Rev. Sci. Instr.*, **42**, 293
124. McDonald, J. D. (1971). Private communication
125. Hirschy, V. L. and Aldridge, J. P. (1971). *Rev. Sci. Instr.*, **42**, 381
126. Martin, L. R. and Goldbaum, R. H. (1971). Private communication
127. Brumer, P. and Karplus, M. (1971). *J. Chem. Phys.*, **54**, 4955
128. Porter, R. N. and Karplus, M. (1964). *J. Chem. Phys.*, **40**, 1105
129. Anlauf, K. G., Maylotte, D. H., Polanyi, J. C. and Bernstein, R. B. (1969). *J. Chem. Phys.*, **51**, 5716
130. Marcus, R. A. (1970). *J. Chem. Phys.*, **53**, 604
131. Kinsey, J. L. (1971). *J. Chem. Phys.*, **54**, 1206
132. Jaffe, S. B. and Anderson, J. B. (1968). *J. Chem. Phys.*, **49**, 2859
133. Nikitin, E. E. (1968). *Ber. Bunsenges Phys. Chem.*, **72**, 949
134. Herschbach, D. R. (1965). *Applied Optics Suppl.*, **2**, 128
135. Polanyi, J. C. (1965). *Applied Optics Suppl.*, **2**, 109
136. Magee, J. L. (1940). *J. Chem. Phys.*, **8**, 687
137. Polanyi, M. (1932). *Atomic Reactions.* (London: Williams and Norgate)
138. Bersohn, R. (1971). *Comments Atomic and Mol. Phys.*, **2**, 156
139. Polanyi, J. C. and Tardy, D. C. (1969). *J. Chem. Phys.*, **51**, 5717
140. Parker, J. H. and Pimentel, G. C. (1969). *J. Chem. Phys.*, **51**, 91
141. Parson, J. M., Schafer, T. P., Tully, F. P., Siska, P. E., Wong, Y. C. and Lee, Y. T. (1970). *J. Chem. Phys.*, **53**, 2123
142. Anlauf, K. G., Kuntz, D. J., Maylotte, D. H., Pacey, P. D. and Polanyi, J. C. (1967). *Discuss. Faraday Soc.*, **44**, 183
143. Persky, A., Greene, E. F. and Kuppermann, A. (1968). *J. Chem. Phys.*, **49**, 2437
144. Greene, E. F., Lau, M. H. and Ross, J. (1969). *J. Chem. Phys.*, **50**, 3122
145. Cowley, L. T., Fluendy, M. A. D., Horne, D. S. and Lawley, K. P. (1969). *J. Sci. Instr. (J. Phys. E)*, **2**, 1021
146. Kempter, V., Kneser, Th. and Schlier, Ch. (1970). *J. Chem. Phys.*, **52**, 5851
147. Raff, L. M. (1966). *J. Chem. Phys.*, **44**, 1202
148. Raff, L. M. and Karplus, M. (1966). *J. Chem. Phys.*, **44**, 1212
149. Truhlar, D. G. and Kuppermann, A. (1970). *J. Chem. Phys.*, **52**, 3841
150. Miller, G. and Light, J. C. (1971). *J. Chem. Phys.*, **54**, 1635
151. Wyatt, R. E. (1969). *J. Chem. Phys.*, **51**, 3489
152. McCullough, E. A., Jr. and Wyatt, R. E. (1969). *J. Chem. Phys.*, **51**, 1253
153. McCullough, E. A., Jr. and Wyatt, R. E. (1971). *J. Chem. Phys.*, **54**, 3578; ibid., 3592

154. Saxon, R. P. and Light, J. C. (1971). *J. Chem. Phys.,* **55,** 455
155. Karplus, M. and Tang, K. T. (1967). *Discuss. Faraday Soc.,* **44,** 56
156. Suplinskas, R. J. and Ross, J. (1967). *J. Chem. Phys.,* **47,** 321
157. Connor, J. N. L. and Child, M. S. (1970). *Mol. Phys.,* **18,** 653
158. Eu, B. C., Huntington, J. H. and Ross, J. (1971). *Can. J. Phys.,* **49,** 966
159. Grice, R. (1970). *Mol. Phys.,* **19,** 501
160. Bernstein, R. B. (1969). *Int. J. Quantum. Chem. IIIS,* 41

7
Ion–Molecule Reactions

J. DUBRIN
Massachusetts Institute of Technology

and

M. J. HENCHMAN
Brandeis University, Massachusetts

7.1   INTRODUCTION

7.1.1   Scope

An anonymous German professor has remarked that ion–molecule reactions should be considered as part of theology, since they do not occur naturally on the surface of the earth. Underlying this rebuke is a viewpoint which is rather prevalent, that the field of ion–molecule reactions is an isolated limb, existing entire in itself, having neither contact nor connection with the vast corpus of chemical kinetics. Here, we argue fervently the contrary opinion with a resolve to convince and convert, even though 5000 years of religious history emphasise the fate awaiting the reforming evangelist, or the heretical zealot.

We have purposely addressed ourselves to a general audience, since the specialist is already well served by some 30 reviews, which have appeared during the past 5 years. Consequently we have attempted to ask what aspects of the field are of more general chemical interest, above and beyond the confines of the particular field itself. The literature is now so vast that a comprehensive review would be impossible in an article of this length. The growth is exponential[1], with over 100 articles currently appearing each year in the *Journal of Chemical Physics* alone.

Accordingly the choice of topics is highly subjective and the coverage of the literature selective, insofar as it has been used to illustrate particular themes. With the lay reader in mind, reference is frequently made to critical review articles rather than to the primary literature, in the hope that the intervening stage will prove a digestive aid. Any reader seeking a general account of the historical evolution of the field should direct himself to the recent, excellent review by Friedman and Reuben[1]. The contemporary scene is developed at length in two recent monographs[2, 3]. The former[2] is noteworthy for its discussion of experimental techniques, theory and the relevance to the chemistry of the upper atmosphere; a useful compilation of rate data for some simple reactions is also included. The latter[3] contains chapters both on particular techniques and fundamental aspects of the field; frequent reference is made to these in the present article.

7.1.2   Techniques

We give a brief indication here of the range of techniques which are presently available. The same beam techniques, which have proved so fruitful for the study of neutral–neutral reactions* have yielded important results on the collision dynamics of ion–molecule reactions[4]. Whereas the neutral-beam studies have been restricted, until recently, to low energies, the full range of translational energy has been explored for ion–molecule reactions. Paradoxically the lowest energy regime, of most interest to chemists, has been the hardest to investigate.

State selection of ionic reactants has been achieved with unrivalled control through the use of photo-ionisation techniques[5]. In contrast, state analysis of products, by means of chemiluminescent studies, has received little attention.

The intense activity in the field of ion–molecule reactions is partly a consequence of the wide variety of techniques available. More than 10 completely different types of apparatus are discussed in a recent review of rate phenomena[6]. Fine control of relative translational energy may be achieved with the merging-beams technique, to give detailed information on the energy dependence of the reaction cross-section[7]. Truly thermal rate constants may be measured over an extended temperature range by the flowing-afterglow technique[8, 9]. Ion cyclotron resonance finds particular use in the characterisation of reactions which are hard to identify by other means and this facility has been imaginatively exploited[10, 11]. Ionic equilibria may be studied in the gas phase by use of ultra-high-pressure mass spectrometry[12].

7.1.3   Preview

The themes developed here presuppose that the study of ion–molecule reactions can throw light on broader kinetic problems. However, several

---

*See Chapter 6 of this volume.

arguments may be used to argue the contrary view and, at this point, it is appropriate to outline them. First, ionic mechanisms are not found in traditional gas kinetics. Secondly, the study of ionic chemistry in the gas phase has no relevance for solutions, because the solvent's role is all-important. Thirdly, most reactions exhibit an activation energy whereas most ion–molecule reactions do not. Fourthly, ion–molecule reactions are dominated by powerful, long-range attractive forces. At appropriate points in this Chapter, the significance of these arguments is examined and found to be less limiting than might be supposed.

The theme, developed in Section 7.2, treats the effect of *energy* upon *reactivity*, this being reflected in the *rate* and the *mechanism*. At the present time, the best control of the reactants' energies (translational and internal) is achieved for ion–molecule reactions and such information is the most detailed and extensive available.

Section 7.3 offers an introduction to a new field, the chemistry of acids, bases and solvated ions in the gas phase, and it is already clear that this is providing implications for solution chemistry.

7.2 ENERGY AND REACTIVITY IN THE GAS PHASE

7.2.1 Reaction rates: excitation functions and rate constants

The notion that most chemical reactions exhibit an activation energy and that translational energy is needed to surmount an energy barrier, dates from Arrhenius and is considered as firmly established from the temperature dependence of rate constants. This is a gross effect, since the rate constant is *not* a *fundamental* measure of the reaction probability, being extensively averaged over the distribution of internal and relative translational energies of the reactants[*]. The fundamental measure of the reaction probability is the reaction cross-section, σ, which should be written as $\sigma(i, j, E)$ for the reaction

$$A_i + B_j \longrightarrow C + D \tag{7.1}$$

where the reactants A and B are in specific internal quantum states, indicated by the subscripts i and j, the relative translational energy or collision energy is E, and $\sigma(i, j, E)$ is a measure of the *total* probability of forming the products C and D in all possible internal quantum states[†]. By analogy with nuclear reactions, we refer to the functional dependence of cross-section upon collision energy as the excitation function[14].

It might be thought that the Arrhenius picture would be firmly established by now, through detailed knowledge of the dependence of cross-section upon collision energy. In fact, very little information indeed is available on this fundamental question for neutral–neutral bimolecular reactions in the gas phase[15]. In large part, this is a consequence of the near impossibility of 'unfolding' information on the excitation function $\sigma(E)$ from knowledge of the temperature dependence of the rate constant $k(T)$[16]. Very recently, this situation has been changed by the production of translationally hot reactants,

---

[*]Nor is it often a true constant. For an excellent recent discussion see reference 13.
[†]For a fuller discussion, see Chapter 6, Section 6.2 of this volume.

by use of chemical accelerators[17, 18] or photochemical recoil[19]. Nevertheless, at the present time, most of the limited knowledge available for neutral–neutral reactions at high translational energies comes from bulk experiments in hot-atom chemistry, where the translational energy is not defined[20, 21] *.

Over forty years ago, it was realised that vibrational excitation of the reactants should be effective in surmounting energy barriers. Polanyi's review[15] emphasises that little is known about this matter for neutral–neutral reactions at the present time†. Less is known about the effect of rotational excitation of the reactants. To give one further example of the present state of ignorance concerning the microscopic behaviour of chemical reactions, only recently has it become widely appreciated that the activation energy, measured in bulk experiments, is not the reaction threshold, i.e. the minimum excitation energy necessary to cause chemical reaction†.

In summary, little is known for neutral–neutral reactions about $\sigma(E)$, yet alone $\sigma(i,j,E)$. The purpose of this prologue is to indicate the extent to which kineticists must rely, at the present time, on the studies of ion–molecule reactions for insight into the effect of energy upon reaction rates. Considerable control of this energy is now possible. First, by use of merging-beam techniques[7], collision energies may be controlled with precision, from energies of 0.1 eV and upwards, to obtain knowledge of the excitation function. Thus the role of translational energy in overcoming energy barriers may be explored in detail. More important perhaps, the role of translational energy may be examined for those reactions, which are not subject to an energy barrier. Secondly, the role of *internal excitation energy*, for reactions with and without barriers, has been examined, using both state-selected ionic reactants (obtained by photo-ionisation[5]) and selectively excited neutral reactants. A range of examples is given in Table 7.1. Section 7.2.2 stresses similarly the important role of ion–molecule studies in exploring the effect of excess energy upon the *dynamics* of reactive collision processes.

Lest this parade of achievement appear to invite complacency, large areas of this subject remain to be charted. Even though accurate excitation functions are now available from merging-beam experiments[7], as is accurate information on the temperature dependence of thermal rate constants from flowing-afterglow measurements[8], no demonstration has yet been given of how the latter may be given by suitable averaging over the former. Information of this type does exist but the distribution of internal quantum states differs for the two experimental situations and this, as may be seen in Table 7.1, can exercise a considerable influence on the reaction rate.

7.2.1.1 *Endoergic reactions: the role of translational, vibrational and electronic energy in surmounting the energy barrier*

As an example of the detailed information available from the investigation of ion–molecule reactions, we consider Chupka's definitive study[5, 24] of the endoergic reaction

$$H_2^+ + He \rightarrow HeH^+ + H \qquad \Delta E = +0.80\,eV \qquad (7.2)$$

---

*See also Chapter 4 of this volume.

†For a discussion of recent beam results, see Chapter 6, Section 6.4.4 of this volume

‡A clear statement is given in reference 22.

Table 7.1 The effect of internal excitation energy of ionic and neutral reactants on reaction rates

	Excited species	Mode of excitation	Reaction*	Result	Reference
(1)	Ion	Rotational	$\underline{H_2^+} + H_2 \rightarrow H_3^+ + H$	$\sigma(J = 0)$ does not differ from $\sigma(J = 1)$ by more than 10%	5, 23
(2)	Ion	Vibrational	$\underline{H_2^+} + H_2 \rightarrow H_3^+ + H$	$\sigma(v = 4)/\sigma(v = 0) = 0.79$	5, 23
(3)	Ion	Electronic	$\underline{Ar^+} + H_2 \rightarrow ArH^+ + H$	$\sigma(^2P_{\frac{1}{2}})/\sigma(^2P_{\frac{3}{2}}) = 1.3$	5, 24
(4)	Neutral	Rotational	$Ar^+ + \underline{H_2} \rightarrow ArH^+ + H$	$\sigma(J = 1)/\sigma(J = 0) = 0.95 \pm 0.025$	25
(5)	Neutral	Vibrational	$O^+ + \underline{N_2} \rightarrow NO^+ + N$	$k(v = 6)/k(v = 0) \approx 350$	2†
(6)	Neutral	Electronic	$O^- + \underline{O_2} \rightarrow O_3 + e^-$	$k(^1\Delta_g) = 3 \times 10^{-10}$ cm^3 molecule^{-1} s^{-1} $k(^3\Delta_g) \approx 0$	26‡

*The excited reactant is underlined.
†Figure 6.15, p. 353.
‡The reaction is endoergic for the ground state ($\Delta E = +0.4$ eV) but exoergic for the excited state ($\Delta E = -0.6$ eV).

More information of this type is available for this particular reaction than for any other. Both relative translational energy and vibrational excitation of the H_2^+ reactant ion are effective in driving the reaction over the energy barrier. Table 7.2 indicates the relative efficiencies of these. In each vertical column, the total energy content, E_t, is held constant and the data, listed therein, record the effect on the reaction probability of partitioning E_t between relative translational and vibrational energy. In general, vibrational

Table 7.2 Relative probability for the reaction $H_2^+ + He \rightarrow HeH^+ + H$ as a function of E_t, the total energy content (translational and vibrational) and of the vibrational quantum number v of the H_2^+ reactant[5]

v	Relative probability*			
	$E_t = 1.0\,eV$	$E_t = 2.0\,eV$	$E_t = 3.0\,eV$	$E_t = 4.0\,eV$
0	0.06	0.10	0.13	0.17
1	0.49	0.35	0.31	0.25
2	1.95	0.93	0.55	0.34
3	—	1.70	0.99	0.56
4	—	2.35	1.22	0.68
5	—	2.49	1.70	0.89

*These values have been corrected to take account of the dependence of collision cross-section with relative translational energy E, assuming $\sigma \propto E^{-\frac{1}{2}}$.

(From Chupka[5], by courtesy of the Plenum Publishing Corp.)

excitation is an order of magnitude more effective than translational energy, this effect becoming less important with increasing total energy content. There would appear to be no convincing explanation as to why the superior effectiveness of vibrational energy should decrease with increasing translational energy for this reaction, which is a direct process at high energy[27].

Reaction (6) in Table 7.1 provides an example of the effectiveness of electronic excitation in overcoming the energy barrier, in this case for an associative detachment process. The reaction, $(N_2^+)_{el}^* + N_2 \rightarrow N_3^+ + N$, provides another example where electronic excitation is undoubtedly very efficient[28] but translational energy is not, because collision-induced dissociation of the N_2^+ takes precedence[29].

7.2.1.2 Exoergic reactions: the role of excess energy in the absence of an energy barrier

Most exoergic ion–molecule reactions exhibit no activation energies or thresholds and thus they permit the examination of a fundamental question — the effect of excess energy upon reactivity in the *absence* of restrictive energy barriers.

(a) *The role of translational energy and the examination of ultra-high-energy chemistry* — Wolfgang has given a particularly clear account of various concepts which control the energy dependence of reaction rates at high energy[14]. Large cross-sections, considerably in excess of gas-kinetic cross-sections, are generally found at low collision energies, due to the powerful attractive force between the ion and the neutral molecule. As competing endoergic channels become energetically attainable, they are generally followed efficiently; these include collision-induced dissociation. The total reactive cross-section, i.e. into all reactive channels, falls with increasing collision energy, since the range of effective interaction decreases correspondingly. While head-on collisions will always occur at any energy (for trajectories with small impact parameters), they will be decreasingly effective for chemical reaction as the collision energy is raised. For stable products to be formed, their internal excitation energy must not exceed the dissociation energy. In general, increase of the collision energy increases the internal excitation energy of the product through the need to conserve momentum and energy.

Perhaps the most important lesson for kineticists, both from these studies and from Wolfgang's hot-atom chemistry[20], is the realisation that a rich and efficient chemistry occurs in the collision-energy range of 10–100 eV, an order of magnitude larger than typical bond energies. Nor is this likely to be the limit. For example, plausible theoretical predictions[30] for the reaction $CH_4^+(CH_4, CH_3)CH_5^+$ indicate a rate constant of $\sim 10^6$ l mol^{-1} s^{-1} at a translational temperature of 10^8 K; this is not a trivial rate constant by traditional kinetic standards*. This is a dramatic demonstration of Wolfgang's point[20] that kineticists gain a distorted view of chemical reactivity by sampling only the thermal regime from the vast range of energy throughout which reaction can occur.

(b) *The role of internal energy* — In the absence of energy barriers, why should internal excitation of the reactants influence reactivity?† Much data are available, indicating a wide variety of effects, but, as yet, no unifying synthesis is apparent.

Rotational excitation of either the ion or the neutral species has little effect on the reaction rate (Table 7.1, Reactions (1) and (4)). The subtle effect in the latter case is ascribed to the anisotropic polarisability of the neutral species; this causing an increase in the range of effective interaction for the $J = 0$ state, whereby the neutral species locks in to the approaching ion[25].

The effect of vibrational excitation is next considered. For the reaction $H_2^+(H_2, H)H_3^+$, the rate constant decreases with increasing vibrational energy of the ion[5, 23] (Table 7.1, Reaction (2)); the reaction $NH_3^+(NH_3, NH_2)NH_4^+$ is similar[5, 31]; but, in other cases, $NH_3^+(H_2O, OH)NH_4^+$ [5, 31] and $CH_4^+(CH_4, CH_3)CH_5^+$ [32], there is little effect. In contrast, vibrational excitation of the neutral reactant in $O^+(N_2, N)NO^+$ (Table 7.1, Reaction (5)) shows a dramatic increase in rate[33] and is much more effective than relative

---

*The nomenclature $A^+(B,D)C^+$, adopted from nuclear chemistry, is a shorthand notation for $A^+ + B \rightarrow C^+ + D$. Where a second channel $A^+ + B \rightarrow E^+ + F$ competes, the nomenclature becomes $A^+\left(B, \dfrac{D}{F} \dfrac{C^+}{E^+}\right)$.

†This discussion disregards cases where such energy would be sufficient to cause decomposition of the products.

translational energy[34]; this finds a reasonable explanation in terms of the available crossings between the potential-energy hypersurfaces[33, 35, 36].

Electronic excitation of the ionic reactant causes an increase in the cross-section of $Ar^+(H_2, H)ArH^+$ [5, 24], a surprising result since the reaction is exoergic and thought to occur for every close-collision (Table 7.1, Reaction (3)).

Such data for these simple systems indicate the crudeness of our present understanding of the most fundamental factors controlling chemical reactivity.

(c) *Statistical models* — It is entirely plausible that reactions involving sufficiently polyatomic reactants should proceed via a long-lived intermediate at low collision energies. Such a mechanism allows the application of simple theoretical models to such processes, considering first the formation of the intermediate and then its statistical decay into products. A start has now been made on testing statistical models for ion–molecule reactions at low energies. The reaction

$$C_2H_4^+ \left(C_2H_4, \begin{matrix} H \\ CH_3 \end{matrix} \right) \begin{matrix} C_4H_7^+ \\ C_3H_5^+ \end{matrix}$$

is a suitable candidate since its product angular distribution supports the hypothesis of a long-lived intermediate at low energies[14, 37]. To test the validity of a statistical model for such a complex system, one is forced to rely on the quasi-equilibrium theory of mass spectra (i.e. activated-complex theory), whose many adjustable parameters make it an insensitive test. Product ratios agree but are not a stringent test; isotope effects in product ratios are also in agreement and this test is more exacting and the agreement more convincing[38]. Surprisingly, however, the translational energy distribution of the products does not appear to agree with the predictions of the statistical model[39].

A much more exacting test of the statistical model involves the application of phase-space theory[40] but its use is restricted effectively to three-atom systems. Regretfully it is precisely for such systems, involving so few atoms, that the statistical model might be expected to fail. For various reasons, discussed at length elsewhere[6], most of the comparisons between experiment and theory which have been made to date are either invalid or questionable, but the three endoergic processes, $H_2^+(He, H)HeH^+$, $Ar^+ + CO \rightarrow Ar + C^+ + O$ and $Ne^+ + CO \rightarrow Ne + C^+ + O$ [41, 24], show encouraging agreement. A more detailed comparison is needed for the first of these[6].

It is clearly important to establish the range of conditions under which a statistical treatment is applicable. For neutral–neutral reactions, the impressive contributions of Rabinovitch and collaborators have come from bulk, macroscopic experiments* and it is encouraging to note the recent application of beam techniques to this question. Thus the reactions of iodine atoms with methyl bromide are found to be not statistical while those of fluorine atoms with a series of alkenes are so[42].

(d) *The failure of statistical models and the use of molecular orbital correlation diagrams* — Several exoergic reactions, such as $He^+(H_2, H)HeH^+$, occur with negligible cross-sections and the failure to observe HeH^+, or its decompo-

---

*These are reviewed in Chapter 1 of this volume.

sition product H^+, has been a source of embarassment to ion–molecule chemists for many years. Phase-space arguments predict a sizable cross-section for this reaction[43] yet, contrary to this, collision merely results in elastic scattering[44]. In a very interesting paper, Mahan has invoked the use of molecular orbital correlation diagrams to show that such behaviour is to be expected in this and other systems[45]. His work reinforces the idea that orbital symmetry is an important constraint which is of fundamental significance in considering the interplay of energy and reactivity.

7.2.1.3 Non-adiabatic processes

Chemical reactions have been known for many years for which the products fluoresce, indicating their electronic excitation and the non-adiabatic nature of the collision. While the formal theory for these processes, involving crossings between potential-energy hypersurfaces, is well established[46], its application to a particular system has been frustrated by our ignorance of the actual hypersurfaces involved. A recent ion–molecule study has avoided this impasse by selecting such a simple system that the hypersurfaces can be calculated with reasonable reliability. Competition between the adiabatic channel $H^+(D_2, HD)D^+$ and the non-adiabatic channel $H^+(D_2, H)HD^+$ shows excellent agreement with the theoretical prediction[47]. Moreover, an important general result emerges from this study, that it is not always meaningful to discuss *a mechanism* for a chemical reaction. Often a sequence of processes is involved in a collision. Formally the non-adiabatic channel may be described as a hydrogen-atom transfer but the study concerned reveals a different picture. The reactants approach along the lower hypersurface, suffer an ion–atom interchange on the lower hypersurface at close separations and undergo a non-adiabatic transition to the upper hypersurface only when the products are well separated.

Similar effects are found for certain charge-transfer reactions. The charge transfer occurs at large separations as the reactants approach and further chemistry may occur at shorter separations, as a consequence of collision-induced dissociation[48].

7.2.1.4 Ion–molecule reactions and neutral–neutral reactions: similar or different?

The notion that there are underlying features of ion–molecule reactions that render their rate behaviour fundamentally different from neutral–neutral reactions is examined briefly here.

(a) *The role of long-range forces* — That long-range forces are operative in ion–molecule reactions at low collision energies is self-evident from the magnitude of measured rate constants and cross-sections. (It should be noted however that certain neutral–neutral reactions exhibit rate parameters of comparable magnitude, an early electron transfer causing powerful coulombic forces to come into play*.) For the simplest case, where the neutral

---

*See Chapter 6, Section 6.4.5.3, of this volume.

reactant is not polar, it has been customary to treat this long-range force as an ion-induced dipole interaction; the corresponding ion-induced dipole potential is then used to compute a close-collision cross-section, summarising those trajectories which cross the centrifugal barrier and lead to an intimate collision. Such an intimate collision is considered to be a necessary condition for chemical reaction and the close-collision cross-section then becomes an upper bound on the reaction cross-section. It is demonstrated at length elsewhere[6] that such a simple electrostatic treatment of the long-range forces *may* only be valid *at thermal energies* and that, even in that energy regime, other contributions cannot, in general, be neglected, i.e. ion–quadrupole, ion–induced quadrupole, etc. The so-called Langevin model, based solely on the ion–induced dipole interaction, is thus a gross over-simplification with respect to quantitative prediction, and is only useful as a qualitative indication of an upper bound to the cross-section at very low energies. We would emphasise particularly that use of *any* attractive potential, which may only be valid at long range, to compute cross-sections at supra-thermal energies is nonsensical and that the continuing widespread use of this procedure is to be deplored.

The presence of a 'basin' in positive ion–molecule potential hypersurfaces is probably universal, the depths of these basins varying from $0.1–10$ eV [6], and constitutes the distinguishing feature, for example, between $D^+(H_2, HD)H^+$ and $D + H_2 \rightarrow HD + H$. In the latter case, the potential energy *increases* as the reactants approach*, the hypersurface exhibiting an energy barrier corresponding to a reaction threshold of *c.* 0.2 eV. Less widely appreciated is the distinction between positive and certain negative ion–molecule reactions in this respect. Many negative ions are unstable, e.g. H_3^-, and as the reactants approach, e.g. $H^- + H_2$, the potential energy will increase*. A comparison between normal neutral–neutral reactions, exhibiting an activation energy, and these negative ion–molecule reactions is instructive. Indeed, the experimental excitation function of $D^-(H_2, HD)H^-$ is similar in shape[49] to that predicted for $T + H_2 \rightarrow HT + T$ [50], rising from a threshold to a maximum and falling again, with increasing collision energy. One may consider the D^- and H_2 reactants, in the negative ion–molecule reaction, riding up the repulsive wall of the hypersurface to reach a region of sufficiently close interaction for the ion–atom interchange to occur, with the products subsequently sliding down again as they separate. For such negative ion–molecule reactions, the similarity with neutral–neutral processes is noteworthy.

Where the potential hypersurface does exhibit a deep basin, ion–molecule reactions will show an individual character at low energies. One feature of this is the comparatively large orbital angular momentum which places constraints on the products by the need to conserve the total angular momentum. At high energies, however, the effect of the basin will be washed out. It is argued in the following section (7.2.2) that the reactions $Cl + H_2 \rightarrow HCl + H$ and $Ar^+(H_2, H)ArH^+$ should show similar behaviour at high energies, if only because it is the repulsive part of the inter-particle potential

---

*The long-range attractive forces will, however, form a shallow basin in the hypersurface at large separations.

which is operative and because these are similar for the two isoelectronic systems under discussion*. Thus, the high-energy chemistry observed for ion–molecule reactions should be of no different kind than that which will be revealed in the corresponding neutral–neutral studies.

(b) *Competing processes* — One feature of ion–molecule collisions is the opportunity for exoergic charge transfer to compete as a reaction channel at the lowest collision energies. This behaviour is less frequently found for negative ion–molecule processes because many neutral molecules exhibit a negative electron affinity. The corresponding processes in neutral–neutral collisions are always endoergic, exhibiting, in general, a threshold in the eV range and a smaller cross-section than that for the ion–molecule reactants[53].

Whenever charge transfer is a possible channel, the channel leading to chemical reaction will be influenced by the competition between the two. Where this charge transfer is exoergic, it may occur over a wider range of impact parameters than can chemical reaction. This may, in principle, yield information on the functional dependence of the reaction probability on impact parameter. Once again, it should be noted from the discussion in Section 7.2.1.3 that charge transfer can occur both before the 'collision'[48] and after it[47], possibilities which may bias the competition strongly in its favour.

A final competitive process must be considered, namely, collisional ionisation to release one electron. For both positive ion–molecule reactions and neutral–neutral reactions, this is highly endoergic and does not compete efficiently in the 'chemical' energy range. For certain negative ion–molecule reactions, it can, however, be exoergic. This process, known as associative detachment, can be exceedingly efficient, occurring perhaps at any early stage during the reactants' approach and forestalling possible competing chemical reactions thereby[49]. Thus the process $H^-(O_2, HO_2)e^-$ occurs to the effective exclusion of other exoergic channels, such as $H^-(O_2, O)OH^-$, at thermal energies[120]; interestingly, the competition becomes important at higher energies[49]. Study of associative detachment reactions is often instructive. Thus the reaction $H^-(H, H_2)e^-$ has a thermal rate constant which is approximately half the Langevin prediction[54], since only half the collisions will follow the attractive potential curve of the H_2^- which results in ejection of the electron[†].

7.2.2 Reaction mechanisms: collision dynamics

Molecular dynamics has as its ultimate aim the detailed (microscopic) description of how molecular systems are transformed from one state to another during a collision. In the last decade, beam and chemiluminescence investigations of reactions between neutral reagents have shown how the product energy and angular distributions can reveal the reaction dynamics[55]. Somewhat more recently, Monte Carlo calculations with semi-empirically calculated surfaces have indicated that the major features of the surface can

---

*The attention which is beginning to be given to the inelastic scattering of He and Li^+ with hydrogen will be instructive in this matter[51, 52]
†A more refined analysis is given in reference 2, p. 203.

often be established from the measured dynamical properties. Unfortunately, only very recently has there been any effort toward the construction of realistic surfaces for ion–molecule processes. Because of this, the chemical and physical implications of the reactive scattering results can only be discussed qualitatively. Models or interpretations based on assumed force fields and their effects must be considered to be largely heuristic. This situation is disturbing since, as we shall see, there exists a wealth of detailed experimental data. The main discussion is preceded by some basic background material, encountered in any description of reactive beam scattering.

As a framework for a brief description of reaction mechanisms (microscopic level) for the general rearrangement collision, it is convenient to consider the two limiting reaction models, which are distinguishable through the measurement of the *c.m. differential-scattering cross-section*[56, 57]. The first is

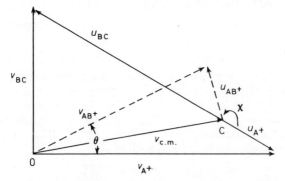

Figure 7.1 'Newton' diagram[56] for the reaction $A^+ + BC \rightarrow AB^+ + C$. The angle between the intersecting beams is 90 degrees. The laboratory velocities of the A^+, BC and AB^+ are v_{A^+}, v_{BC}, and v_{AB^+}; the corresponding centre-of-mass (c.m.) velocities are u_{A^+}, u_{BC} and u_{AB^+}; the c.m. velocity is $v_{c.m.}$. The ionic product is scattered through a c.m. angle χ (with respect to the incoming A^+), which corresponds to the laboratory angle θ. The initial relative-velocity vector is equal to: $u_{A^+} + u_{BC}$, and the final relative-velocity vector is equal to:

$$u_{AB^+} \times \frac{m_{A^+} + m_{BC}}{m_C}.$$ The laboratory and c.m. origins are denoted

by O and C, respectively. The c.m. recoil velocity of the product C is not shown

the 'long-lived complex', 'persistent complex' or 'indirect' mode in which the colliding particles form an intermediate that is bound together long enough to undergo several or more *rotations*. For this case the c.m. angular distribution $I_R(\chi)$ of the products will be symmetric about a plane passing through the c.m. origin and normal to the initial relative velocity vector, u_{BC} (see Figure 7.1). The average period for one complete rotation is typically 10^{-12} s, as compared to a normal vibrational period of 10^{-13}–10^{-14} s. The alternative mode is described by a 'direct', 'direct-interaction', or 'simple' model in which the intermediate decomposes in less than a complete rotational period, and thus $I_R(\chi)$ is asymmetric about $\chi = 90$ degrees. As found by trajectory calculations[50], direct reactions often occur in a time comparable for the reagents to pass unimpeded by one another ($c.$ 10^{-13} s). A forward-scattering or stripping mechanism refers to a direct reaction in which the detected (ionic) product continues in nearly the same c.m. direction as its (ionic) precursor ($\chi \approx 0$ degrees). The other extreme where the product is found at very large c.m. angles ($\chi \approx 180$ degrees) is described by a backward-scattering or rebound mechanism. In a *spectator* stripping reaction of the type $A + BC \rightarrow AB + C$, the characteristic feature is that the incident particle A interacts with only particle B of the target[58]. Since the products separate before there is time for momentum transfer to the remainder of the target, C is merely a spectator. This highly idealised stripping or grazing model may only be important at very high relative velocities, i.e. very short collision times. For the ideal case of negligible target motion, the product will be undeflected ($\chi = 0$, $\theta = 0$ degrees) and will move with a velocity or energy given by: $\dfrac{m_A}{m_A + m_B} \times v_A$ (LAB) or $\times E_A$ (LAB). The remaining energy, $\dfrac{m_B}{m_A + m_B} \times E_A$ (LAB), is deposited in the internal energy of the products.

In a beam experiment, the energy, linear and angular momentum conservation laws are extremely useful in relating the external and internal energy and momenta of the reactants to those of the products[56, 58]. Indeed for reactive encounters at very high collision energies, the dynamics are usually set by the need for disposal of excess internal energy. The conservation law for energy and linear momentum is written as

$$E' + Z' = E + Z - \Delta D_0^0 \tag{7.3}$$

where Z' and Z are the total internal energies of the products and reactants, respectively; E' and E are the final and initial relative kinetic energies; and ΔD_0^0 is the difference in the dissociation energies of the products and reactants, measured from their zero-point vibrational levels. The translational exo-ergicity of the reaction is defined by equation (7.4) and is a measure of the net amount of energy interchanged between translational and internal degrees of freedom (i.e. the energy partitioning).

$$Q(E, \chi) = E' - E \tag{7.4}$$

The maximum value of Q is $Z - \Delta D_0^0$, whereas its minimum value is determined by the requirement that the product be stable w.r.t. unimolecular decay. If the undetected product is an atom and both products are formed in their ground electronic states: $Q > +\Delta D_0^0 + Z - D$, where D is the lowest dissociation energy of the detected product. (The implicit assumption here is $Z'_{vib.} \gg Z'_{rot.}$.)

The total angular momentum is also a constant of the motion and thus

$$L' + J' = L + J \tag{7.5}$$

where L and J are the orbital and rotational angular momentum vectors. For reactions proceeding through a persistent-complex mechanism, this second conservation law predicts a symmetrical distribution about $\chi = 90$ degrees, and the particular shape of $I_R(\chi)$ is governed by the partitioning between L and J, and L' and J' [56, 57]. If $L \gg J$, the usual case, and $L' \gg J'$,

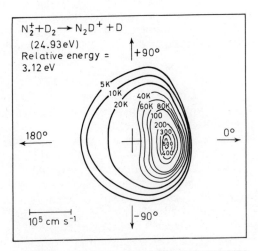

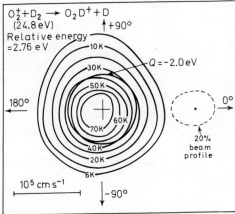

Figure 7.2 Top: An intensity contour map of N_2D^+ from the reaction of N_2^+ with D_2. The distribution is asymmetric about the $\chi = 90$ degrees line, indicating a direct mechanism. Note the dominance of small-angle scattering ($\chi \approx 0$ degrees). Bottom: An intensity contour map of O_2D^+ from the reaction of O_2^+ with D_2. The symmetry about the $\chi = 90$ degrees line is consistent with an indirect mechanism. (From Mahan, B. H.[62], by permission of the American Chemical Society)

strong forward $(\chi = 0$ degrees)–backward $(\chi = 180$ degrees) peaking will occur. If the reverse holds, i.e. $J' \gg L$, the angular distribution will be nearly isotropic.

The measurement of the angular (θ) and kinetic energy distributions of the product, followed by the appropriate LAB-c.m. transformation, gives a differential cross-section or intensity per unit velocity space volume, $I_R(\chi, u)$[59]. This product velocity-vector intensity is often given on a contour plot, and two examples are shown in Figure 7.2. The highly asymmetric distribution about $\chi = 90$ degrees is indicative of a direct reaction and the symmetric distribution in the lower map suggests a persistent-complex mechanism. The more common but less detailed differential cross-section $I_R(\chi)$ is given by:

$$\int_0^\infty I_R(\chi, u) u^2 \, du.$$

For a variety of reasons, in many ion–molecule studies only laboratory energy spectra of the product have been measured at a detector position fixed along the primary beam axis. In the absence of other information, the distinction between direct and indirect modes must be inferred from a comparison of the most probable product speed with the calculated centroid speed ($\equiv$ indirect) and the calculated spectator stripping speed ($\equiv$ direct). From the previous discussion it should be clear that these conditions are neither sufficient nor necessarily correct to unambiguously distinguish these modes. (An additional complication is that wide angle (θ) scattered products are not always detected.) However, for systems in which the spectator and centroid speeds are substantially different, this approach often gives the qualitatively correct answer.

The discussion centres around some of the following well-studied ion–molecule reactions which have been grouped according to whether they proceed via a direct (I) or indirect mechanism (II) at the lowest energy studied. With the exception of (IIa) the reactions are all exoergic and appear to have negligible (<0.1 eV) potential energy barriers. (The approximate relative kinetic energy range is indicated to the right of each reaction.)

(Ia)	$N_2^+ + H_2(D_2, HD) \rightarrow N_2H^+ + H$	
	$CO^+ + H_2(D_2, HD) \rightarrow COH^+ + H$	$(0.1–15 \text{ eV})$ [59–73]
	$Ar^+ + H_2(D_2, HD) \rightarrow ArH^+ + H$	
(b)	$N_2^+ + CH_4(CD_4) \rightarrow N_2H^+ + CH_3$	$(10–50 \text{ eV})$ [74]
	$N_2^+ + D_2O \rightarrow N_2D^+ + OD$	$(1–15 \text{ eV})$ [75]
(c)	$N^+ + H_2 \rightarrow NH^+ + H$	$(2.5–8 \text{ eV})$ [76]
(d)	$N^+ + O_2 \rightarrow NO^+ + O$	$(1–10 \text{ eV})$ [77]
(e)	$O_2^+(^4\Pi_g) + D_2 \rightarrow DO_2^+ + D$	$(2–9 \text{ eV})$ [78]
(IIa)	$O_2^+(^2\Pi_g) + H_2(D_2, HD) \rightarrow HO_2^+ + H$	$(2–10 \text{ eV})$ [79]
(b)	$H_2^+ + H_2 \rightarrow H_3^+ + H$	$(0.1–8 \text{ eV})$ [44, 80, 81]
(c)	$CH_4^+ + CH_4 \rightarrow CH_5^+ + CH_3$	$(1–5 \text{ eV})$ [44, 82]
(d)	$C_2H_4^+ + C_2H_4 \rightarrow C_3H_5^+ + CH_3$	$(0.7–4 \text{ eV})$ [37, 83]

On inspection of this list, one is struck by the great variety of reactions examined and the enormous kinetic energy range explored, in some cases extending from near thermal to collision energies considerably greater than

typical bond dissociation energies. In sharp contrast, the dynamical investigations of neutral systems for the most part have been confined to the
collisions of monovalent atoms at thermal energies, and the prospect for
their investigation at elevated energies in the near future is not promising.
However at c.m. energies greater than c. 0.5 eV, where the effect of long-
range ion–induced dipole forces is presumably unimportant, some chemically
related ion–molecule and neutral–neutral systems (e.g., $Cl + H_2$, $Ar^+ + H_2$)
might possess common dynamical features. It then bears repeating that the
treatment of ionic and non-ionic reaction kinetic systems, as always distinctly
different, is artificial. Emphasis here is placed on the role of reactant kinetic
energy on the ensuing reaction dynamics.

For discussion purposes, it is convenient to divide the chemical reaction
range into three energy regions: 'low', 'intermediate', and 'high'. In the former,
the initial relative kinetic energy is comparable to thermal kinetic energies,
and the long-range attractive potential between reactants and products
may play a dominant role in the dynamics. Above this, and extending to
energies comparable to bond energies (c. 2–4 eV), is the intermediate and,
beyond this, the high-energy region. At high energies, due to the very short
collision time, the chemical interaction is largely of short-range ion–*atom*
character and the creation of new and stable chemical bonds is very difficult.
(Fragmentation and other highly inelastic non-reactive events play a greater
role.) Of course at high energies formation of a persistent complex is very
unlikely and only possible if, for example, there exists a *very* deep basin in
the corresponding potential energy surface.

7.2.2.1   Direct reactions: intermediate and high energies

The pivotal studies of ion–molecule dynamics at intermediate and high
collision energies are due to Henglein and collaborators[69, 70]. From velocity
measurements alone, they were able to correctly show that the reactions of
Ar^+, N_2^+ and CO^+ with H_2 and its isotopic variations (Ia) occurred through
a stripping mechanism, and the most probable product speed was very close
to the predicted spectator or ideal stripping speed. This was the first evidence
in chemical kinetics for a spectator or grazing model. The combined features
of a moderately large reaction cross-section and an extremely short collision
time (c. 10^{-14} s) make this model very plausible here. At high collision
energies, where the ideal stripping model must fail ($Z'_{vib.} > D$), slight positive
deviations from the ideal stripping speed were noted. The forward recoil
of the ionic product and the backward recoil of the freed H-atom provided
the necessary mechanism for lowering the product internal-energy below its
dissociation energy. The finding that the total reaction cross-section is
nearly the same for different isotopes (H, D) at the same energy relative to the
abstracted atom further supports the conclusion that the projectile interacts
almost exclusively with the abstracted atom.

These conclusions were later confirmed by others (Bailey[63] and
Mahan[59–62, 73]), and notably by Mahan *et al.* who extended these studies to
include angular measurements, allowing the construction of the first complete
velocity vector distribution for any chemical reaction (Figure 7.2). From an

inspection of this and other maps, it is clear that the *gross* features of the scattering are best described by a spectator model or one very close to it. Through small-angle scattering is dominant, large-angle events from smaller impact-parameter encounters are evident. The observed rebound contribution might result from a nearly head-on collision having a collinear configuration, N_2^+—H—H. At higher energies the forward-scattered products peak at velocities above the ideal stripping value, as found by Henglein, and in fact they are internally excited to close to their dissociation limit, as evidenced by $Q(\chi = 0$ degrees) becoming more negative with rising energy but eventually reaching a plateau value of $c.\ -2.5\text{eV}$. Hence at high energies the energy partitioning ($\chi \approx 0$ degrees) reflects a modified grazing mechanism in which only the minimum amount of momentum necessary for product stability is efficiently transferred to the unreacted atom. On the other hand, not only are the backward-scattered products usually less internally excited than those scattered forward, but in addition with rising energy the rebound mode increases in importance relative to the stripping mode. These findings together with the absence of a rebound H—D isotope effect are entirely expected from a very small impact-parameter (reactive) collision where the reactant atoms are all strongly coupled.

Suplinskas[84] has applied his kinetic model for chemical reactions to the $Ar^+(D_2,D)ArD^+$ system ($E \approx 3$–8 eV). The product velocity distributions, differential reaction cross-sections, and total reaction cross-sections were all found to be in good agreement with the experimental data. The implication is that, in the relative energy range of several electron volts, the chemical process is dominated by repulsive core interactions, appropriately modified by elastic-scattering potentials. However, the finding that the forward-peaking of the products is primarily due to product attraction (r^{-4}) is difficult to comprehend at these collision energies and almost certainly reflects the assumptions in the kinematic model. Polanyi and Kuntz[85] have carried out trajectory calculations for this reaction by use of the ion–induced dipole potential at large separations of the reactants and the L.E.P.S. (London, Eyring, Polanyi, Sato) hypersurface for the isoelectronic reaction $Cl + D_2 \rightarrow DCl + D$ at all other separations (the attractive potential r^{-4} between the products was neglected). At 2.3 eV c.m. collision energy, the differential cross-section was strongly forward-peaked in agreement with experiment.

The formation of N_2H^+ from the reactions of N_2^+ with hydrogen-containing polyatomic targets (Ib) is of major interest for two reasons: (a) the greater kinematic leverage* relative to molecular hydrogen allows a more sensitive test of the dynamics; (b) the extra degrees of freedom of the neutral reaction fragment can serve as internal energy sinks and thereby affect the energy partitioning. As in the case of molecular hydrogen, the reactions proceed by a stripping mechanism, and, within experimental error, the most probable product speed is identical to the spectator prediction, even at very high collision energies where the model fails for the diatomic target case (e.g. at $E = 47.3\ \text{eV}, Q = -4.0\ \text{eV}$ for $N_2^+(CH_4, CH_3)N_2H^+$). This rather startling result shows that grazing reactions do occur in which substantial amounts

---

*For the same laboratory angular and speed resolution, the resolution of the corresponding centre-of-mass quantities in (Ia) is much less than in (Ib).

of translational energy are converted into vibrational energy of the *neutral* product but without any sizeable momentum transfer to it. At fixed collision energy, Q becomes more negative as χ increases, opposite to that found in reaction Ia where product stabilisation can be achieved only by atom recoil. Indeed at high energies, the extremely negative Q values detected correspond to very violent collisions with a small impact parameter, implying that the methyl radical undergoes decomposition.

At elevated energies it might be expected that a variety of non-reactive inelastic channels would compete with the rearrangement process. The non-reactive scattering of N_2^+ with D_2 [59] and CH_4 [74] and Ar^+ with D_2 [73] have been examined in some detail. Significant inelastic scattering is found for N_2^+ whereas the Ar^+ scattering is essentially elastic. For N_2^+ on D_2 at *c.* 16 eV, there is a negligible elastic component at large angles, and, together with the very small inelastic and reactive contributions, implies that small impact-parameter events lead almost exclusively to simple or dissociative charge-exchange processes. Consistent with this explanation, Vance and Bailey[64] have found sizeable charge-exchange cross-sections for this system at high energies. A comparison of the differential scattering cross-sections for the reactive and non-reactive scattering of Ar^+ by D_2 and He allowed estimates of 0.1 and 0.35 for the zero impact-parameter reaction probabilities at 9.1 and 2.8 eV, respectively[73].

In summary, chemical reactions of the simple abstraction type occur and often quite efficiently at collision energies well in excess of chemical bond dissociation energies or at translational temperatures $> 10^5$ K. The ability to form stable chemical bonds at such high energies and in the presence of a myriad of competing inelastic channels, can be traced to the high probability for atom transfer in grazing collisions. The dependence of the observed isotope effects and energy partitioning on both scattering angle and projectile energy are well understood in terms of energy disposal requirements. On the other hand no very reasonable proposals have appeared that explain on a microscopic level the small-angle forward product recoil* observed at high energies. Collisions having this amount of momentum transfer are difficult to reconcile with our present and perhaps primitive concepts of small-angle reactive scattering. Finally, it should be realised that models other than the spectator can be fabricated, which, though differing from it on a microscopic mechanistic level, may result in very similar macroscopic predictions. Until trajectory calculations are carried out on reliable potential surfaces these questions remain unanswered.

7.2.2.2. *Direct reactions: low energy*

Until only a few years ago, it was commonly thought that at low energies ion–molecule reactions in general involved a persistent-complex mechanism. This belief was based on circumstantial experimental evidence such as the finding of isotope (H/D) effects in reaction Ia at low collision energies and

---

*Forward product recoil is inferred from the observation of ionic product speeds in excess of the value predicted by the spectator stripping model.

the detection of stable ions in mass spectrometry having the same identity as the combined reactants (e.g., H_2CO^+ and $HNNH^+$). The all-too-common misinterpretation of activated complex theory also perpetuated this opinion.

Employing a cross-beam ion–molecule apparatus, Wolfgang *et al.* measured, over the primary beam energy range 0.7–25 eV, the angular and velocity distributions of N_2D^+ and ArD^+ from reaction of N_2^+ and Ar^+ with D_2 [65–67]. It was conclusively shown in this and later studies[71, 72] that these simple atom-transfer reactions are dominated by a direct mechanism at even relative energies as low as 0.1 eV (1000 K). The asymmetric angular distributions are strongly forward-peaked at all energies, and at the highest energy the scattering closely approximates the spectator model, as found by Henglein, Bailey and Mahan. However with decreasing energy, marked deviations from the model become apparent, and for relative energies less than approximately 0.3 eV, the reaction $Q(\chi = 0$ degrees) becomes *positive*, i.e. there is a net conversion of internal into translational energy of the products. Deviations from the spectator model at low energies are totally expected.

In order to explain this behaviour at low energies and, in particular, the inversion in the sign of Q, a model was developed which took into account the attractive r^{-4} potential between the reactants and the products. In its simplest form (polarisation stripping)[65–67] the model states that at large separations the reactants are accelerated toward one another in accordance with the potential: $\phi_{react.} = -\alpha_{D_2}e^2/2r^4$. At some distance of closest approach, r_c, atom transfer occurs and the products recede from one another being decelerated by the potential: $\phi_{prod.} = -\alpha_D e^2/2r^4$. The term Q is expressed in terms of r_c, E, and the polarisabilities; and r_c is obtained by fitting the model to the lowest energy data point. At higher energies (>0.1 eV) reasonably good agreement with the experimental data is found, and for the Ar^+–D_2 system r_c is close to the DCl bond length, 1.27 Å. However it is questionable whether these agreements constitute good evidence that *long-range forces* dominate the *dynamics* at low energies, especially since the necessary assumption of a very strong (*c.* -0.5 eV) ion–induced dipole potential at short separations is incorrect. Indeed a reaction model by Kuntz[86], that is based on a physical picture which is quite different and much more plausible, also fits the data.

Likewise there exists no generally accepted qualitative explanation for why these and other direct reactions at low energies are so strongly forward-peaked. Light and Chang[87] have calculated angular distributions for the reaction $Ar^+(D_2,D)ArD^+$. An impulsive model very similar to the Wolfgang 'polarisation reflection' mechanism[72] (itself, a modification of the polarisation stripping model) is in substantial agreement with the experimental angular distributions over a wide energy range. On the other hand Kuntz's model, mentioned above, also predicts forward-peaking down to very low collision energies.

In anticipation of the description of reactions involving a persistent-complex mechanism, it is essential to state a major point demonstrated by these first low-energy studies[83]. Although a bound intermediate having the composition of the combined reactants is capable of existing for several or more rotational periods, as suggested by a deep basin or well in the hypersurface and confirmed through RRKM unimolecular calculations, it may not be formed in the collision of the reactants. Factors of dynamical nature may make the well inaccessible to the reagents or, stated more generally, the initial relative kinetic energy cannot be easily transformed into internal excitation of the complex. The importance of this second criterion in deciding whether a reaction goes directly or indirectly is nicely illustrated by the $CO^+(H_2,H)$ COH^+ system. The potential energy surface has a deep minimum for at least one configuration of the combined reactants, since the H_2CO^+ ion is known from mass spectrometry. Yet at collision energies much less than the well depth the reaction is still dominated by a direct mechanism. Two important and readily understandable factors can prevent complex formation at low energies; (a) the presence of an entrance barrier to the well; (b) for complex reagents, the necessary nuclear rearrangement may be so extensive that it cannot be achieved during the brief time allowed by the collision. Of course if either or both the criteria are not met, the same products may still be produced by a direct channel.

7.2.2.3 Indirect reactions

The study of persistent-complex ion–molecule reactions provides an important and in some respects a unique opportunity to examine the various statistical models for unimolecular decay. Not only can the average lifetime of the complex be continually changed through the variation of the initial collision energy, but also the translational energy spectra of the products can be easily and accurately measured.

As mentioned in Section 7.2.1.2(c) a start has been made in the use of the measured translational energy spectra[39] to test various statistical theories (see also reference 118). Of perhaps greater significance, the ability to examine and compare in detail a wide variety of reactions, direct and indirect, allows a demonstration of the factors controlling persistent-complex formation. In the past three years, a substantial number of ionic reactions occurring through an indirect route have been uncovered, and in all cases, above some energy, a gradual transition from indirect to direct channels is observed. Our discussion is limited to the specific results of four ion–molecule reactions (equations (7.6–7.9)) below.

(a) For the reaction

$$C_2H_4^+ + C_2H_4 \rightarrow [C_4H_8^+] \rightarrow CH_3 + C_3H_5^+ \rightarrow C_3H_3^+ + H_2 \qquad (7.6)$$

Wolfgang et al.[37, 83] find that both ionic products arise from an indirect mechanism at $E \leqslant 3$ eV, and as written it appears that the bulk of $C_3H_3^+$ comes from the secondary decomposition of a highly excited $C_3H_5^+$ radical. (Total cross-section ratios $C_3H_3^+/C_3H_5^+$ are 0.15, 0.80 and 1.8 at 1.43, 3.25 and 4.1 eV, respectively). With increasing energy the reaction exhibits a continuous transition to a forward-scattering mechanism.

(b) The familar reaction

$$CD_4^+ + CD_4 \rightarrow CD_5^+ + CD_3 \qquad (7.7)$$

and its isotopic variations have been among the most widely studied ion–molecule reactions. Henchman[44] and Henglein[82] have given good experimental evidence for a persistent-complex mechanism at $E < 1$ eV (c.m.). This result is paradoxical since the reaction does not involve a $C_2H_8^+$ intermediate which is completely equilibrated. Scrambling of the hydrogens is not observed; the formal mechanism involves principally the discrete transfer of either a proton or a hydrogen atom and the former occurs with greater frequency than the latter[44]. Thus the intermediate must retain some information as to how it was formed.

(c) For the system

$$H_2^+ + H_2 \rightarrow H_3^+ + H \tag{7.8}$$

the experimental results[44, 80, 81] are consistent with an indirect pathway at energies below 1 eV (c.m.). Due to the relatively few internal degrees of freedom of any H_4^+ complex and the sizable heat of reaction, this conclusion suggests a moderately stable H_4^+, although it has not been observed experimentally elsewhere.

(d) The dynamics of the endoergic reactions:

$$O_2^+(^2\Pi_g) + H_2 \begin{cases} \xrightarrow{\text{(i)}} HO_2^+ + H & \Delta H^0 = 1.96 \text{ eV} \\ \xrightarrow{\text{(ii)}} OH^+ + OH & \Delta H^0 = 1.87 \text{ eV} \\ \xrightarrow{\text{(iii)}} H_2O^+ + OH & \Delta H^0 = 0.66 \text{ eV} \end{cases} \tag{7.9}$$

have been investigated in unusually great experimental detail by Mahan and collaborators[79]. The angular distributions are symmetric and essentially isotropic at collision energies below c. 5 eV. The results for channel (i) are explained on the assumption that the $H_2O_2^+$ complex has a hydrogen peroxide structure with a stability of 2.4 eV with respect to the reactants, and, secondly, there exists a potential barrier to its formation of roughly 2 eV. If there were no barrier or if the barrier were much smaller than the reaction endoergicity, the bulk of the complexes would decay back to the products. Interestingly, the OH^+ angular distribution is symmetrical at even high energies (c. 10 eV), where the formation of a long-lived complex is very improbable and the HO_2^+ and H_2O^+ are formed by a direct mode. The symmetry is probably a consequence of the similarity of the product partners and a short-lived intermediate having a configuration resembling hydrogen peroxide (i.e. dynamically equivalent oxygen atoms) as opposed to say a linear HHOO structure. At lower energies it is very likely that the OH^+ also arises from the decay of the long-lived $H_2O_2^+$ complex. Below 5 eV, the cross-section for the production of HO_2^+ is an order of magnitude greater than those for the OH^+ and H_2O^+ products at the same energy. Quasi-equilibrium theory (QET) has been used to calculate both the product–ion intensity ratios and the lifetimes of the $H_2O_2^+$ complex as a function of the O_2^+–H_2 collision energy[119]. The calculations did not predict the large excess of HO_2^+ over the other two products and thus an additional channel, leading to direct formation of HO_2^+ at all energies, has been suggested[45]. Furthermore, only by inclusion of anharmonicity contributions and vibration–rotation interactions in the QET calculations, can the $H_2O_2^+$ decomposition lifetimes be made to agree with the measured product-ion velocity-angle scattering distributions. The application of QET to such systems is important and currently an area of mu h interest. The apparent failure in this particular case, for which such beautiful data have been obtained, is puzzling.

Some of the results (equations (7.6)–(7.9)) are presently well understood or at least consistent with one's prior expectations; others are not and more work is clearly necessary. In any event, in the near future, from these and other recent beam studies, chemiluminescence and thermal activation studies will emerge a clearer picture of the dynamics of the formation and decomposition of long-lived complexes.

7.2.2.4 *Theoretical and experimental chemical dynamics*

It is obvious that the detailed implications of some of the major scattering results, especially those of the simple atom-transfer reactions at low energies, are not well understood. In the future, more attention must be devoted to the construction of reliable semi-empirical hypersurfaces for these systems followed by quasi-classical or approximate quantum-mechanical treatments[88]. Even for the simplest systems like $Ar^+–H_2$, the difficulties are significant, but at least the theoretical and computational methods are now in a workable form, as testified by recent successes of theoretical dynamic treatments of certain neutral reactions[50, 55].

On the other hand, the exchange reaction $H^+(D_2, HD)D^+$ represents an unparalleled opportunity in chemical kinetics to compare theoretical with experimental chemical dynamics, since the relevant H_3^+ surfaces can be calculated with convincing reliability[89–91]. However, as indicated in Section 7.2.1.3, there is one important complication: above *c.* 2 eV relative kinetic energy, molecular hydrogen ion formation $[H^+(D_2, D)HD^+;\ H^+(D_2, H)D_2^+]$ becomes important, and thus in treating the exchange channel, one must explicitly take into account the non-adiabatic interaction between the lowest and first excited singlet states of H_3^+ [91]. The two lowest singlet surfaces have been computed, and to obtain the dynamics Preston and Tully have devised a 'classical trajectory-surface hopping' approach to non-adiabatic reactions[91, 92]. The method is described in detail and applied to the computation of the cross-sections[92] and angular[47] and velocity[47] distributions of the products from the three channels. The theoretical cross-sections are in good agreement with the very recent experimental measurements (Krenos and Wolfgang[93]; Holliday, Muckerman and Friedman[94]; Maier[95]), and the experimental and theoretical angular and velocity distributions are in good qualitative accord[47]. It is interesting to note that, even now, more is known about the experimental dynamics of the $H^+(D_2, HD)D^+$ reaction than of the simplest neutral exchange reaction $H + D_2 \rightarrow HD + D$, which has been studied for nearly forty years. In the next few years the experimental gap will probably increase, and if the theoretical effort is sustained this, the simplest chemical reaction, may also be the best understood reaction in chemistry.

7.3 THE ROLE OF SOLVATION IN THE LIQUID PHASE: THE RELATIONSHIP OF GAS-PHASE STUDIES TO WET CHEMISTRY

In 1957, the visionary book by Field and Franklin[96] revealed to chemists a new vista of ionic chemistry in the gas phase. One recurrent theme was the exciting prospect that such studies of ionic energetics and reactivity in the gas phase should provide important fresh insights for ionic solution chemistry. Fourteen years later, the extensive contribution to ion energetics is documented in Franklin's recent data compilation[97] but, by and large, it may be said that the contribution towards the understanding of ionic reactivity in solution has been disappointing. There are four main reasons for this: (a) The gas-phase studies have been confined principally to such a high range of translational energy (c. 1 eV and higher) that their relevance to ionic chemistry at thermal energies is not clear. (b) Similarly, the production of reactant ions by electron impact has given them unknown internal energies, substantially in excess of the Maxwell–Boltzmann distribution at thermal energies. (c) The reactions which have been studied most extensively in the gas phase, such as $CH_4^+(CH_4, CH_3)CH_5^+$, have no counterpart in solution, whereas those of most interest in solution, such as $OH^-(CH_3Cl, CH_3OH)Cl^-$, have not been studied in the gas phase. (d) Finally, and most significantly, it can always be argued that the solvent plays an all-important role in solution to the extent that any comparison between the gas phase and solution is highly questionable.

This situation has changed substantially within the past five years. As indicated in the previous section, the study of ion–molecule reactions will surely continue to yield valuable insights for the general field of gas kinetics but we now feel it to be on the threshold of making important contributions to ionic solution chemistry. Admittedly the data available at this time are few, yet they are sufficient to suggest an exciting new prospect for the future. What we can record here is necessarily selective, fragmentary and superficial but, by presenting a survey of some recent results, we seek to interest the general reader of kinetics in the possibilities which lie ahead.

As usual, the situation has been changed by the development of new techniques, specifically ultra-high-pressure mass spectrometry, the flowing afterglow and ion cyclotron resonance, these overcoming to a greater or lesser extent the first three limitations discussed above. For the first two techniques, the reactants possess a truly thermal Maxwell–Boltzmann distribution of translational and internal energy, since they are relaxed by non-reactive collisions within a thermal bath; for ion cyclotron resonance, the distribution is not Maxwell–Boltzmann but can be arranged to be quasi-thermal. Ultra-high-pressure mass spectrometry has permitted the study of solvated ions, their equilibria and hence their thermodynamic properties; these in turn invite interesting speculation concerning the structure of solvation shells. The flowing afterglow yields true thermal rate constants, whose temperature dependence has been examined throughout the range 80–600 K; a very wide range of reactant ions may be prepared, including selectively solvated ions. Finally, comparatively slow reactions may be investigated by use of ion cyclotron resonance, since the double-resonance technique permits an unambiguous identification of the reactant ion responsible for producing a particular ionic product.

7.3.1 Ionic solvation: thermodynamics and structural implications

During the past 6 years[98], a whole new field of 'ionic solvation in the gas phase' has been developed by Kebarle and co-workers. (The reader is directed to a recent comprehensive review by Kebarle[12], from which this brief account is drawn.) By use of an ultra-high-pressure mass spectrometer, either positive or negative ions, designated $A^{\pm}$, are produced in a 'bath' of gaseous solvent molecules B and solvate rapidly to achieve equilibrium, represented by the general equation

$$A^{\pm}B_{n-1} + B = A^{\pm}B_n \qquad (7.10)$$

Measurement of the steady-state ionic intensities of $A^{\pm}B_{n-1}$ and $A^{\pm}B_n$, as a function of temperature and pressure, yield estimates of the changes in enthalpy, entropy and free energy for the successive addition of solvent molecules. Some representative 'enthalpies of solvation', $\Delta H_{n,n-1}$, are given in Table 7.3 to show how this quantity depends on n, the solvation number of the ion. The data are presented in such a form that $\Delta H_{n,n-1}$ represents the enthalpy required to remove a solvent molecule from $A^{\pm}B_n$.

A wealth of interesting information is available from such data. In contrast to studies in solution, this technique focuses upon *individual ions in isolation* to probe the energetics of the close ion–solvent molecular interactions *alone*, in the absence of the bulk solvent. Thus the contribution of such close interactions to the overall behaviour in solution may be examined. Various applications are discussed below.

7.3.1.1 The hydration of positive alkali and negative halide ions

Comparison of the measured values of $\Delta H_{n,n-1}$ with calculated potential energies, based on a model considering electrostatic interactions, shows good agreement but the large difference between $\Delta H_{1,0}$ and $\Delta H_{2,1}$ for the smaller positive ions indicates chemical bonding between the ion and solvent water molecule in these singly-hydrated ionic species[99]. Furthermore, a good correlation is found between the enthalpies measured in the gas phase and total single-ion hydration energies, measured in solution by Randles[100].

7.3.1.2 Structural implications for solvation shells

The species $H_3O^+(H_2O)_3$ has been asserted to possess anomalous stability in solution and, if true for the gas phase, this should be reflected in a discontinuity in the plot of $\Delta H_{n,n-1}$ against n. No such effect is found[101]. The ion $H_9O_4^+$ may yet possess anomalous stability *in water* but the gas-phase results deny the validity of any argument which depends on bonding predictions for the *isolated species*, $H_9O_4^+$. Of interest is the contrasting behaviour in ammonia where anomalous stability is found for the $NH_4^+(NH_3)_4$ ion in the gas phase[102].

7.3.1.3 Competitive solvation

In a mixed-solvent system, is there competitive solvation around particular ions? In the competitive solvation of NH_4^+ by water and ammonia in the gas phase, ammonia is held preferentially in the inner solvation shell but water dominates in the outer one[103]. It is suggested that in the inner shell the

Table 7.3 Enthalpy changes for the successive dehydration of hydrated positive and negative ions in the gas phase

$$A^\pm(H_2O)_n = A^\pm(H_2O)_{n-1} + H_2O$$

$n,n-1$	H^+	Li^+	Na^+	K^+	Rb^+	Cs^+	OH^-	F^-	Cl^-	Br^-	I^-
						$\Delta H_{n,n-1}$/kcal mol^{-1}					
1,0	165	34	24	17.9	15.9	13.7	22.5	23.3	13.1	12.6	10.2
2,1	36	25.8	19.8	16.1	13.6	12.5	16.4	16.6	12.7	12.3	9.8
3,2	22.3	20.7	15.8	13.2	12.2	11.2	15.1	13.7	11.7	11.5	9.4
4,3	17	16.4	13.8	11.8	11.2	10.6	14.2	13.5	11.1	10.9	
5,4	15.3	13.9	12.3	10.7	10.5		14.1	13.2			
6,5	13	12.1	10.7	10.0							
7,6	11.7										
8,7	10.3										

(From Kebarle[12], by courtesy of the Plenum Publishing Corp.)

greater proton affinity of ammonia is important, but in the outer shell the larger dipole moment of the water is the determining factor[12].

7.3.1.4 The influence of solvation number on acid–base behaviour

It is well known that HCl is more dissociated in water than in methanol. In contrast, the gas-phase studies[104] reveal a greater equilibrium constant for (7.11) than for (7.12).

$$HCl + 2CH_3OH \rightleftharpoons CH_3OH_2^+ + Cl^- \cdot CH_3OH \qquad (7.11)$$

$$HCl + 2H_2O \rightleftharpoons H_3O^+ + Cl^- \cdot H_2O \qquad (7.12)$$

As n is increased, the energetics gradually change to yield the reverse behaviour found in solution.

7.3.1.5 Correlation between the solvation energy of a negative ion and the acidity of the solvent

The energy change accompanying the solvation of a negative ion, according to

$$A^- + HB \rightarrow AHB^- \qquad (7.13)$$

may be correlated with the acidity of HB [105]. More generally it depends on the electron affinities of A and B and the H—A and H—B bond strengths[104].

7.3.2 Nucleophilic displacement reactions

7.3.2.1 Reactions of organic anions

Typical of the nucleophilic displacement reactions which can now be studied in the gas phase is the general process

$$X^- + CH_3Cl \rightarrow CH_3X + Cl^- \qquad (7.14)$$

where X^- may be O^-, OH^-, alkoxide, phenyl and benzylic anions[106]. Here again, in the absence of solvent, the measured rate constant reveals the *intrinsic* reactivity. When X^- is the alkoxide ion OR^-, the reaction rate shows a very slight decrease as the extent of alkyl substitution in R is increased. The effect is more marked for a corresponding study of the phenyl and benzylic anions and may be correlated with the extent of charge delocalisation in the anion – a phenomenon familiar in the behaviour of these reactions in solution. In the context of such a model the charge delocalisation may have two consequences in solution – a change in the intrinsic anionic reactivity and a change in its solvation. The gas-phase results permit the study of the first of these in isolation.

7.3.2.2 'Hardness' and 'softness' of acids and bases

The ion-cyclotron resonance technique has been used to study nucleophilic displacement reactions for methyl-substituted species, generalised as

$$X' + CH_3X''^+ \rightarrow CH_3X'^+ + X'' \tag{7.15}$$

where the controlling parameter is the methyl cation affinity of X [107]. Corresponding proton affinities control the order of displacement occurring in the analogous family of reactions

$$X' + HX''^+ \rightarrow HX'^+ + X'' \tag{7.16}$$

In these two examples, H^+ is acting as a 'hard' acid and CH_3^+ as a 'soft' acid in Pearson's nomenclature[108]. If one considers the behaviour of a series of cations, Z^+, with respect to a given neutral X, the cation affinity of Z^+ for X, defined as the enthalpy change for the process

$$ZX^+ \rightarrow Z^+ + X \tag{7.17}$$

becomes a quantitative measure of the 'hardness' of the acid, Z^+, with respect to X. Such measurements, for the isolated reactions in the gas phase, can therefore introduce *quantitative* information into Pearson's empirical theory of 'hard' and 'soft' acids and bases in solution.

It is of further interest to note that molecular nitrogen will undergo this type of easy displacement reaction; for example, displacement of HF from the methyl cation

$$N_2 + CH_3FH^+ \rightarrow CH_3N_2^+ + HF \tag{7.18}$$

to form the methyl diazonium ion [109], with obvious interest for the general problem of nitrogen fixation.

7.3.2.3 Solvent displacement or 'switching' reactions

The discussion in Section 7.3.1 of solvated ions focused upon the experimental measurement of their thermodynamic properties. Attention is now directed to their reactivity, of which solvent displacement,

$$A^\pm \cdot B + C \rightarrow A^\pm \cdot C + B \tag{7.19}$$

is a fundamental process. In effect this is an extreme example of the type of displacement reaction considered in the previous section, whereby the stronger valence forces binding $CH_3N_2^+$ may be replaced, to a lesser or greater extent, by simple electrostatic attraction between the ion and the solvent molecule. An alternative nomenclature refers to 'switching' reactions of 'cluster ions'[110]. This study consists of an investigation of reaction (7.19) for $A^\pm = O_2^+, O^-$ and O_2^- with a variety of diatomic and triatomic inorganic 'solvents'. An order of increasing binding energies for various solvent molecules to a particular ion may be deduced and these differ for the two ions, O_2^+ and O_2^-. For the former, this order is established to be $H_2 \approx N_2 < O_2 < N_2O < SO_2 < H_2O$, whereas for the latter, it becomes $N_2 \approx N_2O \approx CO < O_2 < H_2O < CO_2 < NO$. More limited data for O^- reveal a behaviour similar to that of O_2^-. It is noteworthy that $D(O_2^+ \cdot O_2) < D(O_2^+ \cdot N_2O)$ whereas $D(O_2^- \cdot O_2) > D(O_2^- \cdot N_2O)$.

7.3.3 Relative acidities and basicities in the gas phase

Many of the reactions discussed in Section 7.3.2 are more generally, and more appropriately, considered as acid–base reactions. The ion cyclotron resonance technique has proved to be particularly powerful in identifying such reactions and establishing relative acidities and basicities. We select one dramatic example for which the acidity order in solution is reversed in the gas phase, and we then discuss the determination of proton affinities as a quantitative measure of acidity and basicity.

7.3.3.1 *Relative acidities of water, methanol and ethanol*

Contrary to what is found in solution, the following acidity order has been established in the gas phase: $C_2H_5OH > CH_3OH > H_2O$ [111]. Thus, for the example of reaction (7.20),

$$OH^- + CH_3OH \rightleftharpoons CH_3O^- + H_2O \qquad (7.20)$$

the equilibrium lies to the right in the gas phase and to the left in solution. It is suggested that the polarisable alkyl groups stabilise the negative charge in the corresponding anions by means of an *internal* ion–induced dipole interaction. The situation in solution is clearly different but the gas-phase results have an important negative implication for this. In solution, observed phenomena are frequently interpreted in terms of molecular explanations, e.g. that methanol is a weaker acid than water as a consequence of the inductive effect of the methyl group. Gas-phase studies test this kind of molecular explanation directly and, in this case, show it to be specious. Rather, the explanation must be found in terms of solvent interaction.

7.3.3.2 *Proton affinities*

Ion–molecule chemists have been involved in measuring proton affinities for more than a decade, employing principally a kinetic technique[112]. If the reaction

$$X + YH^+ \rightarrow XH^+ + Y \qquad (7.21)$$

may be shown to be exoergic, then X necessarily has a larger proton affinity than Y. The sensitivity of the double-resonance technique in ion cyclotron resonance has proved to be particularly powerful in these studies. The relevance for acid–base phenomena is obvious: the proton affinity of a neutral molecule X is related to its basicity and that of its anion X^- to its acidity.

Representative values are given in Table 7.4 and are now sufficiently numerous to suggest some trends. Those of the alkali hydroxides correlate with the ionisation potential of the alkali, suggesting that MOH_2^+ is more appropriately $M^+ \cdot H_2O$ [113]. The difference in basicity between NH_3 and PH_3 is similar in the gas phase and solution suggesting surprisingly that the solvent plays an unimportant role in determining this[115]. Where molecules

Table 7.4   Proton affinities (kcal mol⁻¹)

Alkali hydroxides [113]		Hydrides [114, 115]		Hydride anions[116]	
LiOH	241	NH_3	207 ± 3	NH_2^-	389
NaOH	248	PH_3	185 ± 4	PH_2^-	362
KOH	263				
CsOH	270	H_2O	164 ± 4	OH^-	390
		H_2S	178 ± 2	SH^-	350

are functionally similar, then differences in basicity seem to be determined principally by the differences in ionisation potential[117]. On the other hand, the differences in acidity for molecules, within a group, appear to be controlled primarily by the differences in bond energy[116].

7.3.4   The effect of selective ionic solvation on reactivity

The nucleophilic displacement reaction

$$RO^- + CH_3Cl \rightarrow CH_3OR + Cl^- \tag{7.22}$$

is many orders of magnitude faster in the gas phase than in solution, this disparity naturally reflecting the effect of the solvent. In an interesting experiment, Bohme and Young have begun to probe the fine structure of this effect[106]. The anion RO^- is selectively solvated to form $RO^- \cdot ROH$, and the effect of this on the rate of reaction (7.22) is measured. For R = Me, Et, Pr^i and Bu^t, solvation of the anion by one ROH molecule decreases the rate constant of reaction (7.22) by a factor of at least three. Addition of another solvent molecule to form $RO^-(ROH)_2$ decreases the rate constant further. This preliminary study is only a start on what promises to be an exciting controlled simulation of solution kinetics in the gas phase.

7.3.5   Conclusions

Studies in the gas phase naturally cannot reveal directly what occurs in solution but they do yield important insights. All too little is known about the role of the solvent and solvation in solution phenomena and it is therefore customary and natural often to seek explanations in terms of the isolated molecules and ions. Where such phenomena differ in solution and the gas phase, the latter studies rule out such molecular explanations for the former, indicating that the solvent must be playing the dominant role. In all cases, knowledge of the intrinsic molecular parameters, such as proton affinities and electron affinities, is forthcoming from the gas-phase studies and is essential input for the difficult task of seeking explanations for ionic phenomena in solution.

The brevity of this chapter has not allowed us to indicate the full range of current techniques which are being used in these studies; particular omissions are tandem mass spectrometry[1] and chemical-ionisation mass spectrometry[121]. We note, in proof, the recent publication of two reviews of ion cyclotron resonance, each discussing in greater detail several of the topics mentioned in section 7.3[122, 123].

Acknowledgement

We thank Dr. Arthur Werner for a critical review of the manuscript. One of us (J. D.) thanks the United States Atomic Energy Commission and the other (M. H.) thanks the Petroleum Research Fund of the American Chemical Society for financial support of this work.

References

1. Friedman, L. and Reuben, B. G. (1971). *Advan. Chem. Phys.,* **19**, 33
2. McDaniel, E. W., Čermák, V., Dalgarno, A., Ferguson, E. E. and Friedman, L. (1970). *Ion Molecule Reactions.* (New York: Wiley-Interscience)
3. Franklin, J. L. (ed.) (1972). *Ion–Molecule Reactions.* (New York: Plenum)
4. Herman, Z. and Wolfgang, R. (1972). Reference (3), Chapter 12
5. Chupka, W. A. (1972). Reference (3), Chapter 3
6. Henchman, M. J. (1972). Reference (3), Chapter 5
7. Neynaber, R. (1969). *Advan. At. Mol. Phys.,* **5**, 57
8. Ferguson, E. E., Fehsenfeld, F. C. and Schmeltekopf, A. L. (1969). *Advan. At. Mol. Phys.,* **5**, 1
9. Ferguson, E. E. (1972). Reference (3), Chapter 8
10. Baldeschwieler, J. D. and Woodgate, S. S. (1971). *Accounts Chem. Res.,* **4**, 114
11. Henis, J. M. S. (1972). Reference (3), Chapter 9
12. Kebarle, P. (1972). Reference (3), Chapter 7
13. Light, J. C., Ross, J. and Schuler, K. E. (1969). *Kinetic Processes in Gases and Plasmas,* ed. by Hochstim, A. R., 281. (New York: Academic Press)
14. Wolfgang, R. (1969). *Accounts Chem. Res.,* **2**, 248
15. Polanyi, J. C. (1971). *J. Appl. Optics,* **10**, 1717
16. Melton, L. A. and Gordon, R. G. (1969). *J. Chem. Phys.,* **51**, 5449
17. Wolfgang, R. (1968). *Sci. Am.,* **218**, No. 10, 44
18. Wolfgang, R., Branscomb, L. and Zare, R. (1968). *Science,* **162**, 818
19. Gann, R. G., Ollison, W. M. and Dubrin, J. (1971). *J. Chem. Phys.,* **54**, 2304
20. Wolfgang, R. (1965). *Progr. Reaction Kinetics,* **3**, 97
21. Rowland, F. S. (1970). *Molecular Beams and Reaction Kinetics,* ed. by Schlier, Ch., 108. (New York: Academic Press)
22. Menzinger, M. and Wolfgang, R. (1969). *Angew. Chem. Int. Ed. Engl.,* **8**, 438
23. Chupka, W. A., Russell, M. E. and Refaey, K. (1968). *J. Chem. Phys.,* **48**, 1518
24. Chupka, W. A. and Russell, M. E. (1968). *J. Chem. Phys.,* **49**, 5426
25. Sbar, N. and Dubrin, J. (1970). *J. Chem. Phys.,* **53**, 842
26. Fehsenfeld, F. C., Albritton, D. L., Burt, J. A. and Schiff, H. I. (1969). *Can. J. Chem.,* **47**, 1793
27. Leventhal, J. J. (1971). *J. Chem. Phys.,* **54**, 3279
28. Čermák, V. and Herman, Z. (1965), *Collect. Czech. Chem. Commun.,* **30**, 1343
29. Maier, W. B., II (1967). *J. Chem. Phys.,* **47**, 859
30. Bates, D. R., Cook, C. J. and Smith, F. J. (1964). *Proc. Phys. Soc.,* **83**, 49
31. Chupka, W. A. and Russell, M. E. (1968). *J. Chem. Phys.,* **48**, 1527
32. Chupka, W. A. and Berkowitz, J. (1971). *J. Chem. Phys.,* **54**, 4256
33. Schmeltekopf, A. L., Fehsenfeld, F. C., Gilman, G. I. and Ferguson, E. E. (1967). *Planet Space Sci.,* **15**, 401
34. Giese, C. F. (1966). *Advan. Chem. Ser.,* **58**, 20
35. Kaufman, J. J. and Koski, W. S. (1969). *J. Chem. Phys.,* **50**, 1942
36. O'Malley, T. F. (1970). *J. Chem. Phys.,* **52**, 3269
37. Herman, Z., Lee, A. and Wolfgang, R. (1969). *J. Chem. Phys.,* **51**, 452

38. Buttrill, S. E., Jr. (1970). *J. Chem. Phys.,* **52,** 6174
39. Franklin, J. L. and Haney, M. A. (1969). *J. Phys. Chem.,* **73,** 2857
40. Light, J. C. (1967). *Discuss. Faraday Soc.,* **44,** 14
41. Light, J. C. and Lin, J. (1965). *J. Chem. Phys.,* **43,** 3209
42. Lee, Y. T. (1971). Unpublished results
43. Tannenwald, L. M. (1966). *Proc. Phys. Soc.,* **87,** 109
44. Matus, L., Opauszky, I., Hyatt, D., Masson, A. J., Birkinshaw, K. and Henchman, M. J. (1967). *Discuss. Faraday Soc.,* **44,** 146
45. Mahan, B. H. (1971). *J. Chem. Phys.,* **55,** 1436
46. Nikitin, E. E. (1968). *Chemische Elementarprozesse.* ed. by Hartmann, H., 43. (Berlin: Springer-Verlag)
47. Krenos, J., Preston, R., Wolfgang, R. and Tully, J. (1971). *Chem. Phys. Letters,* **10,** 17
48. Masson, A. J., Birkinshaw, K. and Henchman, M. J. (1969). *J. Chem. Phys.,* **50,** 4112
49. Paulson, J. F. (1971). Unpublished results
50. Karplus, M., Porter, R. N. and Sharma, R. D. (1966). *J. Chem. Phys.,* **45,** 3871
51. Gegenbach, R., Strunck, J. and Toennies, J. P. (1971). *J. Chem. Phys.,* **54,** 1830
52. Dittner, P. F. and Datz, S. (1971). *J. Chem. Phys.,* **54,** 4228
53. Lacmann, K. and Herschbach, D. R. (1970). *Chem. Phys. Lett.,* **6,** 106
54. Schmeltekopf, A. L., Fehsenfeld, F. C. and Ferguson, E. E. (1967). *Astrophys. J.,* **148,** L155
55. See Chapters 5 and 6 of this Volume
56. Herschbach, D. R. (1962). *Discuss. Faraday Soc.,* **33,** 149
57. Miller, W. B., Safron, S. A. and Herschbach, D. R. (1967). *Discuss. Faraday Soc.,* **44,** 108
58. Herschbach, D. R. (1966). *Advan. Chem. Phys.,* **10,** 319
59. Gentry, W. R., Gislason, E. A., Mahan, B. H. and Tsao, C. W. (1968). *J. Chem. Phys.,* **49,** 3058
60. Gentry, W. R., Gislason, E. A., Lee, Y. T., Mahan, B. T. and Tsao, C. W. (1967). *Discuss. Faraday Soc.,* **44,** 137
61. Mahan, B. H. (1968). *Accounts Chem. Res.,* **1,** 217
62. Mahan, B. H. (1970). *Accounts Chem. Res.,* **3,** 393
63. Doverspike, L. D., Champion, R. L. and Bailey, T. L. (1966). *J. Chem. Phys.,* **45,** 4385
64. Vance, D. W. and Bailey, T. L. (1966). *J. Chem. Phys.,* **44,** 486
65. Herman, Z., Kerstetter, J., Rose, T. and Wolfgang, R. (1967). *Discuss. Faraday Soc.,* **44,** 123
66. Herman, Z., Kerstetter, J. D., Rose, T. L. and Wolfgang, R. (1967). *J. Chem. Phys.,* **46,** 2844
67. Hierl, P., Herman, Z., Kerstetter, J. and Wolfgang, R. (1968). *J. Chem. Phys.,* **48,** 4319
68. Fink, R. D. and King, J. S., Jr. (1967). *J. Chem. Phys.,* **47,** 1857
69. Henglein, A. (1966). *Ion Molecule Reactions in the Gas Phase,* Adv. in Chem. Series, **58,** 63, ed. by Ausloos, P. J. (Washington: American Chemical Society)
70. Henglein, A., Lacmann, K. and Jacobs, G. (1965). *Ber. Bunsenges. Phys. Chem.,* **69,** 279
71. Kerstetter, J. and Wolfgang, R. (1970). *J. Chem. Phys.,* **53,** 3765
72. Hierl, P. M., Herman, Z. and Wolfgang, R. (1970). *J. Chem. Phys.,* **53,** 660
73. Chiang, M., Gislason, E. A., Mahan, B. H., Tsao, C. W. and Werner, A. S. (1970). *J. Chem. Phys.,* **52,** 2698
74. Gislason, E. A., Mahan, B. H., Tsao, C. W. and Werner, A. S. (1969). *J. Chem. Phys.,* **50,** 142
75. Felder, W., Sbar, N. and Dubrin, J. (1970). *Chem. Phys. Lett.* **6,** 385
76. Gislason, E. A., Mahan, B. H., Tsao, C. W. and Werner, A. S. (1971). *J. Chem. Phys.,* **54,** 3897
77. Tully, J. C., Herman, Z. and Wolfgang, R. (1971). *J. Chem. Phys.,* **54,** 1730
78. Ding, A. and Henglein, A. (1969). *Ber. Bunsenges. Phys. Chem.,* **73,** 562
79. Chiang, M. H., Gislason, E. A., Mahan, B. H., Tsao, C. W. and Werner, A. S. (1971). *J. Phys. Chem.,* **75,** 1426
80. Doverspike, L. D. and Champion, R. L. (1967). *J. Chem. Phys.,* **46,** 4718
81. Durup, J. and Durup, M. (1967). *J. Chim. Phys.,* **64,** 386
82. Ding, A., Henglein, A. and Lacmann, K. (1968). *Z. Naturforsch.,* **23a,** 779
83. Wolfgang, R. (1970). *Accounts Chem. Res.,* **3,** 48
84. George, T. F. and Suplinskas, R. J. (1971). *J. Chem. Phys.,* **54,** 1037
85. Polanyi, J. C. and Kuntz, P. J. (1967). *Discuss. Faraday Soc.,* **44,** 180
86. Kuntz, P. J. (1969). *Chem. Phys. Lett.,* **4,** 129

87. Chang, D. T. and Light, J. C. (1970). *J. Chem. Phys.*, **52**, 5687
88. Gelb, A. and Suplinskas, R. J. (1970). *J. Chem. Phys.*, **53**, 2249
89. Csizmadia, I. G., Polanyi, J. C., Roach, A. C. and Wong, W. H. (1969). *Can. J. Chem.*, **47**, 4097
90. Conroy, H. (1969). *J. Chem. Phys.*, **51**, 3979
91. Preston, R. K. and Tully, J. C. (1971). *J. Chem. Phys.*, **54**, 4297
92. Tully, J. C. and Preston, R. K. (1971). *J. Chem. Phys.*, **55**, 562
93. Krenos, J. and Wolfgang, R. (1970). *J. Chem. Phys.*, **52**, 5961
94. Holliday, M. G., Muckerman, J. T. and Friedman, L. (1971). *J. Chem. Phys.*, **54**, 1058
95. Maier, W. B. (1971). *J. Chem. Phys.*, **54**, 2732
96. Field, F. H. and Franklin, J. L. (1957). *Electron Impact Phenomena and the Properties of Gaseous Ions.* (New York: Academic Press)
97. Franklin, J. L., Dillard, J. G., Rosenstock, H. M., Herron, J. T., Draxl, K. and Field, F. H. (1969). *Ionisation Potentials, Appearance Potentials and Heats of Formation of Gaseous Positive Ions.* (Washington, D.C.: U.S. Dept. of Commerce)
98. Kebarle, P. and Hogg, A. M. (1965). *J. Chem. Phys.*, **42**, 798
99. Džidić, I. and Kebarle, P. (1970). *J. Phys. Chem.*, **74**, 1466
100. Arshadi, M., Yamdagni, R. and Kebarle, P. (1970). *J. Phys. Chem.*, **74**, 1475
101. Kebarle, P., Searles, S. K., Zolla, A., Scarborough, J. and Arshadi, M. (1967). *J. Amer. Chem. Soc.*, **89**, 6393
102. Searles, S. K. and Kebarle, P. (1968). *J. Phys. Chem.*, **72**, 742
103. Kebarle, P. and Hogg, A. M. (1965). *J. Chem. Phys.*, **42**, 798
104. Yamdagni, R. and Kebarle, P. Unpublished results
105. Arshadi, M. and Kebarle, P. (1970). *J. Phys. Chem.*, **74**, 1483
106. Bohme, D. K. and Young, L. B. (1970). *J. Amer. Chem. Soc.*, **92**, 7354
107. Holtz, D., Beauchamp, J. L. and Woodgate, S. D. (1970). *J. Amer. Chem. Soc.*, **92**, 7484
108. Pearson, R. G. and Songstad, J. (1967). *J. Amer. Chem. Soc.*, **89**, 1827
109. Holtz, D. and Beauchamp, J. L. (1971). *Nature Phys. Sci.*, **231**, 204
110. Adams, N. G., Bohme, D. K., Dunkin, D. B., Fehsenfeld, F. C. and Ferguson, E. E. (1970). *J. Chem. Phys.*, **52**, 3133
111. Brauman, J. I. and Blair, L. K. (1970). *J. Amer. Chem. Soc.*, **92**, 5986
112. Talrose, V. L. (1962). *Pure Appl. Chem.*, **5**, 455
113. Searles, S. K., Džidić, I. and Kebarle, P. (1969). *J. Amer. Chem. Soc.*, **91**, 2810
114. Beauchamp, J. L. and Buttrill, S. E., Jr. (1968). *J. Chem. Phys.*, **48**, 1783
115. Holtz, D. and Beauchamp, J. L. (1969). *J. Amer. Chem. Soc.*, **91**, 5913
116. Holtz, D., Beauchamp, J. L. and Eyler, J. R. (1970). *J. Amer. Chem. Soc.*, **92**, 7045
117. Holtz, D., Beauchamp, J. L., Henderson, W. G. and Taft, R. W. (1971). *Inorg. Chem.*, **10**, 201
118. Freund, S. M., Fisk, G. A., Herschbach, D. R., and Klemperer, W. (1971). *J. Chem. Phys.*, **54**, 2510
119. Werner, A. S. (1971). *Ph.D. Thesis,* University of California, Berkeley; Lawrence Radiation Laboratory Report No. UCRL 20363
120. Dunkin, D. B., Fehsenfeld, F. C. and Ferguson, E. E. (1970). *J. Chem. Phys.*, **53**, 987
121. Field, F. H. (1972). Reference (3), Chapter 6
122. Beauchamp, J. L. (1971). *Ann. Rev. Phys. Chem.*, **22**, 527
123. Gray, G. A. (1971). *Adven. Chem. Phys.*, **19**, 141

8
Energy-Transfer Processes

J. I. STEINFELD
Massachusetts Institute of Technology

8.1 INTRODUCTION

In the 3 years since the last reviews on molecular energy transfer[1-5], a large number of experimental and theoretical developments have occurred. In this chapter we shall restrict ourselves to an attempt to answer some of the most salient questions that remained open in this field at the time the last reviews were written. These questions can be phrased as follows: (1) In principle, molecular beam scattering experiments should be capable of yielding the most definitive information on cross-sections. How successful have these methods been in providing values of vibrationally and rotationally inelastic cross-sections? (2) Resonance fluorescence spectrophotometry has proven to be spectacularly successful in generating energy-transfer data in for example, molecular iodine. Is this a general method or is it limited to only a few specialised systems? (3) To what extent have double resonance methods yielded new energy-transfer data? (4) The development of laser sources has been pursued intensively during the last couple of years. Have these devices led to new experimental approaches that were not previously feasible? (5) Can unambiguous relaxation probabilities be extracted from line broadening data? (6) Traditional methods of energy-transfer research cannot resolve processes occurring faster than a few nanoseconds. What prospects are there for extending this range to shorter time scales? (7) A great deal of experimental data has already been accumulated on vibrational and rotational relaxation. Has our theoretical understanding of these processes shown a commensurate improvement?

Let us consider these points one by one.

8.2 MOLECULAR BEAM SCATTERING

In our last review on this subject[1], the only definitive determination of an inelastic cross-section by molecular beam scattering was that by Toennies, who had measured $\sigma (J = 3 \rightleftarrows J = 2)$ in TlF and various gases[6, 7]. Although a great deal of effort has been expended in this field in recent years, results on inelastic scattering are still fairly sparse.

There are two methods by which one would measure directly the energy transferred in a molecular collision. Firstly by use of rotational and vibrational state selected molecular beams, and determining the initial and final states of the scattered molecule; this method was employed by Toennies in his study of TlF. Secondly, by use of both velocity selection and analysis on the beams, and by determination of the inelasticity of the collision as the difference between initial and final relative kinetic energies. (Because of the limitations on beam intensity, simultaneous selection of both internal state and velocity is not deemed to be practical, given the current state of the art.)

State selection has been demonstrated by Bernstein and co-workers[8], who used a ten-pole electrostatic field to select states with $J = 0$, 1 and 2 for CsF and RbCl. Scattering experiments have not yet been reported from this system. Rather more effort has gone into velocity selection in which ionic species are handled with ease by virtue of their net charge, and results are available for a variety of systems. For example, Datz and co-workers[9] have measured vibrational excitation in collisions of K^+ with D_2, Mahan and co-workers[10] in NO^+ and $O_2^+ + He$, and Cosby and Moran[11] in $O_2^+ + Ar$ and $O^+ + O_2$. A sample of the data obtained for the latter system is shown in Figure 8.1. Although these results are of both practical and theoretical interest, it is difficult to compare them directly with results obtained by other methods, since one does not usually deal with ionic species in experiments involving

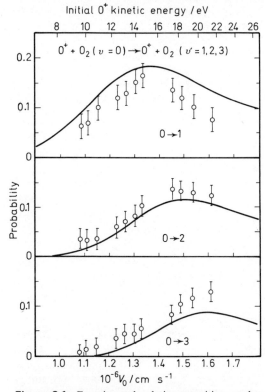

Figure 8.1 Experimental relative transition probabilities (Φ) for the vibrational transitions $v = 0 \rightarrow 1, 2$ and 3 in the system $O^+ + O_2$, for $\theta_{lab.}$ between $\pm 1^\circ$. Solid curves are probabilities calculated for the corresponding transitions using an approximate semi-classical 3-dimensional theory based on work by Shin (1969). *J. Phys. Chem.*, **73**, 4321. (From Cosby and Moran[11], by courtesy of the American Institute of Physics)

bulk gases. Data of this type for neutral–neutral collisions should soon be available from the beam apparatuses of Lee *et al.* and Greene *et al.* but this work is still in an early stage.

8.3 RESONANCE FLUORESCENCE SPECTROPHOTOMETRY

8.3.1 Iodine

Although an extensive literature exists on fluorescence and energy transfer in iodine[148], this system is still capable of yielding new information, as our recent results demonstrate[12, 13]. I_2 was excited to the $J' = 11$ and 15 levels of the $v' = 43$ level of the $B(^3\Pi o_u^+)$ state by the 5145 Å radiation from an argon laser, and vibrational relaxation was observed from $v' = 43$ to $v' = 48$ through 35. Total vibrational relaxation cross-sections (in units of 10^{-16} cm^2) were measured for collisions with H_2 (4.8), He(7.4), Ne(16.8), Ar(23.0), CO_2(17.5), Kr(29.8), Xe(22.6), and I_2(19.0). The dependence of energy-transfer probability on collision reduced-mass is thus similar to that found in $v' = 25$ [14]. Vibrational level changes by $\Delta v = +5$ to -8 were observed; the relative probability for a given value of Δv is essentially the same for all collision partners studied. In subsequent work, the rotational inelasticity accompanying these collisions was resolved in detail[110]. Rotational quantum state changes of $\Delta J'$ up to ± 12 were obtained; this is much less than complete rotational thermalisation. Apparently, for molecules with high moments of

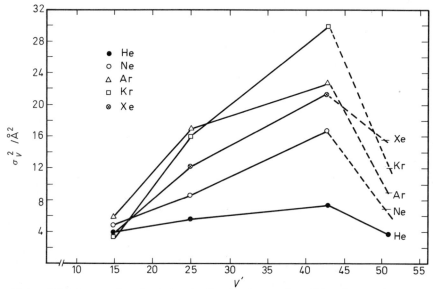

Figure 8.2 Cross-sections for change in vibrational energy in collisions between I_2^* and the indicated rare-gas atom, as a function of vibrational level. For $v' = 50$, only upper limits could be established for Ne through Xe. (From Steinfeld and Schweid[13], by courtesy of the American Institute of Physics)

inertia, the rotational velocity is so slow that energy transfer takes place more rapidly though vibrational $\rightleftarrows$ translational exchange than through rotational $\rightleftarrows$ translational exchange; a similar phenomenon was noted in studies of Na_2 fluorescence[22]. The rotational energy transfer probability from $J' = 11$ and 15 is also clearly asymmetric to positive values of ΔJ, as required to facilitate an approach toward equilibrium.

Fluorescence from $v' = 50$ of iodine, which lies 380 cm^{-1} below the molecular dissociation limit, was excited by a Cd 5086 Å source[13]. As Figure 8.2 shows, vibrational energy transfer probabilities are *less* in this level than in lower-lying levels of the same molecule. This decrease in probability is accompanied by a marked increase in quenching of the excited state. The mechanism proposed to account for this involves direct collisional dissociation of the molecule, namely:

$$I_2^* \, (v' > 50) + M \rightarrow I(^2P_{\frac{3}{2}}) + I^* \, (^2P_{\frac{1}{2}}) + M,$$

which competes successfully with the more conventional process

$$I_2^* \, (v' = 50) + M \rightarrow I_2^* \, (v' = 50 \pm \Delta v) + M.$$

This mechanism ('collisional release'), had been suggested previously by Callear[15] to account for his observation of excited I atoms following irradiation of I_2 below its molecular dissociation limit.

8.3.2   Hydrogen*

The most interesting system to study by monochromatically excited fluorescence should be molecular hydrogen, because this is a sufficiently simple system to suppose that an interaction potential between the molecule and simple rare-gas atoms might be calculated with reasonable accuracy, and also used in manageable scattering calculations to predict the observed cross-sections. However, the absorption and emission of H_2 lies in the vacuum ultraviolet. Recently, Moore and his co-workers at Berkeley have succeeded in carrying out this experiment[16, 17]. They used the 1048 Å and 1066 Å lines of an argon lamp to excite the $(v' = 3, J' = 2)$, $(v' = 5, J' = 2)$, and $(v' = 6, J' = 5)$ levels of the $B(^1\Sigma_u^+)$ state of HD. Their measured cross-sections are summarised in Table 8.1.

Table 8.1 $\sigma_{eff.}/Å^2$ with stated collision partner

Process	HD	D_2	^{3}He	^{4}He	Ne	
$J' = 2 \rightarrow J' = 0$	1.3	1.8	4.7	3.8	3.5	
$J' = 2 \rightarrow J' = 1$	3.0	1.5	1.9	1.8	1.8	$v' = 3$ is
$J' = 3 \rightarrow J' = 2$	1.2		2.7	2.3	2.6	
$J' = 4 \rightarrow J' = 2$	1.7		4.2	2.9	2.1	unchanged
$J' = 5 \rightarrow J' = 2$			1.3	0.8		
$J' = 5 \rightarrow J' = 3$			2.2	2.0		
$J' = 5 \rightarrow J' = 4$			2.6	2.7		
$J' = 6 \rightarrow J' = 5$			2.4	2.3		$v' = 6$ is unchanged
$J' = 7 \rightarrow J' = 5$			1.2	1.0		
$v' = 3 \rightarrow v' = 2$			0.89	0.76	0.81	all $\Delta J'$

8.3.3   Nitrogen dioxide†

For some time, attempts have been made to apply the technique of monochromatically excited fluorescence to the study of vibrational energy-transfer in small polyatomic molecules. In such systems, it becomes very difficult to identify transitions between specific quantum levels. The results obtained with NO_2 may serve to illustrate the difficulties encountered. Kaufman et al.[18, 19]

*See Note (1), added in proof.
†See Note (2), added in proof.

studied the fluorescence of NO_2 as a function of exciting and emitted wavelengths and of NO_2 pressure. They concluded that an average of 1000 cm^{-1} of vibrational energy was removed from each excited NO_2 molecule upon collision with a ground-state NO_2 molecule. A similar conclusion was reached by Schwartz and Johnston[20]. Energy transfer of this magnitude must surely proceed by near-resonant transfer between excited and ground-state molecules, i.e.

$$NO_2^* \,(^2B_1, \text{ high } v') + NO_2 \,(^2A_1, \text{ low } v'')$$
$$\rightarrow NO_2^* \,(^2B_1, \text{ low } v') + NO_2 \,(^2A_1, \text{ high } v'').$$

Data on vibrational relaxation of NO_2^* by rare-gas atoms would be useful in elucidating the energy transfer mechanism, but such data are not yet available. The electronic quenching cross-section, i.e., for the process

$$NO_2^* \,(^2B_1) + NO_2 \,(^2A_1) \rightarrow 2\,NO_2 \,(^2A_1)$$

is reported[20] to be of the order of 0.01 times the gas-kinetic value, which is in agreement with Broida's earlier[21] determination of $0.9 \,\text{Å}^2$.

8.3.4   Other systems

A few more examples of fluorescence studies of energy transfer should suffice to demonstrate the versatility of this technique. Na_2 is excited to the ($v' = 6$, $J' = 43$) level of the B $^1\Pi_u$ electronic state by the 4880 Å line of the Ar$^+$ laser[22]. For this molecule, a rotationally inelastic cross-section of 65 Å^2 has been measured. In vibrationally inelastic collisions, a persistence of rotational state was noted; $v' = 5$ and $v' = 7$ tended to be populated preferentially around $J' = 43$ in each level. In a similar study[23], several different levels of Li_2 (B $^1\Pi_u$) were excited with an argon-ion laser and cross-sections for rotational and vibrational transitions were obtained in detail. Broida[24] carried out a similar experiment in BaO, using the same laser line to excite the ($v' = 8$, $J' = 49$) level of the A $^1\Sigma$ state. Rotational and vibrational relaxation was observed; cross-sections for vibrational energy transfer were 0.1 with He, 0.4 with Ar and 1 Å^2 with N_2.

A study of the fluorescence of OH [25], produced in the $A^2\Sigma^+$ state by photodissociation of H_2O at 1236 Å, gave evidence for the near-resonant ($\Delta E = 27$ cm^{-1}) vibration-rotation transfer process:

$$OH(v' = 0, K' = 20) + M \rightarrow OH(v' = 1, K' = 15) + M$$

with a cross-section of c. 1.4 Å^2 for M = Ar and N_2. The C $^3\Pi_u$ state of N_2 was excited by fission fragments from a californium-252 source[26]; vibrational deactivation cross-sections with ground-state N_2 are quoted as $\sigma(v = 1 \rightarrow 0) = 2.5 \pm 0.7\,\text{Å}^2$, $\sigma(v = 2 \rightarrow 1) = 1.6 \pm 0.4\,\text{Å}^2$. Some question must be raised about these last results since all the standard theories predict that the vibrational deactivation probability should be proportional to the higher vibrational quantum number; thus, $\sigma(v = 2 \rightarrow 1)$ is expected to be about twice $\sigma(v = 1 \rightarrow 0)$.

Finally, an interesting kinetic absorption study on HBr by Donovan and co-workers[27] should be mentioned. Vibrationally excited HBr was produced by the reaction sequences:

$$HBr + hv \rightarrow H\,(1\,{}^2S_{\frac{1}{2}}) + Br\,(4\,{}^2P_{\frac{1}{2}})$$
$$Br\,(4\,{}^2P_{\frac{1}{2}}) + HBr\,(v'' = 0) \rightarrow Br\,(4\,{}^2P_{\frac{3}{2}}) + HBr\,(v'' = 1)$$
$$H\,(1\,{}^2S_{\frac{1}{2}}) + Br_2 \rightarrow HBr\,(v'' = 0, 1, 2, \dots) + Br.$$

Relaxation of the excited HBr by collision with He, H_2, HBr, HCl, N_2, CO, CH_4, SF_6, Br_2 and Br is reported. In a similar study[28], vibrationally excited CS was produced by the flash-initiated reaction:

$$NO_2 + hv(\lambda > 3200\,\text{Å}) \rightarrow NO + O(2\,{}^3P)$$
$$O(2\,{}^3P) + CS_2 \rightarrow SO + CS\,(v'' = 0, 1, 2 \dots)$$

The relaxation of the CS† by NO_2, CS_2, Ar, He, H_2, D_2, N_2O, CO_2 and $CH_3 \cdot CH_2 \cdot CH{=}CH_2$ was monitored by kinetic absorption spectroscopy. In summary, it appears that the methods of resonance fluorescence and kinetic absorption spectrophotometry still have plenty to offer the chemist interested in molecular relaxation processes.

See Note (3), added in proof.

8.4 DOUBLE RESONANCE METHODS

The perturbation of the equilibrium population of a pair of molecular energy levels by intense pumping radiation, followed by absorption at another frequency to determine the recovery of the populations, is known as double resonance spectroscopy. While this method is now widely used in both nuclear and electron spin resonance spectroscopy, its application to the 'optical' part of the spectrum has awaited the development of sources both sufficiently intense and monochromatic to saturate selectively a particular pair of rotational or vibrational energy levels. The first developments along these lines came in the area of microwave–microwave double resonance, in part due to the availability of suitable pumping sources (klystrons); it has recently been extended to the infrared region.

8.4.1 Microwave–microwave double resonance studies of rotational energy transfer

The first use of microwave double resonance for relaxation studies was a steady-state experiment on OCS and $(CH_2)_3O$ transitions by Wilson[29] who observed a relaxation time for the saturation of 2.5 µs from intensity ratios vs. pressure studies. This was followed[30, 31] by modulated microwave double-resonance experiments, in which levels connected by collision-induced transitions were identified and relative rates for $\Delta J = 0, 1, 2$, and 3 transitions measured. More recently, Flygare and co-workers[32] carried out both time-resolved direct-decay and phase-comparison experiments on double resonances in OCS interacting with itself and with Ar, He, and O_2, obtaining relaxation times and rotational relaxation cross-sections. Other work on ammonia[33] and SO_2 [34] has been reported; in the latter work, the $6_{15} \rightarrow 5_{24}$ transition in

SO_2 at 23.44 GHz was pumped, and the $6_{06} \rightarrow 6_{15}$ transition monitored; a very fast relaxation time of 14 ns torr was found.

The most prolific developments in this area are by Oka. He reported[35] pumping of the $2_{12} \rightarrow 2_{21}$ transition in $(CH_3)_2O$, and observation of the $3_{21} \rightarrow 3_{30}$ and $3_{12} \rightarrow 3_{21}$ transitions. It was found that the 3_{21} level is preferentially coupled to the 2_{12} by collisions, and the 3_{30} and 3_{12} levels to the 2_{21}, in accordance with dipole-like selection rules. This was extended to studies[36] in H_2CO, HCN and H_2CCO; in these systems the 'propensity rules' $\Delta J = 0, \pm 1$ and parity $+ \rightleftarrows -$ was found. In NH_3–He collisions, however[37], the collision appears to be 'quadrupole-type'; that is, parity going $+ \rightleftarrows +$ or $- \rightleftarrows -$ when $\Delta J = \pm 1$. This is in contrast to NH_3–Ar collisions, in which 'dipole-type' (i.e., parity $+ \rightleftarrows -$) propensity rules are observed. In NH_3–NH_3 collisions[38], dipole-type collisions lead to the propensity rules $p(\Delta J = 0) > p(\Delta J = 1)$ when $J \simeq K$, and $p(\Delta J = 0) \simeq p(\Delta J = 1)$ when $J \gg K$; also $p(\Delta J > 1)$ is much smaller than $p(\Delta J = \pm 1)$, and $p(\Delta K \neq 0)$ is much smaller than $p(\Delta K = 0)$. Further investigation[39] of NH_3–rare gas collisions showed that when the quantum number K did change by predominantly three units at a time, which indicated an 'octupole-type' interaction.

The problem with steady-state measurements of this type is the difficulty of distinguishing between, for example, a single $\Delta J = 2$ transition and two successive $\Delta J = 1$ transitions. Oka solved this problem by employing 'triple resonance' in studies[40] on HCN, NH_3, and CH_3OH to clamp the population of the intermediate level at a fixed value. In CH_3OH, dipole-like propensity rules were again observed[41]. Other studies included modulated double resonance experiments in CD_3CN and PF_3 [42], in which the relaxation time for rotational saturation was different from that deduced from pressure broadening and a comprehensive study of NH_3 relaxation[43] by H_2, HD, D_2, O_2, N_2, CH_4 and SF_6. All of these systems showed dipole-like propensity rules, suggesting that this is probably the dominant mechanism in rotational relaxation.

8.4.2   Infrared–microwave double resonance

The success of microwave–microwave double resonance in revealing rotational relaxation pathways led naturally to attempts to use the intense, monochromatic light of a molecular laser to saturate vibrational energy levels, and follow their relaxation by microwave absorption. The first experiment of this type to be reported used a CO_2 laser incident on CH_3Br [44]. Since pumping was presumed to be on the $^PP_1(9)$ line of the v_6 band, and $K = 1$ lines decreased in intensity while $K = 0$ lines remained unchanged, it was felt that a true double resonance was being observed. This result was subsequently confirmed in detail by workers at Lille[45]. However, since a number of other laboratories have since failed to reproduce this experiment, it must be concluded that this was not an authentic effect.

True infrared–microwave double resonance has been observed by Oka, however, in NH_3 pumped by an N_2O laser[46, 47]. The P(13) laser line at 927.739 cm^{-1} coincides with the $^QQ_-(8,7)$ line of the v_2 band in $^{14}NH_3$, while the P(15) line at 925.979 cm^{-1} is 332 mHz lower than the $^QQ_-(4,4)$

line of the same band in $^{15}NH_3$; in the latter case, the molecule had to be Stark-tuned into resonance with the laser. This technique should be applicable to a large number of other molecular systems; results have been obtained, for example, with CH_3Cl pumped by a CO_2 laser[48].

Infrared–infrared double resonance experiments, in which infrared monitoring is coupled to infrared pumping, is discussed in the next Section.

8.5 LASER RELATED METHODS

The molecular laser, operating on stimulated vibrational infrared emission, is intimately dependent on the dynamics of vibrational energy-transfer in gases for its existence. It is only fitting then, that such lasers have proven extremely valuable as research tools in investigating these same processes. In this section, we will consider several of the most important recent applications.

8.5.1 Laser-excited infrared fluorescence

The first important use of an infrared laser in energy transfer studies was the excitation of selected molecular vibrational levels, and the analysis of the decay of the ensuing fluorescence to yield relaxation times. The first such system was the 3.39 μm line of the He–Ne laser used to excite the v_3 vibration of CH_4 [49, 50]. The relaxation of this vibration into other modes of the methane ('V–V transfer') and into translation ('V–T transfer') was studied in collisions with O_2 [51] and with He, Ne, Ar, Kr, Xe, H_2, HD, D_2, CO, N_2, O_2 and C_2H_6 [52].

With the availability of the high-powered CO_2 laser, many new possibilities presented themselves. The first to be exploited was the use of the laser radiation (at 10.6 μm) to excite the lower laser level of CO_2 gas (10^00, at 1388 cm^{-4}) to the upper level (00^01, at 2349 cm^{-1}), which then proceeded to fluoresce at 4.3 μm. This feature made detection of the fluorescence and rejection of scattered exciting light particularly easy. Relaxation of the excited CO_2, by both V–V and V–T processes, was studied in the presence of CO_2 [53], He, Ne, Ar, Kr, Xe, H_2, D_2, H_2O, N_2, SF_6, BCl_3, CH_3, CH_3F, C_2H_4, CH_3I, CH_3Br and CH_3Cl [54], isotopically substituted CO_2, N_2O, N_2 and CO [this to observe the effect of near-resonant vibrational exchange][55], and H_2O, HDO and D_2O over a range of temperatures[56]. The temperature variation of the relaxation of CO_2 by CO_2, N_2, He, O_2, H_2O, and other gases has been re-investigated[57, 58, 108], because of the importance of these processes in the operation of the CO_2 laser itself. An analogous experiment, by use of the N_2O laser to excite N_2O, has been carried out in several laboratories[59-61]. Flynn finds $p\tau = 1.95$ μs atm for N_2O–N_2O collisions, while Yardley finds $p\tau = 1.74$ μs atm for the same process, and also reports relaxation measurements with He, Ne, Ar, Kr, H_2, D_2 and N_2. HCl excited by the HCl chemical laser[62, 63] has received attention. Some of the measured cross-sections for transfer of the HCl vibrational quantum to other molecules, in Å^2[10^{-16} cm^2] are: rare gases, $< 10^{-5}$; H_2, 0.0003; HCl, 0.004; DCl, 0.017; H_2O, 2; HBr, 0.2; HI, 0.035; CO, 0.013; N_2, 0.004; CH_4, 0.37. These data are interpreted primarily in terms of vibration–rotation energy transfer, rather than as a near-resonant V–V' phenomenon.

While the introduction of laser excitation has caused a massive thrust in the field of infrared fluorescence decay, the more traditional method of molecular excitation by a resonant emission source[122] still yields new data. Miller and Millikan[123] extended their CO relaxation studies down to 100 K, and found significant departure from the Landau–Teller relation obeyed at higher temperature. They attribute this to the effect of scattering resonances arising from the attractive part of the intermolecular potential.

All the molecular laser sources cited have been used to pump molecules of the same species as exist in the laser. The use of an accidental resonance between a laser line and an absorption line of some other molecule has also been exploited. The most widely studied system of this sort, namely, the fluorescence of SF_6 excited by the CO_2 laser, is discussed in the following section in the context of the relevant double-resonance experiments.

8.5.2   Non-linear effects: saturation and double resonance

In the laser-excited fluorescence experiments described above, it does not matter to the interpretation whether or not the system is linear, in the sense that a given amount of radiation input is presumed to produce a proportional amount of fluorescence output. The high power available from molecular lasers, however, makes it possible to explore the non-linear behaviour of systems in which the radiation flux produces a non-equilibrium distribution in molecular energy levels, the relaxation of which can then be monitored by either emission or absorption spectroscopy. The SF_6 molecule, excited by the CO_2 laser, has been investigated thoroughly in this manner, and its modes of relaxation are now quite well understood.

The first indication of non-linear response observed in this system was saturation of the laser line absorption, that is a decrease of the extinction coefficient at high powers[64–67]. This work, and subsequent pulse trans-mission measurements[68], indicated that the non-equilibrium populations produced by the intense radiative pumping were being relaxed at essentially a gas-kinetic collision rate, presumably by rotational energy transfer. How-ever, some infrared fluorescence measurements reported at about the same time[65, 69] were interpreted in terms of the vibrationally excited state having a relaxation time of 45 ms torr, which implies a cross-section only 10^{-6} times the gas kinetic value. This disagreement was resolved by double resonance experiments[70, 71], in which decay of the excited-state absorption was moni-tored. This technique is similar to that used earlier to monitor the decay of the 10^00 level of CO_2 [72].

The relaxation scheme in SF_6 is shown in Figure 8.3. Essentially, the satura-tion of a particular rotational line of the $v_3 \leftarrow 0$ transition is relaxed by a combination of rotation–translation and vibration–vibration exchange processes, occurring on essentially every kinetic collision. This produces a distribution of vibrational states characterised by a vibrational temperature $T_{vib.}$ in excess of the ambient translational-rotational temperature. The two temperatures then relax toward each other with a time constant of $(p\tau_{vib.}) =$

0.12–0.16 ms torr, implying a cross-section σ_{v-t}^2 approximately 1/500th of the gas-kinetic value. This interpretation was confirmed by subsequent fluorescence experiments[73, 74], in which the short fluorescence decay was observed directly, and also fluorescence at 16 μm (occurring by vibration–vibration transfer from the v_3 to the v_4 mode of SF_6) was observed to appear with an extremely rapid rise time, and decay at the same rate as the 10.6 μm

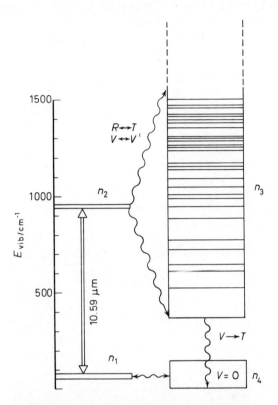

Figure 8.3 Relaxation model for laser-pumped SF_6. Immediately following saturation of levels n_1 and n_2 by the laser pulse, rapid (i.e. gas-kinetic) relaxation processes distribute the excitation over the n_3 levels, which relax back to equilibrium by a slower $V–T$ process. (From Steinfeld *et al.*[71, 75], by courtesy of the American Institute of Physics)

fluorescence and double-resonance signals. It appears as if the earlier fluorescence measurements missed the short collisional V–T decay times, and saw only the long thermal-conduction decay of the coupled vibrational and translational temperatures. In fact, analysis of the SF_6 data shows that infrared fluorescence is easily observable if $T_{vib.}$ is as little as 20 K higher than $T_{transl.}$. A further confirmation of the proposed model for SF_6 relaxation is its success in predicting the self Q-switching behaviour when this gas is incorporated into the laser cavity itself[75, 76]. Other Q-switching results are in substantial agreement with the model as well[77]. Double resonance studies of other systems (C_2H_4 [153], BCl_3 [154]) are under way.

8.5.3 Molecular relaxation effects on laser operation

Since molecular infrared lasers operate on inverted populations of vibrational and rotational levels, the formation and decay mechanisms of these levels are important in understanding the operation of such systems. An example of the investigation of such processes, using laser behaviour as a measurement probe, is the determination of the rotational relaxation rate in CO_2 from laser pulse recovery rates. Using gain-saturation measurements, Caroll and Marcus[78] obtained a rate constant of 0.66×10^7 s^{-1} torr^{-1} for the process, and concluded that $\Delta J = \pm 2$ was the preferred transition.

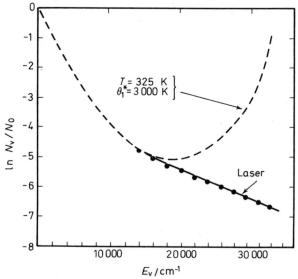

Figure 8.4 Vibrational distribution in CO according to Treanor's model (see text), and observed laser transitions. (From Legay et al.[84], by courtesy of the National Research Council of Canada)

With an improved technique, Cheo and Abrams[79] found a value of 1.1×10^7 s^{-1} torr^{-1}, and concluded that all rotational levels are statistically thermalised in a single collision. They also[80] obtained rates of 0.7×10^7 and 1.0×10^7 s^{-1} torr^{-1} for relaxation by He and N_2, respectively. Recent work[81] tends to support the latter results. Other investigations of this system[82] have tended to focus on the vibrational relaxation rates for CO_2 with N_2, He and H_2O, which are the main components of the laser system. These rates have been reviewed and summarised by Taylor and Bitterman[83].

Another system of considerable interest is the carbon monoxide laser, which has been interpreted[84-86] in terms of a vibrational relaxation model due to Treanor[87] in which the population of any vibrational level of an anharmonic oscillator is given by:

$$N_v = N_0 \, e^{-v \cdot E_1/k\theta^*} \, e^{(vE_1 - E_v)/kT}$$

where θ^* is the effective vibrational temperature determined from the N_1/N_0 ratio. If the molecule is anharmonic, so that $vE_1 > E_v$, and $\theta^* \gg T$ by virtue of slow vibration-translation relaxation rates, then this formula predicts an inverted population above a certain vibrational level v^*, as shown in Figure 8.4. It is notable that the transitions observed in the CO laser are just those at or above this initial value v^*, as indicated in Figure 8.4. Just recently, however, the relative population of CO excited vibrational levels were measured directly, by a flow-tube technique[109]. In this technique, a stream of active nitrogen (containing large amounts of molecules with $v'' \geqslant 1$) is fed into a CO atmosphere, and infrared emission from CO levels is observed. It was found that, while N_v was decidedly non-Boltzmann, it did not fit Treanor's model (as in Figure 8.4) very well; in particular, no actual population inversion was found. While the excitation conditions in this case differ from those previously used[84, 85] it is true that Treanor's model does omit some important factors, such as radiative decay of excited vibrational states. When this factor is taken into account[149], a population distribution is predicted which is much closer to that found experimentally.

Vibrational relaxation rates are also important in determining the operating efficiency of the chemical laser, which operates by producing more molecules in excited than ground vibrational levels by an appropriate chemical reaction. Pimentel[88] has discussed the relationship between these rate constants and laser activity. Most of these lasers have used hydrogen halides in the operating medium, but a carbon monoxide chemical laser has been reported[89-91] in which the mechanism is reminiscent of Smith's kinetic spectroscopy work described earlier[28]. The laser is initiated by flashing a mixture of CS_2 and O_2 (to produce CS radicals), and the reaction:

$$CS + O_2 \rightarrow CO^\dagger + SO$$

is the presumed source of vibrationally excited CO.

Compelling evidence for vibration–vibration energy transfer is also seen in the chemically-pumped CO_2 laser[92-94]. In this system vibrational energy is first produced in DF (by a $D + F_2$ or $F_2O + D_2$ reaction) or HBr (by a $H + Br_2$ reaction). The vibrational quanta of 2998 and 2649 cm^{-1}, respectively, are transferred in a near-resonant exchange to the 00^01 level of CO_2 at 2349 cm^{-1}, from which laser action at 10.6 μm is observed.

8.6 RELAXATION INFORMATION OBTAINED FROM LINE BROADENING

As described in our earlier review[1], a great deal of effort has been expended in determination of pressure broadening of spectral lines, primarily in the microwave and selected infrared regions. The measured frequency broadening is related to a molecular relaxation cross-section by the expression

$$2\pi\Delta v_{\frac{1}{2}} = 2n\langle v\rangle\sigma^2.$$

The problem really comes down to just what σ^2, determined in this way, is the cross-section for. Most treatments of line broadening are usually carried out in terms of an impact approximation[95, 96]. More exact treatments are now in use; an example of recent work from the Leiden Molecular Physics group should suffice to demonstrate the scope of the information obtainable from line broadening studies.

The availability of high-power, single-frequency lasers has made the measurement of Rayleigh scattering linewidths much more convenient than previously. In a recent study of this sort by Knaap and co-workers[97], broadening cross-sections of 65, 94 and 117 Å^2 were found for CO_2, OCS and CS_2, respectively. This technique appears to be very generally applicable to molecular systems; other techniques, such as microwave lineshape-measurement or nuclear-spin relaxation, are restricted to limited classes of molecules. The cross-section measured by this experiment is that for reorientation of the tensor polarisation in a collision and is the same as is measured by an entirely different phenomenon, namely, the Senftleben–Beenakker magnetic-field effect on the viscosity[98]. In the case of CO_2, the cross-section measured by this latter technique is 69 Å^2, in excellent agreement with the light-scattering determination.

8.7   RELAXATION AT PICOSECOND TIME SCALES.

Nearly all the work described in the preceding five sections was carried out under low collision-density conditions in molecular beams or low-pressure gases in order to ensure that the rate of relaxing collisions did not exceed the frequency response of the detector used to monitor the property of the system that was returning to equilibrium, which typically had a time constant not much less than a few nanoseconds. This restriction appears to rule out any study of relaxation processes in liquids in which the time between molecular encounters is of the order of:

$$\frac{\text{(mean distance between molecules)}}{\text{(mean thermal velocity)}} = \frac{10^{-8}\ \text{cm}}{10^4\ \text{cm s}^{-1}} = 10^{-12}\ \text{s}.$$

Certain kinds of intramolecular relaxation processes, such as anharmonic coupling of vibrations, also appear to occur on this timescale. Techniques have recently appeared, however, which enable us to probe some of the processes occurring at this time scale, namely the use of picosecond pulses from mode-locked lasers as optical pumping and probing sources.

If the cavity of a Q-switched solid-state laser was lengthened sufficiently, the normal 10–20 ns Q-switched pulse broke up into a train of sub-pulses which were observed to have a width of 5×10^{-10} s [99]. This was the time constant of the photoelectric detector used to monitor the pulses and subsequently it was found that the actual duration was a few picoseconds by use of the two-photon fluorescence technique[100, 101]. In this technique, the pulse train is passed through a solution of a dye which is transparent at the probe frequency ω_0, but absorbs to a fluorescing state at $2\omega_0$. The pulse train is doubled back upon itself by a mirror in the dye cell; where succeeding pulses overlap, the doubled light intensity at ω_0 causes two-photon absorption at

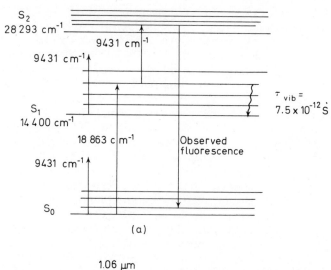

(a)

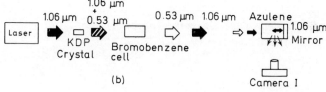

(b)

(c)   (d)

Figure 8.5 Principles of vibrational relaxation measurements in azulene solutions by use of two-photon fluorescence (a) Energy levels of azulene. The fundamental frequency photon of 9431 cm^{-1} is not absorbed by the molecule, but the double-frequency photon at 18 863 cm^{-1} populates an excited vibration level of S_1, which is nearly non-fluorescent. This level can then absorb a second photon of 9431 cm^{-1} to reach S_2, which fluoresces; but if the vibrational energy is relaxed prior to the arrival of the second photon, the 0–0 gap between S_1 and S_2 is too large for absorption to occur. (a) Apparatus for measuring decay of vibrationally excited S_1 azulene. (c) Pulse display of mode-locked laser photon using conventional two-photon fluorescence; the 2 ps pulse width sets the lower time resolution limit of the experiment. (d) Fluorescence display of azulene, as observed in the apparatus of Figure 8.5b. (From Rentzepis[106, 107], by courtesy of the American Association for the Advancement of Science and the American Chemical Society)

$2\omega_0$, the probability of which is proportional to the square of the local intensity. The spatial extent of the fluorescence spot, as recorded by a camera, is a direct measure of the pulse duration. Since the spots are less than 1 mm in length, the pulses must be less than $l/c = 3 \times 10^{-12}$ s in duration.

Recently, several new and more convenient methods for resolving picosecond light pulses have been reported. One involves placing a fast Kerr shutter[102], consisting of a cell filled with CS_2 between crossed polarisers, in front of the camera. The shutter is opened by a portion of the light pulse itself, so that a direct photograph of the pulse can be taken. A variation on this method[103] involves passing the gating pulse transversely through the Kerr cell in front of the camera; the spatial resolution of the resulting film image provides time resolution of the event being photographed. An even simpler way of accomplishing this[104] involves passing the light beams through an etalon consisting of a tilted pile of glass microscope slides; the delay introduced by the refractive index of the glass is sufficient to time-resolve different parts of the wavefront reaching the camera.

The two-photon fluorescence technique has been used to probe molecular reorientation, intramolecular relaxation and vibrational relaxation time in solution; several reviews have appeared[105-107]. As an example of the kind of information obtainable, let us consider the determination of the vibrational relaxation time of azulene dissolved in methanol[147]. Azulene is remarkable in that the lowest excited singlet state, S_1, at 14 400 cm^{-1}, is almost totally non-fluorescent, while the second excited singlet, S_2, at 28 293 cm^{-1}, is strongly fluorescent. The fundamental output of a neodymium laser at 1.06 μm (9431 cm^{-1}) is thus not absorbed by azulene molecules, but its second harmonic at 0.53 μm (18 863 cm^{-1}) can produce a vibrationally excited level of S_1 which can then absorb a second photon of 1.06 μm radiation to reach S_2, which then fluoresces. If the vibrational excitation is removed, the 9431 cm^{-1} energy is insufficient to reach S_2. The energy levels are summarised in Figure 8.5a). Figure 8.5b shows the apparatus used to measure the vibrational relaxation of azulene; the 0.53 μm pulse is delayed with respect to the 1.06 μm by passage through a medium (bromobenzene) which has larger index of refraction at 0.53 μm than at 1.06 μm. The appearance of the fluorescence track in azulene is shown in Figure 8.5d; from the spatial decay profile, a relaxation time of $\approx 7 \times 10^{-12}$ s can be established.

8.8   DEVELOPMENTS IN ENERGY TRANSFER THEORY

Up to this point, the emphasis in this review has been on the development of a number of new techniques which permit us to measure molecular energy-transfer efficiencies that were hitherto unobtainable. We must now consider the most important question of whether the data that are thus accumulated can be systematised in terms of an appropriate theoretical framework, i.e. do we understand what we are doing? There are two interlinked problems[1]. Firstly the assignment to particular experiments of the appropriate microscopic cross-sections and/or transition probabilities that are actually being measured. This area is under investigation at the present time and we hope to detail results in a future review. Secondly, the relation of these cross-sections and transition probabilities to those parts of the intermolecular potential that are effective in coupling various molecular degrees of freedom with each other, via a tractable inelastic scattering-theory. Developments in the latter area will form the main emphasis of this section.

8.8.1 Vibrational excitation of the harmonic oscillator

The exchange of vibrational energy between a harmonic oscillator and a particle colliding in a collinear (one-dimensional) configuration and interacting by a potential:

$$V = V_0 e^{-Z/L},$$

where Z is the distance from the third particle (A) to the center of mass of the oscillator (BC) has been investigated. After a number of approximate solutions to the scattering problem Secrest and Johnson developed the amplitude-density-function results, which were highly accurate numerical solutions to the quantum-mechanical equations[111]. Subsequential, classical[112] and other quantum-mechanical[113] treatments of the same model systems appeared, which were in good accord with the 'exact' results. All of these treatments have recently been compared by Rapp[114].

Heidrich, Wilson and Rapp[115] found that a simple semi-classical analytic expression reproduces to very high accuracy all the previous detailed classical and quantum-mechanical results. Their expression for the oscillator transition probability is:

$$P_{if} = i! f! (<\Delta E>_{fi})^{(i+f)} e^{-<\Delta E>_{fi}} \left[\sum_{l=0}^{\min(if)} \frac{(-\Delta E_{fi})^{-l}}{l!(i-l)!(f-l)!} \right]^2,$$

where

$$<\Delta E>_{fi} = 2 \left\{ \frac{\pi}{\alpha(1+1/m)} \operatorname{cosech} \left[\frac{\pi}{\alpha} \left(\frac{m}{2E_0} \right)^{\frac{1}{2}} \right] \right\}^2,$$

$$(E_0)^{\frac{1}{2}} = \frac{1}{2} [(E_t - i - \frac{1}{2})^{\frac{1}{2}} + (E_t - f - \frac{1}{2})^{\frac{1}{2}}],$$

and E_t is the constant total energy (sum of kinetic and oscillator energies) of the system, the mass parameter

$$m = M_A M_C / M_B (M_A + M_B + M_C),$$

and the potential information is contained in the parameter

$$\alpha = L^{-1} [\hbar M_C / \omega_{\text{vib.}} M_B (M_B + M_C)]^{\frac{1}{2}}.$$

Thus, the problem of vibrational energy transfer in a collinear collision involving a harmonic oscillator can be considered an essentially solved problem in mechanics. Unfortunately, actual molecular collisions involve anharmonic oscillators, non-exponential (and non-monotonic) potentials, and non-collinear configurations, so that the application to real systems still involve a number of approximations.

8.8.2 Rotational energy transfer: close-coupled versus classical calculations

At the time of the last review[1], one of the areas of most intensive calculation was the solution of the close-coupled rotational excitation problem[116, 117]. Nearly exact forms for the $S(j', l'; j, l; J)$ matrix could be obtained, but the computations were laborious and perhaps prohibitively so for total angular momentum, $J \gg 10\text{–}12\,\hbar$. Accordingly, some alternative approach of approximating the close-coupled results was sought. One approach was to determine, from inspection of the $|S|^2$ (or transition probability) matrices, when a statistical approximation in which all dynamically accessible final states were equally probable[118] could be employed; this approximation appears, by this analysis, to be valid over only a limited range of collision energies and mass combinations. It appears, finally, that the most efficient way of getting a quick estimate of rotational energy transfer efficiencies for a variety of inititial conditions may be a judicious selection of classical collision trajectories by use of the efficient numerical integration methods now available[119]. For heavy systems, this is certainly the approach to be preferred: it is surprising that these calculations are stated in Ref. 119 to be 'readily interpretable and of predictive value', even for hydrogenic systems.

It will be interesting to see whether a similar classical approach can be equally applicable to the vibrational energy transfer problem, in the case of more general molecular (i.e., anharmonic) and intermolecular (i.e., non-exponential) potentials. Calculations of this type have been carried out already for Br_2–Ar [120] and I_2–M [121] collisions, the latter for the purpose of direct comparison with the detailed vibrationally and rotationally inelastic cross-sections measured for the excited molecular state. The most serious question in this approach is how light, i.e. quantised, a system has to be before a properly phase-averaged classical calculation fails to reproduce the exact quantum-mechanical results.

8.8.3 Other developments in inelastic scattering theory

A vast quantity of work in inelastic scattering theory has appeared during the past two years, and we can only cite a few examples here. In addition to the classical and quantum-mechanical analyses discussed above, a good deal of effort has been directed toward semi-classical treatments of the same problems[124–126], with good results. Zare[127] has accounted for the different rotational 'propensity rules' in collisions involving the Λ-components of a Π-state in terms of the torque on the molecular electron density produced by a collision.

In the area of vibrational energy transfer, Thomson[128] has solved a quantum-mechanical problem for interaction via a Morse potential, taking account of the scattering resonances introduced by the attractive well.

A similar problem is the near-resonant exchange of CO_2–N_2 which has technical importance. Sharma and Brau have analysed[129] this system by use of semi-classical calculations, with the interaction between the dipole moment of CO_2 and the quadrupole moment of N_2 as the dominant source of the exchange. The problem of coupling between vibrational, rotational and translational modes has been discussed by Nikitin[130].

8.9 CONCLUDING REMARKS

As the study of molecular energy transfer is an extremely active field, both experimentally and theoretically, space limitations preclude even the mention of many facets of research in this area, beyond those posed for discussion in the Introduction. For example, the traditional methods for examining vibrational energy transfer, namely, ultrasonic attenuation and shock-tube relaxation, have continued to yield new results. In the former area, the extension of measurements to higher and lower temperatures than previously has been one of the main areas of activity[131]. Recent shock-tube studies include the systems:

$$CO_2 + N_2 \text{ }^{[132]}, HCl \text{ }^{[133,134]}, HBr \text{ and } HI \text{ }^{[135]}, DBr \text{ and } DI \text{ }^{[136]}, CO + N_2,$$
$$O_2, D_2 \text{ and } H_2 \text{ }^{[137]}, N_2 + N_2O \text{ }^{[138]}, CO_2 + Ar \text{ }^{[139,140]} \text{ and } N_2O + Ar \text{ }^{[141]}.$$

A recent novel study[142] involved the relaxation of CO by iron atoms, in which both species were produced by shock-decomposition of $Fe(CO)_5$. In many of the cases cited, the spate of activity was a result of the use of new fast sensitive infrared detectors to monitor vibrational emission, which represented a considerable improvement on methods involving the monitoring of density changes or the like used previously. This is the case, for example, in a recent shock-tube study[143] of HF relaxed by HF, N_2, He and D_2 in which it was concluded that 'sticky collisions', involving the attractive part of the HF–HF potential, may be more important in the relaxation process than the usually invoked impulsive V–T or V–R type of interaction.

In contrast to these classical methods, we have neglected to mention one of the most exciting new techniques, namely, that of stimulated Raman excitation[144]. In this approach, a substantial fraction of the molecules are brought to an upper vibrational state by an intense laser pulse and the population of the state probed either by anti-Stokes emission with a second laser pulse or by light scattering from the refractive index fluctuations caused by local temperature changes in the system. Since the pulses are from a Q-switched laser, the time resolution is sufficiently short to permit the use of fairly dense samples. Further discussions of this field will be forthcoming[145,146].

Acknowledgements

I thank A. N. Schweid, J. Levy, P. Houston, D. Frankel and G. Holleman for the literature search, the National Science Foundation (Grants GP-6504, GP-9318, and GP-23266), the Air Force Office of Scientific Research (Grant C-9-218), the U.S. Army Research Office (Contract No. DAHCO4-70C-0015), the Petroleum Research Fund of the American Chemical Society (Grant 2523-AC5), and the North Atlantic Treaty Organization (Grants SA.5-2-05B(405)717(69)AG and SA.S-2-05B(405)1158-(70)AJ) for financial support and the Alfred P. Sloan Foundation for a Research Fellowship.

Notes added in proof

(1) The relatively large cross-sections for vibrational relaxation of electronically excited H_2 should be contrasted to those determined for collisions involving ground-state H_2, using the laser-Raman technique[144]. These have been determined as $4 \times 10^{-6} Å^2$ for $H_2(v'' = 1) + H_2 \rightarrow 2H_2$ $(v'' = 0)$, and $7 \times 10^{-7} Å^2$ for $H_2(v'' = 1) + He \rightarrow H_2(v'' = 0) + He$. The difference in vibration frequency (1357 cm^{-1} for the B state, as compared with 4395 cm^{-1} for the ground state) can account for only a part of the six orders of magnitude difference in vibrational deactivation efficiencies.

(2) A resonant-transfer mechanism, of the sort proposed in Section 8.3.3 to account for efficient relaxation in NO_2^*–NO_2 collisions, has recently been invoked in a study of vibrational energy transfer in NO [150]. Using monochromatic excitation of single isotopic species, Melton and Klemperer found cross-sections for the processes

$$^{14}NO(A^2\Sigma^+, v' = 1) + {}^{15}NO(X^2\Pi, v'' = 0) \rightarrow$$
$$^{15}NO(A^2\Sigma^+, v' = 1) + {}^{14}NO(X^2\Pi, v'' = 0): \sigma = 20 \pm 5 Å^2$$

$$^{14}NO(A^2\Sigma^+, v' = 0) + {}^{15}NO(X^2\Pi, v'' = 0) \rightarrow$$
$$^{15}NO(A^2\Sigma^+, v' = 0) + {}^{14}NO(X^2\Pi, v'' = 0): \sigma = 10 \pm 7 Å^2$$

and

$$^{14}NO(A^2\Sigma^+, v' = 1) + {}^{15}NO(X^2\Pi, v'' = 0) \rightarrow$$
$$^{15}NO(A^2\Sigma^+, v' = 0) + {}^{14}NO(X^2\Pi, v'' = 1): \sigma = 1.4 \pm 0.2 Å^2$$

A calculation[151] based on a Förster-type electronic energy transfer model gave 18, 16, and $1.4 Å^2$, respectively, for these three processes.

(3) As if in confirmation of this last statement, kinetic absorption spectroscopy has been used very recently by Porter and Formosinho[152] to study a very much larger molecule than the sort mentioned thus far, namely, vapour-phase anthracene excited by a Q-switched laser pulse. They were able to see both vibrational deactivation of the triplet molecules – which appears to proceed *via* an exponential distribution of energy losses per collision – and 'back-crossing' from the vibrationally hot triplet back to the singlet manifold.

References

1. Gordon, R. G., Klemperer, W. and Steinfeld, J. I. (1968). *Ann. Rev. Phys. Chem.*, **19**, 215
2. Borrell, P. (1967). *Advan. Mol. Relaxation Processes*, **1**, 69
3. Moore, C. B. (1969). *Account. Chem. Res.*, **2**, 103
4. Stevens, B. (1967). *International Encyclopedia of Physical Chemistry and Chemical Physics: Topic 19. Gas Kinetics: Volume 3. Collisional Activation in Gases* (Oxford: Pergamon Press)

5. Burnett, G. M. and North, A. M. (eds.) (1969, 1970). *Transfer and Storage of Energy by Molecules: Volume 2: Vibrational Energy; Volume 3: Rotational Energy* (London: Wiley-Interscience)
6. Toennies, J. P. (1962). *Discuss. Faraday Soc.*, **33**, 96
7. Toennies, J. P. (1965). *Z. Physik*, **182**, 257; (1966). ibid., **193**, 76
8. Waech, T. G., Kramer, K. H. and Bernstein, R. B. (1968). *J. Chem. Phys.*, **48**, 8978
9. Dittner, P. F. and Datz, S. (1968). *J. Chem. Phys.*, **49**, 1969, (1971), *ibid.*, **54**, 4228
10. Cheng, M. H., Chiang, M. H., Gislason, E. A., Mahan, B. H., Tsao, C. W. and Weiner, A. S. (1970). *J. Chem. Phys.*, **52**, 6150
11. Cosby, P. C. and Moran, T. F. (1970). *J. Chem. Phys.*, **52**, 6157
12. Kurzel, R. B. and Steinfeld, J. I. (1970). *J. Chem. Phys.*, **53**, 3292
13. Steinfeld, J. I. and Schweid, A. N. (1970). *J. Chem. Phys.*, **53**, 3304
14. Steinfeld, J. I. and Klemperer, W. (1965). *J. Chem. Phys.*, **42**, 3475
15. Broadbent, T. W., Callear, A. B. and Lee, A. K. (1968). *Trans. Faraday Soc.*, **64**, 2320
16. Akins, D. L., Fink, E. H. and Moore, C. B. (1970). *J. Chem. Phys.*, **52**, 1604
17. Fink, W. H., Akins, D. L. and Moore, C. B., to be published
18. Keyser, L. F., Kaufman, F. and Zipf, E. C. (1968). *Chem. Phys. Lett.*, **2**, 523
19. Keyser, L. F., Levine, S. Z. and Kaufman, F. (1971). *J. Chem. Phys.*, **54**, 355
20. Schwartz, S. E. and Johnston, H. S. (1969). *J. Chem. Phys.*, **51**, 1286
21. Sakurai, K. and Broida, H. P. (1969). *J. Chem. Phys.*, **50**, 2404
22. Bergmann, K. and Demtröder, W. (1971), to be published
23. Ottinger, Ch. and Poppe, D. (1971). *Chem. Phys. Lett.*, **8**, 513
24. Sakurai, K., Johnson, S. E. and Broida, H. P. (1970). *J. Chem. Phys.*, **52**, 1625
25. Welge, K. H., Filseth, S. V. and Davenport, J. (1970). *J. Chem. Phys.*, **53**, 502
26. Calo, J. M. and Axtmann, R. C. (1971). *J. Chem. Phys.*, **54**, 1332
27. Donovan, R. J., Husain, D. and Stevenson, C. D. (1970). *Trans. Faraday Soc.*, **66**, 2148
28. Smith, I. W. M. (1968). *Trans. Faraday Soc.*, **64**, 3183
29. Cox, A. P., Flynn, G. W. and Wilson, E. B. (1965). *J. Chem. Phys.*, **42**, 3094
30. Ronn, A. M. and Wilson, E. B. (1967). *J. Chem. Phys.*, **46**, 3262
31. Gordon, R. G., Larson, P. E., Thomson, C. H. and Wilson, E. B. (1969). *J. Chem. Phys.*, **50**, 1388
32. Unland, M. C. and Flygare, W. H. (1966). *J. Chem. Phys.*, **45**, 2421
33. Roussy, G., Demaison, J. and Barril, J. (1969). *Compt. Rend. Acad. Sci.*, **Ser. C269**, 1080
34. Macke, B., Messelyn, J. and Wertheimer, R. (1969). *J. Phys.*, **30**, 665
35. Oka, T. (1965). *J. Chem. Phys.*, **45**, 754
36. Oka, T. (1967). *J. Chem. Phys.*, **47**, 13
37. Oka, T. (1967). *J. Chem. Phys.*, **45**, 4852
38. Oka, T. (1968). *J. Chem. Phys.*, **48**, 4919
39. Oka, T. (1968). *J. Chem. Phys.*, **49**, 3135
40. Oka, T. (1968). *J. Chem. Phys.*, **49**, 4234
41. Lees, R. M. and Oka, T. (1969). *J. Chem. Phys.*, **51**, 3027
42. Oka, T. and Shimizu, T. (1970). *Phys. Rev.*, **A2**, 587
43. Daly, P. W. and Oka, T. (1970). *J. Chem. Phys.*, **53**, 3277
44. Ronn, A. M. and Lide, D. R. (1967). *J. Chem. Phys.*, **47**, 3669
45. Lemaire, J., Houriez, J., Bellet, J. and Thibault, J. (1969). *Compt. Rend. Acad. Sci.*, **Ser. B368**, 922
46. Shimizu, T. and Oka, T. (1970). *J. Chem. Phys.*, **53**, 2530
47. Shimizu, T. and Oka, T. (1970). *Phys. Rev.*, **A2**, 1177
48. Frenkel, L., Marantz, H. and Sullivan, T. (1971). *Phys. Rev.*, **A3**, 1640
49. Yardley, J. T. and Moore, C. B. (1966). *J. Chem. Phys.*, **45**, 1066
50. Yardley, J. T. and Moore, C. B. (1968). *J. Chem. Phys.*, **48**, 14
51. Yardley, J. T. and Moore, C. B. (1968). *J. Chem. Phys.*, **49**, 1111
52. Yardley, J. T., Fertig, M. N. and Moore, C. B. (1970). *J. Chem. Phys.*, **52**, 1450
53. Yardley, J. T. and Moore, C. B. (1967). *J. Chem. Phys.*, **46**, 4222, 4491
54. Stephenson, J. C., Wood, R. E. and Moore, C. B. (1968). *J. Chem. Phys.*, **48**, 4790
55. Stephenson, J. C., Wood, R. E. and Moore, C. B. (1968). *J. Chem. Phys.*, **48**, 4790
56. Heller, D. F. and Moore, C. B. (1970). *J. Chem. Phys.*, **52**, 1005
57. Rosser, W. A., Wood, A. D. and Gerry, E. T. (1969). *J. Chem. Phys.*, **50**, 5996
58. Rosser, W. A. and Gerry, F. T. (1969). *J. Chem. Phys.*, **51**, 2286
59. Bates, R. D., Flynn, G. W. and Ronn, A. M. (1968). *J. Chem. Phys.*, **49**, 1432

60. Yardley, J. T. (1968). *J. Chem. Phys.*, **49**, 2816
61. Arditi, I., Morgottin-Maclon, M., Goregnen, H. and Doyennette, L. (1970). *Compt. Rend. Acad. Sci.*, **B270**, 477
62. Chen, H.-L., Stephenson, J. C. and Moore, C. B. (1968). *Chem. Phys. Lett.*, **2**, 593
63. Chen, H.-L. and Moore, C. B. (1971). *J. Chem. Phys.*, **54**, 4072, 4080
64. Burak, I., Nowak, A. V., Steinfeld, J. I. and Sutton, D. G. (1969). *J. Quant. Spectroscopy and Rad. Transfer.*, **9**, 959
65. Wood, O. R., Gordon, P. L. and Schwarz, S. E. (1969). *I.E.E.E. J. Quantum Electronics*, **QE-5**, 502.
66. Djeu, N. and Wolga, G. J. (1971). *J. Chem. Phys.*, **54**, 774
67. Brunet, H. (1970). *I.E.E.E., J. Quantum Electronics*, **QE-6**, 678
68. Sutton, D. G., Burak, I. and Steinfeld, J. I. (1971). *I.E.E.E. J. Quantum Electronics*, **QE-7**, 82
69. Wood, O. R. and Schwarz, S. E. (1970). *Appl. Phys. Lett.*, **16**, 518
70. Burak, I., Nowak, A. V., Steinfeld, J. I. and Sutton, D, G. (1969). *J. Chem. Phys.*, **51**, 2275
71. Steinfeld, J. I., Burak, I., Sutton, D. G. and Nowak, A. V. (1970). *J. Chem. Phys.*, **52**, 5421
72. Rhodes, C. K., Kelly, M. J. and Javan, A. (1968). *J. Chem. Phys.*, **48**, 5770
73. Bates, R. D., Flynn, G. W., Knudtson, J. T. and Ronn, A. M. (1970). *J. Chem. Phys.*, **53**, 3621
74. Bates, R. D., Knudtson, J. T., Flynn, G. W. and Ronn, A. M. (1971), to be published
75. Burak, I., Houston, P. L., Sutton, D. G. and Steinfeld, J. I. (1971, *I.E.E.E. J. Quantum Electronics*, **QE-7**, 73
76. Houston, P. L., Sutton, D. G. and Steinfeld, J. I. (1971). *Appl. Phys. Lett.*, to be published
77. Lee, S. M., Gamn, L. A. and Ronn, A. M. (1970). *Chem. Phys. Lett.*, **7**, 463
78. Carroll, T. O. and Marcus, S. (1968). *Phys. Lett.*, **27A**, 590
79. Cheo, P. K. and Abrams, R. L. (1969). *Appl. Phys. Lett.*, **14**, 47
80. Abrams, R. L. and Cheo, P. K. (1969). *Appl. Phys. Lett.*, **15**, 177
81. Crafer, R. C., Gibson, A. F. and Kimmitt, M. F. (1969). *Brit. J. Appl. Phys.* **2**, 1135
82. Carbone, P. J. and Witteman, W. J. (1969). *I.E.E.E. J. Quantum Electronics*, **QE-5**, 442
83. Taylor, R. L. and Bitterman, S. (1967). AVCO Everett Research Laboratory Report No. 282, ARPA Contract F33(615)–68–C–1030
84. Legay, F., Legay-Sommaire, N. and Taïeb, G. (1970). *Can. J. Phys.*, **48**, 1949
85. Taieb, G. and Legay, F. (1970). *Can. J. Phys.*, **48**, 1957
86. Legay-Sommaire, N. and Legay, F. (1970). *Can. J. Phys.*, **48**, 1966
87. Treanor, C. E., Rich, J. W. and Rehm, R. G. (1968). *J. Chem. Phys.*, **48**, 1798
88. Parker, J. H. and Pimentel, G. C. (1969). *J. Chem. Phys.*, **51**, 91
89. Arnold, S. J. and Kimball, G. H. (1969). *Appl. Phys. Lett.*, **15**, 351
90. Pollack, M. A. (1966). *Appl. Phys. Lett.*, **8**, 237
91. Gregg, D. W. and Thomas, S. J. (1968). *J. Appl. Phys.*, **39**, 4399
92. Cool, T. A., Falk, J. and Stephen, R. R. (1969). *Appl. Phys. Lett.*, **15**, 318
93. Gross, R. W. F. (1969). *J. Chem. Phys.*, **50**, 1889
94. Cool, T. A. and Stephen, R. R. (1970). *J. Chem. Phys.*, **52**, 3304
95. Cooper, J. (1967). *Rev. Mod. Phys.*, **39**, 167
96. Murphy, J. S. and Bozzo, J. E. (1967). *J. Chem. Phys.*, **47**, 691
97. Keijser, R. A. J., Jansen, M., Cooper, V. G. and Knaap, H. F. P. (1971). *Physica*, **51**, 593
98. See Ref. 1, p. 231, for a discussion
99. De Maria, A. J., Stetser, D. A. and Heynau, H. (1966). *Appl. Phys. Lett.*, **8**, 174
100. Giordmaine, J. A., Rentzepis, P. M., Shapiro, S. L., Wecht, K. W. and Duguay, M. A. (1967). *Appl. Phys. Lett.*, **11**, 216, 218
101. Duguay, M. A., Shapiro, S. L. and Rentzepis, P. M. (1967). *Phys. Rev. Lett.*, **19**, 1014
102. Duguay, M. A. and Hansen, J. W. (1971). *I.E.E.E. J. Quantum Electronics*, **QE-7**, 37
103. Rentzepis, P. M., Topp, M. R., Jones, R. P. and Jortner, J. (1970). *Phys. Rev. Letters*, **25**, 1742
104. Topp, M. R., Rentzepis, P. M. and Jones, R. P. (1971). *Appl. Phys. Letters*, to be published
105. De Maria, A. J., Glenn, W. H., Brienza, M. J. and Mack, M. E. (1969). *Proc. I.E.E.E.*, **57**, 2
106. Rentzepis, P. M. (1970). *Science*, **169**, 239
107. Rentzepis, P. M. and Mitschele, C. J. (1970). *Anal. Chem.*, **42**, 20A

108. Stephenson, J. C., Wood, R. E. and Moore, C. B. (1971). *J. Chem. Phys.*, **54**, 3097
109. Horn, K. P. and Oettinger. P. E. (1971). *J. Chem. Phys.*, **54**, 3040
110. Kurzel, R. B., Steinfeld, J. I., Hatzenbuhler, D. and Leroi, G. E. (1971). *J. Chem. Phys.*, **55**, 4822
111. Secrest, D. and Johnson, B. R. (1966). *J. Chem. Phys.*, **45**, 4556
112. Kelley, J. D. and Wolfsberg, M. (1966). *J. Chem. Phys.*, **44**, 324
113. Gutschick, V. P., McKoy, V. and Diestler, D. J. (1970). *J. Chem. Phys.*, **52**, 4807
114. Rapp, D. and Kassal, T. (1969). *Chem. Rev.*, **69**, 61
115. Heidrich, F. E., Wilson, K. R. and Rapp, D. (1971). *J. Chem. Phys.*, **54**, 3885
116. Lester, W. A., Jr. and Bernstein, R. B. (1967). *Chem. Phys. Lett.*, **1**, 207, 347
117. Lester, W. A., Jr. and Bernstein, R. B. (1968). *J. Chem. Phys.*, **48**, 4896
118. Lester, W. A., Jr. and Bernstein, R. B. (1970). *J. Chem. Phys.*, **53**, 11
119. Bernstein, R. B. (1970). *Proc. Conf. on Potential Energy Surfaces in Chem.*, (Santa Cruz, California), p. 27
120. Razner, R. (1969). *J. Chem. Phys.*, **51**, 5602
121. Feldman, D., Sutton, D. G. and Steinfeld, J. I. (1971). *Proc. VIIth Intl. Conf. on Phys. of Electronic and Atomic Collisions.* (Amsterdam: North-Holland Publishing Co.), p.638
122. Millikan, R. C. (1963). *J. Chem. Phys.*, **38**, 2855
123. Miller, D. J. and Millikan, R. C. (1970). *J. Chem. Phys.*, **53**, 3384
124. Cross, R. J., Jr. (1969). *Proc. Intl. School of Physics "Enrico Fermi", Course XLIV*, 50 (New York: Academic Press)
125. Cross, R. J., Jr. (1968). *J. Chem. Phys.*, **48**, 4838
126. Roberts, R. E. (1970). *J. Chem. Phys.*, **53**, 1937
127. Ottinger, Ch., Velasco, R. and Zare, R. N. (1970). *J. Chem. Phys.*, **52**, 1636
128. Thompson, S. L. (1968). *J. Chem. Phys.*, **48**, 3400
129. Sharma, R. D. and Brau, C. A. (1968). AVCO Everett Research Laboratory Report 303, on Contract F33615–68–C–1030
130. Nikitin, E. E. (1970). *Comments Atom. Mol. Phys.*, **2**, 59
131. See, for example: Bass, H. E., Winter, T. G. and Evans, L. B. (1971). *J. Chem. Phys.*, **51**, 644
132. Taylor, R. L. and Bitterman, S. (1969). *J. Chem. Phys.*, **50**, 1720
133. Bowman, C. T. and Seery, D. J. (1969). *J. Chem. Phys.*, **50**, 1904
134. Borrell, P. and Gutteridge, R. (1969). *J. Chem. Phys.*, **50**, 2273
135. Kiefer, J. H., Breshears, W. D. and Bird, P. F. (1969). *J. Chem. Phys.*, **50**, 3641
136. Breshears, W. D. and Bird, P. F. (1970). *J. Chem. Phys.*, **52**, 999
137. Sato, Y., Tsuchiya, S. and Kuratani, K. (1969). *J. Chem. Phys.*, **50**, 1911
138. Roach, J. F. and Smith, W. R. (1969). *J. Chem. Phys.*, **50**, 4114
139. Simpson, C. J. S. M., Chandler, T. R. O. and Strawson, A. C. (1969). *J. Chem. Phys.*, **51**, 2214
140. Kamimoto, G. and Matsui, H. (1970). *J. Chem. Phys.*, **53**, 3910
141. Kamimoto, G. and Matsui, H. (1970). *J. Chem. Phys.*, **53**, 3987
142. Von Rosenberg, C. W. and Wray, R. C. (1971). *J. Chem. Phys.*, **54**, 1406
143. Bott, J. F. and Cohen, N. (1971). *J. Chem. Phys.*, **55**, 3698
144. Ducuing, J., Joffrin, C. and Coffinet, J. P. (1970). *Opt. Commun.*, **2**, 245
145. Moore, C. B. (1971). *Ann. Rev. Phys. Chem.*, **22**, in the press
146. Moore, C. B. (1971). *Advan. Chem. Phys.*, in the press
147. Rentzepis, P. M. (1968). *Chem. Phys. Lett.*, **2**, 117
148. Steinfeld, J. I. (1971). Molecular Spectroscopy: Modern Research. Commemorative Volume of Silver Jubilee Symposium on Molecular Structure and Spectroscopy [Rao and Mathews, Eds.] in the press (New York: Academic Press)
149. Brau, C. A., Caledonia, G. E. and Center, R. E. (1970). *J. Chem. Phys.*, **52**, 4306

151. Gordon, R. G. and Chiu, Y.-N. (1971). *J. Chem. Phys.*, **55**, 1469
152. Formosinho, S. J. (1971). *Ph.D. Thesis, University of London*
153. Flynn, G. W. (1971). *Proc. Cambridge Conf. on Molecular Energy Transfer*
154. Houston, P. L., Nowak, A. V. and Steinfeld, J. I. (1972), to be published

9

Reactions of Solvated Electrons

F. S. DAINTON
University of Oxford

9.1 BASIC CONCEPTS: FREE AND LOCALISED ELECTRONS

9.1.1 Free and quasi-free electrons

An electron of thermal energy in a vacuum can be described as 'truly free' with a velocity determined solely by its kinetic energy and mass. If placed in an assembly of molecules with which it does not react chemically, the electron will experience short-range repulsive forces and relatively long-range attractions due to the electronic polarisation of the surrounding molecules which it causes. When the concentration of molecules is increased, the dimensions of the primitive cell diminish, the polarisation potential will rapidly change and local field effects due to screening influences of neighbouring molecules will become important. As a result the energy of this 'quasi-free' electron will depend markedly upon the density of the medium which, in the case of molecules of low polarisability, e.g. He and Ne, is likely to be greater than the *in vacuo* value and similarly, if the polarisability is high, this energy may be less than the *in vacuo* value.

In this description it has been assumed (a) that the molecules are uniformly distributed and randomly oriented, and (b) that this distribution and orientation is unaffected by the presence of the electron. Neither assumption is true for the vast majority of real systems. Thus in liquids and glassy solids there exist 'holes' of a wide range of sizes. In liquids the volumes of individual voids will fluctuate at rates dependent on the temperature whilst it is reasonable to assume that in glasses this fluctuation rate will be very small if not actually zero. Consequently a 'quasi-free' electron, either because of its own motion or, in a liquid, because of fluctuation in hole size, has a finite probability of finding itself at a point where the short-range repulsive forces of the molecules confine it to this particular volume element sufficiently long for molecular motions to occur which bring about changes which minimise the free energy. In this process the cavity volume may change and the orientations of the molecules forming the cavity wall and to a lesser degree those in outer layers of solvent molecules may alter. The energy ultimately attained is then the sum of the electronic energy (E_e) of an electron confined to a cavity of a particular shape and size and the energy (E_m) necessary to cause the rearrangement of the molecules to form the cavity. The electron is then 'localised' in the sense that its wavefunction is largely confined to the vicinity of the cavity, in contrast to that of the 'quasi-free' electron which is spread over many molecules. For such a localised electron-state to persist sufficiently long to be regarded, in any meaningful way, as a discrete entity with identifiable physical and chemical properties, two conditions must be satisfied. First, $E_e + E_m$ must be smaller than the energy of the 'quasi-free' electron in the same medium and, secondly, the spontaneous decomposition or reaction of the localised electron must be sufficiently slow.

9.1.2 Solvated and trapped electrons, e_s^- and e_t^-

In this chapter we are concerned with electrons in liquids and amorphous solids* for which both these conditions are satisfied and which are frequently referred to as 'solvated electrons' though it is perhaps better to reserve this term and the corresponding symbol, e_s^-, for electrons localised in liquids but which can diffuse through liquids whilst retaining the characteristics of e_s^- and to use the term 'trapped electron', designated by e_t^-, for localised electrons in solids which are normally immobile and can only be mobilised (i.e. become e_m^-) by gross disruption of the cavity and hence loss of the character of e_t^-. Before discussing the chemical reactions of such localised electrons and their relevant physical properties it will be useful to describe theoretical models a little further. Much recent work has been carried out in this field but fortunately the reader can be referred to two comprehensive reviews which have recently been published by Jortner[2].

9.1.3 Factors regulating the cavity size and the energy of e_s^-

It is important to recognise that both E_e and E_m depend on the size of the cavity and the number and arrangement of solvent molecules which comprise the wall. First, if a spherical cavity is assumed and if localisation involves increase of the radius from r to R, there will be an increase in the surface free energy $= 4\pi\gamma(R^2 - r^2)$, where γ is the surface tension and an increase of free energy $= \frac{4}{3}\pi P(R^3 - r^3)$ for the work done against the external pressure P. Secondly, the very fast electronic polarisation of the solvent molecules in the cavity wall induced by the central electronic charge will be followed by a slower rotation of the solvent molecules tending to orient them with the positive ends of their dipoles pointing inwards. Therefore, whilst this rotation contributes to the stabilisation of the system by changing the repulsive forces between the electron and the solvent molecules, it also destabilises it by increasing repulsions between solvent molecules. Obviously the magnitude of these effects, all of which influence E_m, will also depend on the number of solvent molecules in the first shell (N). Calculations have been made by Jortner et al. for various values of N in liquid and supercritical

---

* We are not here concerned with electrons trapped at some specific site in a crystalline lattice, e.g. in a defect such as a negative ion vacancy comprising an F centre, or confined between molecules which have specific relative orientations in molecular crystals as, e.g. between two parallel but opposed CN groups in acetonitrile crystals[1].

gaseous ammonia and for liquid water[3]. Thirdly, E_m will be affected because it is regulated by the form of the potential defined by the short-range forces (attraction and repulsion) and long-range attractive polarisation-forces between the electron and its enclosing molecules. Finally, there is the effect of the electron on the molecules outside the first shell and for which it is generally assumed that the potential $= -(D_{op}^{-1} - D_s^{-1})e^2/d$ where D_{op} and D_s are the high frequency and static dielectric constants and d is the distance from the centre of the cavity. This is equivalent to treating the molecules outside the first shell as a dielectric continuum.

Many calculations have been carried out by Jortner and his collaborators[3] which incorporate all these factors and they have been able to evaluate the shape of the potential energy-cavity radius curves for the 1S and 2P states of the localised electrons in cavities containing different numbers of solvent molecules in assemblies of either NH_3 or H_2O molecules of different densities and to place these curves on a vertical scale relative to that of the quasi-free electron in the same medium. Typical curves are shown in Figure 9.1,

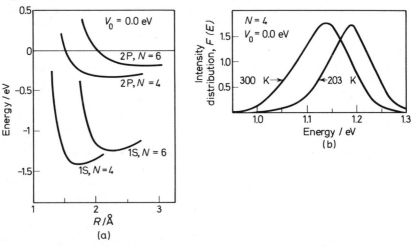

Figure 9.1 (a) The total energies ($E_e + E_m$) for the ground (1S) and excited (2P) states as a function of the distance (R) from the centre of a spherical cavity to the 'hard core' of either four ($N = 4$) or six ($N = 6$) symmetrically disposed NH_3 molecules assuming zero energy of the quasi-free electron in liquid ammonia ($V_0 = 0$) (b) The expected form of the spectrum for 2P ← 1S transition at 203 and 300 K for (a) above with $N = 4$. Note (i) that the line width is only 30% of that observed and (ii) the spectrum is asymmetric to the *low* energy side contrary to experiment (From Copeland et al.[3], by courtesy of The American Institute of Physics)

and the general conclusions to be drawn from this work are:
  (a) The transition energy, $\Delta E(2P \leftarrow 1S)$, can be calculated and decreases with increasing R.
  (b) The optical absorption for this transition is an asymmetric broad line with a low-energy tail and has a high oscillator-strength.
  (c) The energy for photo-detachment of an electron from its hole into the vacuum state which is equal to the energy of the 1S level plus that of the quasi-free electron can be calculated.

(d) If assumptions are made about the level of the conduction band, the photoconductivity threshold can be calculated and is expected to be less than the photoelectric threshold in (c) above.

(e) The average radius of the 1S wave-function is greater than R.

Solvated and trapped electrons are important in chemistry for a variety of reasons. By definition they are, next to the 'quasi-free' and 'truly free' electrons, at once the simplest free radical, the simplest reducing agent and the simplest nucleophile. Moreover, since they are, in general, only lightly bound to the medium, most of their reactions with molecules, ions and other free radicals will be considerably exothermic and exergodic and are expected to have high rate-constants. They therefore constitute a powerful tool for the investigation of many chemical problems including rates and mechanisms of solvation, diffusion-controlled reactions, the chemistry of hyper-reduced ions, and charge neutralisation processes.

9.2 THE FORMATION OF e_s^- AND e_t^-

There are several ways in which e_s^- or e_t^- may be prepared:

9.2.1 By spontaneous reactions

There are at present two kinds of *spontaneous reaction* in liquid media which can lead to the formation of e_s^-. First the dissolution of a metal (equation (9.1))

$$M_s \rightarrow M_s^+ + e_s^-$$ (9.1)

of low ionisation-potential for which ΔG_1^0 is negative because of the relatively large solvation energy of M^+ and e^-. This is true for alkali metals, either pure or in amalgam with mercury, when brought into contact either with liquid ammonia or amines[4] or molten salts when e_s^- is relatively long lived or with protic solvents such as water or alcohols when e_s^- is relatively short lived. By use of the rotating cryostat, e_t^- in the amorphous solid can be formed[5]. Mere dissolution of the metal does not establish that Reaction (9.1) has taken place, because, as Dye has shown[6], other anions and ion pairs may be found. For instance, in ethereal solutions of alkali metals, M^+M^- ion pairs as well as free M^- are produced. In this method the metal may either be predeposited as a film on the vessel wall or deposited electrolytically at a cathode. The second kind of spontaneous reaction is based on the idea that in a protic solvent, HA, a hydrogen atom is the acid of which e_s^- is the conjugate base (equation (9.2))

$$H_s \rightleftharpoons H_s^+ + e_s^-$$ (9.2)

and therefore reaction of a hydrogen atom with the anion of the solvent (equation (9.3)) should produce the solvated electron.

$$H_s + A_s^- \rightleftharpoons HA_s^- \equiv e_s^-$$ (9.3)

This has been shown to be true in water and liquid ammonia where $A_s^- = OH_{aq}^-$ and $NH_{2,am}^-$ respectively. The hydrogen atom need not be a free hydrogen atom; one which is lightly bound to a molecule, e.g. as in a free radical, will suffice as equation (9.4) illustrates.

$$CH_2OH + MeO_{alc}^- \rightleftharpoons CH_2O + e_{alc}^- \tag{9.4}$$

9.2.2 By injection into the medium

If the medium is an insulator, electrons can be *injected* by field emission from a cathode. Although it has some merits, this method is unlikely to be of great utility in exploring the reactivity of electrons and it also suffers from the disadvantage of requiring high electrical fields and being an intensely local source of electrons[7].

9.2.3 Photochemically from plates, films, cathodes and solutes

For kinetic investigations there are considerable advantages to be gained from the use of methods in which e_s^- can be produced at controlled rates which can be started and stopped at will. There are several ways in which this can be achieved by use of light to stimulate electron formation. Photo-electric emission from metals in the form of metal plates[8] or thin films of alkali metals[9] leading to e_s^- or e_t^- in liquid and solid hydrocarbons has been used with success, as has photo-emission from a mercury cathode[10], but in these cases the effect is to generate the e_s^- or e_t^- within a very short distance of the surface. Despite this disadvantage, techniques of this kind are useful for measuring, for example, the potential of 'quasi-free' electrons[8] and can also be modified to determine diffusion[10] and rate constants of e_s^-.

Light stimulated formation of e_s^- is most valuable when the electrons are uniformly distributed throughout the solvent. Many solutes, D, exist from which electrons can be detached by a monophotonic process (equation (9.5)).

$$D + h\nu \rightarrow D^+ + e_s^- \tag{9.5}$$

For example e_{aq}^- may be formed by photo-detachment from hydrated cations (e.g. Fe^{2+}, Eu^{2+}, Cr^{2+}), anions (e.g. $Fe(CN)_6^{4-}$, I^-, Br^-, Cl^-, OH^-, SO_4^{2-}) and even neutral molecules (e.g. C_6H_5OH, $C_6H_5NH_2$, tryptophan)[11]. Other solutes can be ionised by biphotonic processes; a much studied example is TMPD (tetramethyl *p*-phenylene diamine) dissolved in hydrocarbons[12]

$$TMPD + h\nu \rightarrow TMPD^* + h\nu' \rightarrow TMPD^+ + e_s^- \tag{9.6}$$

These methods allow the employment of flash photolytic techniques combined with kinetic spectroscopy to determine rates of reaction of e_s^- but they do have several demerits: (a) with the exception of the anions in aqueous solution the quantum yields are rather small, (b) D and D^+ may have absorption spectra which overlap that of e_s^- and (c) D^+ may be reactive towards the solute forming a light absorbing species, e.g. $I + I^- \rightarrow I_2^-$, $\lambda_{max} = 385$ nm and almost invariably reacts with e_s^- in the reverse of equation (9.5). The disadvantages (b) and (c) can in some circumstances be removed or reduced by addition of a reagent which will destroy D^+. In the case of $D = OH_{aq}^-$ the use of hydrogen is especially convenient because $D^+ = OH$ reacts with H_2 according to equation (9.7)

$$OH + H_2 \rightarrow H_2O + H; \, k_7 = (5 \pm 1) \times 10^7 \, M^{-1}s^{-1} \qquad (9.7)$$

and if the solution is sufficiently alkaline reaction (9.8) will occur ensuring that the only product is e_s^-.

$$H + OH_{aq}^- \rightarrow e_{aq}^-, \, k_8 = 1.8 \times 10^7 \, M^{-1}s^{-1} \qquad (9.8)$$

9.2.4 Radiolytically

The most widely applicable method is that of radiolysis in which the molecules of the medium are ionised by exposure to α-, β- or γ-rays from natural or artificial radio-isotopes or to high-energy charged particles, e.g. e^-, $_1^1H$, $_1^2D$, $_2^4He$, which emerge from an accelerator. Accelerators not only deliver high currents but can usually be operated in pulsed modes so that the (somewhat more expensive) analogue of flash photolysis, namely, pulse radiolysis can be used to obtain information about the optical and kinetic properties of e_s^- [13]. The very considerable advantages of this method of generating solvated or trapped electrons are: (a) no photoionisable solute, D, is required and therefore disadvantages (b) and (c) of the photochemical method are automatically eliminated, (b) the ionisation process is unaffected by the state of aggregation so that the method can be used with gases, liquids and crystalline or amorphous solids with equal facility, and (c) in multi-component systems each component contributes electrons roughly in proportion to the product of the mole fraction and the number of electrons per molecule so that the solute in a dilute solution is virtually unaffected directly by the incident radiation, all the e_s^- being derived from ionisation of solvent molecules (equation (9.9)).

$$S \rightsquigarrow S^+ + e^- \qquad (9.9)$$

Although S^+ may persist briefly as a positive hole, it frequently decomposes or reacts to form a stable ion and a radical which, in the case of aqueous systems, is OH and can therefore be converted to e_{aq}^- by reactions (9.7) and (9.8) as in the photochemical case.

There are three features of the ionisation process in radiolysis which make this method more complicated than photoionisation. In the first place the ionisation events do not occur uniformly throughout the volume but take place in the wake of the primary or secondary fast charged particle. This is the more marked the higher the Linear Energy Transfer (L.E.T. = energy deposition per unit length of track) of the radiation but even for radiation of low L.E.T. such as ^{60}Co γ-rays or 3 MeV electron beams c. 50% of the ionisations occur in clusters of on average 5 ionisations within a small volume element. In the second place the electron which emerges from reaction (9.9) has much higher kinetic energy than that formed by either process (9.5) or (9.6) and it therefore travels some distance from the site of its parent molecule before it has lost sufficient energy to permit solvation. Finally, ionisation may be accompanied or followed by formation of excited states of solvent molecules and these or the free radicals which they may form by dissociation may cause side reactions. However, these three complications are well understood and there are various devices which can be adopted either to minimise their effect or to make allowance for them and the radiation chemical method is still the most effective means of generating solvated and trapped electrons.

Representative examples of these methods and of the principles underlying their use are to be found in references 4–13.

9.3 RELEVANT PHYSICAL PROPERTIES OF e_s^- AND e_t^-

To understand fully the factors which govern the extent and rate of reaction into which solvated electrons enter, it is necessary to know the values of its thermodynamic functions, its effective charge and certain kinetic properties which regulate its movement through the system. Additionally, rate constants should be measured with as high a precision as possible over a wide variation of experimental conditions. For the hydrated electron and to a slightly lesser extent the ammoniated electron this information is more complete than for very many common reagents. In this Section we summarise these properties for the former species on the assumption that this will be of greater interest to chemists concerned with reaction kinetics.

9.3.1 Charge

The *charge* of the solvated electron in several solvents has been found to be -1 by studying the effect of changing ionic strength on the rate constant for the reaction of e_s^- with a solute A of ionic charge Z_a,

$$e_s^- + A^{Z_a} \rightarrow \text{Product a} \qquad (9.10)$$

since, as Brønsted and Bjerrum showed, this should be given by equation (9.11) for low ionic strengths and assuming that the distance of closest approach to e_s^-, A and the activated complex are all equal to a.

$$\ln k_{10} = \ln k_0 - Z_a \frac{\alpha'}{(DKT)^{\frac{3}{2}}} \sqrt{\mu}/(1 + \beta a \sqrt{\mu}) \qquad (9.11)$$

Here α' is a universal constant, $\beta\sqrt{\mu}$ is a measure of the inverse of the thickness of the ion atmosphere and $\beta = (4\pi/DKT)^{\frac{1}{2}}$. In general $\ln k_{9.10}$ has been found to be linearly dependent on $\sqrt{\mu}/(1 + \sqrt{\mu})$ and to have the expected slope which is of course larger the lower the dielectric constant (D) of the medium[14].

9.3.2 Ion-atmosphere relaxation time

The validity of equation (9.11) rests on the assumption that the solvated electron has a fully developed ion atmosphere and that the activated complex has time to form its ion atmosphere. However, in practice when photochemical or radiation chemical methods of generating e_s^- are used, a solution of A in the solvent is prepared and either illuminated or irradiated so that the half-life of electrons before reaction is $0.69/k_{10}[A]$, and as $[A]$ is increased, a point may be reached at which this time becomes smaller than the ion–atmosphere relaxation time, τ, when the effective activity coefficient of the solvated electron will be unity and independent of ionic strength, whilst that of the activated complex will correspond to that of an ion with a central charge, $Z_a - 1$, but the unperturbed ion atmosphere of the original ion of charge Z_a. For reactions in water at 25 °C where $\alpha'/(DKT)^{\frac{1}{2}}$ $= 2.303 \times 0.510$ this simple theory predicts that $\log_{10} k_{10}$ should change from the value given in equation (9.11) by an amount equal to $\Delta \log_{10} k_{10}$ given in equation (9.12) as the concentration of A is changed to carry $0.69/k_{10}[A]$ through τ.

$$\Delta \log_{10} k_{10} = 0.51 \left\{ \frac{Z_e(Z_a - 1)\mu^{\frac{1}{2}}}{1 + a_+ \beta \mu^{\frac{1}{2}}} + \frac{\mu^{\frac{1}{2}}}{1 + a_e \beta \mu^{\frac{1}{2}}} \right\} \tag{9.12}$$

Figure 9.2 shows that this effect has been observed for the fast reaction of e_{aq}^- with $I_{3,aq}^-$ by use of the reaction between e_{aq}^- and N_2O as a reference reaction at constant μ. The values of τ which data of this kind suggest are in reasonable agreement with the Debye–Hückel–Falkenhagen theory and lead to $\tau \approx 10^{-10} D/\mu \Lambda_\infty$ at 25 °C [15]. However the theory outlined here is highly

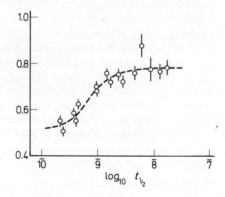

Figure 9.2 The variation of $\log_{10} k(e_{aq}^- + I_3^-)/k(e_{aq}^- + N_2O)$ with the half life $(= 0.69/(\Sigma k_i[A_i])$ of the hydrated electron at 25 °C in water at a constant ionic strength of 0.23 (From Dainton and Logan[15], by courtesy of The Royal Society)

approximate and for a more refined treatment reference 16 should be consulted. Very recently, Hunt et al., by use of pulse-radiolysis techniques of very high time resolution, have been able to observe this effect directly, i.e. without the use of a competing reference reaction[17].

9.3.3   Optical spectrum

The *spectrum* of the solvated electron has been measured for a variety of solvents and is independent of the method of formation provided only that when method 9.2.1 is used to produce stable solvated electrons the concentration of e_s^- is sufficiently low. It always consists of a broad continuum (width at half height much greater than predicted by theory, e.g. in water it is 0.9 eV) with a surprisingly long high-energy tail*. λ_{max} varies from c. 500 to >2000 nm as the polarity of the medium decreases as illustrated in Figure 9.3 and is in qualitative agreement with theory. The origins of the

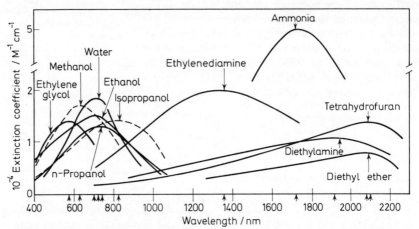

Figure 9.3 The absorption spectrum of e_s^- in various polar solvents (From Dorfman et al.[18(a)], by courtesy of Verlag Chemie GmbH)

shape of this spectrum are discussed below. For the reaction kineticist the most significant feature is that the extinction coefficient is always sufficiently large at λ_{max} for micromolar concentrations of e_s^- to be measured with path lengths of a few centimetres. The spectrum of e_t^- in glassy solids is of similar shape and extinction coefficient but displaced to shorter wavelengths than that of the corresponding liquid. Apart from its theoretical significance, this fact is extremely useful because thermal annealing of these glasses, containing solutes with which e_s^- at room temperature would react very rapidly, allows these reactions to be carried out 'in slow motion' and the details of their mechanisms to be more thoroughly explored. Additionally, as described below, a study of spectral changes in warmed or illuminated glasses containing e_t^- provides information about the trapping and mobilisation processes.

9.3.4   E.S.R. spectrum

The *e.s.r. spectrum* of a free electron is a narrow singlet. A narrow singlet is observed for e_{aq}^- [19] and a broader singlet is characteristic of e_t^- in a variety

---

*There is also evidence of a second absorption band with $\lambda'_{max} < 200$ nm for e_{aq}^- [18b].

of glasses, although in a few cases there is some limited evidence of weak satellites which afford clues as to the number and orientation of the molecules in the cavity wall[5]. More striking evidence on this point is provided by the effect on the line-width of isotopic substitution in solvent molecules. Thus in aqueous glasses based on H_2O the line width is 14 G whereas in D_2O-based glasses it is only 6 G, clearly indicating that the inside of the cavity wall is, as expected from the polarity of the O—H dipole, lined with H rather than O atoms. This is even more convincingly shown in alcohol glasses where the line width of ROD glasses is less than that of ROH glasses irrespective of whether the R group contains C—H or C—D bonds[20]. The degree of interaction of the electron and the spin of the nuclei forming the cavity wall will depend upon the cavity size and both the line width and the spin-lattice relaxation times are generally expected to diminish the smaller the cavity (vide infra). That the relaxation time of the microwave excited electrons is longer than that of other free radicals trapped in the glass, is shown by the lower microwave power at which the e_t^- singlet signal saturates.

The e.s.r. signal is also useful as an indicator of the changes undergone on illumination or thermal annealing of e_t^- or on reaction of e_t^- with a solute when the product can often be uniquely characterised and its optical spectrum correctly assigned.

9.3.5   Equivalent conductivity, Stokes radius and diffusion constant

Of great importance to the rates at which electrons in liquid react is their *mobility*. For more than half a century physicists have measured the steady-state currents which develop between two charged plates immersed in non-conducting liquids when the latter are continually irradiated. By studying the effects of radiation intensity and electrical-field strength on these currents it is possible to estimate both the volume-recombination coefficients of e^- with positive ions and the electrical mobilities of the charge carriers. An alternative method is to use high-intensity pulsed-radiation sources and study the rise and post-pulse decay of the current. The results are clear-cut but in some respects perplexing. Mobilities ($cm^2 V^{-1} s^{-1}$) in the range 10^0–10^{-2} are found for helium, hydrogen and hexane and it must be concluded that e_s^- in these systems is localised whereas for Ar, Kr, Xe and liquids comprised of large spherical molecules, such as CMe_4 or $SiMe_4$, the mobility is in the range 50–2000 and the electron is probably in the 'quasi-free' state[8, 21].

The values for the hydrated and ammoniated electrons are of special interest because of the relationships between electrical mobility, u and diffusion constant, D, and Stokes-law radius, r_s, embodied in equations (9.13) and (9.14)

$$u' = 1.6 \times 10^{-12} D/kT = 1.04 \times 10^{-5} \lambda/cm^2 ohm^{-1} equiv^{-1} \qquad (9.13)$$

and $$r_s = F^2/6\pi N\eta\lambda = \frac{0.82/\text{Å}}{\eta/\text{P} \times \lambda/cm^2 ohm^{-1} equiv^{-1}} \qquad (9.14)$$

and because of their significance for the rates of the diffusion controlled reactions into which these species can enter. The stability of dilute metal–liquid ammonia systems permits direct measurement of Λ_∞ [22] and at $-37\,°C$ this is $c.$ 1.0×10^3 and by combining this with the known viscosity of ammonia (0.0025 P) and the measured cationic transport number it is found that for e^-_{amm}, $u' \approx 10^{-2}$ cm^2 V^{-1} s^{-1}, $D \approx 2 \times 10^{-4}$ cm^2 s^{-1} and $r_s \approx 0.35\,Å$. The hydrated electron even in the purest water has a half-life of about 0.5 ms so that in this case it is necessary to use a pulse-conductivity method and the counter ion is H^+_{aq}. The values for e^-_{aq} at $25\,°C$ are $u' = 1.8 \times 10^{-3}$ cm^2 V^{-1} s^{-1}, $D = 4.8 \times 10^{-5}$ cm^2 s^{-1} [23] and hence $r_s = 0.5\,Å$. Evidently both e^-_{aq} and e^-_{amm} are localised but the (unmodified) Stokes-law radii are much smaller than the theoretically predicted size of the cavity and therefore also smaller than the average size of the wave functions. It must be concluded that when both these species diffuse, the solvation shell is not preserved but disintegrates and reforms rapidly. The walls of the cavities must therefore be in a continual state of fluctuation and at any moment there will be a distribution of cavity-wall configurations (*vide infra*). The frequency of these fluctuations is probably very high ($> 10^{12}$ s^{-1}) and is expected to be related to the solvation time (*vide infra*).

9.3.6 Thermodynamic properties

The thermodynamic properties of e^-_s are known with any certainty only for the aquated and ammoniated electron. In principle, those of the latter are obtainable by the application of standard methods to stable solutions of alkali metals but in practice the existence of various equilibria make the analysis of the data rather complicated[24]. The thermodynamic data for e^-_{aq} are obtained indirectly from pulse-radiolysis measurements of rate constants as the following illustrates[25].

By measuring (a) the rate of development of the characteristic spectrum of e^-_{aq} in hydrogen saturated alkaline solutions in which $c.$ 10^{-5} molar H atoms are produced during the pulse by reaction (9.7) and (b) the variation of this rate with $[OH^-]$ $k_{9.15}$ is found to be 2×10^7 M^{-1} s^{-1}.

$$H_{aq} + OH^-_{aq} \rightleftharpoons e^-_{aq} \tag{9.15}$$

Since the half-life of e^-_{aq} in the purest water is 880 μs, $k_{-9.15} = 810\,s^{-1}$ and hence $K_{9.15} = 2.5 \times 10^4$ M^{-1} and $\Delta G^0_{9.15} = -25.2$ kJ mol^{-1}. Combining $K_{9.15}$ with the ionic product of water we obtain $\Delta G^0_{9.16} = +54$ kJ mol^{-1}.

$$H_{aq} \rightleftharpoons H^+_{aq} + e^-_{aq} \tag{9.16}$$

and, since $\Delta G^0_{9.17} = 210$ kJ, we derive $\Delta G^0_{9.18} = 265$ kJ and hence the

$$\tfrac{1}{2}H_{2,\,g,\,1\,atm} \rightarrow H_{g,\,1M\,ideal} \tag{9.17}$$

$$\tfrac{1}{2}H_{2,\,g,\,1\,atm} \rightarrow H^+_{aq} + e^-_{aq} \tag{9.18}$$

standard oxidation potential on the hydrogen scale of the aquated electron is 2.75 V and e^-_{aq} is therefore a more powerful reducing agent than H atoms.

The free energy of hydration of an electron, i.e. $\Delta G^0_{9.19}$, is equal to $\Delta G^0_{9.16} -$ (the free energy of hydration of H $(= -8 \text{ kJ}) - \Delta G^0_{9.20}$ $(= 230 \text{ kJ})$

$$e^-_g \rightarrow e^-_{aq} \tag{9.19}$$

and has the value 170 kJ

$$H_g \rightarrow H^+_g + e^-_g \rightarrow H^+_{aq} + e^-_g \tag{9.20}$$

which is smaller than that for many univalent anions. Furthermore if Noyes' semi-empirical formulae relating free energies, enthalpies and entropies of hydration to univalent anionic radii are applied $\Delta S^0_{9.19} \approx 12 \text{ J K}^{-1} \text{ mol}^{-1}$ and $\Delta H^0_{9.19} = 164 \text{ kJ}$ which is close to the energy of the photon of wavelength $= \lambda_{max}$ in the absorption spectrum of e^-_{aq} [26].

Hentz[27] and his collaborators have found that pressures up to 6000 bar have no effect on the relative rates of reactions (9.15) and (9.21) where $RH = Me_2CHOH$, HCO^-_2 or $PhCH_2OH$

$$H + RH \rightarrow H_2 + R \tag{9.21}$$

and therefore the volumes of activation of these two reactions are identical. Since previous work had shown $\Delta V^{\ddagger}_{9.21} = -5.9 \text{ cm}^3$ and Hentz et $al.$ had already demonstrated that reaction ·9.15 is accelerated by increasing pressure to an extent which corresponds to $\Delta V^{\ddagger}_{9.15} = -14.2 \text{ cm}^3$ it follows that $\Delta V_{9.15} = 8.3 \text{ cm}^3$. The partial molal volumes of OH^-_{aq} and H_2O are well established and making a reasonable estimate of $\overline{V}_H$ they obtained a partial molal volume for e^-_{aq} between -1.7 and $+2.7 \text{ cm}^3$. If allowance is made for electrostriction, this is not inconsistent with a cavity radius of $c.$ 1 Å. Although there is still considerable argument about the concentration dependence of the molar volume of metal–ammonia solutions there is little doubt that the size of the cavity of e^-_{amm} is considerably larger than that of e^-_{aq} and recent measurements by Böddeker of the isothermal compressibility of these solutions accord with the notion of e^-_{amm} as a 'large soft rubber ball' [28].

It is easily shown that the thermal expansion coefficient of the cavity, α, is equal to

$$\frac{3}{R} \cdot \frac{d(ch/\lambda)/dT}{d(ch/\lambda)/dR}$$

and similarly its isothermal compressibility, β, is equal to

$$\frac{3}{R} \cdot \frac{d(ch/\lambda)/dP}{d(ch/\lambda)/dR}.$$

Taking the values of $d(ch/\lambda)/dR$ and R calculated by Jortner et $al.$[3] for liquid ammonia and for water and the observed blue shift of λ_{max} for e^-_{amm} and e^-_{aq} caused by lowering the temperature[36] or increasing the pressure[59] to calculate the numerator in these expressions, values of α and β are obtained for both media which are much larger than those of the pure liquids. These results may be taken to imply that the cavity containing the electron is more isothermally compressible and thermally expansible than the holes in the bulk liquid, i.e. is 'softer' than the solvent.

9.4 EVIDENCE FOR A BROAD DISTRIBUTION OF CAVITY SIZES AND ENERGIES

9.4.1 In liquid ammonia

Attention has already been drawn to the discrepancy between the predicted shape of the absorption spectrum and the observed much broader and asymmetrical spectrum of e_s^- (see Figure 9.3). This discrepancy might be accounted for by assuming that there is a distribution of cavity sizes, each cavity having a discrete value of R, $\Delta E(2P \leftarrow 1S)$, and a transition probability which can in principle be calculated from the wavefunctions of the 1S and 2P states evaluated by Jortner and others. Rusch, Koehler and Lagowski[29] have shown that the distribution which resolves this discrepancy for the 1500 nm band of e_{amm}^- is given by equation (9.22) indicating that about 96% cavities have a radius between 2 and 4 Å and 20% lie between 3.20 and 3.45 Å.

$$n_i = N(R = 3.0 \text{ Å}) \exp \left\{ -4\ln 2(r_i - 3.00/1.15)^2 \right\} \tag{9.22}$$

9.4.2 In aqueous systems

Such calculations have not yet been made for e_{aq}^-, but there is adequate experimental evidence to show that e_t^- in various glasses exists in holes of different sizes and it is a likely presumption that this is also true for e_s^-. The proof consists simply in perturbing the equilibrium distribution under conditions where the recovery rate is very low.

9.4.2.1 Photoredistribution of e_t^-

Early investigations had revealed that illumination with white light of electrons trapped in organic glasses caused some bleaching of the characteristic colour of e_t^-. Although this was partly due to photodecomposition, especially in alcoholic glasses where reaction (9.23) occurs; by use of filtered light it was shown that selective photobleaching of the red end of the e_t^- spectrum could be achieved[30].

$$e_t^- + h\nu \rightarrow RCHO^- + H_2 \tag{9.23}$$

When similar techniques were applied to alkaline aqueous glasses at 77 K, it was possible to achieve a fully compensatory condition in which no e_t^- were destroyed (i.e. no loss of spins) and the spectrum merely changed shape whilst $\int$O.D. dν remained constant[31]. Figure 9.4 shows typical results for e_t^- formed either photochemically from $[Fe(CN)_6]^{4-}$ ions or by γ-irradiation and illustrates that by using red light the spectrum is shifted to the blue, i.e there is a loss of absorption at the red end and again at the blue end. By use of blue light the process is reversed. This 'photo-shuttle' may be represented by equation (9.24).

$$e_{t,i}^- \underset{h\nu_j}{\overset{h\nu_i}{\rightleftharpoons}} e_{t,j}^-; \, \nu_i < \nu_j \tag{9.24}$$

It is presumed that red light causes redistribution of cavities to favour narrower and deeper holes whilst blue light causes the reverse. That there

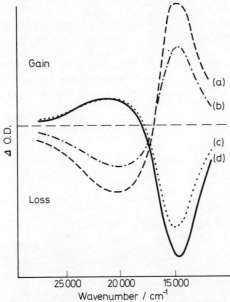

Figure 9.4 Spectral changes produced by illumination of e_t^- in 10 M aqueous hydroxide glass at 77 K using $\lambda > 640$ nm (curves c and d) and $\lambda < 500$ nm (curves a and b). e_t^- produced by γ-irradiation for curves a and d and by photodetachment from $[Fe(CN)_6]^{4-}$ for curves b and c (From Buxton et al.[31], by courtesy of The Faraday Society)

is a real change in hole distribution is shown by the fact that the e.s.r. signal of $e_{t,i}^-$ saturates at a lower microwave power than that of $e_{t,j}^-$ [31].

9.4.2.2 *Photomobilisation*

The question arises as to whether a photon merely changes the shape of the cavity whose electron it excites or whether it ejects the electron which then moves through the medium and is subsequently trapped at another site. Figure 9.5 depicts this latter process and since these systems show

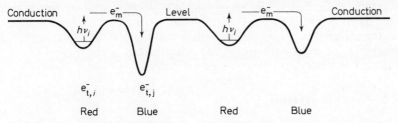

Figure 9.5 Potential energy diagram to illustrate photoshuttling and photoconductivity in alkaline aqueous glasses.

photoconductivity when illuminated with light of wavelength falling within the e_t^- spectrum[32] equation (9.24) should be replaced by (9.25),

$$e_{t,i}^- + hv_i \rightarrow e_m^- \rightsquigarrow e_{t,j}^-$$

$$(9.25)$$

$$e_{t,j}^- + hv_j \rightarrow e_m^- \rightsquigarrow e_{t,i}^-$$

where e_m^- denotes a photo-ejected and therefore free or quasi-free but certainly mobile electron. Confirmation of this mechanism is provided by the fact that the quantum yield of permanent photobleaching diminishes with time but the shape of the optical density/I_{abs} curve is independent of the concentration of e_t^- initially present, as shown in Figure 9.6. Evidently

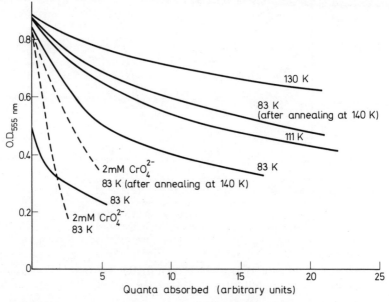

Figure 9.6 Photobleaching of e_t^- in 10 M aqueous hydroxide glass with light of $\lambda = 500–620$ nm at various temperatures and in presence or absence of a solute (CrO_4^{2-}) reactive to e_m^- (From Buxton et al.[31], by courtesy of the Faraday Society)

in a steady light flux electrons 'hop' from trap to trap and those which are nearest to a solute with which they can react (either the conjugate positive hole or radical O^- or a deliberately added solute) are destroyed first so that the quantum yield for photobleaching progressively diminishes. Clearly the greater the possibility of retrapping a mobile electron the smaller will be the mean 'hopping distance' and the lower the bleaching quantum yield. The typical results displayed in Figure 9.6 suggest that this retrapping probability increases with increasing temperature, a conclusion which is supported by other data.

9.4.3 In other systems

Although photolysis of the traps may be a complicating feature which is absent from aqueous systems, evidence is accumulating that mechanisms similar to those described for aqueous glasses can also operate in non-aqueous glasses. Thus one or more of the phenomena of photoconductivity, photobleaching, photo- and electrophoto-luminescence and photo-perturbation of the e_t^- spectrum have been observed in glasses based on alcohols, ethers and hydrocarbons[33]. There is no evidence to contradict the conclusions that the observed asymmetric spectrum is partly due to a wide trap-distribution, that, if it also corresponds to 2P ← 1S transition, the 2P level is very close to the conduction band, and that mobile electrons can react with solutes or be retrapped in the medium with an efficiency which increases with increasing temperature.

9.5 RATE PROCESSES NOT INVOLVING CHEMICAL REACTION: SOLVATION, RELAXATION AND MOBILISATION

We have mentioned that the observed Stokes radius is too small to be consistent with the idea that each e_s^- diffuses complete with its own intact solvation shell. Equally, since processes which depend upon finite diffusion rates have frequently been observed at temperatures where $\exp(-ch/\lambda_{max} kT) \leqslant 10^{-50}$, translational displacement cannot occur by a mechanism in which the electron becomes thermally excited to the 2P level. Most probably electrons can move by escape from one cavity to a neighbouring potential binding-site by some rotational motion of the solvent molecules which destabilises the donor cavity and stabilises the potential recipient in the course of which the energy levels of both are temporarily matched, thus permitting tunnelling. For a better understanding of this process it becomes important to know something of the nature and rate of thermally induced configurational changes of the cavity. There are two distinguishable ways in which this can be done. First, trapped electrons in glasses can be warmed to a temperature at which their initial configuration is not the equilibrium one and the rate of change is sufficiently slow to be studied by optical and/or e.s.r. spectroscopy. Secondly, the actual solvation process of free electrons can be studied at any temperature.

9.5.1 Optical and e.s.r. spectroscopic changes produced by warming e_t^-

Although the localisation process outlined in Section 9.1 is exothermic, the final configuration of an electron trap in a low-temperature rigid glass will be a function of the initial shape of the unoccupied hole and the competition between the molecular motion induced by the exothermicity of the trapping process and the quenching of this motion by the rate of heat dissipation. It would be surprising if the resultant trap depth distribution initially observed in glasses were the equilibrium distribution appropriate to all temperatures of the glass and it would therefore be expected that warming of the glass sufficient to allow limited slow molecular motion would cause a redistribution of trap sizes and consequently a change in the absorption spectrum.

Changes of this kind have been observed. Thus when an alkaline aqueous glass γ-irradiated at 77 K is warmed there is a loss of absorption at the wings and a gain in the centre which, because the loss is preferentially from the red end, results in a slight blue shift of λ_{max} which is accompanied by the expected change in the e.s.r signal-saturation characteristics[31]. Since an increase of temperature in a liquid system always causes a red shift of λ_{max} this effect must be due to a slow adjustment of the metastable distribution to the equilibrium distribution. It is significant that if the spectrum at 77 K is perturbed by, for example, photoshuttling or selective photobleaching, the spectrum ultimately attained in the warmed glass is always the same. The most dramatic red-shifts are observed when the glass is irradiated at 4.2 K which is a sufficiently low temperature to ensure trapping of electrons in cavities which would be too shallow to confine them for long periods at 77 K. Warming even to 77 K then causes a blue shift of considerable magnitude. Higashimura and his colleagues[34] have demonstrated a shift of λ_{max} for e_t^- in methanol glasses from 700 to 530 nm when the temperature is raised from 4 to 77 K. A similar effect but in a different wavelength region was observed with 2-methyltetrahydrofuran. In ethanol and equimolar glycol–water glasses, the 4 K glass shows a marked infrared band centred about 2000 nm which disappears on warming the sample to 77 K and

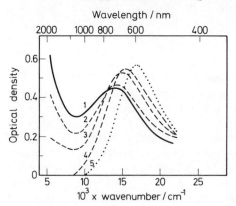

Figure 9.7 The temperature induced irreversible change in the optical absorption spectrum of e_t^- generated in glycol–water glass at 4 K. Spectra measured at 4 K (1), 20 K (2), 30 K (3), 40 K (4) and 77 K (5) (From Hase et al.[34], by courtesy of The American Institute of Physics)

augments the existing visible absorption (see Figure 9.7). This change is accompanied by a very striking broadening of the e.s.r. signal; in the glycol–water case the change is from 3 to 14.3 G. Neither the optical nor e.s.r. spectral changes can be reversed on re-cooling the glasses to 4 K.

9.5.2 Isothermal optical changes

Attempts have been made to use the technique of pulse radiolysis to measure the rate of development of the spectrum after the pulse. This is easily achieved in glasses with apparatus with time resolution of only 10–100 ns and marked spectral changes have been observed in hydrocarbon, ether, alcohol and aqueous glasses[35, 36]. When allowance is made for instrumental artifacts, it seems that these changes involve a blue shift, i.e. loss of absorption from the red end of the spectrum in accord with expectations based on the discussion in Section 9.5.1.

Of course liquids are of much greater interest than glasses in this connection. So far even time resolution[17] of 10 ps has failed to reveal any changes in the spectrum of e_s^- in water, glycol or C_1–C_4 alcohols at room temperature. Also Walker and Kenney-Wallace[37] by use of an ingenious technique of subjecting a known concentration of e_{aq}^- to a very intense flux of light of $\lambda = 694$ nm for 20 ns saw no change in the spectrum and therefore concluded that if photons of this wavelength, which is close to λ_{max}, were desolvating electrons according to the analogue of reaction (9.25) then the half-life of the resolvation process was < 5.5 ps.

Very recently the solvation process has been seen in cooled liquids; the most striking results are those of Baxendale and Wardman[38]. The end-of-pulse spectrum in n-propanol at 152 K (curve (a) in Figure 9.8) changes with

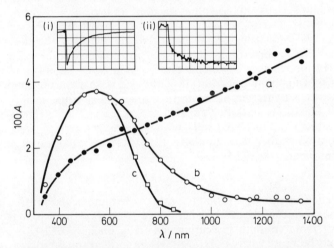

Figure 9.8 The absorption spectra observed after a 5 ns pulse of 12 MeV electrons into n-propanol at 152 K. Curve (a) end-of-pulse spectrum. Curve (b) 200 ns after pulse. Note that dielectric relaxation time at this temperature is 750 ns estimated from reference 55(c) (From Baxendale and Wardman[38], by courtesy of Macmillan Journals Limited)

a half-life of 60 μs to a spectrum closely resembling that of e_s^- appropriate to liquid propanol at this temperature (curve (b)) and they conclude (a) that at least half of the traps which are ultimately formed exist at 10^{-8} s, (b) that the red traps are unstable and that the further solvation of these is faster than would be expected from the extrapolated value of the dielectric relaxation time of propanol at this temperature[55c]. The second conclusion leads these authors to suggest that, as mentioned in Section 9.5.1, the exothermicity of the trap-deepening process permits local heating which accelerates the rotation of the solvent molecules.

9.6 CHEMICAL REACTIONS OF THE SOLVATED ELECTRON

e_s^- will react with a wide range of solutes. The rate constants of these reactions may be measured for stable e_s^-, e.g. in ammonia and amines by use of flow techniques with optical detection. The immediate products may be identified by e.s.r. and optical spectral studies on warmed irradiated glasses containing the solute of interest. Flash photolysis and pulse radiolysis can be utilised to measure the rates of reaction of short-lived e_s^- and also to identify the products of the reaction and their subsequent fates. Literally hundreds of reactions of e_s^- have been investigated[39] and of this large body of knowledge which has been accumulated there is space to discuss only the salient features[40].

9.6.1 Types of reaction

These are classified in Table 9.1. Reactions (9.26) and (9.27) have been written as unimolecular processes but given second-order rate-constants. Their mechanisms are unknown but the fact[41] that $E_{9.26} = 36$ kJ mol^{-1} and $\Delta S_{9.26}^{\ddagger} = -131$ J deg^{-1} mol^{-1} suggest that reaction (9.26) involves a motion of the molecules of the cavity which involves shrinkage of the cavity and brings some molecules into a position favourable to hydration of the OH$^-$ ion which is produced. The fact that $\Delta G_{9.26}^0$ is only 25 kJ mol^{-1} indicates that for many reactions in which e_{aq}^- is an electron donor it may also serve as a hydrogen atom donor with almost equal facility. As the tabulated equations indicate, most of the reactions are electron capture processes but reaction (9.39) may be an example where e_{aq}^- surrenders a hydrogen atom rather than an electron, i.e. perhaps equation (9.39) should be written as equation (9.54).

$$e_{aq}^- + C_2H_3 \cdot CONH_2 \rightarrow C_3H_6ON + OH^- \qquad (9.54)$$

The evidence for this view is that the rate of the bimolecular mutual destruction of the product of the reaction between e_{aq}^- and acrylamide is unaffected by change of ionic strength indicating that it is uncharged[42]. However, this may merely be a case where $C_3H_5ON^-$ is first formed and rapidly protonated by the solvent before it combines with another radical.

Solvated electrons react rapidly with other odd-electron species including the parent molecular ion (reaction (9.28)) when an excited molecule is produced which Bullot and Albrecht[33] have shown may then luminesce. Reactions (9.53) and (9.53(a)) are important in the radiolysis of water and it is still an open question as to whether the former involves the spin-paired dielectron $e_{2,aq}^{2-}$ as an intermediate although there is little doubt that e_2^{2-} can be trapped in solid matrices[43].

In all media, solvated electrons are rapidly converted to hydrogen atoms by reaction with H_s^+ (cf. reactions 9.29 and 9.33). The rate constant for reaction of e_{aq}^- with an acid, HA $\rightarrow$ H + A$^-$ (cf. reactions (9.30), (9.31) and (9.32)), decreases with increase of pK_{HA} according to the Brønsted equation, $k = G K_{HA}^{0.51}$ [44]. Non-dissociative capture processes by molecules, e.g. reactions (9.34)–(9.37), may well have relatively simple mechanisms but capture by cations, e.g. reactions (9.43) and (9.44), are probably more complex because of the considerable energy changes of the orbitals about a cation where it accepts the electron and the resultant effect on the ligands[40]. Not surprisingly, therefore, there is considerable selectivity in these reactions and many of them, e.g. with Zn^{2+}, Mn^{2+}, Lu^{3+}, are probably

Table 9.1   Types of reaction of solvated electrons

Description	Equation number	Example	Medium	Rate constant at room temp. $(M^{-1}s^{-1})$
Decomposition of e_s^-	9.26	$e_{aq}^- \rightarrow H + OH_{aq}^-$	Water	1.4×10^1
	9.27	$e_s^- \rightarrow H + EtO_s^-$	Ethanol	7×10^3
Recombination with parent ion	9.28	$TMPD^+ + e^- \rightarrow TMPD + h\nu$	3MP Glass	
Reaction with acids	9.29	$e_{aq}^- + H_{aq}^+ \rightarrow H$	Water	2×10^{10}
	9.30	$e_{aq}^- + H \cdot CO_2H \rightarrow H + HCO_2^-$	Water	8×10^7
	9.31	$e_{aq}^- + HOAc \rightarrow H + AcO^-$	Water	8×10^7
	9.32	$e_{aq}^- + H_2PO_4^- \rightarrow H + HPO_4^{2-}$	Water	$\sim 10^6$
	9.33	$e_s^- + H_s^+ \rightarrow H$	Methanol	$4 \pm 1 \times 10^{10}$
Non-dissociative	9.34	$e_s^- \phi_2 \rightarrow \phi_2^-$	Ethanol	4×10^9
capture by a molecule	9.35	$e_{aq}^- + O_2 \rightarrow O_2^-$	Water	2×10^{10}
	9.36	$e_{aq}^- + CO_2 \rightarrow CO_2^-$	Water	8×10^9
	9.37	$e_{aq}^- + I_2 \rightarrow I_2^-$	Water	5×10^{10}
	9.38	$e_{aq}^- + C_2H_4 \rightarrow ?$	Water	$\sim 5 \times 10^6$
	9.39	$e_{aq}^- + C_2H_3CONH_2 \rightarrow ?$	Water	1.8×10^{10}
	9.40	$e_{aq}^- + C_6H_6 \rightarrow ?$	Water	$< 7 \times 10^7$
Non-dissociative	9.41	$e_{aq}^- + NO_3^- \rightarrow NO_3^{2-}$	Water	$\sim 10^{10}$
capture by an ion	9.42	$e_s^- + NO_3^- \rightarrow NO_3^{2-}$	Methanol	4×10^7
	9.43	$e_{aq}^- + Cd^{2+} \rightarrow Cd^+$	Water	5×10^{10}
	9.44	$e_s^- + Cd^{2+} \rightarrow Cd^+$	Methanol	1.5×10^{10}
Dissociative capture by a molecule	9.45	$e_{aq}^- + H_2O_2 \rightarrow OH + OH^-$	Water	1.2×10^{10}
	9.46	$e_{aq}^- + N_2O \rightarrow N_2 + O^-$	Water	9×10^{10}
	9.47	$e_{aq}^- + C(NO_2)_4 \rightarrow$ $NO_2 + C(NO_2)_3^-$	Water	5×10^{10}
	9.48	$e_{aq}^- + \phi Cl \rightarrow \phi + Cl^-$	Water	5×10^8
	9.49	$e_{aq}^- + \phi Br \rightarrow \phi + Br^-$	Water	4×10^9
	9.50	$e_{aq}^- + \phi I \rightarrow \phi + I^-$	Water	1.2×10^{10}
Dissociative capture by an ion	9.51	$e_{aq}^- + BrO_3^- \rightarrow BrO_2 + O^-$	Water	4×10^9
	9.52	$e_{aq}^- + S_2O_8^{2-} \rightarrow SO_4^- + SO_4^{2-}$	Water	$\sim 10^{10}$
Combination with an	9.53	$e_{aq}^- + e_{aq}^- \rightarrow H_2 + 2OH^-$	Water	5×10^9
odd electron species	9.53(a)	$e_{aq}^- + OH \rightarrow OH^-$	Water	3×10^{10}

activation rather than diffusion controlled. However, they are all sufficiently fast and the products frequently have sufficiently distinctive spectra to enable the exploration of the chemistry of highly unstable hyper-reduced ions.

9.6.2   Reaction rates

9.6.2.1   *Activation and diffusion control*

The forces between molecules, the interplay of which determines whether bonds are broken and made, are of short range. Bimolecular chemical reactions are therefore regarded as occurring between entities which, outside a reactive distance, r, move independently, unless the reagents are charged when allowance must be made for electrostatic interactions. It is assumed that at the distance, r, the molecules will react with a probability which depends on their energies, the relevant potential energy profile, and to a lesser degree, their mutual orientation. The time for reaction is therefore made up of two parts, namely t_d = time to diffuse to the distance r and t_r = time for reaction. Correspondingly we may write

$$k_{obs}^{-1} = k_r^{-1} + k_d^{-1} \qquad (9.55)$$

Two cases may be distinguished. First, when $k_d \gg k_r$, $k_{obs} = k_r$ and the reactions are not limited in rate by diffusion. For reactions of the solvated electrons with molecules of closely similar structure which come into this category the variation of k_{obs} with structure should therefore be similar to the variation of k_{obs} for reaction of a different nucleophile with the same molecules. This has been found for the reactions of e_{aq}^- with monosubstituted benzene (reactions (9.40), (9.48), (9.49), (9.50)) and o-, m- and p-substituted phenoxide ions[45], where ln k is linearly dependent on the Hammett σ-function. It is important to recognise that because a reaction is not diffusion controlled it is not necessarily activation-controlled. The reactions of e_{aq}^- with benzene derivatives is in fact one in which the changes in k_{obs} and hence k_r are not due to changes in the activation energy, which is always in the range 12–16 kJ mol^{-1}, but is caused by changes in the pre-exponential term[46]. The origins of this are not well understood and these reactions would repay further experimental and theoretical study.

9.6.2.2   Diffusion control

The second case is when $k_d \ll k_r$ and, therefore, the reaction is diffusion controlled. Three consequences flow from this; first, if electrons are instantaneously generated in solutions of a solute A, the observed rate constant should diminish from the initial 'true' value given by k_r to a steady state value $k_{ss} = k_d$. Secondly, the value of k_{ss} is given by the Debye–Schmoluchowski equation (9.56), in which r is the reactive distance, D is the relative diffusion-coefficient of A and e_s^-, i.e.

$$k_{ss} = 4\pi r D \delta / (e^\delta - 1) \qquad (9.56)$$

$D = D_A + D_e$ and $\delta = -Z_A/\varepsilon r k T$, where ε is the static dielectric constant. Thirdly, it follows that the temperature dependence of the rate will be regulated by the way in which r, D and ε vary with temperature.

Almost all the data hitherto reported are values of k_{ss} at room temperature. If r is taken to be 5 Å, the term $\delta/(e^\delta - 1)$ at 25 °C in water has values of 0.17, 0.45, 1, 1.9 and 3.0 for $Z_A = -2, -1, 0, +1$ and $+2$ respectively and, allowing for this, the value of D calculated from k_{ss} for most of the reactions which, *prima facie*, are diffusion controlled, is not far removed from 5×10^{-5} cm^2 s^{-1}. Better agreement can be obtained by adjusting r and D_A to take account of the size of the solute A. Moreover, there is reasonably good agreement between theory and experiment for the limited number of reactions of e_s^- with ionic solutes in solvents for which ε is markedly less than in water and for which comparison is possible.

9.6.2.3 Temperature dependence of k_d

Not many studies of the effect of temperature on k_{ss} have been made and most measurements refer to aqueous systems over a limited range of temperature. Interpretations of the results commonly assume that k_{ss} can be represented in Arrhenian form and that $E_a = RT^2(\text{d ln } k_{ss}/\text{d}T) = E_d = RT^2$ (d ln D/dT) has a constant value*. For water[47] and methanol[48] the agreement between the measured activation energies of reaction and diffusion is fairly good but it is obvious that for liquids which form glasses there is a finite temperature at which diffusion becomes infinitely slow (T_0) and therefore the Arrhenius equation is inapplicable to k_{ss}. Not surprisingly, the application of the Arrhenius equation to diffusion-controlled reactions just above T_0 leads to absurd values. For example in 2-methyltetrahydrofuran (MTHF) in the range 16–22° above T_0 $k_{ss} = 10^{34}\exp(-59 \ kJ/RT)$ [49] and in alkaline aqueous solution Ershov and Pikaev[50] found $k_{ss} = 5 \times 10^{21}$ exp

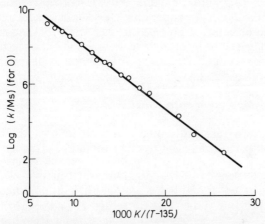

Figure 9.9 Test of equation (9.57) for the reaction of e_{aq}^- with nitrate ion over the temperature range 172 to 277 K (From Buxton et al.[36], by courtesy of The Faraday Society)

$(-70 \ kJ/RT) \ M^{-1} \ s^{-1}$ at 160–180 K, whereas at room temperature the expression would be $\approx 10^{12} \exp(-15 \ kJ/RT) \ M^{-1} \ s^{-1}$. Measurements over a wide temperature range[36, 51] show that ln k_{ss} plotted against T^{-1} is not a straight line but is concave to the origin and, in fact, ln k is linearly dependent on $(T-T_0)^{-1}$. Figure 9.9 illustrates these points. The data lead to constant and reasonable values of A_0, T_0 and B in equation (9.57) which are characteristic of the solvent as the data for 10 M aqueous hydroxide solutions in

$$k_{ss} = A_0 \exp[-B/(T-T_0)] \tag{9.57}$$

---

*In these treatments the term $RT^2\text{d ln}[\delta/(e^\delta-1)]/\text{d}T$ has been neglected. This is justified at low temperatures where it is much less than E_a but less justified at room temperature.

Table 9.2 show*. It is interesting to note that Jarrousseau and Valensi[52] find that a similar equation applies to the temperature dependences of the diffusion of gaseous oxygen, the equivalent ionic conductance and fluidity (inverse viscosity) in this medium, and additionally that there is independent evidence

Table 9.2   Values of A_0, T_0 and B in equation (9.57) for 10 M hydroxide solution

Reaction	$\text{Log}_{10} A_0/M^{-1} s^{-1}$	T_0/K	B/K
$e^-_{aq} + NO^-_{3\,aq}$	12.02 ± 0.04	135	870 ± 10
$e^-_{aq} + CrO^{2-}_{4\,aq}$	12.00 ± 0.08	135	760 ± 9
$e^-_{aq} + NO^-_{2\,aq}$	12.05 ± 0.09	135	840 ± 5
$e^-_{aq} + WO^{2-}_{4\,aq}$	10.97 ± 0.11	135	780 ± 16

derived from differential thermal analysis[53] and thermoluminescence[54] that 135 ± 5 K is a temperature at which aqueous systems become rigid solids. Equation (9.57) is also known to apply to the inverse viscosity (Tammann–Hesse–Vogel equation) of many hydrocarbons and polyhydric alcohols where it is often also valid for the inverse dielectric relaxation time[55]. The theoretical significance of the equation is a matter of argument but equations of this form can be deduced provided it is assumed that the liquid contains 'holes' and B may well be inversely related to the volume thermal-expansion coefficient of the medium[56].

9.6.2.4   Time-dependent rate constants

Noyes[57] has shown that the diminution with time of the observed rate constant of a diffusion controlled reaction should be given by

$$k_{obs} = (1 + k_r e^{x^2} \text{ erf } x/k_{ss}) k_{ss}/(1 + k_{ss}/k_r) \tag{9.58}$$

where $x = (Dt)^{\frac{1}{2}}(1 + k_r/k_{ss})/r$. For $x > 4$, which is a condition satisfied for glass-forming aqueous systems when $T < 250$ K and $t > 10^{-8}$s, e^{x^2} erf x is close to $(\sqrt{\pi} x)^{-1}$. Consequently, when a small concentration of e^-_{aq} is generated in such a medium containing an excess of A, the optical density (O.D.) of absorption spectrum of e^-_{aq} should decline with time according to

$$\ln(\text{O.D.}_0/\text{O.D.}) = [A] k_{ss}\{t + k_{ss}t^{\frac{1}{2}}/2(\pi D)^{\frac{3}{2}}\} \tag{9.59}$$

Figure 9.10 displays an example of this behaviour. Such graphs are useful not merely as a test of the theory[58] but because they allow r to be evaluated since

$$r = \{(\text{intercept})^2/16[A] \text{ slope}\}^{\frac{1}{3}} \tag{9.60}$$

Preliminary data suggest that r is medium dependent, being large in MTHF and changing with temperature in aqueous systems. In both systems conditions can be established for which r may be several tens of Å suggesting that at low temperatures electron tunnelling may play an important role[58].

---

*See footnote on previous page.

9.7 CONCLUSIONS

Because the electron is the simplest free radical, a strong reducing agent and a nucleophile, it is of intrinsic chemical interest and can be used to produce novel chemical species. Because of its charge and nature, it can be confined within a group of solvent molecules, it becomes localised and behaves like a rather mobile anion, but with an intense absorption spectrum which enables low concentrations to be measured. Because it can be produced in very short periods of time by pulse radiolysis or flash photolysis and can be trapped in glassy matrices, it is a uniquely valuable tool for the exploration of many rate processes including those rearrangements of solvent molecules involved

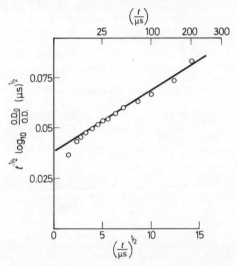

Figure 9.10 Test of equation (9.59) for the reaction of e_{aq}^- with nitrate ion at 189 K in 9.5 M aqueous lithium chloride solutions[58]

in its localisation or delocalisation and transport. Moreover, it enters into many diffusion controlled reactions, the study of which can enlarge our understanding of liquids, of transport through them and of electron tunnelling. In the 10 years since the first unambiguous identification of solvated electrons much has been learnt, probably more than about any single chemical compound in the same period of time, and it seems likely that it will continue to be an unusually useful tool for probing many problems in physical, inorganic, organic and industrial chemistry.

Acknowledgement

The author gladly acknowledges his indebtedness to many students and colleagues with whom he has been privileged to have discussions over the last few years, notably, Drs Buxton, Cattell and Salmon and without whom many researches mentioned here would have been impossible.

References

1. Bonin, M. A., Takeda, K. and Williams, F. (1969). *J. Chem. Phys.*, **50**, 5243
2. Jortner, J. (a) (1971). *Ber. Bunsenges. physik. Chem.*, **75**, 696; and (b) (1970). *Actions Chimiques et Biologiques des Radiations*, **14**, 7 (Paris: Masson et Cie)
3. Copeland, D. A., Kestner, N. R. and Jortner, J. (1970). *J. Chem. Phys.*, **53**, 1189
4. See, e.g. (a) *Solutions Metal–Ammoniac*, (1964), ed. by Lepoutre, G. and Sienko, M. J. (New York: Benjamin); and (b) *Metal–Ammonia Solutions*, (1970), ed. by Lagowski, J. J. and Sienko, M. J. (London: Butterworth)
5. Bennett, J. E., Mile, B. and Thomas, A. (1964). *Nature (London)*, **201**, 919
6. Dye, J. L., in Ref. 4(b), p. 1
7. Schmidt, W. F. and Schnabe . W. (1971). *Ber. Bunsenges. physik. Chem.*, **75**, 654
8. Holroyd, R. A. and Allen, M. (1971). *J. Chem. Phys.*, **54**, 5014
9. Froben, F. W. and Willard, J. E. (1971). *J. Phys. Chem.*, **75**, 35
10. Barker, G. C. (1971). *Ber. Bunsenges. Physik. Chem.*, **75**, 714
11. For a review see Stein, G. (1969). *Actions Chimiques et Biologiques des Radiations*, **13**, 119 (Paris: Masson et Cie); and Joschek, H. I. and Grossweiner, L. I. (1966). *J. Amer. Chem. Soc.*, **88**, 3261
12. Johnson, G. E. and Albrecht, A. C. (1966). *J. Chem. Phys.*, **44**, 3162, 3179
13. For illustrations of this technique see (a) *Pulse Radiolysis* (1965), ed. by Ebert, M., Keene, J. P., Swallow, A. J. and Baxendale, J. H. (London: Academic Press); and (b) Pikaev, A. K. (1967). *Pulse Radiolysis of Water and Aqueous Solutions* (London: Indiana University Press)
14. See e.g. Czapaki, G. and Schwarz, H. A. (1962). *J. Phys. Chem.*, **66**, 471; Collinson, E., Dainton, F. S., Smith, D. R. and Tazuke, S. (1962). *Proc. Chem. Soc.*, 140; Buxton, G. V., Dainton, F. S. and Hammerli, M. (1967). *Trans. Faraday Soc.*, **63**, 1191; Gordon, S., Hart, E. J., Matheson, M. S., Rabani, J. and Thomas, J. K. (1963). *J. Amer. Chem. Soc.*, **85**, 1375
15. Dainton, F. S. and Logan, S. R. (1965). *Proc. Roy. Soc. A*, **287**, 281
16. Logan, S. R. (1967). *Trans. Faraday Soc.*, **63**, 3004, 3009
17. Bronskill, M. J., Wolff, R. K. and Hunt, J. W. (1970). *J. Chem. Phys.*, **53**, 4201
18. (a) Dorfman, L. M., Jon, F. Y. and Wageman, R. (1971). *Ber. Bunsenges. physik. Chem.*, **75**, 681; (b) Nielson, S. O., Pagsberg, P., Hart, E. J., Christenson, H. and Nilsson, G. (1969). *J. Phys. Chem.*, **73**, 3171
19. Avery, E. C., Remko, J. R. and Smaller, B. (1968). *J. Chem. Phys.*, **49**, 951
20. Zeldes, H. and Livingston, R. (1959). *J. Chem. Phys.*, **30**, 40
21. Table 4, Ref. 2 (a) and Schmidt, W. F. and Allen, A. O. (1970). *J. Chem. Phys.*, **52**, 4788
22. Dewald, R. R., Ref. 4(b), p. 497
23. Barker, G. C., Fowles, P., Sammon, D. C. and Stringer, B. (1970). *Trans. Faraday Soc.*, **66**, 1498
24. (a) Jolly, W. L. (1965). *Solvated Electron*, 30 (Washington: American Chemical Society); (b) Lepoutre, G. and Demortier, A. (1971). *Ber. Bunsenges. physik. Chem.*, **75**, 647
25. Tables of rate constants of e_{aq}^- are to be found in Anbar, M. and Neta, P. (1965). *Int. J. Appl. Rad. Isotopes*, **16**, 227
26. Jortner, J. and Noyes, R. M. (1966). *J. Phys. Chem.*, **70**, 150; and Noyes, R. M. (1968). *Advan. Chem. Series*, **81**, 65
27. Hentz, R. R. and Brazier, D. W. (1971). *J. Chem. Phys.*, **54**, 2777
28. Böddeker, K. W. and Vogelsegang, R. (1971). *Ber Bunsenges. physik. Chem.*, **75**, 638
29. Rusch, P. F., Koehler, W. H. and Lagowski, J. J. (1970). Ref. 4(b), p. 41
30. Dainton, F. S., Salmon, G. A. and Zucker, U. F. (1968). *Chem. Commun.*, 1172
31. Buxton, G. V., Dainton, F. S., Lantz, T. and Sargent, P. (1970). *Trans. Faraday Soc.*, **66**, 2962
32. Eisele, I. and Kevan, L. (1970). *J. Chem. Phys.*, **53**, 1867
33. Dyne, P. J. and Miller, D. A. (1965). *Can. J. Chem.*, **43**, 2696; Skelly, D. W. and Hamill, W. H. (1966). *J. Chem. Phys.*, **44**, 2891; Baverstock, K. F. and Dyne, P. J. (1970). *Can. J. Chem.*, **48**, 2182; Fujii, S. and Willard, J. E. (1970). *J. Phys. Chem.*, **74**, 4313; Bullot, J. and Albrecht, A. C. (1969); *J. Chem. Phys.*, **51**, 2220; and Habersbergerova, A., Josimović, L. and Teply, J. (1970). *Trans. Faraday Soc.*, **66**, 656, 669
34. Hase, H., Noda, M. and Higashimura, T. (1971). *J. Chem. Phys.*, **54**, 2975; Smith, D. R. and Pieroni, J. J. (1967). *Can. J. Chem.*, **48**, 2723; and Yoshida, H. and Higashimura, T. (1970). *Can. J. Chem.*, **48**, 504

35. Richards, J. T. and Thomas, J. K. (1970). *J. Chem. Phys.*, **53**, 218
36. Buxton, G. V., Cattell, F. C. and Dainton, F. S. (1971). *Trans. Faraday Soc.*, **67**, 687
37. Kenney-Wallace, G. and Walker, D. C. (1971). *Ber. Bunsenges. physik. Chem.*, **75**, 634
38. Baxendale, J. H. and Wardman, P. (1971). *Nature (London)*, **230**, 448
39. See, e.g. the list given by Anbar, M. and Neta, P. (1967). *Internat. J. Appl. Rad. Isotopes*, **18**, 493
40. See *Solvated Electron*, (1965). (Washington: American Chemical Society)
41. Fielden, E. M. and Hart, E. J. (1967). *Trans. Faraday Soc.*, **63**, 2975; (1968). *Trans. Faraday Soc.*, **64**, 3158
42. Chambers, K., Collinson, E. and Dainton, F. S. (1970). *Trans. Faraday Soc.*, **66**, 142
43. Khodzhaw, O. F., Ershov, B. G. and Pikaev, A. K. (1967). *Acad. Nauk. SSSR Ser. Khim.*, 1882, 2253; Zimbrick, J. and Kevan, L. (1967). *J. Amer. Chem. Soc.*, **89**, 2483
44. Rabani, J. (1965). In Ref. 40, p. 242
45. Anbar, M. and Hart, E. J. (1964). *J. Amer. Chem. Soc.*, **86**, 5633
46. Anbar, M., Alfassi, Z. B. and Bregman-Reisler, H. (1967). *J. Amer. Chem. Soc.*, **89**, 1263
47. Ref. 46 and Thomas, J. K., Gordon, S. and Hart, E. J. (1964). *J. Phys. Chem.*, **68**, 1524
48. Cattell, F. C. and Janovsky, I., unpublished results
49. Dainton, F. S. and Salmon, G. A. (1965). *Proc. Roy. Soc.*, **285A**, 319
50. Ershov, B. G. (1971). *Progress and Problems in Contemporary Radiation Chemistry*, **2**, 261 (Prague: Czech. Academy of Sciences)
51. Kawabata, K., Okabe, S., Fujita, S., Horii, H. and Taniguchi, S. (1968). *Ann. Rep. Radiation Center Osaka Prefecture*, **8**, 70
52. Jarrousseau, J-C. and Valensi, G. (1969). *J. Chim. Phys.*, 33
53. Angell, C. A. and Sare, E. J. (1970). *J. Chem. Phys.*, **52**, 1058
54. Moan, J. (1971). *Ber. Bunsenges. physik. Chem.*, **75**, 668
55. See (a) Haward, R. N. (1970). *Rev. Macro. Chem.*, **C4(2)**, 191; (b) Haranadh, C. (1963). *Trans. Faraday Soc.*, **59**, 2728; and (c) Davidson, R. H. and Cole, D. W. (1952). *J. Chem. Phys.*, **20**, 1389
56. Cohen, M. H. and Turnbull, D. (1959). *J. Chem. Phys.*, **31**, 1164
57. Noyes, R. M. (1961). *Progress in Reaction Kinetics*, **1**, 129
58. Buxton, G. V., Cattell, F. C. and Dainton, F. S., unpublished results
59. Vogelsgesang, R. and Schindewolf, U. (1971). *Ber. Bunsenges. physik. Chem.*, **75**, 651 and Schindewolf, U., Kohrmann, H. and Lang, G. (1969). *Angew. Chem.*, **8**, 512

10
Relaxation Techniques

J. E. CROOKS
BP Chemical International Ltd., Surrey

10.1 INTRODUCTION

The advent of relaxation techniques, about 20 years ago, as heralded in a Discussion of the Faraday Society[1,2], was a breakthrough in chemical kinetics. Up to that time, the rate at which a chemical reaction could be followed was limited by the rate at which the reagents could be mixed. Even with the most sophisticated mixing techniques, it is very difficult to mix solutions in less than a millisecond. Relaxation techniques avoid the mixing problem, as the reactants are already mixed, and in equilibrium with their products. The system is perturbed by a rapid change in an external parameter, and the rate at which the system relaxes, i.e. moves to the new equilibrium position, gives a measure of the rates of the chemical reactions involved. Relaxation methods can be classified according to the nature of the external parameter which is perturbed, and also according to whether a single step or a continuously varying perturbation is applied.

A single-step perturbation can readily be made in temperature, pressure or electric field intensity. Nearly all equilibria are sensitive to temperature and pressure, the sensitivity being measured by ΔH^0 and ΔV^0 respectively, and ionic equilibria are also sensitive to electric field intensity by virtue of the second Wien effect. The experiments are so designed that the perturbation is as rapid as possible, generally occurring in a time very much less than that for the chemical relaxation. For small perturbations, the reagent concentrations after the perturbation, followed by, for example, spectrophotometry or conductimetry, obey a first-order law as the system moves to the new equilibrium, and the time constant of the exponential change is the relaxation time of the reaction, τ.

A continuously varying perturbation of sinusoidal form can also be applied to a system in equilibrium. An ultrasonic wave in a liquid gives a simultaneous variation of temperature and pressure, and an electromagnetic wave gives a variation in electric-field intensity. At low frequencies the equilibrium moves in phase with the applied perturbation, whereas at high frequencies the equilibrium has not enough time to adjust in either direction, and so is unaffected. However, if the frequency of the applied perturbation is of the same order of magnitude as the reciprocal relaxation time, the chemical reaction interacts with the applied perturbation. The phase of the transmitted wave lags behind the phase of the imposed sine wave, and energy is absorbed by the system, the maximum absorption occurring when the applied frequency is equal to $(2\pi\tau)^{-1}$.

A rather different kind of relaxation technique involves the interaction between a chemical reaction and the spin relaxation time of a nucleus in a magnetic field. A line in a·nuclear magnetic resonance spectrum is usually quite sharp, but is broadened if due to a nucleus affected by a chemical reaction in such a way that the nucleus may have more than one chemical shift during its relaxation. The width of the line gives a measure of the rate of the reaction.

These techniques were introduced at about the same time, but some have proved much more popular than others. The single-step temperature-jump and the n.m.r. line-broadening techniques account for most of the recent published work using relaxation techniques, the ultrasonic technique accounting for much of the rest. The commercial availability of apparatus might be one reason for this.

Fast reaction studies published up to 1967 have been extensively reviewed by Eyring and Bennion[3], and the biological applications of relaxation techniques have been reviewed by Schechter[4]. Detailed accounts of experimental methods are given in a recently published text[127], which, although it will not replace 'Weissberger'[116] as the Bible of fast reaction techniques, is a very useful supplement. It is perhaps in the biological fields that the most interesting and significant advances will be made, and it is for this reason that some such applications are discussed in this Review. This Review is intended to emphasise the variety of problems which can be tackled by the use of relaxation techniques.

10.2 SINGLE-STEP PERTURBATION TECHNIQUES

10.2.1 Temperature-jump

10.2.1.1 Experimental methods

Joule heating, caused by the rapid discharge of a capacitance at high voltage through an electrolyte solution, is employed in the vast majority of temperature-jump studies. The possibility of using a laser for heating, with its potentially enormous power output of over 100 MW, delivered over a few nanoseconds, has often been discussed, and some recent papers discuss such an application. Caldin, Crooks and Robinson[5] give a critical account of the drawbacks of laser heating. The main problem is the efficient conversion of the power output of the laser in the form of light in the visible or near infrared regions of the spectrum into thermal energy in the sample solution. If the laser is not Q-switched, so that it delivers its energy over several hundred microseconds, the light can be readily converted to thermal energy by a minute concentration ($\sim 10^{-6}$ mol l^{-1}) of inert dye. Each dye molecule is repeatedly raised to an excited state by a photon, then falls to the ground state by deactivating collisions with solvent molecules. If however the full potentialities of the laser are to be exploited, the laser must be Q-switched, so that it delivers its energy in a microsecond or less. Conversion of optical to thermal energy by the dye is no longer efficient, as each dye molecule has time to convert only one or two photons during the light pulse. Direct absorption of the laser light in the solvent is difficult, as all solvents are transparent to light of the wavelength emitted by the ruby laser (694 nm). Even water only feebly absorbs light of the wavelength emitted by the neodymium laser, having an optical density of 0.15 at 1.06 μm. In certain special circumstances, one of the reagents may be used to absorb the laser light, e.g. as in the work of Hoffmann, Yeager and Stuehr[6] on the complexation of Ni^{2+}. Hoffman, Yeager and Stuehr used conductimetric detection, more sensitive but less specific than the more common spectrophotometric detection. A most ingenious application of laser heating has been made by Rigler, Jost and De Maeyer[7]. The light from a Q-switched neodymium laser is conducted to a microcell through a light-pipe. The fibres of the light-pipe are arranged regularly around the circumference of the disc-shaped cell, so that the laser light is multiply reflected and passes many times through the aqueous sample solution. The solution is thus uniformly heated, and all the light is absorbed. The sample cell is small enough to be placed on the stage of a microscope, so that temperature-jump studies of single biological cells can be undertaken. There is always the danger that the concents of a biological cell might be damaged by the high power levels found in a laser pulse. Lee and Chance[8] have described a simple apparatus for subjecting a biological system to a temperature-jump by exposing the outside of a silver-walled sample cell to a 200 J light flash. The temperature of the cell contents was raised 5 °C in 400 ms, an extremely long time by conventional temperature-jump standards, but short enough to enable transient behaviour to be seen in the metabolism of yeast cells.

The tedious labour of evaluating relaxation times from exponential traces displayed on an oscilloscope screen may be reduced by comparing these traces, stored photographically[9] or electronically[10] with synthetic exponentials of known relaxation time.

There is a constant effort on the part of designers of apparatus for single-step techniques to increase the signal-to-noise ratio. Continuous perturbation methods are intrinsically less troubled by noise, since noise, being random tends to cancel itself out over an extended time. Rüppel and co-workers[11, 12] have described various ways in which a single-step perturbation, repeated at regular intervals (e.g. a temperature-jump produced by a microwave discharge from a pulsed magnetron) may be sampled repetitively so as to average out the noise. The signal voltage is sampled at a range of fixed intervals from the start of the pulse, and the voltages at each interval are added up to give a display of signal as a function of time, integrated over a large number of pulses. The improvement in the signal-to-noise ratio is theoretically proportional to the square root of the number of pulses in the integration, and in practice an improvement of the order of 100 times can be achieved. The main disadvantage of this technique is that the signal which appears out of the noise may b. some transient other than the chemical relaxation required, e.g. electrical pick-up, or optical transients caused by changes in the refractive index of the sample consequent on the temperature jump. Most workers in the field rely on reducing the noise on a single trace, e.g. by working at as low a bandwidth and as high a light level as possible.

10.2.1.2 Theoretical methods

The problems of the user of relaxation techniques do not end with the determination of relaxation times, as rate constants have still to be evaluated and the relationship between these quantities are generally complex. Czerlinski[13] has given a detailed analysis of many of the possible complex reaction schemes which may be encountered, with especial reference to enzyme–substrate systems. The most complex scheme considered was a cycle containing four bimolecular steps, as shown in Figure 10.1. Czerlinski gave expressions for the slowest and second slowest relaxation times for this and eight related systems in terms of equilibrium and rate constants for the separate steps. The equations were solved by making various assumptions about which steps were the slowest, and by taking advantage of the symmetry of the reaction scheme. Although such compilations have their uses, most workers in the field seem to prefer to derive their own relationships, even if considerable approximations and simplifications have to be used,

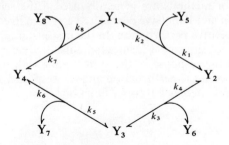

Figure 10.1 General reaction scheme for a four-step cyclic process

as no compilation as yet is sufficiently comprehensive. In these derivations it is generally assumed that the ratio of the forward and backward rates of the reactions setting up an equilibrium gives the equilibrium constant. In a paper by Koren and Perlmutter-Hayman[14], the breakdown of this relationship sheds new light on the mechanism of a reaction. For the hydrolysis of dichromate

$$Cr_2O_7^{2-} + H_2O \underset{k_b}{\overset{k_f}{\rightleftharpoons}} 2HCrO_4^-$$

the relaxation time τ should be given by

$$\tau^{-1} = k_f + 4k_b[HCrO_4^-] \tag{10.1}$$

As expected, temperature-jump studies gave a single relaxation time, whose concentration dependence fitted equation (10.1). However, the value of k_f/k_b found, namely 1×10^{-2} mol l^{-1} was very different from the value of K determined spectrophotometrically, namely 2.7×10^{-2} mol l^{-1}. The value of k_f was determined separately by stopped-flow experiments and found to be in agreement with the value from the temperature-jump experiments. The explanation is that the reaction proceeds by two paths, the second being autocatalytic

$$Cr_2O_7^{2-} + HCrO_4^- + H_2O \underset{k_b'}{\overset{k_f'}{\rightleftharpoons}} 3HCrO_4^-$$

so that the observed relaxation time is given by

$$\tau^{-1} = k_f + 4[HCrO_4^-](k_b + k_f'/4) + 4k_b'[HCrO_4^-]^2 \tag{10.2}$$

As for the simpler scheme, the intercept on a plot of τ^{-1} v. $[HCrO_4^-]$ gives k_f, in agreement with the value from stopped-flow studies, but the slope is not equal to k_b. The curvature to be expected on this plot from equation (10.2) is too slight to be noticed in the concentration range used.

Arbesman and Kim[15, 16] have produced a mathematically rigorous treatment of the problem of the two-step system, e.g.

$$A_1 + A_2 \rightleftharpoons A_3 \rightleftharpoons A_4$$

for which the two relaxation times are of similar magnitude. If the relaxation times differ by at least an order of magnitude their separation is simple, but trial-and-error methods are generally used if the relaxation times are similar. Arbesman and Kim have considered a system in equilibrium subject to a regularly repetitive perturbation. In the first instance, this perturbation was chosen to be of sinusoidal form, of angular frequency ω. Not only did the amplitude of the response of the system vary with frequency, but also the phase lag between the perturbation and the response. The response may in general be represented as a transfer function

$$G(j\omega) = \frac{a(j\omega)}{v(j\omega)} \tag{10.3}$$

where $v(j\omega)$ is the amplitude and phase of the perturbation as a function of frequency, and $a(j\omega)$ is the amplitude and phase of the response as a function of frequency. The values of $j\omega$ at which the denominator of $G(j\omega)$ becomes zero, i.e. the poles of the transfer function, and the value of $j\omega$ at which $G(j\omega)$ itself is zero, i.e. the zero of the transfer function, can be found experimentally. These quantities, the poles and the zero, are related simply and unambiguously to the individual rate and equilibrium constants. For some relaxation techniques, e.g. ultrasonics, $G(j\omega)$ is determined directly. Arbesman and Kim have extended this method of calculation to include data for which the perturbation is a repetitive step function, or square wave e.g. a temperature jump produced by a microwave discharge from a pulsed magnetron. The variation of response with frequency was determined by numerical Fourier analysis of the experimental response to the step perturbation. Arbesman and Kim have not yet applied their technique to real experimental data, but have produced synthetic data, degraded by the addition of small errors to represent noise, using arbitrary relaxation times. These relaxation times are recovered to an accuracy of $\pm 10\%$ even though τ_2/τ_1 is as low as 2.3.

10.2.1.3　Proton-transfer reactions

In the early relaxation studies a great deal of attention was paid to rates of proton transfer to and from atoms other than carbon, as these rates had hitherto been too fast to study. This work culminated in the definitive paper by Eigen[17], in which he showed that in the proton-transfer reaction

$$AH + B \underset{k_b}{\overset{k_f}{\rightleftharpoons}} A + HB$$

the reaction in the exothermic direction occurs at the diffusion-controlled rate ($\sim 10^{10}$ mol^{-1} l s^{-1}) unless prevented from doing so by special causes. These causes include intramolecular hydrogen-bonding and the need for solvent reorganisation. Most reactions between carbanions and hydroxonium ions are slow in aqueous solution because the negative charge on the carbanion is on an oxygen atom. Before the C—H bond can be formed, the charge must shift to the carbon at the site of attack of the proton, and this involves extensive reorganisation of the hydration sheath. More recent papers from Eigen's laboratory provide additional evidence. Ahrens and Maass[18] have studied the proton-transfer reactions between a wide range of oxygen, nitrogen and sulphur acids and bases, and found that the reaction rate reached the diffusion-controlled limit when ΔpK, the difference in pK_a between AH and HB, exceeded one or two units. Even at much greater values of ΔpK, the values of k_f or k_b for proton transfer to or from the typical carbon acid acetylacetone, in its ketonic form, did not approach the diffusion-controlled rate[19]. The rate constant for the reaction

$$E^- + H_3O^+ \rightarrow KH + H_2O$$

(where E^- and KH are the enolate and ketonic forms respectively) was only 1×10^7 mol^{-1} l s^{-1}, although ΔpK is 10.6, and the rate constant for

$$KH + OH^- \rightarrow E^- + H_2O$$

was as low as 4×10^4 mol^{-1} l s^{-1}, even though ΔpK is -5.1.

Further evidence that a proton-transfer reaction is retarded if the proton is held in an intramolecular hydrogen bond has been provided by Rose and Stuehr[20], in a study of a range of indicator acids possessing intramolecular hydrogen bonding, including alizarin yellow, tropaeolin O, Clayton yellow and quinazarin-2-sulphonic acid. The rate of the reaction

$$AH + OH^- \rightarrow A^- + H_2O$$

was found to be substantially below the diffusion-controlled limit, being as low as 3×10^4 mol^{-1} l s^{-1} for Clayton yellow, in which the proton is held in a N—H...O hydrogen bond.

As might be expected, proton-transfer reactions in aprotic solvents show different features to those in aqueous solution. These reactions must be studied by microwave or laser temperature-jump, as the solutions are non-conducting. Whereas the reaction between bromophenol blue and aliphatic amines in chlorobenzene is a simple diffusion-controlled reaction[21], the reaction between bromophenol blue and pyridine bases in chlorobenzene is more complex[22]. The overall rate of formation of the ion-pair product is very much less than for a diffusion-controlled reaction, and the overall activation energy for 2,6-dimethyl pyridine as base is negative. The mechanism was therefore postulated to be two-step, with a fast pre-equilibrium in which a hydrogen-bonded complex is formed, followed by a slow intramolecular rearrangement to form the ion-pair.

$$AH + B \underset{\text{fast}}{\rightleftharpoons} AH...B \underset{\text{slow}}{\rightleftharpoons} A^-...HB^+$$

Ivin and co-workers[23] have found a similar negative activation energy for the reactions between 2,4-dinitrophenol and aliphatic amines in chlorobenzene, but have not interpreted their results in terms of an intermediate complex.

10.2.1.4 *Metal-complex formation*

A large number of papers have been published in the last few years applying the temperature-jump technique to determine the rates of formation of metal complexes, notably those of Ni^{2+}, Co^{2+}, Cu^{2+} and Fe^{3+}. The Eigen–Wilkins mechanism is generally taken as the starting point of the discussion:

$$M \cdot H_2O + L \underset{K_{os}}{\rightleftharpoons} M \cdot H_2O \cdot L \overset{k_{exch}}{\rightleftharpoons} M \cdot L$$

The replacement of H_2O by L proceeds via the fast pre-equilibrium formation of the outer-sphere complex $M \cdot H_2O \cdot L$, with an equilibrium constant of K_{os}. Values of K_{os} may be calculated using a simple electrostatic model. The rate-determining step is the loss of H_2O from this complex, the rate of which is equal to k_{exch}, the rate of exchange of solvation with H_2O in the bulk of the solvent, as determined by n.m.r. spectroscopy. The overall rate is thus given by

$$k_{obs} = K_{os} \times k_{exch} \tag{10.4}$$

Some doubt has been expressed, however, as to whether this is an adequate explanation, especially for studies in non-aqueous solvents[24].

The work of Jones and Margerum[25] on the addition of a third ethylenediamine (denoted by en) ligand to diaquobis(ethylenediamine)Ni^{II} may be taken as a typical example. The reaction is accompanied by a pH change, as protonated en dissociates to replace en complexed by the Ni^{2+}. The use of an indicator, phenol red, enabled this pH change to be followed spectrophotometrically. Only a single relaxation time, varying with pH and reagent concentration, in the range 1–10 ms, was observed, and this fitted the equation

$$\tau^{-1} = k_f\{[Ni(en)_2^{2+}] + [en] + [H(en)^+] + [H_2(en)_2^{2+}]\} + k_r \tag{10.5}$$

This can be interpreted in terms of the mechanism shown in Figure 10.2. The observed pH dependence suggests reaction paths $(1) \to (4)$ and $(2) \to (3)$ are significant. As there is only one observed relaxation time, intermediates (3) and (4) are present in only negligible quantities, otherwise separate relaxation times would be observed for $(1) \to (4) \to (5)$ and $(2) \to (3) \to (4) \to (5)$. The relationship between the relaxation time and the rates and equilibrium constants for the individual steps may be derived assuming steady-state conditions for (3) and (4), and the equation was solved with the aid of a regression computer programme. The value of k_{14} found, 5.5×10^6 mol^{-1} l s^{-1}, was very large for Ni^{2+}, being about 6000 times that predicted by the Eigen–Wilkins mechanism. This acceleration was attributed to the labilisation of coordinated H_2O by the en ligands already on the Ni^{2+}. The rate of reaction between en and $Ni(H_2O)_6^{2+}$ was also faster than predicted by the Eigen–Wilkins mechanism, as discussed by Rorabacher[26]. Rorabacher postulated an internal conjugate-base mechanism, whereby the departure of a water molecule from the hydration sheath is accelerated by hydrogen-bonding to a nitrogen on the approaching ligand. Whereas for the reaction of en with $Ni(H_2O)_6^{2+}$ the chelate ring closure $(4) \to (5)$ is much faster than the loss of monodentate ligand intermediate $(4) \to (1)$, the rates are similar for the reaction of en with $Ni(en)_2^{2+}$ ($k_{41} = 1.2 \times 10^5 s^{-1}$, $k_{45} = 2.2 \times 10^5 s^{-1}$) so that the rate of chelate ring closure to a certain extent limits the rate of formation of Ni(en)$_3$. The two phenomena of labilisation of a metal ion by ligands already bound, and slow chelate ring closure have been the subject of several papers. Kirschenbaum and Kustin[27] found similar results for the reaction between Cu^{2+} and en. The forward rate constant, of value 4×10^9 mol^{-1} l s^{-1}, was about 10 times greater than that predicted by the Eigen–Wilkins mechanism, attributable to the internal conjugate base mechanism. The rate of substitution by H(en)$^+$ was only 1×10^5 mol^{-1} l s^{-1}, since the amine must be deprotonated, an exothermic step, before the chelate ring can close. Similarly

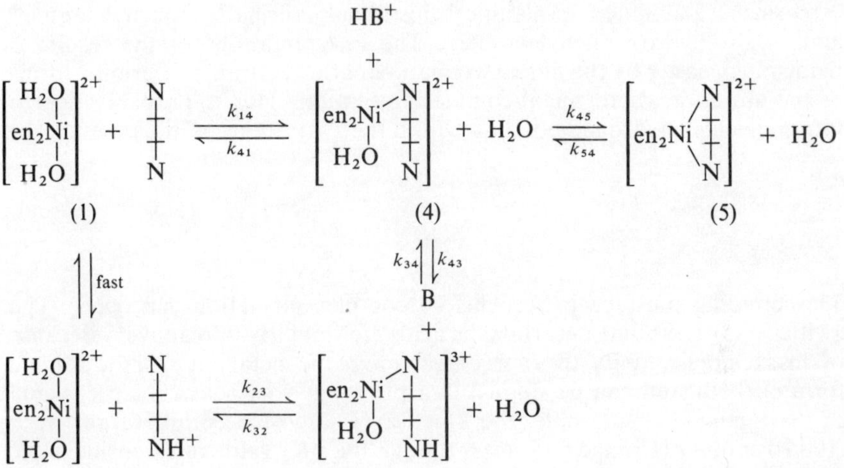

Figure 10.2 Scheme for the reactions between Ni^{2+} and ethylenediamine

the rate of complexation of Cr^{2+} and Cu^{2+} by mono-protonated $\alpha\alpha'$-bipyridyl ($k = 6 \times 10^3$ mol^{-1} l s^{-1} and 3×10^5 mol^{-1} l s^{-1} respectively) is much slower than the rate of complexation by neutral bipyridyl[28] ($k = 4 \times 10^7$ mol^{-1} l s^{-1} and 4×10^7 mol^{-1} l s^{-1} respectively). Whereas the rate of reaction between Ni^{2+} and 9-methylpurine is 7×10^3 mol^{-1} l s^{-1}, as expected from the Eigen–Wilkins mechanism, the rate for adenine complexation is much lower[29], having a value of 3×10^2 mol^{-1} l s^{-1}. Chelation of adenine is sterically restricted, so that the complexation rate is limited by the slow chelation rate. The acceleration of rate of substitution by ligands already bound to the metal ion is clearly seen in work done by Margerum and co-workers[30] on the reaction

$$Ni^{II}L(H_2O)_x + NH_3 \underset{k_{21}}{\overset{k_{12}}{\rightleftharpoons}} Ni^{II}L(H_2O)_{x-1}NH_3 + H_2O$$

For a series of ligands, L, ranging from H_2O to triethylenediamine, such that the number of coordinated H_2O molecules falls from six to two, the value of k_{12} rose from $3 \times 10^3 \ mol^{-1} \ l \ s^{-1}$ to $1 \times 10^5 \ mol^{-1} \ l \ s^{-1}$.

Much work has been done on the reactions of Cu^{2+}, Ni^{2+} and Co^{2+} with amino acids. In all, work on 13 amino acids has been reported, namely glycine[31], L-cysteine[32], L-carnosine[33], arginine[34], glycylsarcosine[35], leucine[36], sarcosine[37], α-alanine[38], β-alanine[38], histidine[38], serine[39], L-phenylalanine[40] and L-3,4-dihydroxyphenylalanine[40]. The interpretations of the results is made much easier by the unreactive nature of the neutral, zwitterionic forms of the amino acids to metal-complex formation. This is probably due to intramolecular hydrogen bonding within the zwitterions, of the type

$$\begin{array}{c} C-C \\ / \quad \backslash \\ N^+ \quad O^- \\ \backslash H \cdots \end{array}$$

This bonding must be broken before complex formation can occur. The zwitterions of pyridine decarboxylic acids are similarly unreactive[41]. Because of this lack of reactivity, the rates of reaction of the metal ion with the anionic form of the ligand can be studied in a pH range well below the pK_a value of the ligand. For example, the kinetics of sarcosine complexation were studied in the pH range 6–7, even though the pK_a value of sarcosine is 10. This has the advantage that a large Ni^{2+} or Co^{2+} concentration can be used without the relaxation times being too short to measure, so that only the mono-substituted complexes are present in appreciable quantities. Furthermore, the kinetics of complexation of Cu^{2+} can be studied even though the second-order rate constants approach the diffusion-controlled limit. In general, the kinetics fit the Eigen–Wilkins mechanism but, as previously discussed, Ni^{2+} and Co^{2+} become increasingly labile with substitution, and the rate of complex formation is reduced if chelate ring closure is sterically unfavourable.

The kinetics of complexation of Fe^{3+} follow a rather different pattern, in that hydroxyferric species are significant. $FeOH^{2+}$ is much more reactive than Fe^{3+}, so that the observed relaxation times are usually highly sensitive to pH. The general reaction scheme[42] is shown in Figure 10.3. Only one relaxation time was observed, despite the complexity of the scheme, as $[FeOH^{2+}]$ is in a steady state, and the protolytic equilibria are rapidly established. For such a system

$$\frac{d}{dt}[FeA^{2+}] = k_{obs}[Fe^{3+}][A^-] - k'_{obs}[FeA^{2+}] \tag{10.6}$$

where k_{obs} and k'_{obs} are the overall association and dissociation rates. These values of k_{obs} and k'_{obs} can be determined directly by stopped-flow experiments if the conditions are such that the reaction is slow enough. The relaxation time is given by

$$\tau^{-1} = k_{obs}B \tag{10.7}$$

where B is a complicated function involving known equilibrium constants and concentrations only. A direct comparison between stopped-flow and temperature-jump results was thus possible, and good agreement was found. Results for HA being HN_3 [42], HCl [43, 44], HBr [43, 44], $HCNS$ [43], $H_2C_2O_4$ [43], CH_3CO_2H [45], $C_2H_5CO_2H$ [45] and CH_2ClCO_2H [45] show that k_{34}, the rate

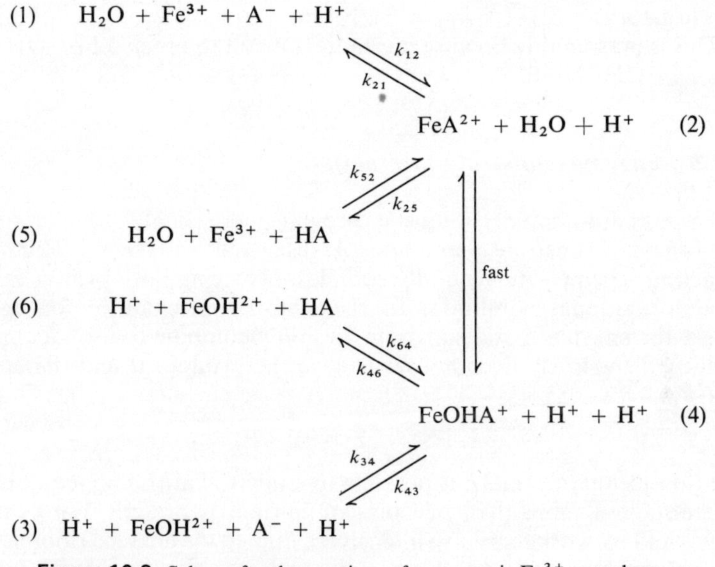

(1) $H_2O + Fe^{3+} + A^- + H^+$

k_{12}

k_{21}

$FeA^{2+} + H_2O + H^+$ (2)

k_{52}

k_{25}

(5) $H_2O + Fe^{3+} + HA$

fast

(6) $H^+ + FeOH^{2+} + HA$

k_{64}

k_{46}

$FeOHA^+ + H^+ + H^+$ (4)

k_{34}

k_{43}

(3) $H^+ + FeOH^{2+} + A^- + H^+$

Figure 10.3 Scheme for the reactions of monomeric Fe^{3+} complexes

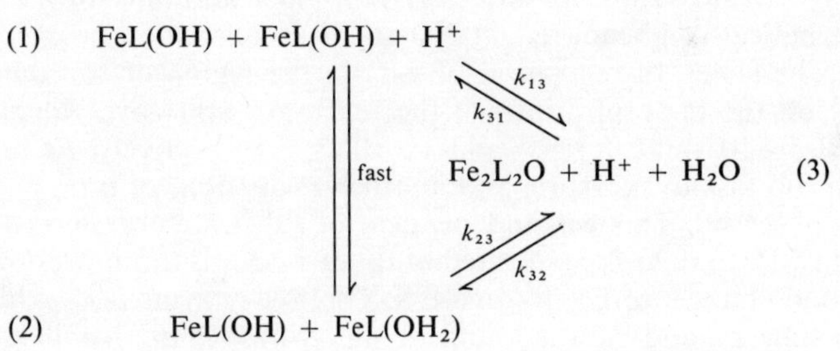

(1) $FeL(OH) + FeL(OH) + H^+$

k_{13}

k_{31}

fast $Fe_2L_2O + H^+ + H_2O$ (3)

k_{23}

k_{32}

(2) $FeL(OH) + FeL(OH_2)$

Figure 10.4 Scheme for the formation of dimeric Fe^{3+} complexes

of complexation of the hydroxyferric species lies in the range $1-3 \times 10^4$ mol^{-1} l s^{-1}, independent of the ligand in accordance with the Eigen–Wilkins mechanism.

The kinetics of Fe^{3+} complexes is further complicated by the formation of dimeric species containing Fe—O—Fe bridges, as shown in Figure 10.4.

Systems in which L is ethylenediaminetetra-acetate[46, 47], N-(2-hydroxyethyl) ethylenediaminetriacetate[47] and *trans*-1,2-cyclohexanediaminetetra-acetate[47] have been studied. Only a single relaxation time has been found, fitting the equation

$$\tau^{-1} = 4A[FeL(OH)] + B \tag{10.8}$$

where A and B are functions of concentration and equilibrium constants. By plotting A and B against $[H^+]$, values of individual rate constants may be obtained. The dimer was found to form more easily from the aquo species and one molecule of hydroxy species than from two molecules of the hydroxy species (e.g. for L = EDTA, $k_{13} = 6 \times 10^2$ mol^{-1} l s^{-1}. $k_{23} = 2 \times 10^4$ mol^{-1} l s^{-1}). This is presumably because it is more difficult to break a Fe—OH bond than a Fe—OH$_2$ bond.

10.2.1.5 Enzyme–substrate interactions

The temperature-jump technique is proving increasingly popular for the study of enzyme–substrate interactions. As discussed in a review by Hammes[66], the reaction schemes are generally complex, involving both proton transfer and conformational equilibria. In the simplest mechanism for enzyme catalysis, the enzyme E and substrate S rapidly combine to form a complex ES, which then slowly decomposes to give the product P and the enzyme back again.

$$E + S \underset{k_{-2}}{\overset{k_2}{\rightleftharpoons}} ES \xrightarrow{k_f} E + P$$

Relaxation techniques make it possible to study the initial pre-equilibrium, and determine if more than one pre-equilibrium is present. For example, Hammes and co-workers[48] have intensively studied the enzyme ribonuclease, with such model substrates as cytidine and uridine 3'-monophosphates. The extent of formation of ES during the relaxation was followed either by its absorption at 260 nm, or indirectly by the pH, determined by the absorption of the indicator, phenol red, at 550 nm. Two relaxation processes were observed. The faster, of relaxation time τ_2 in the region of 40–100 μs, was dependent on the concentrations of the reactants, and gave values for k_2 and k_{-2} of 5×10^7 mol^{-1} l s^{-1} and 1×10^4 s^{-1} respectively. At high [S], τ_2 became too fast to measure, and another relaxation, of time τ_3 around 1 ms, was observed. This was independent of substrate concentration, but varied with pH, and so is associated with an isomerisation, presumably a conformation change, of ES, to give E'S. The free enzyme also exhibited a relaxation time τ_1, and τ_1 was found to vary with pH in a similar fashion to τ_3, suggesting the same ionising group takes part in both processes. It is likely that this group is the imidazole group of histidine-48, which is known to be at the hinge of the active site.

One difficulty in the application of relaxation techniques to the equilibrium between E and S is that ES might irreversibly decompose to give the products of the reaction before the relaxation experiment can be performed. This difficulty is most usually avoided by using a model substrate, i.e. a substance, similar to a real substrate, which binds to the enzyme but is not decomposed by it. For example, in the studies described above[48], uridine 3'-monophosphate was not hydrolysed by ribonuclease, being itself the product of the ribonuclease-catalysed hydrolysis of uridine 2':3'-cyclic phosphate. Guillain and Thusius[49] have discussed the use of proflavin as a model substrate. Proflavin is a strong competitive inhibitor (i.e. it binds but does not react) for trypsin and chymotrypsin, and the binding is associated with a large shift in the visible spectrum, so that [ES] may be readily determined. Guillain and Thusius found the rate of binding of proflavin to chymotrypsin to be $8 \times 10^7 \, mol^{-1} \, l \, s^{-1}$. An alternative solution to the problem of decomposition of ES is to use a combination stopped-flow temperature-jump apparatus[50, 51], in which E and S are mixed only a few milliseconds before being subjected to a temperature-jump. In this way del Rosario and Hammes[50] studied the interaction of ribonuclease with uridine 2':3'-cyclic phosphate, and found that the kinetics follow a similar pattern to that for uridine 3'-monophosphate as substrate.

Aspartate transcarbamylase is a well-known regulatory enzyme, catalysing the first step in the pathway of pyrimidine biosynthesis. A final product, cytidine triphosphate, strongly inhibits the enzyme, so that the pathway is subject to feedback control. The enzyme is allosteric, i.e. it has separate, distinct, but interacting binding sites for substrates and inhibitors. There are two models for allosteric control, that of Monod[52], involving two conformational states of the enzyme, and that of Koshland[53], involving sequential changes as each substrate or inhibitor is bound. Relaxation data unequivocally show[54] that the mechanism involves two different conformations of the enzyme, to one of which substrate or inhibitor binds preferentially, i.e. the mechanism is of Monod type. A general sequential mechanism of the Koshland type would exhibit a spectrum of relaxation times, which is not observed.

10.2.1.6 Miscellaneous processes

The versatility of the temperature-jump technique is illustrated by the range of miscellaneous applications collected in this Section, which do not fit into the more common categories.

The kinetics of hydration of formaldehyde[55] and 2-methyl butyraldehyde[56] have been studied, using the carbonyl absorption at 294 nm as a measure of aldehyde concentration. The reaction is subject to both acid and base catalysis. Meisenheimer complexes have often been postulated as intermediates in aromatic substitution, e.g.

Conventional kinetics only give a value of k_{12}, and that only if $k_f \gg k_{21}$. By applying the temperature-jump technique to the nucleophilic substitution reaction of amines with trinitrobenzene, Bernasconi[57] has derived values for the individual rate constants, and thus also produced equilibrium constants for Meisenheimer complex formation. Further equilibrium constants involving protonated species have also been found. Plane and co-workers[58] have followed the dimerisation of a water-soluble porphyrin by observation of the Soret band at 390 nm. The association rate was found to be 8×10^7 $mol^{-1} l s^{-1}$, similar to that found for the dimerisation of acridine orange[59]. This is presumably a diffusion-controlled reaction, the rate constant being much lower than the generally quoted figure of $10^{10} mol^{-1} l s^{-1}$ because of the high stereospecificity of the interaction. The activation energy is similar to that for diffusion through water ($\sim 4 kcal mol^{-1}$).

Two studies have been made of the rate at which surfactant molecules cluster to form micelles. Eyring and co-workers[60] have used light scattering as the mode of detection of micelle concentration, which is more specific for micelles than light transmission, as used by Krescheck and co-workers[61]. Despite the apparent complexity of the aggregation process, only a single relaxation time was observed. The expression relating this to the rates of the processes involved may be derived by assuming that the first molecule to leave a micelle does so much more slowly than the others. This rate has been found to be $5 s^{-1}$ for sodium lauryl sulphate[60] and $50 s^{-1}$ for dodecylpyridinium iodide[61].

Fast reactions of nucleic acids have been studied by the temperature-jump technique. The kinetics of recombination of separated strands of synthetic, oligomeric, nucleotides of adenine and uridine have been the subject of a classic study by Porschke and Eigen[62]. The presence of the double-helical form was shown by absorption at 260 nm. It was found that the mechanism of renaturation involves nucleation by the formation of three hydrogen-bonded base pairs in a row, followed by a rapid 'zipping up' process, at a rate of the order of 10^4 base pairs a millisecond. A similar process was observed[63] in naturally-occurring yeast phenylalanine specific transfer RNA. Proflavin may be inserted, or 'intercalated' between the base pairs in DNA, and so alter the reading of the genetic code. The rate of insertion of proflavin has been measured[64]. Two relaxation times were found, showing that there are two forms of the proflavin–DNA complex. In the less strongly bound form, proflavin is attached externally to the double helix. The insertion reaction, to produce the more strongly bound form, occurs in the millisecond time range.

10.2.2   Electric-field jump

10.2.2.1   Experimental methods

A significant drawback of the temperature-jump method is that a perturbation of shorter duration than a few microseconds is extremely difficult to arrange. The electric-field jump technique is much faster, as an electric field of sufficient intensity to produce an adequate perturbation ($\sim 10^5 V cm^{-1}$) can conveniently be applied by a delay line, giving a pulse of 50 ns rise-time and a few microseconds duration. The electric-field jump technique is thus excellent for the observation of very short relaxation times, but cannot be used to study slower reactions. A major practical difficulty is the shielding of the detector from the high-voltage pulse.

10.2.2.2 Proton-transfer reactions

The use of the electric-field technique to observe the ionisation of 2,6-dinitro-phenol has been described by Eigen[62]. Eyring and co-workers[67] have studied the rate of ionisation of the indicator acid, methyl red. Relaxation times as short as 0.62 µs were observed, and the re-protonation of the anion was found to be diffusion-controlled, having a rate constant of $4 \times 10^{10} \, mol^{-1} \, l \, s^{-1}$.

10.2.2.3 Metal ion hydrolysis reactions

Eyring and co-workers have studied the hydrolysis reactions of Al^{3+} [68], Sc^{3+} [69], Cr^{3+} [70], In^{3+} [71] and Ga^{3+} [71]. The hydrolyses of these ions proceed according to the equations

$$M^{3+} + H_2O \underset{k_{-1}}{\overset{k_1}{\rightleftharpoons}} MOH^{2+} + H^+ : MOH^{2+} + H_2O \underset{k_{-2}}{\overset{k_2}{\rightleftharpoons}} M(OH)_2 + H^+$$

Although two relaxation times may be expected, only one, of a few microseconds duration, was observed, probably because the amplitude of the other relaxation is too small. For a coupled system of two equilibria, one amplitude is proportional to the sum of two large terms in reactant concentrations, specific rates, relaxation times and derivatives of the equilibrium constants with respect to the electric field intensity, whereas the other amplitude is proportional to the difference of these terms. The observed relaxation time has been assigned by various arguments to the first hydrolysis. For example, if the observed relaxation time for the hydrolysis of Cr^{3+} refers to the second step, the values of k_2 and k_{-2} are found to be $1.1 \times 10^9 \, mol^{-1} \, l \, s^{-1}$ and $1.5 \times 10^5 \, s^{-1}$ respectively. These values are not in themselves implausible, but their ratio, $2.4 \times 10^{-4} \, mol \, l^{-1}$, is very different from the value of K_2 of $2.8 \times 10^{-6} \, mol \, l^{-1}$ determined by spectrophotometry. Values of k_{-1} ranging from $8 \times 10^8 \, mol^{-1} \, l \, s^{-1}$ for Cr^{3+} to $1 \times 10^{10} \, mol^{-1} \, l \, s^{-1}$ for Sc^{3+} have been found.

10.2.3 Pressure-jump

10.2.3.1 Experimental methods

In most pressure-jump apparatus the perturbation is caused by the bursting of a thin disc, which reduces the pressure on the sample from about 50 atm to 1 atm in about 100 µs. The reaction is generally followed conductimetrically, as pressure disturbances set up transient changes in the refractive index of the sample solution which make spectrophotometric detection difficult. Conductimetric detection is easier to arrange for pressure-jump than for temperature or electric-field jump, as the detector electrodes do not have to be protected from a high-voltage pulse. The main disadvantage of the pressure-jump technique is the comparatively long time required for the perturbation. This can be considerably reduced, down to a microsecond or so, by use of a shock-wave technique. Since the velocity of sound in a liquid increases with pressure, the leading edge of a pressure pulse steepens as the pulse is transmitted along a column of liquid. Jost[72] has described an apparatus in which a pressure pulse produced by the bursting of a brass disc at 1000 atm is sharpened by transmission down a tube of water 1.5 m long. The resulting perturbation may be detected spectrophotometrically. In apparatus described by Hoffman and Yeager[73], ethanol was used as the driver liquid, so that a sharp shock-front was produced by a lower driver pressure, detection being by conductimetric methods. The minimum observable relaxation time was found to be around 1 μs for both apparatus.

10.2.3.2 Metal-complex formation reactions

The pressure-jump technique has been used mainly to study the rate of metal-complex formation. Discussion of the results has been generally based on the Eigen–Wilkins model, as discussed in Section 10.2.1.4. Strehlow and Knoche[74], using a simple electrostatic model, have computed an activation energy for the replacement of H_2O by SO_4^{2-} on Be^{2+} in aqueous solution. The activation energy for the second, rate-determining, step (the expulsion of H_2O from $Be^{2+} \cdot H_2O \cdot SO_4^{2-}$) was calculated to be of the order of 90 kcal mol^{-1} for the simple S_N1 mechanism. The value of the activation energy measured using the pressure-jump technique was only 13 kcal mol^{-1}, showing that the reaction is concerted. Be^{2+} was chosen as a test of the theory as the small size of the ion means that large differences between the S_N1 and the concerted mechanism may be expected. Macri and Petrucci[75] have studied the formation of complexes of Mg^{2+}, Mn^{2+}, Ni^{2+}, Co^{2+} and Zn^{2+} with m-benzenedisulphonate in anhydrous methanol. The rates were much smaller, by factors of 15–100, than those predicted by the Eigen–Wilkins model, although rates of these reactions in aqueous solution were only smaller than those predicted by a factor of five. The Eigen–Wilkins model appears to give better predictions for lanthanide complexes in aqueous solution. The rates of formation of the oxalate complexes of 10 lanthanides[76] have been found to be very similar to the rates for the murexide complexes, ranging from 8×10^7 $mol^{-1}\,l\,s^{-1}$ for La^{3+} to 6×10^6 $mol^{-1}\,l\,s^{-1}$ for Tm^{3+}. There is a rapid decrease in the rate between Eu^{3+} and Dy^{3+}, suggesting that there is a change in the number of water molecules in the first hydration sphere as the ionic radius decreases.

10.3 CONTINUOUS PERTURBATION TECHNIQUES

10.3.1 Ultrasonics

10.3.1.1 Experimental methods

The pulse method, as described by Brundage and Kustin[78], has been generally employed. An ultrasonic pulse is attenuated by passage through the sample solution, and the attenuation is measured by comparison with a standard signal from a calibrated attenuator. The sample cell must be constructed and aligned to extremely small tolerances, of the same order of magnitude as the wavelength of the ultrasonic wave. The minimum observable relaxation time is limited by the accuracy of construction, but ultrasonic frequencies up to 200 MHz have often been employed. The main drawback of the ultrasonic technique is that the detection is very non-specific. All that is observed is an absorption peak, and assignment of that peak to a particular chemical reaction is often difficult. Furthermore two or more peaks may coalesce, so that the detection of only one peak does not necessarily mean that only one reaction is involved[79, 80].

10.3.1.2 Hydrogen bond formation

The ultrasonic technique is especially suitable for the study of systems in which the rates are fast and the equilibrium constants low. Fairly concentrated solutions (> 0.1 mol l^{-1}) must be used to obtain an adequate signal amplitude, but this is an advantage for such systems as negligible quantities of products are formed in dilute solutions. For all relaxation techniques, appreciable quantities of both reactants and products must be present in the system at equilibrium for the signal to have an appreciable amplitude. The kinetics of formation of hydrogen-bonded complexes have been mainly studied by this technique. The formation of the hydrogen-bonded dimers of benzoic acid[81], 2-pyridone[82], 1-cyclo-hexyluracil[83], and various complexes containing N—H...O, N—H...S and N—H...N bonds[84] have been found to be all diffusion-controlled, with rates of the order of 10^9 mol^{-1} l s^{-1}. An example of the caution with which ultrasonic data must be treated is shown in a paper by Rassing and Jenson[80] on the hydrogen-bonded association of benzyl alcohol in cyclohexane. Although only one relaxation was observed, the infrared spectrum of the solution showed that considerable quantities of oligomeric species were present.

10.3.1.3 Proton-transfer reactions

The rates of deprotonation by hydroxyl ion of the alkylammonium cations of triethylamine[78], serine[85], threonine[85] and piperidine[86] have been found to be 2×10 mol^{-1} l s^{-1}, as predicted by Eigen[17].

10.3.1.4 Metal-complex formation

The ultrasonic technique is fast enough to enable the formation of outer-sphere complexes to be studied, e.g. the reaction between cobalt–amine complexes and sulphate ions[87]. These reactions have been found to proceed at rates of around 10^{11} mol^{-1} l s^{-1}, in accordance with the predictions of the Debye–Smoluchowski equation for the diffusion-controlled reaction of ions. The rates of inner-sphere complex formation of the lanthanides with sulphate[88, 89] and acetate[90] ions have been studied. The relaxation around 40 MHz has been assigned to the step

$$La^{3+} \cdot H_2O \cdot L^{2-} \rightarrow LaL^- + H_2O$$

rather than the step

$$La^{3+} \cdot H_2O + L^{2-} \rightarrow La^{3+} \cdot H_2O \cdot L^{2-}$$

on the grounds that the latter, the formation of the outer-sphere complex, gives rise to a relaxation around 200 MHz for divalent transition metal sulphates[91], and some indication of this relaxation in the lanthanide system has been similarly observed above 200 MHz. The rates of complex formation found are similar to those found by the pressure-jump technique[76].

10.3.1.5 · Helix–coil transformations

The rate of interconversion between the α-helical and randomly-coiled conformations of polypeptides is known to be very fast ($\tau \sim 10$ ns) despite erroneous deductions from n.m.r. measurements[92], which gave rates slower by a factor of 10^5. As the conformation change is accompanied by a molecular volume change, it may be observed by the ultrasonics technique. Values of 11 ns and 15 ns have been determined for the relaxation times of poly-L-glutamic acid[93] and poly-L-ornithine[94] respectively.

10.3.2   Dielectric relaxation

10.3.2.1   Experimental methods

Any reaction in which there is a change in dipole moment, e.g. the association of dipolar molecules, can in principle be studied by the dielectric relaxation technique. The intensity of the signal is proportional to the square of the electric field strength, so that it is customary to superimpose the oscillatory electric field on a static field of intensity of the order of 10^5 V cm^{-1}. The sample solution must thus be of low conductivity. The dielectric relaxation technique resembles the ultrasonics technique in that reagent concentrations of 0.1 mol l^{-1} and applied frequencies of 100 MHz are typical.

10.3.2.2   Hydrogen bond formation

The rate of dimerisation of benzoic acid[107], ε-caprolactam[62], 2-amino-pyridine[62] and the rate of formation of the ε-caprolactam–2-aminopyridine complex[62] have been measured by the dielectric relaxation technique. Association rates of around 10^9 mol^{-1} l s^{-1} have been found, as found by the ultrasonic technique for similar systems.

10.3.2.3   Helix–coil transformations

Schwarz and Seeling[108] have found a relaxation time of 500 ns for the helix–coil transformation of poly-γ-benzyl-L-glutamate. A large signal amplitude has been observed as the co-operative nature of the transformation makes it very sensitive to electric field intensity.

10.4 NUCLEAR MAGNETIC RESONANCE

10.4.1 Experimental methods

The mean lifetime, τ, of a nucleus on a particular site in a molecule may be calculated from the transverse longitudinal relaxation time, T_2. For example, for slow exchange, where nuclei on the two possible sites can still be distinguished,

$$\tau^{-1} = 1/T_2' - 1/T_2 \tag{10.9}$$

where T_2' is the measured transverse relaxation time, and T_2 is the transverse relaxation time due to all processes except exchange. Values of T_2' may be measured by choosing conditions in which exchange does not occur, or may be calculated. Values of T_2 are calculated from the width Δ, of the line broadened by exchange:

$$\Delta = 1/\pi T_2' \tag{10.10}$$

In more sophisticated calculations, line-shapes calculated for various values of T_2' are fitted to the observed line-shapes with the aid of a computer programme. Alternatively T_2' may be measured directly, as in the spin-echo technique[95]. Although line-widths in the proton magnetic resonance spectrum are most commonly observed, resonances due to other nuclei, e.g. ^{7}Li, ^{11}B, ^{17}O, ^{31}P and ^{55}Mn have also been used.

10.4.2 Proton-exchange processes

The n.m.r. method has been much used to study proton-transfer processes. By contrast with other techniques, all processes giving rise to proton exchange are detected including symmetrical processes in which there is no net chemical reaction. The results are thus more difficult to interpret, but give information about solvent–solute interactions which cannot be obtained in other ways. Ralph and Grunwald have given a review[105] of earlier work on the kinetics of hydrogen-bonded solution complexes of amines in water and hydroxylic solvents. The mechanism of ionisation of imidazole[45], shown in Figure 10.5, shows how detailed a picture can be built up. The noteworthy feature of this mechanism is that it is possible to detect experimentally an intermediate in which the proton is in the outer hydration sphere of the imidazole. Proton exchange on imidazole can also occur by transfer of the type

$$\text{Im} + (\text{H}_2\text{O})_n + \text{ImH}^+ \rightleftharpoons \text{ImH}^+ + (\text{H}_2\text{O})_n + \text{Im}$$

For most such processes, n is unity, so that the reaction is termolecular, but for imidazole[96] n is about 1.4. In aqueous t-butanol the process is termolecular. The intervening solvent molecule can be H_2O or Bu^tOH, and the kinetics of these two processes can be resolved. The ratio of the rates is similar to the $[H_2O]/[Bu^tOH]$ mole ratio, suggesting that the relative abundance of the two encounter complexes is nearly statistical. Proton exchange between water and t-butanol has been studied by Ralph and Grunwald[98], using water enriched with $H_2^{17}O$ to split the proton resonance. The reaction was found to be termolecular, with concerted proton addition and subtraction.

The line-broadening technique has been used to measure rates of proton exchange on purine in aqueous solution[99], tertiary benzylamines in aqueous[100] and alcoholic [101] solutions, guanidine in aqueous dimethylsulphoxide[102]

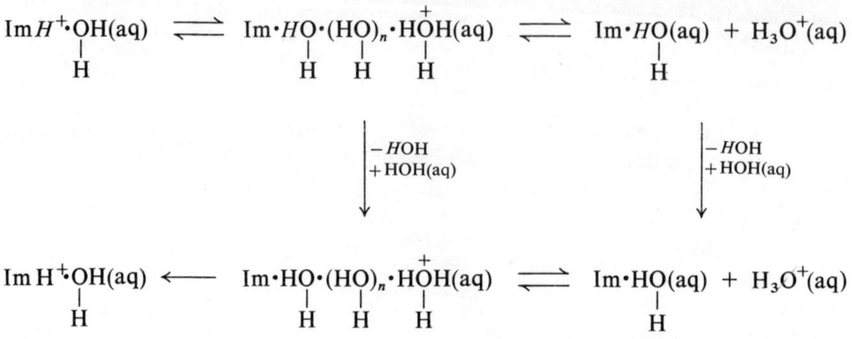

Figure 10.5 Scheme for proton exchange on imidazolium ion

and in aqueous N,N-dimethylacetamide[103] and methylamine in aqueous formic acid[126]. Reaction schemes of the type proposed by Grunwald and Ku[104], in which every possible motion of a proton is considered, are generally put forward, but only some of the many rate constants have been evaluated.

Proton exchange on the biochemically-significant mercapto group has been studied by Whidby and Leyden[106]. The exchange on mercaptoacetic acid and ester in acetic acid solution was found to be base-catalysed. The rate of proton transfer from the mercapto site to the acetate catalyst was found to be quite slow, the rate constant being 7×10^2 mol^{-1} l s^{-1} and the activation energy being 12 kcal mol^{-1}. The rate of proton loss from $Al(H_2O)_6^{3+}$ in aqueous solution was found[115] to be 8×10^4 s^{-1}, in good agreement with the value of 1.1×10^5 s^{-1} determined by the electric-field jump method[68].

10.4.3 Ligand exchange on metal ions

Rates of solvent exchange on transition metal ions are of interest as they provide essential data for the calculation of rates for the Eigen–Wilkins mechanism. Rates of H_2O exchange are generally measured using water enriched by $H_2^{17}O$. Studies on ethylenediamine[109] and ammonia[110] complexes of Ni^{2+} show that the exchange rate increases with the number of amine ligands, as previously discussed in Section 4.2.1.3. However, this effect is slight for complexes of Ni^{2+} with terpyridyl[111] and dipyridyl[112], and for complexes of Mn^{2+} with phenanthroline[113]. The rate of water exchange on

Ni^{2+} complexed by a fused-ring tridentate ligand[114] shows that rigidity of the chelating ligand is not an important kinetic factor.

The rate of ligand exchange in nitrilotriacetate (NTA) complexes of Cd^{2+}, Zn^{2+} and Pb^{2+} has been found from the lifetime of complexed NTA, as measured by the proton resonance line-width[117]. The rates found were 10^3 times slower than those predicted by the Eigen–Wilkins mechanism, and it was suggested that the rate is controlled by the rate of proton migration from the nitrogen on the free ion $(HN^+TA)^{3-}$. The measurement by n.m.r. spectroscopy of rates of ligand exchange in organometallic systems, e.g. the rate of exchange of trimethylamine coordinated to trimethylgallium, has been reviewed by Brown[118].

Discrepancies between n.m.r. spectroscopic and temperature-jump results for the rates of binding of transition metal ions to adenosine triphosphate have been discussed by Kustin and co-workers[119]. These discrepancies were

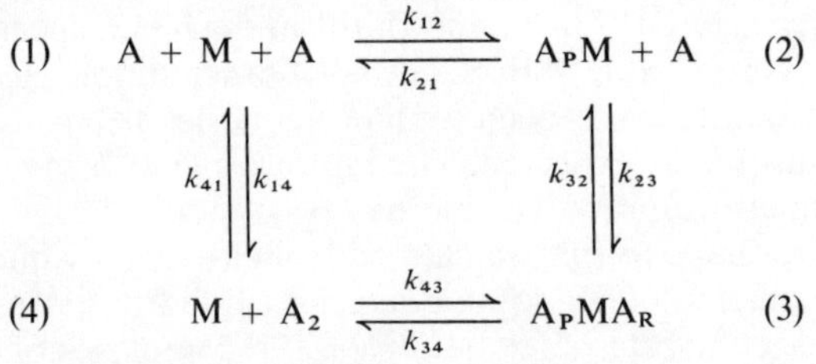

(1) $A + M + A$ $\underset{k_{21}}{\overset{k_{12}}{\rightleftharpoons}}$ $A_PM + A$ (2)

$k_{41} \uparrow\downarrow k_{14}$ $k_{32} \uparrow\downarrow k_{23}$

(4) $M + A_2$ $\underset{k_{34}}{\overset{k_{43}}{\rightleftharpoons}}$ A_PMA_R (3)

Figure 10.6 Scheme for the reactions between divalent transition metal ions and adenosine triphosphate

shown to be caused by the use of dilute solutions for the temperature-jump studies and concentrated solutions, in which 1:2 complexes predominate, for the n.m.r. spectroscopic studies. Kustin and co-workers have made a careful study of the system using line-broadening of the ^{31}P resonance as a measure of the lifetime of free adenosine triphosphate. The complete reaction scheme is shown in Figure 10.6 where A_P and A_R represent adenosine triphosphate bound by its phosphate groups and its ring nitrogens respectively. At high [ATP] the equilibrium $(4) \rightleftharpoons (3)$ is accessible to study by n.m.r. spectroscopy, but at low [ATP] only $(1) \rightleftharpoons (2)$ can be studied. Temperature-jump studies, on the other hand, give values for the rates $(2) \rightleftharpoons (3)$ as well as for $(1) \rightleftharpoons (2)$, so that the two techniques are complementary.

10.4.4   Enzyme–substrate interactions

The n.m.r. spectrum of a macromolecule is complex, but much information can be obtained from the n.m.r. spectrum of a small molecule which is able to penetrate the environment of the macromolecule. If an enzyme is under study, the probe molecule must be a model substrate or inhibitor, as a real substrate would react before the spectrum could be observed (cf. Section 10.2.1.5). An increase in line-width for nuclei in the probe indicates rotational restriction of the probe within the active site, or an interaction with the nuclei in the active site. It is convenient to measure the transverse relaxation time directly, as a single resonance can be picked out of a complex spectrum. In the spin-locking, or rotating-frame technique described by Sykes[123], the nuclei are taken by rapid adiabatic passage with the centre of the resonance, and the rate of decay of the consequent magnetisation gives a measure of T_2'. The rates of reaction of the inhibitor trifluoracetyl-D-phenyl-alanine with chymotrypsin[123], of the inhibitor N-acetyl-D-glucosamide with lysozyme[124], and of the inhibitor succinate with aspartate transcarbamylase[125] have been measured. The rates are in the region of 10^4–10^5 mol^{-1} l s^{-1}, much slower than the diffusion-controlled rate, showing that conformational changes in the enzyme must occur before the inhibitor is firmly bound.

10.4.5 Miscellaneous processes

The rate of electron transfer between Mn^I and Mn^{II} through an isonitrile ligand has been measured[120] from linewidths in the n.m.r. spectrum of ^{55}Mn, and found to be comparable with the rates of other simple electron-transfer processes. Ethyl isonitrile has been found to conduct electrons more readily than t-butyl isonitrile, as expected. The interchange of bridge and terminal protons in μ-dimethylaminodiborane has been studied[121] by matching the observed ^{11}B line-shape to that calculated from assumed values of the rate.

A somewhat different technique, deuterium quadrupole relaxation, has been used to study the rate of formation of π–π complexes of 1,3,5-trinitro-benzene with aromatic hydrocarbon bases. The deuteron relaxation time was found from the shape of the triplet fine structure of the —CHD— resonance. Very short relaxation times, down to a few nanoseconds, can be measured by this technique.

References

1. Eigen, M. (1954). *Disc. Faraday Soc.*, **17**, 194
2. Ogg, R. A. (1954). *Disc. Faraday Soc.*, **17**, 215
3. Eyring, E. M. and Bennion, B. C. (1968). *Ann. Rev. Phys. Chem.*, **19**, 129
4. Schechter, A. N. (1970). *Science*, **170**, 273
5. Caldin, E. F., Crooks, J. E. and Robinson, B. H. (1971). *J. Phys. E.*, **4**, 165
6. Hoffman, H., Yeager, E. and Stuehr, J. (1968). *Rev. Sci. Instr.*, **39**, 649
7. Rigler, R., Jost, A. and De Maeyer, L. (1970). *Exp. Cell Research*, **62**, 197
8. Lee, I. Y. and Chance, B. (1969). *Anal. Biochem.*, **29**, 331
9. Crooks, J. E., Zetter, M. Z. and Tregloan, P. (1970). *J. Phys. E.*, **3**, 73
10. Eggers, F. (1971). *Nature (London)*, **229**, 89
11. Brumm, P., Kilian, F. P. and Rüppel, H. (1968). *Ber. Bunsenges. Phys. Chem.*, **72**, 1085
12. Buchwald, H.-E. and Rüppel, H. (1971). *J. Phys. E.*, **4**, 105
13. Czerlinski, G. H. (1968). *J. Theor. Biol.*, **21**, 408, and papers cited therein
14. Koren, R. and Perlmutter-Hayman, B. (1970). *Isr. J. Chem.*, **8**, 1
15. Arbesman, R. W. and Kim, Y. G. (1969). *Ind. Eng. Chem. Fundam.*, **8**, 216
16. Arbesman, R. W. and Kim, Y. G. (1969). *Chem. Eng. Sci.*, **24**, 1627
17. Eigen, M. (1964). *Angew. Chem. Int. Ed. Engl.*, **3**, 1
18. Ahrens, M. L. and Maass, G. (1968). *Angew. Chem. Int. Ed. Engl.*, **7**, 818
19. Ahrens, M. L., Eigen, M., Kruse, W. and Maass, G. (1970). *Ber. Bunsenges. Phys. Chem.*, **74**, 380
20. Rose, M. C. and Stuehr, J. (1968). *J. Amer. Chem. Soc.*, **90**, 7205
21. Crooks, J. E., Sheridan, P. J. and O'Donnell, D. (1970). *J. Chem. Soc. B*, 1285

22. Crooks, J. E. and Robinson, B. H. (1970). *Trans. Faraday Soc.*, **66**, 1436
23. Ivin, K. J., McGarvey, J. J., Simmons, E. L. and Small, R. (1971). *Trans. Faraday Soc.*, **67**, 104
24. Caldin, E. F. and Bennetto, P. (1969). *Chem. Commun.*, 599
25. Jones, J. P. and Margerum, D. W. (1970). *J. Amer. Chem. Soc.*, **92**, 470
26. Rorabacher, D. B. (1966), *Inorg. Chem.*, **5**, 1891
27. Kirschenbaum, L. J. and Kustin, K. (1970). *J. Chem. Soc. A*, 684
28. Diebler, H. (1970). *Ber. Bunsenges. Phys. Chem.*, **74**, 268
29. Karpel, R. L., Kustin, K. and Wolff, M. A. (1971). *J. Phys. Chem.*, **75**, 799
30. Jones, E. P., Billo, E. J. and Margerum, D. W. (1970). *J. Amer. Chem. Soc.*, **92**, 1875
31. Pearlmutter, A. F. and Stuehr, J. (1968). *J. Amer. Chem. Soc.*, **90**, 858
32. Davies, G., Kustin, K. and Pasternack, R. F. (1968). *Trans. Faraday Soc.*, **64**, 1006
33. Pasternack, R. F. and Kustin, K. (1968). *J. Amer. Chem. Soc.*, **90**, 2295
34. Davies, G., Kustin, K. and Pasternack, R. F. (1969). *Intern. J. Chem. Kinet.*, **1**, 45
35. Kustin, K. and Pasternack, R. F. (1969). *J. Phys. Chem.*, **73**, 1
36. Pasternack, R. F., Gibbs, E. and Cassatt, J. C. (1969). *J. Phys. Chem.*, **73**, 3814
37. Pasternack, R. F., Kustin, K., Hughes, L. A. and Gibbs, E. (1969). *J. Amer. Chem. Soc.*, **91**, 4401
38. Makinen, W. B., Pearlmutter, A. F. and Stuehr, J. (1969). *J. Amer. Chem. Soc.*, **91**, 4083
39. Karpel, R. L., Kustin, K. and Pasternack, R. F. (1969). *Biochim. Biophys. Acta*, **177**, 434
40. Karpel, R. L., Kustin, K., Kowalak, A. and Pasternack, R. F. (1971). *J. Amer. Chem. Soc.*, **93**, 1085
41. Kowalak, A., Kustin, K. and Pasternack, R. F. (1969). *J. Phys. Chem.*, **73**, 281
42. Accasina, F., Cavasino, F. P. and D'Alessandro, S. (1967). *J. Phys. Chem.*, **71**, 2474
43. Cavasino, F. P. (1968). *J. Phys. Chem.*, **72**, 1378
44. Yasunaga, T. and Harada, S. (1969). *Bull. Chem. Soc. Jap.*, **42**, 2165
45. Accasina, F., Cavasino, F. P. and Di Dio, E. (1969). *Trans. Faraday Soc.*, **65**, 489
46. Gilmour, A. D. and McAuley, A. (1970). *Inorg. Chim. Acta*, **4**, 158
47. Wilkins, R. G. and Yelin, R. E. (1969). *Inorg. Chem.*, **8**, 1470
48. Hammes, G. G. and Walz, F. G. (1969). *J. Amer. Chem. Soc.*, **91**, 7179, and papers cited therein
49. Guillain, F. and Thusius, D. (1970). *J. Amer. Chem. Soc.*, **92**, 5534
50. del Rosario, E. J. and Hammes, G. G. (1970). *J. Amer. Chem. Soc.*, **92**, 1750
51. Hammes, G. G. and Haslam, J. L. (1969). *Biochemistry*, **8**, 1591
52. Monod, J., Wyman, J. and Changeux, J.-P. (1965). *J. Mol. Biol.*, **12**, 88
53. Koshland, D. E., Nemethy, G. and Filmer, D. (1966). *Biochemistry*, **5**, 365
54. Eckfeldt, J., Hammes, G. G., Mohr, S. C. and Wu, C.-W. (1970). *Biochemistry*, **9**, 3353
55. Schecker, H.-Gg. and Schulz, G. (1969). *Z. Phys. Chem. (Frankfurt)*, **65**, 221
56. Ahrens, M. L. and Maass, G. (1971). *Angew. Chem.*, **83**, 80
57. Bernasconi, C. F. (1970). *J. Amer. Chem. Soc.*, **92**, 129
58. Das, R. R., Pasternack, R. F. and Plane, R. A. (1970). *J. Amer. Chem. Soc.*, **92**, 3312
59. Hammes, G. G. and Hubbard, C. D. (1966). *J. Phys. Chem.*, **70**, 1615
60. Bennion, B. C., Tong, L. K. J., Holmes, L. P. and Eyring, E. M. (1969). *J. Phys. Chem.*, **73**, 3288
61. Krescheck, G. C., Hamori, E., Davenport, G. and Scheraga, H. A. (1966). *J. Amer. Chem. Soc.*, **88**, 246
62. Eigen, M. (1967). *Fast Reactions and Primary Processes in Chemical Kinetics*, 358 (New York: Wiley)
63. Romer, R., Riesner, D., Maass, G., Wintermeyer, W., Thiebe, R. and Zachan, H. G. (1969). *Fed. Eur. Biochem. Soc. Lett.*, **5**, 15
64. Li, H. J. and Crother, D. M. (1969). *J. Mol. Biol.*, **39**, 461
65. Eigen, M. and Wilkins, R. G. (1965). *Advan. Chem. Ser.*, **49**, 55
66. Hammes, G. G. (1968). *Accounts Chem. Res.*, **1**, 321
67. Lloyd, L. P., Silzars, A., Cole, D. L., Rich, L. D. and Eyring, E. M. (1969). *J. Phys. Chem.*, **73**, 737
68. Holmes, L. P., Cole, D. L. and Eyring, E. M. (1968). *J. Phys. Chem.*, **72**, 301
69. Cole, D. L., Rich, L. D., Owen, J. D. and Eyring, E. M. (1969). *Inorg. Chem.*, **8**, 682
70. Rich, L. D., Cole, D. L. and Eyring, E. M. (1970). *J. Phys. Chem.*, **73**, 713
71. Hemmes, P., Rich, L. D. and Eyring, E. M. (1970). *J. Phys. Chem.*, **73**, 2859
72. Jost, A. (1966). *Ber. Bunsenges. Phys. Chem.*, **70**, 1057

73. Hoffman, H. and Yeager, E. (1968). *Rev. Sci. Instr.*, **39,** 1151
74. Strehlow, H. and Knoche, W. (1969). *Ber. Bunsenges. Phys. Chem.,* **73,** 427
75. Macri, G. and Petrucci, S. (1970). *Inorg. Chem.,* **9,** 1009
76. Graffeo, A. J. and Bear, J. L. (1968). *J. Inorg. Nucl. Chem.,* **30,** 1577
77. Geier, G. (1965). *Ber. Bunsenges. Phys. Chem.,* **69,** 617
78. Brundage, R. S. and Kustin, K. (1970). *J. Phys. Chem.,* **74,** 672
79. Gusenkov, G. M. (1969). *Zh. Fiz. Khim.* **43,** 107
80. Rassing, J. and Jenson, B. N. (1970). *Acta Chem. Scand.,* **24,** 855
81. Borucki, L. (1967). *Ber. Bunsenges. Phys. Chem.,* **71,** 504
82. Hammes, G. G. and Spivey, H. O. (1966). *J. Amer. Chem. Soc.,* **88,** 1621
83. Hammes, G. G. and Park, A. C. (1968). *J. Amer. Chem. Soc.,* **90,** 4151
84. Hammes, G. G. and Park, A. C. (1969). *J. Amer. Chem. Soc.,* **91,** 956
85. White, R. D., Slutsky, L. J. and Pattison, S. (1971). *J. Phys. Chem.,* **75,** 161
86. Applegate, K., Slutsky, L. J. and Parker, R. C. (1968). *J. Amer. Chem. Soc.,* **90,** 6909
87. Elder, A. and Petrucci, S. (1970). *Inorg. Chem.,* **9,** 19
88. Fay, D. P., Litchinsky, D. and Purdie, N. (1969). *J. Phys. Chem.,* **73,** 544
89. Fay, D. P. and Purdie, N. (1970). *J. Phys. Chem.,* **74,** 1160
90. Fay, D. P. and Purdie, N. (1970). *J. Phys. Chem.,* **74,** 275
91. Eigen, M. and Tamm, K. (1962). *Z. Elektrochem.,* **66,** 93
92. Ullman, R. (1970). *Biopolymers,* **9,** 471
93. Saksena, T. K., Michels, B. and Zana, R. (1968). *J. Chim. Phys.,* **65,** 597
94. Hammes, G. G. and Roberts, P. B. (1969). *J. Amer. Chem. Soc.,* **91,** 1812
95. Ralph, E. K. and Grunwald, E. (1969). *J. Amer. Chem. Soc.,* **91,** 2422
96. Ralph, E. K. and Grunwald, E. (1968). *J. Amer. Chem. Soc.,* **90,** 517
97. Ralph, E. K. and Grunwald, E. (1969). *J. Amer. Chem. Soc.,* **91,** 2429
98. Ralph, E. K. and Grunwald, E. (1969). *J. Amer. Chem. Soc.,* **91,** 2426
99. Marshall, T. H. and Grunwald, E. (1969). *J. Amer. Chem. Soc.,* **91,** 4541
100. Leyden, D. E. and Morgan, D. R. (1969). *J. Phys. Chem.,* **73,** 2924
101. Grunwald, E., Lipnick, R. L. and Ralph, E. K. (1969). *J. Amer. Chem. Soc.,* **91,** 4333
102. Tewari, K. C., Li, N. C. and Kurland, R. J. (1969). *J. Phys. Chem.,* **73,** 2853
103. Tewari, K. C., Schweighardt, F. K. and Li, N. C. (1971). *J. Phys. Chem.,* **75,** 688
104. Grunwald, E. and Ku, A. Y. (1968). *J. Amer. Chem. Soc.,* **90,** 29
105. Grunwald, E. and Ralph, E. K. (1971). *Accounts Chem. Res.,* **4,** 107
106. Whidby, J. F. and Leyden, D. E. (1970). *J. Phys. Chem.,* **74,** 202
107. Bergmann, K., Eigen, M. and De Maeyer, L. (1963). *Ber. Bunsenges. Phys. Chem.,* **67,** 819
108. Schwarz, G. and Seelig, J. (1968). *Biopolymers,* **6,** 1263
109. Desai, A. G., Dodgen, H. W. and Hunt, J. P. (1969). *J. Amer. Chem. Soc.,* **91,** 5001
110. Desai, A. G., Dodgen, H. W. and Hunt, J. P. (1970). *J. Amer. Chem. Soc.,* **92,** 798
111. Rablen, D. and Gordon, G. (1969). *Inorg. Chem.,* **8,** 395
112. Grant, M., Dodgen, H. W. and Hunt, J. P. (1970). *J. Amer. Chem. Soc.,* **92,** 2321
113. Grant, M., Dodgen, H. W. and Hunt, J. P. (1971). *Inorg. Chem.,* **10,** 71
114. Letter, J. E. and Jordan, R. B. (1971). *J. Amer. Chem. Soc.,* **93,** 864
115. Fong, D.-W. and Grunwald, E. (1969). *J. Amer. Chem. Soc.,* **91,** 2413
116. Friess, S. L., Lewis, E. S. and Weissberger, A. (Editors) (1963). *Technique of Organic Chemistry, Vol. VIII–Part II; Investigation of Rates and Mechanisms of Reactions* (New York: Interscience)
117. Rabenstein, D. L. and Kula, R. J. (1969). *J. Amer. Chem. Soc.,* **91,** 2492
118. Brown, T. L. (1968). *Accounts Chem. Res.,* **1,** 23
119. Sternlicht, H., Jones, D. E. and Kustin, K. (1968). *J. Amer. Chem. Soc.,* **90,** 7110
120. Matteson, D. S. and Bailey, R. A. (1969). *J. Amer. Chem. Soc.,* **91,** 1975
121. Schirmer, R. E., Noggle, J. N. and Gaines, D. F. (1969). *J. Amer. Chem. Soc.,* **91,** 6240
122. Brevard, C. and Lehn, J. M. (1970). *J. Amer. Chem. Soc.,* **92,** 4987
123. Sykes, B. D. (1969). *J. Amer. Chem. Soc.,* **91,** 949
124. Sykes, B. D. (1969). *Biochemistry,* **8,** 110
125. Sykes, B. D., Schmidt, P. G. and Stark, G. R. (1970). *Chem. Commun.,* 1180
126. Delpuech, J. J., Ducom, J. and Michon, V. (1970). *Chem. Commun.,* 1187
127. Kustin, K. (Editor) (1969). *Methods in Enzymology, Vol. XVI; Fast Reactions* (New York: Academic Press)

THE SEA OF STONES

THE

Sea of Stones

A NOVEL IN THREE PARTS

CHICAGO

JERUSALEM

BEIRUT

BY

FLORENCE CHANOCK COHEN

PUSHCART PRESS
WAINSCOTT, NEW YORK 11975

WINNER OF THE TWELFTH ANNUAL EDITORS' BOOK
AWARD
*Sponsoring editors for the Editors' Book Award are Simon Michael
Bessie, James Charlton, Peter Davison, Jonathan Galassi, David Go-
dine, Daniel Halpern, James Laughlin, Seymour Lawrence, Starling
Lawrence, Robie Macauley, Joyce Carol Oates, Nan A. Talese, Faith
Sale, Ted Solotaroff, Pat Strachan, Thomas Wallace. Nominating ed-
itor for this novel: Morris Philipson.*

Distributed by W.W. Norton & Co., New York, N.Y.

This is a work of fiction and no reference is intended to real persons,
places or circumstances, unless otherwise indicated.

Manufactured in The United States of America by

RAY FREIMAN & COMPANY
*184 Brookdale Road
Stamford, Connecticut 06903*

ACKNOWLEDGEMENTS

There are others to whom I remain deeply indebted, particularly to the Director of the Chicago University Press, Morris Philipson, who had an enduring belief in this work and nominated it for the Editors' Award. My thanks to Mary Douglas, Distinguished Professor of Anthropology and Philosophy at Princeton University, who nudged me to write the book in the first place. My editors, Alan Friedman, Martha Hall, and Fred Shafer who so fully grasped this story, helped me see what was needed to make it work. My literary friends, writers themselves, Tilde Sankovitch, Fania Weingartner, Sondra Fargo, Sondra Gair, and Kate Rollins, to whom I so frequently turned for advice, I owe thanks for their patience and enlightening comments in reading and rereading the manuscript in all of its formative stages. To my son, Aaron, I owe loving thanks for many hours of his time for technical assistance in preparing the manuscript and for providing the Hebrew text. Thanks also to my Israeli friend, Ruth Hedvat Shorr, who read for authenticity.

My gratitude to Chaim Shur for permission to use the story of the death of his son, Avida, in Lebanon. May his death stand as a memorial to all the fallen.

I owe profound thanks to the Rockefeller Foundation for granting me a residency to write at the Villa Serbelloni in Bellagio, Italy, where the cordial atmosphere of the staff and my fellow residents together with the bracing air of the Alps lifted my spirits and cleared my head for working to my best capacity. I owe thanks and appreciation to Joan Cohen for her deep interest, and for her drawings to suggest a concept for the book jacket.

To my valued Jewish friends in Israel and no less my Palestinian friends, I owe the privilege of knowing them as distinct individuals in the family of humankind.

THE SEA OF STONES

CHICAGO

CHAPTER ONE

A<small>T THE END OF THE</small> underpass the facades of apartment houses stood like old men in shadows. The rainbow painted in the underpass faded into what she knew existed inside the buildings: kitchen tables with cracked porcelain tops, crucifixes hung over beds, lovers tangled with each other. Every time Freya passed these buildings on her way to visit her father, something in herself was aroused by imagining the lovers.

In the park near the apartment house where Isaak Pushkin lived were an oak and a cottonwood. Inside Poppa's building, at least the wallpaper didn't bulge and crust from too many layers, the plumbing worked, the floors in the entrance hall and on the stairs were clean. She stopped to shift her shopping bag at the place where she had set up a lemonade stand and where Poppa put her out of business. "What kind of profit making *hundler* are you anyway? Who put you up to it? Your cousin Evelyn?" Issak Pushkin gave her to understand that she was a dupe for capitalist imperialism. The lemonade stand had been dismantled.

The stairway was dark: inevitably a light bulb was out and she had to feel her way up to the second floor. The change in smells caught her off guard every time. Olive oil instead of chicken fat, tomato paste instead of sweet noodle pudding. When she was five steps up, Freya remembered holding the lilacs for Dora who lived in the basement apartment. The lilacs could wait.

She lingered at the door before she used her key to let herself in. Since Momma died, she had trouble opening the door knowing that her mother would not be on the other side, waiting, her face awash with relief when she saw Freya. Momma's face was always full of wonder, even when Poppa repeated and repeated the exact same stories about the meetings of the threadbare remnants of the Workmen's Circle that she was rarely well enough to attend. Momma's excitement was in her living room where she could sit like a shadow, observe people arguing fine points, finding morale, finding themselves. Close friends, even friends like Dora, seemed too possessive, as though to win people to herself was to misuse Poppa's cause.

Freya opened the door and put the light on in the kitchen. There were dishes in the sink and two slices of stale halah bread on the table.

"Poppa!" she called out.

His holler came from the other side of the apartment. "You know where I am!"

She put the groceries she brought far back on the bottom shelf of the refrigerator and she filled the empty coffee pot with water for the lilacs. She searched her purse to see if she remembered the cigarettes before she went to his room down the little corridor off the kitchen.

He was lying on the bed like a long shadow over the white sheets. "Poppa," she said. "Get up! I'll raise the blinds."

He didn't answer which was his usual consent and when light opened up the room, he squinted, shivered, lifted himself heavily, one foot eased to the side of the bed, then the other. He was fully dressed in blue serge pants and a shirt buttoned to the neck, shirt sleeves rolled to the elbow, the whole look of him awkward, bony. Arthritis had made claws of his hands but no one had known about the arthritis until it had crippled him. Issak Pushkin was not a man to complain of pain. For a man like Pushkin, pain was the mark of an honest life.

"Get up, get *up*, Poppa. I'll make you some tea."

He looked at her from the corners of his black and heavy-lidded eyes. "Today is Friday. You come on Saturday."

She felt herself shrugging and tried to stop the gesture before it was complete. She never shrugged in Beaulac but here her

6

body went its own way. "I know," she said. "I took leave from the Museum."

He glanced at her obliquely. "Give me a cigarette."

"You shouldn't smoke."

"Give me a cigarette. Who told you I shouldn't smoke? Karl said so? You still listen? You already divorced him!"

"Dr. Kaplan said you shouldn't smoke."

"Give me a cigarette."

"I'm not going to bring them anymore. You'll *see*." She saw his face turn red, his mouth twist with frustration as though he couldn't make his slightest wish work for him. She knew his litany: the world refused him anything, everything.

She opened the pack of cigarettes and took one out, lighted it and inserted it between his fingers, watching while he brought the cigarette to his mouth and inhaled and blew out the first long drag. A cough sputtered out. He took the cigarette out of his mouth and caught his breath, held it while the deep wheeze filled the air around her.

"I brought a suitcase," Freya said.

"What for you brought a suitcase? There was an earthquake in Boo-Luc?"

"Beaulac," she corrected him. "It's only three weeks since Momma died. I thought just maybe you would need me." Without looking at him she picked up socks and underwear and two shirts lying across an over-stuffed chair that was part of this room for as long as she could remember. The chair needed a dusting; the whole apartment needed cleaning out. Momma's clothes were still in the closet.

"How do you feel, Poppa?" she asked, turning toward him. She would never bring cigarettes again.

"How does it feel to die?" he said, looking at her directly, narrowing his eyes while he handed her the cigarette to crush in the ashtray.

Pretending to be flip, she said, "You'll never die."

"I'm not human? I'm human."

"Poppa, why don't you sit in the park while I make lunch? It's so hot, the streets are cracking."

7

Pushkin shrugged. "The streets are cracking? Let the government worry for their property. I haven't got no property."

She stared at him. How was it possible for Pushkin to be different from everyone else? Not to mellow one brain cell's worth. Whatever Pushkin's personality was made of, it had come with a lifetime guarantee.

"That's right," she said. "Pushkin hasn't got no property. He's independent."

He lifted his eyebrows and snorted. "Independent? If I have to take a piss, it depends on how much the alderman owes me." He motioned her closer with his elbow and she sat down on the edge of the bed. "Give me another cigarette," he said.

"You want to kill yourself?"

"I got my way, you got yours."

"No cigarettes."

He opened his eyes wide and made a feeble attempt at a whistle. "Go to Boo-Luc."

"I'm sorry Beaulac doesn't suit you."

He looked at her sideways. "It suits you?" he said.

"Poppa, if you don't stop, I'll go home as soon as I give you lunch."

"Don't get excited. I didn't mean nothing."

When she looked at him again, there were tears in his eyes; she saw his hand quiver and she felt a drawing in her fingers. Why did he have to show that he couldn't be touched or wounded, that he couldn't be shaken? But that was his pretense. From the time she was a child he protected himself from feelings as though his last defense would be ripped away.

While she stared at him, she remembered his strong, sure footstep in the hallway, nervousness on Momma's face when she heard him at the door, and then her eyes brightening when he burst in with a look of triumph—the union had six new members! The world was getting better. And Momma would laugh with pride and bring out the honey cake Dora baked for them. The world would be better, a world of hope, because the union had six new members. But Poppa had been sliding to the Left of the Socialist party. More and more he admired Lenin and spoke in-

8

cessantly of "the ten days that shook the world" and "the dictatorship of the proletariat" as compatible with American democracy. He was exalted by the heroism and sacrifice of the Russian people. How many times had she heard his monologue about his early years in America, how he had defied the profusion of splits in the Jewish Socialist Federation and stuck to his own version of secular Jewish trade unionism in a Workmen's Circle. It was Poppa who had insisted that she go to the eclectic little school for children of workers, the *Abraham Lincoln Schule,* that would incorporate secular Jewish socialism, universal communism, and selective American history—Moses, Lincoln, and Lenin merged to chaos. It was only Jewish nationalism, Zionism, that was anathema.

His eyes were brimming over and Freya pulled out another cigarette and lit it for him. She put the pack on the table next to his bed. "Poppa? Do you want me to stay or not?"

"It makes no difference," he said.

"And that's why you call me every morning? Because it makes no difference?"

She had caught him off guard. He turned his head to look at her and turned away again. She felt herself softening; Poppa was no longer sure of his world; he seemed to have no place in it anymore. For Poppa, having been was his being.

She went for the Bissel carpet sweeper in the closet. There was a boundary of bread crumbs around his bed. "Last week I left cooked meals in the freezer. Are you eating them?"

He turned on his side, away from her. "I eat only kosher," he said. "Like your Uncle Saul."

Her face went hot. "Kosher!" she cried. "Is that your idea of a joke? When Momma wanted to eat kosher you said she was a dupe of the priests!"

He was pursing his lips. "Kosher," he mocked. "So long as the Jews ate kosher what did it matter that their heads got skinned after the meal?"

"Terrific," she said, attacking the floor around his bed, remembering how Momma ate unkosher meat as though it carried the plague.

9

"Freya!" She heard his voice boom and she saw his eyes, narrowed and enraged. "I don't need no Kapos here. Go home!" He jerked his elbow at her to motion her away from the bed and then he lowered himself to the sheets and sank into the mattress, a muscle twitching at his temple when he closed his eyes. "Shut the light."

She straightened up; she was clutching the long handle of the Bissel and she stood there watching his frame heave, the dark hulk of him moving up and down. She started toward him but it was no use; he had tuned her out.

She was washing dishes that had been gathering dust in the cupboard when the phone rang. A joyous hissing came out of the ear piece when she said hello. "You're here!" Dora cried.

"I'm here, Dora. How are you?"

"God bless you. I wanted just to make the telephone call. A telephone is so good. I should come upstairs to see you, maybe yes?"

"Sure, Dora, come up."

"Dora is coming up," she called out to Pushkin. She put the sweeper in the closet and went in to look at him. His look back at her was downcast and grieved. "I know what you got there inside you," he murmured. "You blame me. I did something wrong to you? I did something wrong to Noah?"

Somehow he had swung his legs over the side of the bed; he was sitting up but his hands were shaking. The door clicked and Dora came in. She sang out, "Hello Pushkin, hello Freya! God should bless you."

"What do you want?" he hollered back at her, wiping his face with the edge of the sheet. He lifted himself up and, slowly, he crept through the doorway of his bedroom toward Dora's voice.

She carried a large jar. Her hair was disarranged as though she had been rushing about. She only smiled at his yell.

"What should I want, Pushkin?" she said. She held up the jar. "Compote! I made it myself. I'll put up tea. You see, Pushkin? If a dish of compote with tea is good then Pushkin will come to table."

Pushkin groaned. He motioned Dora away with his hand. But he had seated himself at the table, smoothing his tangled hair. Dora took small bowls out of the kitchen cabinet and began filling them with apple and raisin compote that she stored in her cold basement apartment.

"It smells wonderful, Dora," Freya said.

"You didn't taste it yet," Pushkin said.

"It looks delicious."

"When you taste it, you'll give a compliment."

"He's right," Dora said. "Pushkin is right." She was smiling. Her face flushed with excitement and the little curls around her head seemed to dance. Freya stood there, looking at Dora, artless and brimming over with what everyone mistook for simplemindedness. Dora was not simple-minded. Dora was not secretive like Lena Pushkin who thought she had an ally in God, someone to listen and forgive her for pretending atheism to please Poppa. Dora's belief in God had nothing to do with supplications; her belief celebrated the existence of the world.

They drank tea and ate compote spoon by spoon; the silence was as though each of them had vanished into another world. Pushkin sipped and Dora smiled at him. He ate a spoonful of compote and Dora nodded at him. Adoration thickened the air.

Freya looked at her watch and remembered. It was almost noon; Karl held morning office hours and he might still be there. Pushkin and Dora sat quietly sipping tea.

The nurse answered. Yes, he was still in but almost out the door; yes, she would get him.

There was silence. "Karl?"

"Freya? What's going on?" he said.

"I wanted to ask a favor . . . I would appreciate it, really, I would."

"I'm practically out the door."

"It won't take a minute. Can you take Hammer for two weeks?"

11

There was a silence. Then, quietly, almost in a whisper: "For two *weeks, why?*"

"I've decided to stay with my father, you know, set him up. And it will take two weeks to get this place clean. Can you take Hammer?"

"You set Pushkin up in two weeks and you'll get the Nobel Peace Prize. Anyway, Henry won't walk Hammer for two weeks. He's not the only dog in the building."

"Just this once? I'll get Henry something special."

"That's not the point."

"Karl! We share custody of Hammer! Why can't you do this for me?"

"I'll talk to Henry."

"Otherwise, I'd have to put Hammer in a kennel."

"You don't put Hammer *anywhere* until you check with me."

"I just checked with you."

"He gets depressed in kennels. Last time he lost five pounds."

"I wouldn't send him to Auschwitz," she said.

"Which two weeks?" he asked.

"Starting today."

"I know what's going on, Freya. Didn't I tell you the old man would squeeze you?"

She cupped her hand over her mouth and began mumbling so Pushkin and Dora wouldn't hear. "The truth is, Karl, Dr. Kaplan said my father needs me here for a couple of weeks. I should have stayed after the funeral. That's what he said. . ."

"Freya? All of a sudden your voice sounds like there's a bag over your head. Where are you?"

"Karl, yes or no?"

"Dr. Kaplan is it? Some oily story."

"Kennel? You decide," she said.

"I have the key to the house. I'll get him tomorrow."

"Hammer needs to be fed! I *can't* go back to Beaulac tonight. I was so sure you'd get him, I didn't leave a key with Claire."

"Why so sure? You never think ahead, do you? That's what makes you so damn vulnerable."

"This is no time for a lecture, Karl!"

12

"You are a pain in the butt . . . I'll get Hammer tomorrow. That's the best I can do."

Pushkin was looking at her from narrowed eyes. "You brought a suitcase for nothing?"

"I have to go home tonight, Poppa," she said. "I'll come back tomorrow."

Dora's face glowed. "If not for Freya, who then? I'll make a pudding with apples! Not so good as your Momma could make, when she had strength, may she rest in peace and be blessed by God, but a pudding just the same."

Pushkin looked up at Freya through bulging red-streaked eyes. His white mane was disheveled over his head. He hadn't shaved and the coarse gray stubble spread over his face and under his chin and crept around his neck like an unchecked growth. His pockmarks were muted but his face was sunken and hollow with a bewildered expression like someone who had stumbled out of a dark jungle into a clearing where the sun was too bright. He held his twisted hands close to his sides so they looked clipped and useless. For the first time Freya saw how Lena's long illness had torn him apart. Every now and then his body convulsed in a shudder as though he were reliving all those years with a woman who was always dying and wouldn't die.

"You needed Karl's permission?" Pushkin said.

"No."

Dora's jar of compote was empty. She held it up. "See? *God* helped Pushkin come to table."

Tears filled Dora's eyes, a network of wrinkles creased her forehead.

"Don't cry!" Pushkin hollered at her. "You wanted I should come to table? So I came. Don't be a fly on my head."

Freya watched. All those years Momma listened and listened to his yarns about the revolution in Russia, year in, year out. Shots fired from the Aurora on the Winter Palace still sounded in this apartment. Smoke still lingered from Lenin's triumphant return to the train station in Leningrad. Momma didn't dare invoke the name of God. And now he winked at Dora's God. It was for Dora that he got out of bed.

13

There was no letter from Noah in the mailbox when she got home. Inside, the heat had sucked up the air and Hammer was curled under the table in the kitchen. She opened windows and turned on the fan and then she filled a bowl with kibble and meat and shoved it under the table. "Why don't you eat?" she said, bending down to look at the old Irish Setter. "What's wrong?" Hammer sniffed and turned his head. She sat down and kicked off her shoes. "Come on out good boy. Don't do this to me." But Hammer didn't move. She got up and poured herself a glass of ice tea and reached into a jar for a cookie. She bent over and reached out to Hammer. "You win," she said, and Hammer snatched the cookie from her hand. Hammer always won. Like Noah always won. Noah was an athlete, a star.

She turned the fan toward her and took a long sip of tea. There was a six weeks old letter from Noah on the table, the last letter she had, the letter he wrote back when she told him she was working again at the Natural History Museum. Her project was making labels for desert dioramas. She remembered boasting to her cousin, Evelyn. Noah's letter was poetry! *See?* Was she wrong to push poetry? And Evelyn had answered, "What scares me about your kid is that he might find what he's looking for."

Noah was trying to please her; his letter was florid. Not like Noah, captain of the Beaulac High School swimming team.

I camped for a week in the Sinai and all night I could hear the wind like voices of the prophets. It was SO still. In some ways it made me feel spooky—there was such force and power in the striped mountains and the white dunes and dust devils all around me, that for the first time in my life I felt that God was more than an old wives' tale.

Noah was bewitched, taken over, as though he had never grown up in Beaulac, as though it had taken but one summer to dissolve his identity. She read the letter twice again, though she knew every word by heart. When was he coming home?

Momma, I know you didn't want me to come here but that's because you didn't know that this magical land is forested with

14

pine, cypress, eucalyptus, and casurina trees. In the spring the
hills go wild with almond blossoms, cyclamen, narcissus and
anemone.

She should phone the Haases to see if they had a letter from
Michael. It was Michael's idea that he and Noah should go to
Israel for a summer, to work on a kibbutz—Michael was the one
who had turned Noah's head. Michael was the son of Holocaust
survivors, a living miracle. But Michael Haas was only a sports-
minded kid from Beaulac whose father had survived Treblinka to
make good in real estate. The Haases had survived. But underneath their smooth manner, Freya sensed a barbed defensiveness
toward those who knew their history and shook their heads in
specious sympathy or disbelief, or simply did not appreciate how
success was triumph in a scarred life. Though their sons were
best friends, the Haases had always been cool toward her as
though she had failed in some expectation or that she, far more
than they, was inappropriate living in a Gentile suburb. They,
who had already decoded the mechanisms of survival, were entitled to live where they pleased, but she, the Jewish wife of a
Jewish doctor, had pretenses about being so very American! Did
they think she walked about as though she were immune to the
never cooling fires of anti-semitism? She could not seem to establish a comfortable presence with them or understand their intense interest in Noah; it seemed that they had taken Noah on as
a Jewish responsibility. What good would it do to phone the
Haases?

She looked at Noah's letter again. What was he trying to tell
her?

You can't make friends unless you go to the army. Everyone
goes here. Girls, too. No one sits on their ass waiting to get
blown away. I know a beautiful girl here from Morocco—her
name is Carmella and you won't believe it, Momma, but she's so
much like YOU; she has long black hair and a dark complexion
and grey eyes. Carmella has been helping me with my Hebrew
but she turned eighteen and she went to the army, like everyone
else her age. I feel left out—I'm nineteen and I'm not exactly a

15

physical wreck but I walk around feeling like a cheat, living off this land and doing nothing to protect it from our enemies. Did I tell you that Michael Haas volunteered? I'm not sure he did the right thing, but I'm not sure he didn't, either.

Her neck was running with sweat when she slapped the letter to the table. Jewish history had taken on new meaning; the morning paper still lay on the kitchen table: June 4, 1982. Israeli tanks were in South Lebanon.

She mopped her neck while she stared at the letter . . . a girl who was now in the army . . . From Morocco? Noah wouldn't volunteer for the army because of a girl from Morocco; Noah was smart; he was cool. Michael Haas was different.

The fan was blowing at her eyes and distorting her vision so that she saw the room in parallax. The house hung around her like a loose robe; it was a fishnet to squirm in. She examined the kitchen, while Hammer tried to nudge her bare feet away from his bowl. The kitchen, she told herself, lacked nothing except a celebration of itself, neither high tech nor traditional; it was a setting, a habitat like the house itself, no one style prevailing, stucco. English Cotswold maybe. Percival Clayton, the original owner's dream house spiffed up to qualify as Beaulac fake historical. She wasn't worthy of old Clayton's legacy. Everyone in Beaulac knew the story about how Clayton plunged into the Lake every day of his life. In winter he would walk to the Lake with a raccoon coat over his body, which was bare except for swimming trunks and galoshes. Then on the beach two blocks away, he threw off his coat and galoshes at the snow-banked shore and dived in.

Percival Clayton himself had rewarded them for buying his house with a Swiss music box that tinkled out *The Moldau.* But someone more deserving should have his house, folks who would appreciate a public trust, huge rooms with wrap around windows that had an openness, Karl said was truly American—disclosure and unification, Karl's formula for assimilation into the mainstream. And that's where Noah had been reared, in the mainstream without dust devils or whispers of prophets or soldiers that quoted Isaiah.

16

She took her empty glass to the sink and when she passed the window, she stopped in front of the small mirror she kept on the sill and ran her fingers through her long, tangled hair. She could hear Momma's thin string of a voice, *God gave me such a beauty, a Russian princess. But Freya did not have merry eyes. Freya's eyes were too wide and gray—too deep.* And Poppa, the boss, was answering back from his wall of scorn: *Who did Momma think they were raising? Catherine the Great? If the noble Rosa Luxemburg had worried about the shape of her ass, would she have given her life for the People?*

Freya stared back at herself: no, she did not have merry eyes. She was not Catherine the Great. There was no use fooling herself; Karl had sacrificed himself to a woman who did not have merry eyes. Karl would win the Superior Person Award. Easily. Karl could always be trusted to do what he said was right, without annulment or repeal: second thoughts were weakness; his judgement was immutable. She, on the other hand, was left fragmented from considering all sides, groping for some slit of ambiguity to breathe through.

It was after midnight. She bent down to peer at Hammer under the table. "If you don't eat this minute," she said, "The state of Illinois will explode."

Hammer didn't eat. He followed her to the bedroom and settled beside her while she lay at the edge of the mattress, her hand resting on Hammer's head, while she imagined the morning paper lying on the kitchen table like a grenade. Israeli tanks had gone into Southern Lebanon. *What if and what if and what if. . . .*

Freya moved closer to the edge of the mattress. Since Karl moved out, she didn't feel entitled to full occupancy: the cool side of the bed and the unused pillow beside her existed in a space that wasn't hers. It was part of the silence and darkness of the room. On the wall opposite the bed was a photograph of Noah, a huge blowup she had made from a snapshot, when they were walking on the beach together, the spring before he went away. Noah in a T-shirt and bare feet holding out a stick for Hammer to chase, his tall, muscular body supple and animated.

What was lacking in his life? Love? She loved him fiercely. But what of the atmosphere of love that she brought with her from Frick Street, the feeling she had of defeat and lack of faith? On Frick Street, who spoke of love? Love had been stamped out by the corpses of Auschwitz. She was hardly five years old when Poppa had come home waving a Marxist magazine in Momma's face. "This is the work of the God you love so much?"

The corpses burned in Poppa like fire and he cried out that any of the pictures could be his own mother or Tante Luba or Labe the furrier. Momma began to wail as though the pictures truly were Poppa's mother or Tante Luba or Labe the furrier, and she, Freya, watching, had felt a hot stream of urine splash down her legs.

Poppa said her generation was stonewalled, stained with the blood of the Holocaust. No Jew would recover. For generations to come, every Jew, in one way or another, would suffer the damage. The effect on Freya was as though she had been dropped into poison; she expected only sorrow, never joy; if her life was of no importance why should she have wishes and dreams? Through Pushkin, through her, how much damage had rubbed off on Noah?

After Noah left, Karl badgered her for a full year. Over and over he told her that Noah, part Pushkin, was predisposed to *causes*, like diabetes. He was not surprised that Noah had taken himself off to Israel—could Pushkin exist without a crusade? His attacks were low-voiced but relentless, aimed, she was sure, at her, her father's daughter, her mother's daughter—genetically unfit to raise happy children. At first she was bewildered and wounded, then, one morning, months later, after yet another night filled with hurt and sleeplessness, she blurted out to Karl that he should move out. Afterward, she was struck with wonder at her strength.

Hammer was snoring on the floor. She got out of bed and sat down on the rocker, chilled in her nakedness and chilled in the void in her body that once she believed Karl might fill. Fair haired and angular, he seemed to have had an amulet quality

about him; if she let him love her she would be transformed—
she would become a Gould. Freya Pushkin, who could only love
Asher, would disappear.

She got out of the rocking chair and went to the bureau to put
Noah's letter away in a small metal box she kept locked, the same
box where she stored the copy of the last letter she wrote to
Asher twenty years ago. She unlocked the box and slipped Noah's
letter in while she scratched to the bottom and lifted out the letter she had not read since she wrote it twenty years ago. She took
the letter to the window where she stood, clutching it, not moving, looking out. The silhouette of her fallen tree lay on the
ground near the hedges because she had been stubborn and ignored the tree surgeon's advice that the roots of the spruce
wouldn't go deep enough. The tree had fallen over a week ago
without having been stressed by wind or struck by lightning;
over it went, still alive and bluegreen, not one frond decayed,
heaving and hurling itself onto the ground as though in a howl of
loneliness.

The letter was crumpled, damp from her hands, when she
went back to sit in the rocker. After twenty years the ink hadn't
faded.

Dearest Asher, my beloved,

*You did what you said you would do, you came back to get me
after a year, after I didn't keep my promise to live with you in
Israel. This time you hardly said good-bye. And I'm no closer to
leaving here now. Maybe I could have lived if I hadn't seen you
again; now, I'm not sure it matters. This last week we spent,
Asher, is what I'll live on. I know you didn't believe I couldn't
break away and I know you won't forgive me. But you saw
Momma with your own eyes. You saw how gruff Poppa is with
her. There's no one but me, Asher! Dr. Gould says that Momma
can live on for many years if she has good care. I'm the only one
to take care of her!*

*I ask myself over and over what a child owes a parent. Are
children not entitled to freedom? Ever? Are children not entitled to follow their dreams? I would never be such a parent,*

Asher, that would make my child want me to die so they could be free. I want Momma to die! I pray she'll die! How can I live with myself? How can I live without you? Our last week here was like the first time we made love on the beach when I was fifteen. We had six years. Last week we had seven nights. There isn't anymore for us, Asher. I feel it, somehow, like a hot coal in my heart. You musn't write to me again. I won't answer. You musn't come to get me again. I won't see you. Unless Momma dies, there's no way out! I love you, Asher, I'll love you all of my life.

Freya put the letter on the bed and sat there looking at it, her heart racing. She had not decided to leave him; it had been decided for her. A sense of Asher began to fill up inside her, fragments left over and lingering, the memory of his body pressing her into wet leaves and the whispers of promises.

She sat there, staring at the letter, remembering, finally, that Poppa knew, oh he *knew* how much she loved Asher. His way of trying to soften the blow of their separation was to diminish Asher in her eyes: Asher was especially stonewalled. *Asher had dreamer ideas . . . dreamer ideas . . . a universalist-nationalist? What sense did it make that he could be both? It was like sitting with one ass on two chairs.*

It was two o'clock when Freya replaced Asher's letter in the metal box and went back to bed; she lay there half awake, the tedium of her thoughts sinking into numbness.

She bolted up to a shrill ring. The telephone! It was still dark. Four o'clock. A second ring sirened through her.

"Freya?" Karl's voice was calm, bedside mannerish, like when he told patients they had cancer. She looked at the clock again tensing at his voice, holding on to the phone as though it would support her.

"What's wrong, what happened?" she asked. Something was happening. Something about Noah.

"Nothing's *wrong*," Karl was saying. "I wanted to confirm that I'll pick Hammer up today."

20

Freya sank back to the pillow; she caught her breath. She could feel a fast, hard, pulse in her throat. "Are you crazy? Do you know what time it is?"

"Freya. Take a deep breath. You don't sleep, anyway, when it's hot. And I've been worried about you since your mother's funeral. If you go to your father for two weeks, he'll squeeze you dry. For the old man, you're a pushover! I know how you are."

She sat up and tightened her grip on the phone. "I don't believe this," she said.

"Freya, for God's sake. I know how you are."

"Stop telling me you know how I am! You *don't* know how I am. Not anymore."

"Did I say you were a bad person?"

She felt herself stiffen in the moment of silence.

"Freya? I'll pick up Hammer but I want you to know. I'm worried about him. How does he seem to you?"

"You're worried about *Hammer*? On a Saturday morning before dawn, you're worried about Hammer?"

"We're sharing custody, right? And I don't think he's well."

"You may *not* have sole custody of Hammer."

"I didn't say that. I'm concerned."

"Since you've been up all night worrying. Hammer is obnoxious. There's a bitch in heat somewhere. He doesn't eat and there's something wrong with the way he pees. A trickle at every tree."

"He might have a prostate problem. Take him to the vet."

"He doesn't pee like that when he visits you?"

"The doorman walks him. He didn't mention."

"It's me, then. I give him angst."

"Take him to the vet."

"Listen, I've had three hours sleep or doesn't that keep you up nights?"

"You weren't listening. I'm worried. About the Audi too. When was your last tune up? The Audi is an expensive piece of property. When I said you could have it, I thought you'd take care."

Freya wiped the sweat from her face with an edge of the sheet. "I'll let you know if the Audi needs you," she said.

"Listen," Karl said, "Get Hammer to eat kibble instead of cookies. There's shit in those cookies you buy at Casey's."

"Anything else to pave our way to friendship?"

"What's wrong now, Freya? You asked me to do you a favor so you could stay with your father and I'm doing it, so why are you angry? I can hear it in your voice."

"That's why people get divorced. Right?"

"We said we'd stay cordial. I've kept my part of the bargain for a year. What will you do about Pushkin?"

He took her off guard. "Oh, my *father,*" she said. "*He* worries you. Now that's what I call a new leaf."

"Not your father. You. And I know what you're thinking, Freya. I'm worried about Hammer and the Audi but I didn't even ask if there was a letter from Noah. Our own son. I'm a rotten father. Right?"

"My God, Karl, what do you want from me?"

"And that's what happens when I mention Noah. You regress. You turn Pushkin on me."

"Why are you calling me names at four in the morning?"

"Listen, Freya, our divorce wasn't supposed to end up in nuclear winter. And when was your last tune-up? You should check every 5000 miles. . ."

"I heard you. Now can I get out of this bed to pee?"

"Which reminds me of Hammer," he said.

She hung up.

CHAPTER TWO

THE SKY WAS STILL overcast when she went downstairs after Karl's call. At half-light she and the house didn't fight; the forty windows didn't gawk at her. Through the window she could see hot pockets of blue beginning to form. Later she would try Noah's kibbutz once more, but the answer would be the same. *Yes; yes, he still lived there. They would give Noah Gould a message. No, she couldn't give any information about the army. Did she want to speak to someone else?*

The phone was ringing. Noah?

"It's George Sargeant," the voice was saying. "Tree service. Sorry to call so early on a Saturday morning, but we need to schedule. How about Monday for hauling away the tree?"

"Monday?" she said. "This coming Monday?"

"Something wrong, Mrs. Gould?"

"No. I thought you were someone else."

"I'm the tree service. Sargeant."

"Do I have to be here when you do it?"

"It's not exactly euthanasia, Mrs. Gould."

"Monday," she said. "I won't hang around. I don't want to see. . ."

"I didn't mean to hurt your feelings, Mrs. Gould. I know you loved your tree and all that. Monday then?"

"Okay."

"But a live tree committing suicide?" she had said to Sargeant when he came out to look at the tree. "Why should it want to do that when it had everything to live for?"

23

"I wouldn't put it that way, Mrs. Gould," he had said, and he looked at her as though she were deranged.

He didn't have to tell her: monkeys, sheep, and trees didn't commit suicide. But to her, the tree was like Wordsworth's poetry. *One impulse from a vernal wood, May teach you more of man, of moral evil and of good than all the sages can.*

She was at the kitchen sink squinting against the glare shafting through the window when the hot light caught in her eyes. Fragments from Yeats settled in her head: *I will arise and go now, and go to Innisfree . . . Nine bean rows will I have there . . . A hive for the honey bee . . .* What was left to keep her in Beaulac?

Hammer's high-pitched yelp got her up. He wanted out. She bent down to the old Irish setter and nuzzled his ear. Hammer panted in her face.

"Something wrong, good boy? Why didn't you eat? I'll take you to the vet."

She opened the back screen door and let him out. "No street!" she yelled after him. "No *street*." And then she remembered it was hardly seven in the morning and that her neighbor, Claire, was disgusted with her for letting Hammer out on his own.

"Hammer!" she called out the door. "Hammer, *come!*"

But it was Claire coming toward her, looking like Christ Church and golf at the Lake Shore Club with her short blonde hair that never got messed no matter what. At seven o'clock in the morning Claire looked balanced; she was always balanced; she exposed what she was and what she wanted. She never explained, she never apologized. Her Christmas card lined the family up against the fireplace and took aim; even the cat looked merrily dumb-struck.

"You'll wake up the dead," Claire said, and then she put her hand over her mouth. "I put my foot in it. . ."

"It's okay," Freya said, smiling.

"Where were you last night?" Claire said. "I phoned you."

"I went to my father's in the city."

"One of these days you'll get mugged in the parking lot. You should have an escort. You should get married again."

"So I can visit my father?"

24

"I thought you might need company last night," Claire said. "And I wanted to share my thoughts about Jewish funerals. It was strange. . . ."

She smiled at Claire. *Share her thoughts.* "You weren't sure Jews die, Claire?" she said. "Jews go on eating up the world's resources? Forever? Like my mother?"

"Freya! You're lucky I didn't hear that," Claire said. "What's wrong with you, anyway? You're a beautiful woman! You drive men to drink. What are you saving yourself for?"

"For Christ's sake, Claire!"

"Okay," Claire said. "Enough." Her feelings were hurt. She turned to go and glanced back. "Get that tree hauled away before it rots your grass. And get your damn dog out of my swimming pool."

Freya called out to her. "You're my best friend, Claire!" The remark about the Jewish funeral was only Claire's dip into Jewish outer space.

"Hammer!" she called. "Hammer, you better come!"

Hammer raced over through a break in the hedges between the yards and stopped at her feet shaking water off his fur. "What's the matter with you, anyway? You've got no manners, no class. You're not fit to live in Beaulac." She nuzzled Hammer's ear and looked over at Claire's yard.

She let him into the house and glanced back at the garden. Thank God it rained before the new flower seeds died. Today would be a scorcher. There were red streaks in the sky left over from sunrise and heat and pollution that drifted up from the Gary mills fifty miles away. The air had a thin, misty appearance that distorted space at the back of the yard where the fallen tree lay green and dead. A quirk. Like she was a quirk in Beaulac, putting on airs so she wouldn't be marked as a child of bricks and cracked pavements, a stranger, a transplant like the tree. The verse obsessed in her mind. *I will arise and go now and go to Innisfree . . . and go now and go now and go now . . .*

She looked at her watch when the phone rang. My God, seven-thirty, on the dot. She caught the phone on the fifth ring and Pushkin's voice barked out. Every morning, call to battle, catching her at seven-thirty to the second before she went to work. Poppa's hello and good-bye was her second waking, a cold plunge into concrete.

"It's Saturday," he said, without a hello.

"I know."

"SO?"

"I'll be *there*, Poppa. I told you I would be back today. For God's sake, Poppa. First Karl. Now you."

"*Karl?* The big doctor began his rounds on you?"

She didn't answer. Swipes at Karl was Pushkin's sport. There was a symbol in Karl that proved something about a world that had landed him such a son-in-law. He barely tolerated him as Momma's doctor. Upper-class German Jews were a knife-edge away from butchery. And in Pushkin's opinion there was something cruel and carnal about Karl's long legs. The world was knifed by long legs. Polish Counts. English lords. Pushkin even made up a story about a demonic Pope with long legs. Pushkin was high on his sport. When Karl told him once that people who suffer and suffer and suffer must be in love with suffering, Pushkin closed his eyes as though blacking out something foul. "Hungry people like to suffer?" he asked, accentuating hungry as though he would swallow Karl on the next word. "*Slaves* like to suffer?" Pushkin had put his heart into the s-l-a-v-e-s as though reading from Exodus. Karl shot back, "If you're not hungry and you're not a slave and still you go on suffering, you should look into it." Pushkin had risen from his chair, his face pale. "You *hear* what the big doctor says?" he murmured to Freya. Then he blew his nose.

"Are you there or not?" he was saying now.

"I'll see you, Poppa. Soon."

But he wasn't finished. "Boo-luc," he muttered. "People ask me where does your daughter live? Boo-luc, I say. That's a *place* they ask me? Boo-luc is a place or is it the name of your daughter? I'm hanging up," he said. And he did.

There was no use talking to him. Even his brother, Saul, five years younger and a widower for 33 years, couldn't persuade Pushkin to stay with him for a few weeks after the funeral. Love

26

did not abound between the brothers Pushkin. When Uncle Saul got rich he acquired religion and a $350 prayer shawl.

Freya hung up the phone but Pushkin's drumbeat was in her ear. What was his own and what belonged to someone else melded into one. Lena Pushkin existed now because she didn't exist; after a twenty year illness she was the new fact in Poppa's life. He carried on. He railed. Turn off the lights. Pull down the blinds. *His wheel had come full circle.* One morning he reported that he woke up to a ringing in his head like a dentist's drill, another time the smell of candle wax was in his nose; shadows were in his eyes the last time she went to see him, shadows with every light in the apartment turned on.

And then he would inflect his voice like a shrug: Did *she* care? She, meaning Freya. As for Noah, when he turned seventeen, he went to test his mettle in Israel. Pushkin didn't have to say it. The divorce from Karl was his only sign that his daughter could be redeemed.

She gathered up the trash and took it out to the back alley, past the fallen tree. When she turned back to the house, her cousin, Evelyn, was coming up the driveway, swinging toward her, blue denim jumpsuit smudged with mixtures of artist's pigments which made her look bespangled. Evelyn had enough swish to destabilize the Beaulac Historical Society.

She was carrying a plastic bag marked, *Bumpers*, and Freya sighed. Evelyn had come by every day since her mother died. "I brought you some fish," Evelyn said.

"At this hour in the morning? Doesn't anyone *sleep?*"

"Bumpers had a sale. Yesterday."

"Just what I need. Fish."

"Shark steaks. How about that?" Evelyn said.

"You cook them on a lance?"

"Aren't you going to put the shark away?" Evelyn asked, setting up a lawn chair that had flipped over near the vegetable patch.

Freya was standing over her and Evelyn looked up in a way that announced her intention to sit.

"So," Freya said. "Coffee. Right?"

27

"Right. I have to talk to you."

"You've been talking to me every day since Momma died."

"I thought of something I haven't said before," Evelyn said. She nodded her head toward the tree. "I notice what quick action you're taking on that dead phallic symbol."

"Let's go inside," Freya said, loosening her hair from the rubber band that held it. "I have to get ready. If I don't get to Pushkin before noon, he'll call the KGB."

"So let him," Evelyn said. She had settled herself in the lawn chair, her legs stretched out in front. She took a crushed visor cap out of her pocket and perched it to shield her eyes.

"Evvy! I have to go. . !"

"It's hot, hot, hot," Evelyn said, puffing up her cheeks and blowing out. She looked up at Freya. "Why don't you wear a hat? Just because you have indestructible looks doesn't mean you won't blanch out. You're forty-one. Too bad you never figured it out with Lottie."

"It's a good thing Lottie had the sense to die," Freya said. "Lottie the Good hadn't heard a word you said for years. How, tell me, did an 82-year-old deaf psychoanalyst stay in business until they stretchered her out? What did you need it for?"

"You exaggerate. Naturally. That's you. Lottie wasn't deaf for more than two years. By then she didn't need to hear what I said. She knew. And there wasn't much left to say. It took me two years to acknowledge her deafness and the fact that she'd probably die on me. Separation. You *cope*. You walk around like Mrs. Miniver."

"I wouldn't have had the time for Lottie. Didn't you tell me it took Lottie three years to get you to cry and three years to get you to stop? Two years to acknowledge her deafness and two her death adds up to eight. Right?"

Evelyn sat up straight. "Now that I'm kicked in the ass, should I leave?"

Freya kissed the top of Evelyn's head. "I'll make coffee. So I'll be late. Why should you care?"

"Get a hat!"

Outside, she handed Evelyn a mug of coffee and watched her take it between her hands in her deliberate, posturing way.

Freya dragged over another lawn chair and sat down.

"So?" Freya said.

"Where's your hat?"

Something in Freya resisted. After Karl, Sargeant, Pushkin, and the shark steaks, she had had enough for a day. And the morning wasn't half begun.

"You better find a way to work it through with Pushkin," Evelyn said. "Or maybe you want to stay paralyzed for the rest of your life, pretending. Your mother did that."

"My mother did not pretend. Not *my* mother. She accepted."

"Sure she did. The only place she could pray was sitting on the toilet."

"Okay, Evvy. That's it." She felt jarred, intruded upon." I hate that expression, *work through*. It makes me think of how Momma used to pull elastic through my underpants with a safety pin."

"I don't care what you hate. You never got one chance in Hell . . . !"

"For Christ's sake, Evvy!"

"I mean it," Evelyn said. "You better get things straight with the old man before he dies."

"Don't worry. He won't die. They'll have to chip him out of the Albany Park with a jack hammer. Anyway, how long did you agonize with Lottie? Eight years? And the last two gave you chronic laryngitis from screaming into her ear. That's working through?"

"Go ahead. Heap scorn," Evelyn said, looking away.

She felt a sudden fury at Evelyn who had taken off her cap again and was settling it on her knee, examining her nails.

"Why are you fighting me?" Evelyn said. "I know what's eating you. Don't you think I love Noah too? Noah wouldn't get caught in a war. Noah is special . . . he's smart . . . Noah is a lamb. . ." Evelyn's voice was breaking; she looked away. "Unless he had to prove something to the Herr Doctor Gould."

"What do you mean, prove something?" Freya murmured.

"Forget it," Evelyn said. "I have a textbook mind. You've *got* to work it through with your father," Evelyn said. "First of all, he needs you." Evelyn reached out for Freya's hand. "Don't look at me that way. You were nailed to your mother! It killed me to watch. All those years. So let yourself get exploited for two more weeks. Find out what he needs and get out. When you've emptied the garbage, you'll know what to do."

"Do about *what?* The divorce is final. Noah is thousands of miles away."

"You won't just sit here. You'll act. You'll do something." She broke off and looked at Freya.

"What's that supposed to mean?" Freya said. "Poppa wouldn't let me work anything through. He wouldn't give me a moment's peace."

"Because you won't give him a moment's credit!"

"For what?"

"He's not such a bad guy. He loves you, Freya . . . just because he's been through so much in his life he can't show it? Think about that."

"No, Evvy girl! *You* think about it. To Pushkin you're a *moojik*. He thinks you get gussied up for the landlords. And he didn't like your hat at the funeral."

Evelyn laughed. "So I'm a *moojik*. So what?"

"Evvy! Stop lecturing me! I'm going there! I was there yesterday. I'm going back today for two weeks. So don't you dare say another word. God, when I imagine actually sleeping in that apartment, I can't, I get the shakes. You should have heard him on the phone this morning."

"The shakes are better than crawling into a hole. You were a promising writer until you started writing labels for the museum. How about writing fortune cookies? That should keep your mind off your life."

"Is that *all*, Evvy?" She wasn't listening anymore.

"You have to forgive. . ." Evelyn said, rubbing circles into her knee.

"Oh sure. Pushkins are big forgivers like the forgiving brothers, Saul and Issak Pushkin. Neither of them can start a sentence without, 'I'll never forget how my own brother, Issak,' or, 'If I live to be a hundred, I'll never forget how my own brother, Saul. . .' " Freya said.

Evelyn kept rubbing her knee, not looking at Freya. "Have you seen Karl?" she asked.

"He gets Hammer every other weekend."

"How does he look?"

"How does Karl usually look? Overbooked."

"Is he having an affair?"

Freya narrowed her eyes. "What you need, Evvy, is a part in a play, some nineteenth-century drama. How should I know what Karl does in bed?"

Evelyn shrugged. "It's better you divorced him. He loved you for your body."

Freya smiled and kissed the top of Evelyn's head again. "Never mind. And listen, you're no *meeskeit*. Not to me. Go home already!"

She stood there after Evelyn had gone; she was drained, her thoughts racing until she let herself go loose and sank down on the lawn chair. Then she got up again to pick lilacs for Dora.

Her watch said ten o'clock. Pushkin would be revving up. She'd take him the shark for dinner. Perfect. Poppa was in the apartment now, alone in grief, desolate, frightened, maybe. Pushkin? *Frightened?* Was she inventing sweet songs about Poppa? Freya started quickly into the house, Poppa's limerick running in her head. *Gorky and Lenin were listening to Beethoven: Lenin was in ecstasy and said to Gorky, "This music puts me off guard; it makes me want to pet dogs and children's heads. But I know if I did they would bite my hand."*

CHAPTER THREE

Every day for two weeks while Freya stayed with her father, she told herself to cook the shark steaks. But Dora had cooked something Poppa liked better. During the day Pushkin sat up in bed with *Das Kapital* in front of his face; the book was ragged, pages hung from the binding. Freya found her old library card in the kitchen drawer and went to the branch library on the boulevard to get Poppa books from the New Deal: *Roosevelt and Hopkins*. He took one look and told Freya that if she knew what he knew about Roosevelt she wouldn't read his lies. When Jews begged him to bomb the concentration camps, did he? *Did he?* She took the book back and got him Shirer's, *Rise and Fall of the Third Reich*. "You don't think I know how Hitler rose and how he fell?" Poppa scowled. "The capitalists put him up, we tore him down."

Every morning she called Claire to check for a letter from Noah. Every morning she hung up the phone dispirited and went out with Dora's shopping list; she walked to Mack's lot to check on the Audi, then down the length of Albany Boulevard to sit in a coffee shop where she could sink into a booth and read the paper. *Israeli tanks had gone beyond the Litani River. Casualties were high.* Every day the headlines printed maps, pictures of bombing and crazed Arab women running with children. The headlines, maps, pictures terrified her. Where was Noah? She could not think about Israel. She knew the Bible purely as

32

literature. Modern Israel had been a gap in her education; it was not indifference she felt but a lack of questions. She accepted modern Israel because she was a Jew, like the color of one's eyes that no one could change; whether you liked them or not, you went on using the same eyes. But her opinions vacillated with what she was reading at the time. Palestinians were right. Israel was right. Still, the Holocaust haunted her . . . pictures of corpses in Poppa's newspaper . . . urine splashing down her legs. And Momma telling her: "Love your own, not the world." *Had Momma whispered that to Noah?*

Freya went back to the apartment to scrub: kitchen furniture one day, living room next; she scrubbed walls and floors and finally she scrubbed down her own little room where she had slept as a child. All week she lay awake there for hours, the street light beating against the window blinds.

Evelyn phoned every day and every day she offered to come and help out, and every day Freya said, no. "Don't start on me, Evvy. I'll see you soon. When I set Poppa up, I'll get out. Like you told me to do. Anyway, he's getting set up with Dora."

"What's that supposed to mean?" Evelyn answered, astonished.

Karl phoned to tell her that, no, he hadn't had a letter from Noah, either. And Hammer had been to the vet and there was nothing special. He pees like an old dog pees.

"It's almost two months since we haven't heard from Noah," Freya cried into the telephone. "I should phone the State Department."

"Don't get worked up. Noah is busy; he's using himself up so he won't have to face medical school."

Dora was filling in everywhere; she had even persuaded Poppa to talk to his brother, Saul. Oh, how Momma wanted to bring the brothers together. She had prayed to God so fervently for the brothers Pushkin. But Dora was doing business with a God Pushkin could understand. Momma was sick all those years while downstairs sat Dora.

Freya sat in the coffee shop stirring the spoon in her coffee, back and forth, back and forth, a sense of displacement coming over her as she wandered over the crags of her life.

After ten years at the Museum, Freya didn't mind the brackish odors that wafted over from the taxidermy lab down the hall, acerbic, cloying. The Museum was always a perfect refuge, a stillpoint, a certainty of cycles and survival, where she was encircled by natural history, immersed in a world of creatures and habitats and ecological chains that were traceable, predictable: layers on layers of ancient rocks, sediment from peaceful accumulations of creatures' skeletons in the sea to violent eruptions of lava flows. Cause and effect laid the ground rules: nature knew its boundaries. The papier-machè existence of the Museum was safe haven; it kept her life balanced in a world where she could write labels about where beetle eggs hatched and spiders worked their shifts. The Museum fortified her against her bankrupt marriage to Karl, the relic of her love for Asher, and the silences left in herself after growing up on the steppes of Poppa's revolution. Natural laws were not programmed for revolutions. In the museum atmosphere she could take pleasure in the omniscient where she could experience summer or winter, wilderness of jungle villages, and see the world created with styrofoam and exotic materials secured in perpetuity with epoxy. The flaw was that in the human world people were not like slime molds who reassembled themselves from one stage to another in order to survive. Slime molds were more careful to survive than Noah was. Slime molds were not prey to ideologies.

She was running her fingertips over the soothing formica of the table, staring straight ahead, her legs crossed, and, then, with a start, she uncrossed her legs and stood up. If she didn't get out of here she would sink under the weight of her mind. If slime molds tried to survive why shouldn't she?

It was during the second week of her stay with her father that her mother's death took on reality. Momma's absence was palpable in the dusky, cramped living room where Freya had taken to drinking coffee by herself. Nothing of Momma lingered, not a whiff of antiseptic or bath salts or lemon skin lotion Freya used to keep Momma's skin moist and the odor of waste from seeping into her pores. The dent in the overstuffed couch Momma had occupied had disappeared.

Freya felt herself shiver. Oh how she had wished her mother would die. At night, when she slept in her little room, yearning for Asher, she wished her mother dead. *Good will come back to you, Freyala. You'll see.* But it was too late. Asher was lost to her.

Pushkin began reading the *Chicago Tribune* again, his style picking up from where he had once denounced the Mayor, the aldermen, the foreign correspondents. They were nothing more than tools of a rotting system—they were liars, conspirators. The real truth was the same truth. He knew. The rats in the street were initialled C. I. A.

For the past week, when Dora turned on the radio, he listened, but it wasn't until Thursday of her second week that Pushkin mentioned Noah. Until then, he had said nothing, asked nothing; he pretended not to hear when she phoned Claire to check for mail. There were more news broadcasts. He heard.

Her two weeks would be up Saturday and the thought of going home to Beaulac was heavy inside her. Everything she wanted to escape came leering back at her. If she sold the house, then what? More labels at the Museum? A schoolgirl's affair? In two days she would leave her father's house. Except for Momma's boxes in the closet everything was done. The boxes had been stored high on the closet shelf, side by side, unopened and accumulating dust for years. When she was a little girl the closet was scary and irresistible, Momma's private space like Ali Baba's jars.

Pushkin called to her from his bedroom. "You bought the cigarettes?" he said. She came to stand over him. "Cigarettes," he repeated while she stood there looking at him, not wanting to cross him. Words she never meant to speak streamed out of her mouth: "Dora won't like it." He looked at her sideways. "Something is funny with you?"

She shrugged. "She worries about you."

"That's what you meant?" he said.

She nodded, avoiding his eyes.

"You didn't have to stay with me," Pushkin said. "I didn't need no Kapos."

"I'm sorry. I didn't think."

"You're Freya the Thinker? You think about Noah? You don't know what's going on over there?"

"It's *you* who doesn't want to know! It's killing me that you won't even talk about Noah. Why can't we talk like father and daughter? Noah is your grandson!"

He caught his breath and eyed her intensely. "Listen to me!" he boomed. "Listen," he repeated, his voice falling as he stared out to the room as though visions were glaring at him from the walls.

"Listen. You hear? When the Red soldiers needed food my father didn't sit at home with swollen eyes to cry that his brothers are starving. He went for hundreds of miles on top of a boxcar with two sacks of bread and a sack of rice. My mother told him he was crazy, he was a *shlemeil* to risk his life—for what? How many could he save and for how long with two sacks of bread and a sack of rice? My father answered her back that it was not how many; it was not for how long. It was the idea of going to rescue. To go and rescue, my father said, is also to rescue yourself."

And suddenly something burst in her. She sank down, her forehead resting on the frame at the foot of the bed. "Poppa!" she cried out. "Don't do this to me! You know I haven't heard from Noah. I don't know what's happened to him!" She felt her forehead press into the wood frame while Pushkin, propped at the head of the bed, let out his wheezing.

She bolted up. "Poppa, there's something in Noah . . . he doesn't want to be rescued."

Pushkin frowned. His face was softening, he leaned toward her. "He doesn't *want?* He doesn't want is his business. You go to rescue. That's your business."

The pain on his face was not from gnarled fingers. She reached out to take his hand; instead, she went to the kitchen to bring the pack of cigarettes she brought from the supermarket. She took one out of the pack and lit it for him and put it between his lips.

"Okay?" she said, and turned out of the room so he wouldn't see the tears on her face.

When he called out to her again there was something in his voice that turned her back quietly, a tremor, a scratch, a catch of confusion. He was sitting there with his head down, disillusionment hunched into his shoulders as though he had lived long enough to see his effect on the world.

"Don't look like that, Poppa."

"What did I do wrong to your life?"

"You didn't do anything wrong. Don't work yourself up before Dora gets here."

"Don't Dora me! I did something wrong to you. I didn't make it easy for you to run away with your Prince? Your life was so terrible? Even you took care of Momma, you went to college. A full scholarship. You studied. I remember your books."

"You don't remember my books. I never showed them to you."

"I don't remember? From Donne to Dryden? From Blake to Byron? You think I forgot what you dragged home?"

She felt herself wince, regressed to a child caught out. He had *known* and said nothing. She was supposed to have been a sociology major but hidden under her underwear like fugitives were Donne to Dryden and Blake to Byron and assorted paperback books on how to write novels and short stories.

"You knew. You never said anything."

"I wanted you should be something special for the world."

"Right. Rosa Luxemburg."

Her words were going past him but he looked up with a jerk of his head; he coughed and held his mouth until he could speak again. "It was the *Schule*," he said. "That's what I did wrong to you."

"The *Schule?* My God, Poppa!" She felt her breath catch. Pictures began to flash. The cold room. Albany Boulevard. *Abraham Lincoln Schule* painted across a gray unwashed storefront window where children of Jewish Socialists went for their real education—after regular school.

It had taken her years to see their teacher, Berman, for what he was, a man of few pleasures, stolid and alone, driven toward any sensation that might alleviate the boredom of his life at the *Schule*. He didn't like his students. He didn't like their parents who paid him a pittance while they monitored his ideology with fierce regularity. Often, he would sit and stare at his pupils, especially at her, in a penetrating rage of silence that terrified her and made her feel that it was she, only she, that caused him to look clenched and blotchy. She would look down at her papers, with the same helplessness and unworthiness that she felt when Poppa told her about the Holocaust, ashamed that Berman had succeeded in fondling her so many times, afraid of the heat of his gaze. Berman's anger seeped into her as if separating the blameless from the culpable. Berman was her symbol of penance.

Though times changed, diehards like Pushkin didn't give up. She was sent to the *Schule* even though only a few children of the old C.P. still went. She remembered the signs that were plastered across every wall inside . . . A Meeting of the Young Pioneers . . . A parade even when parades had petered out. The parade of diehards. White blouses and red arm bands. Mr. Berman running from one end of the room to the other. This one in front, that one in back. Mr. Berman always looking for a nub end of chalk. The dog-eared books. Reciting stories from Sholom Aleichem in Yiddish. The Internationale. Songs from the Spanish Civil War . . . *Los Quatros Generales* . . . *Los Quatros Generales, Mamita* . . . And *Meadowlands* in Russian . . . *Polushka Polya* . . . And the cold, the bitter cold of the place because the sun was going down and they had to come here after the full day at the "English" school. The sun was going down and there was no warmth even though the radiators hissed and choked and splattered steam up toward the tin painted and embossed ceiling, or at the rows of wooden desks. The lift-top and adjoining seat desks were always too small and someone was always falling out of one; the desks made of wood that had never been alive. There were holes where the ink wells should have been, and the tops were rough and splintered with gouged out carvings. And the stinking toilets in the cellar. The dark toilets where you peed as quickly as you could because someone would come in, maybe a boy, maybe a prowler. She was afraid.

"Freya?" Pushkin said. "The *Schule* was so terrible for you?"

"The *Schule* is like a dream," she said.

"A bad dream? Tell me!"

What could she tell?

"You was smart," he said. "Mr. Berman said you was proof that a beautiful girl didn't have to be dumb."

She felt herself shiver at the sound of his name. Mr. Berman. She stifled the impulse to tell him how Mr. Berman had run his hands up her legs. He wouldn't believe her. She would take his breath away; she would destroy something more in him.

Poppa was chuckling, wheezing. "You never gave him peace. But you had a good time? You danced, I saw. Such songs you sang! Berman used to say to me, 'Pushkin, why is your nose running? They're having a good time like bandits.' And I would tell him, 'Berman! One thing is private property, Berman, my nose.'"

She remembered the singing and the dancing, the celebrations after the May Day Parade, the hot-dogs and the pop, the wild chases in the park, and the ice cream, the Cracker-Jacks, the salts Poppa gave her to settle her stomach, and relief when the salts dissolved and the taste disappeared.

Mr. Berman would bellow at them: *Who else in this world had such a thing as May Day, such a good time, so free, so wild—like bandits. Sing, you stinking little bandits!* But something magical had happened to her only because she went to the *Schule*. It was on May Day, the year she was fifteen, that she and Asher Frank fell in love.

After the speeches were over and the songs were sung and the ice cream was passed, a very tall boy came over to where she was standing. He had black wavy hair and deep blue eyes; he was looking at the name tag she had pinned to her white middy blouse.

"I've never known anyone with the name Freya before," he said in a voice that sent a vague tremor of excitement through her. "I'm Asher Frank."

"I've never seen you here," she said.

39

"I've never been. I live on the South Side. I'm a student at the U. of Chicago."

His eyes were drifting over her. "You have unusual hair," he said.

She was pulling off her red arm band, frowning. "I don't like wearing this. . ." She drew back, aware of how he was looking at her. "Are you thinking that I look like Rosa Luxemburg?"

Asher Frank ran his hands through his hair. "Rosa Luxemburg?" He was laughing and so was she.

Their lives began on that May Day. For Freya, at fifteen, and for Asher, a freshman at the university, six years followed of sleeping together on the beach or going camping on the Dunes, or listening to music. Asher helped her with her math and taught her to play tennis and took her to poetry readings. He told her about his great-great-grandfather who was the founder of the Vicksburg synagogue in Mississippi and a Colonel in the Confederate army.

Asher was a student of philosophy, but philosophy was too sedentary to make a life; he was an athlete, and for studies he preferred Middle East languages, politics. He was a Socialist and a Zionist; he wanted to go to Israel.

Freya would lie in Asher's arms while they planned and made promises. She would go to Israel with him; they would be married. Nothing on earth could separate them. They would be lovers forever.

Poppa was shaking his head. "In the Schule you drew a picture of Lenin," he said. "Something wonderful. . ."

She remembered lettering it. *Vladimir Illyitch Lenin.* Carefully, not letting the red crayon slip, her index finger tight, her wrist cramped because the lettering was the crucial aspect of the picture, the touch that would give him his honor and her honesty. Cramps in her fingers bringing forth slowly, painstakingly, *Vladimir Illyitch Lenin.* Berman gave out little medals of Lenin as a baby to worthy students. The baby Lenin. Had Poppa forgotten that she never got one?

"What was wrong?" Pushkin said. "Tell me." She saw his eyes widen, the pain gone out of them, anger welling up.

"Poppa! No one went to the *Schule*. But I went to the *Schule*. And who wanted a collective state anymore? Everyone was getting rich. All I knew about sweat shops was what you told me. I never saw a worker get shot. Everything changed but people like us. We didn't know how to change."

"Change from what?" he hollered back. "From a bad world to a good world? Was it only the sweat shops and the workers getting shot you learned from me? There was no blood in Europe? There was no flesh rotting? What changed? *What?*"

"Poppa! Drawing pictures of *Vladimir Illyitch Lenin* didn't make a life for me. It didn't make a life for Momma. Your heroes fell, one by one, like rotten wood. For God's sake, Poppa! What did you learn? That change doesn't matter? Even Russia changed. The Jews are leaving."

"Jews," he said. "That's right, Jews are going out. To where? To be safe in their Homeland? You can be sure Noah is carrying a gun."

"Maybe Jews are fighting for their lives! It would be different, wouldn't it, if unionists were fighting for their lives. Then you wouldn't mind Noah carrying a gun!"

She saw his face blanch and she was seized with another irony—all these years of assimilation, waltzing through the mainstream at Saks Fifth Avenue, fishing at Bumpers in golf shirts embroidered with little alligators, she was nothing more than a Jewish mother whose son had plunged into another Jewish IDEA, a Jewish earthquake, not Marx this time, not Freud, but Zionism.

"Poppa," she said. "How long can you hate history?"

Pushkin was pushing himself back on the bed, half rising, his eyes grown large and red-rimmed, the pock marks of his face alive as though they had been lit. He was pointing his finger at her.

"What do you know from *history?*" he said. "For you there is no more history. Truth don't exist anymore. Everything is *re-la-*

tive. Start from something you get nothing; start from nothing you get nothing. A new arithmetic came out of you. With one ass you could sit on two chairs. And everyone should love everyone. So I'll tell you. Someone who loves everyone is a fool like Dora, or a liar like the priests, or a lunatic. I hated plenty! I hated the killers. No, I don't love the victims, no, I don't do nothing for them. I see how victims become killers the first chance they get. Victims, despots. . ."

"Poppa," she said. "Let it go." She felt glued to his vitriol; she couldn't take her eyes off him. She felt cold. She saw his eyes twitch but he seemed to be sitting straighter; something in him was coming alive again, a spark ignited from when their tiny living room was crowded with comrades. The fights with the Trotskyites. Marsalka with his fat behind spilling over the chair. Factionists. Deviationists. Splinters. Who was right? Who was wrong? Who was the true orator of the golden predictions and the singer of the fiery songs?

"Why did you throw it away?" he said, shrugging. Then he sighed. "Maybe you lost too much."

That night, Freya lay in the little room that had been her bedroom as a child; two nights more and she would never sleep in this room again. The walls were clean and scrubbed; the floors were polished and the flowered wallpaper was washed. The room lay in readiness for vacancy.

She lay there awake. When had she found peace in this apartment? But there had been times . . . she, Momma and Poppa. She picked up the framed photograph from the bedside table. Poppa, Momma, herself in the grass strip called park outside the apartment house. Momma was half blurred and Poppa sat there like a commissar hovering over the little girl, Freya, who sat between them wearing a white dress and an arm band. It must have been May Day. The old men were there. Always the old men, lounging on benches, never looking up, always the same in the way their knees crossed, in their shirts unbuttoned, in hands held to faces shutting out the light, the way it was when the three of them sat in the park on airless nights. They would sit there waiting for a breeze to pick up. Momma would make herself a fan out of a brown paper bag and wave it languidly; how carefully she had torn the brown paper and folded it accordionlike so that it was elongated and graceful and she could move it against the darkness with her eyes shut. On nights like that in the park, Poppa would bring walnuts and crack the shells. His knuckles were large and the bones protruded like bald summits. The crack of the walnuts would break the silence. He would hand her a nut. She would chew. Momma's fan stirred the air. They were inside themselves. Once, she had seen an old man hobble over to a bench opposite them; he sank down on it, his arm tucked under his head for a pillow, a crumpled newspaper dropping from his hand. The wind had blown and pages had fluffed up around his feet. "Who is he, Poppa? Where does he live?"

Poppa had answered solemnly. "He is a King. He is the King of Hope."

There had been muffled echoes from far off and leftover sounds and the rustling of newspapers as the breeze came up. And then, a loud and prolonged fart from the King of Hope. Momma fanned the air. Poppa cracked a walnut. And Freya had sat there making her first file card for her collection.

The next morning she began her last task in the apartment before she went home the next day.

"What are you doing out there?" he hollered from the bedroom as soon as the sounds of boxes being lifted down reached him.

"Straightening up, Poppa. Don't worry."

"What are you? A spy from the board of health? Let it stay!"

The shelf of the closet was piled box upon box, dust over everything. On the floor of the closet was a packing case and she pulled it out into the center of the small living room. Dust blew into her eyes and nostrils. The boxes on the shelf were tied with cord and knotted securely. Ten boxes, some the size of a dress box, some more like shoe boxes, all of them aged and stressed, one with the name of a department store Freya never heard of. She stood there in the middle of the boxes and the dust; what would she do with the stuff? How many years had they lain there without sifting or pruning? Had Momma remembered what was in them? What would be missed if she sent the lot unopened to the incinerator?

But the boxes had to be opened. She untied the knot of one small box and lifted off the lid. Buttons, buckles, hooks and eyes, crochet hooks, bits of yellowed crocheted edging snipped from the border of some larger piece, empty spools, and dozens of dress snaps. She set the box aside and began on what seemed a companion box of the same size except that it was tied with a faded green ribbon knotted without an inch to spare. Judging from the condition of the threads at the end of the knot it had been tied and untied over and over. Freya dug with her finger-nails until the knot came loose and then, lifting off the lid, she heard herself gasp. A million colored lights shot out at her; she put her hand in and sifted through hundreds of loose glass beads of all shapes and sizes, some opaque, some opalescent: transparent crystals, glittering rhinestones, luminescent ambers, rose pearls, strings of black sequins, turquoise octagons, milky balls flecked orange and yellow, crimson teardrops, all the beads shimmering together, glistening, so that she felt caught and held in a multi-colored radiance. Her fingers kept sifting through hundreds of little prisms and she agitated and stirred them to new brilliances, hypnotized apart from any connection they might have had to Lena Pushkin.

And then she remembered. Momma never approved of ornaments or jewelry. The beads were part of necklaces or bracelets; the sequins must have been snipped from dresses, the crimson glass teardrops cut away from some garish lampshade, the rhinestones pried loose from some buckle or clasp, all of them saved over decades—there were hundreds of them. She was sifting and sifting and slowly she withdrew her hand; the friction of the cold glass beads sent shivers through her fingers. If Lena Pushkin had disapproved of frivolousness why hadn't she thrown the beads away? Had she only pretended to disapprove so Pushkin wouldn't come down on her? But she had tied them up in a box so part of herself would smother while she walked the earth disguised as the humble wife of a worker.

She examined the faded green ribbon again, fingering the ends that had had the hardest use; the threads were loose from endless repetitions of tying and untying, tying and untying. Momma in her drabness, Momma in her monotone, had known where to find light.

She put the green ribbon inside the box and before she replaced the lid she lifted out three dazzling beads, then, she replaced the lid. The three beads were her inheritance.

The rest of the boxes would go to the incinerator. She had seen enough.

She sat there staring out into the living room, at the brown sofa, the lamp with the parchment lampshade jaunty and tilted like a dancer on a stick. What could anyone know about a person from leftover possessions? What if she died tomorrow and someone came to clean out her drawers and cupboards? What would they know about her? Her books? Pablo Neruda and Anna Achmatova and Shakespeare? *How-to* books on writing? What could be known from her possessions? Perfume from Bonwits? Noah's baby shoes? The black pearl Karl gave her? The yellowed birthday card she got from Asher when she was seventeen? *I love you, my sweetheart. I love you.*

She was sitting cross-legged, the box between her knees, a wild restlessness tugging at her. She sifted and sifted, beads falling over and through her fingers; there might have been an aroma emanating from the box in the way she drew in breath. She sat there for how long she didn't know, thinking. Momma had been the quantity of X. But for two weeks she had been exposed to Poppa. Isaak Pushkin was not the quantity of X. Isaak Pushkin was visible; he had courage. He would have ridden the boxcars.

Looking out into the room, the three glass beads warm in her fist, Freya knew what she would do.

CHAPTER FOUR

FREYA DRESSED HASTILY in a sleeveless blue gauze dress and tied her long hair to the nape of her neck with a red scarf; as always when at her father's house, she felt vulnerable and regressed. She had gathered up her overnight case and a shopping bag of laundry. It was eight blocks to Mack's lot where she kept her car. She lifted the phone to call a taxi but she put it down again. Walking would clear her head even with the suitcase and shopping bag—a last mile of contrition.

She was at the front door when she put her bags down and went back to the bedroom, where she stood staring at the bottom drawer of her bureau where she had kept the large, black envelope. She opened the drawer and slid the envelope out, peering in to what she knew was there, photocopies of letters written about the Siege of Vicksburg by Asher's great-great grandmother, Lillian Mary Frank, a grand lady in Mississippi during the Civil War. His great-great grandfather had been a Colonel in the Confederate Army. Asher had wanted her to know what was in his blood.

She put the envelope back and went to the door again, then she turned back, lifted it out and took it with her.

Traffic moved over the Outer Drive as on a slow conveyer belt and she could see the beach, a wide expanse of sand stretching back from the lake's edge. Bodies were running, a red and white beach umbrella blew over and rolled in the sudden gust of wind that swept it to the shore's edge. The sunlight suddenly switched off and the sky over the lake blackened to small shocks of lightning that galloped over the surf like a weird fiesta. It was only another few miles to Karl's apartment building; maybe she would beat the storm and she could walk into the building cool and unshaken. Her hands tightened on the wheel; she switched on the radio to an all news station while she startled at a crackle, then a huge clap of thunder. *Israeli tanks were moving through South Lebanon.* She switched the radio off. "Go *already . . . Go already.*" Poppa's ashen face had left a clammy chill inside her. *Maybe you lost too much.* And his look of astonishment when she told him she was going to ride the boxcars in Israel. "Go already. Go!" he had said. She was being nudged, pushed, out of everyone's life.

Karl's apartment house was a solid twelve story building on the inner drive off Lincoln Park, an older building reminiscent of the splendor of Sullivan's time before the Bauhaus school caught the imagination of the city in sheer plate glass and steel. She had driven past the building many times since Karl moved there, remarking to herself that at least he hadn't moved into a building shaped like a trapezoid or a cauliflower with a huge spaceship of a chandelier suspended in the lobby.

The storm hadn't broken after all and her hands were steady when she drove into the underground garage and waited for the attendant to take her keys. She wiped her face with a tissue and tightened the red scarf around her hair before she took the elevator up to where Henry sat behind a desk in the air conditioned lobby. He was older than she had imagined him, in his sixties maybe, balding, heavy jowled and somber, an expression to match his job of screening comings and goings. There were two phones on his desk and a yellow pad; he kept his hand on one phone while she looked around her before she spoke to him, taking in the quiet decor of the parquet floor and oriental area rugs and wood paneling. How had she let herself get stuck in Beaulac? "*Xanadu!*" she said aloud, and Henry's head snapped up with a suspicious look, his hand on the phone.

"I'm Freya Gould, Henry," she said. "The ex-missus. Dr. Gould said he'd leave me a key."

He was taken off guard, eyeing the straps of her blue gauze dress that had slipped over one shoulder. "You planning to take your dog home?"

"No," she said.

"A pretty good-looking animal," Henry said, eyeing her. "Spoiled rotten."

"Thanks for helping out," she said.

"Spoiled," Henry said. "You bet."

"Did Dr. Gould leave the key?"

"I'm not supposed to let anyone in."

She took her driver's license out of her bag and held it out. "See? I am who I say I am." She smiled at him. "If I promise not to pinch the silver?"

Henry handed her the key, perplexed. "Don't say I didn't tell you," he said. "Ten sixteen."

"Thanks," she said, and she walked toward the elevator.

She turned the key in the door and walked in to a small foyer in Karl's apartment. There was a console table against the wall with a black ceramic bowl sitting on top. The bowl was stuffed with mail. Maybe there was a letter from Noah. She withdrew her hand quickly and stared at the mail for a possible green form in the pile. She was spying. Henry was right.

She walked past the foyer and into a living room of polished hardwood floors and a thick oval area rug of solid taupe. There was no sign of Hammer. The picture window in the center of the room looked out to the lake; she could see a fleet of sailboats and when she sank down in the armchair to watch them bobbing in the harbor, the sense of loss descended again. All those years of living in Beaulac. She had raised Noah to fit in where she herself didn't fit, to conform when she didn't conform. Noah was left to tough out reality in Beaulac.

She got out of the chair. A clock began to chime and she looked up to the mantle over the fireplace to the French antique clock. Where did he get it? Where had he found the ceramic bowls, the lamps, the antique zither on another wall? If he hadn't used a decorator, there had to be a woman living here. Who else would

48

have such taste and restraint? But where was Hammer? Did the woman take her dog?

She found the bedroom. A king-sized bed. A quilted charcoal spread. Everything elegant. Karl probably wore satin pajamas. And the woman wouldn't fall into bed naked like a gypsy girl; she would have a sheer and lacy peignoir. *In Xanadu, did Kubla Khan.* . . . Where was her dog?

And he *had* made the bed! Or maybe *she* had made the bed and left so they wouldn't collide. She sat down on the edge of the bed wishing she hadn't come. It had taken Karl one year to create a beautiful life for himself. And she still moldered in Beaulac wrapped in a Gogol overcoat.

She got up and went to a closet that stretched across the length of a wall. She slid open the doors, her curiosity piqued to see the woman's style in clothes. There was a cough behind her and she swung around, her face hot while she slid the closet door shut.

"You left the front door wide open," Karl said. He was standing there in beige slacks, his hospital ID clipped to the pocket of his short sleeve beige shirt; he looked cool, his sandy hair had a new kind of cut—he looked like a TV anchorman. His eyes were fixed on her.

She stuttered, "I wasn't going to steal the silver. I promised Henry."

"You made a big hit with Henry," Karl said.

"I'll bet."

"Finished in the closet?"

"Finished. It isn't ten o'clock yet," she said.

"I should have waited until you rinsed out a few things."

"I'm sorry. I was looking for Hammer. Sometimes he gets into a closet and can't get out . . . you know how he is."

Karl lifted an eyebrow. "Blitzie takes Hammer shopping with her on Saturdays. Henry didn't tell you?"

"*Blitzie!* Is that a reindeer? I hope *Blitzie* knows that Hammer is half my dog."

"Blitzie helps with my half."

She shrugged. "I want to talk about Hammer," she said. "I don't have much time."

"How's Pushkin?" Karl said.

"Excellent. He applied for a pilot's license."

Karl must have turned up the air-conditioning; she was cold. The woman had no right to her dog. Hammer was hers!

She felt as though she would shiver but she clenched her hands. She had to see this through; no matter how arrogant Karl became, she wouldn't react; she would do what Momma always said she should do, catch flies with honey.

"Do I owe you full disclosure about Blitzie?" Karl said, grinning.

"Certainly not! It's Hammer I care about."

"Never mind," Karl said. "Blitzie is really Blitzstein. Inge Blitzstein. Tante Sunnenzimmer's distant relative. Remember Tante Sunnenzimmer? She's ninety-seven years old."

"I know," Freya said. "And she still mows the grass."

He looked at her obliquely. "You never appreciated Tante, did you?" he said.

"I'm going to Israel," she blurted out, watching his face, his long fingers suddenly flexing.

"Since when?" he said.

"Since yesterday."

"A hidden agenda. I knew it."

"He's your *son*, Karl! If you read the papers . . . I never heard you worry. Not once did I hear you worry!"

"I don't worry out loud."

"Let's make this friendly, Karl. I need you to take Hammer. It would make everything easier. There's Blitzie . . ."

"You want to give Hammer to Blitzie?"

"Karl! Why can't we grow up?"

"I thought I *was* grown up until I found you tearing up my closet."

"I said I was sorry. I'm *sorry*. Oh Karl, Noah is on some kibbutz and there's a war and all you ever say about him is, he disappointed you. Noah is our son. Whatever he does . . . You have to *help!*"

She followed Karl out to the living room. He sat down on the sofa, folded his arms over his chest and stared at her.

"Why are you looking at me like that?" she said.

"You get more beautiful every year, Freya," he said. "When you came to the hospital with your mother, I couldn't take my eyes off you. I still can't. What turned me off was your bloody Pushkin streak. You could have had the best clothes money could buy, but you insisted on wearing blue jeans and a sweater. Why such a pretty dress now? Is it money you want? I settled a lot of cash on you. What do you do with it?"

She had sat down opposite him, struggling to stay cool, not to show her humiliation, not to cry. "You make me feel like a ward of the state. Maybe that's how Noah felt. Karl! You need to *share* this!"

She was getting up, choking back tears, her voice quivering. "It's not right that you should make me beg for our son. Where's my purse? I shouldn't have come here."

He got up too, his hands on her shoulders. "Okay. I worry about Noah, too, but I don't kid myself about him."

"Why didn't you talk to him about medical school, I mean *talk* like a father who cares."

"I didn't hide what I felt. Noah will come back when he's damn good and ready. If he wants to be a hero, why stop him? He can use a little discipline. It's better than wasting film."

"There's a war!" she cried.

"I don't think he's marched off to war," Karl said. "More likely he's taking pictures of sand dunes and prophets. I read his last letter. Remember?"

Karl went to his study and came back with a check in his hand.

"I don't need five thousand dollars," she said, glancing at the check.

There was a grin of self love on his face. He had been generous. "Keep accounts," he said. "Pay back what you don't use. Interest free. Anyway, you might have to pay someone off."

He came closer to her and reached for her shoulder. She turned away and tried to move but he held her arm. "You plan to see that genius you were so crazy about?"

She didn't answer and held out the check. "I don't need five thousand dollars."

He forced a laugh. "I wouldn't want you to sell the pearl I bought you. Let's do the right thing. A united front."

She stood there, hesitating, then she took the check. She could be proud, refuse it, but Noah was Karl's son.

"When do you leave for Israel? How long do you plan to stay?" Karl said.

"As soon as I get a reservation. Monday, I hope. I don't know for how long."

"Don't lose the check. Get travelers checks. And put the Audi in the garage and lock . . ."

"My purse," she interrupted him.

"Maybe you left it in the foyer. And with the door open. See what I mean?"

It was there. Next to the black bowl stuffed with mail. Had she actually left the door open?

She put the check in her purse and turned to leave. "I'll call you when I have a reservation," she said.

He was close to her again and she moved back. "Don't get hijacked," he said. "I wouldn't want you turning up in a harem."

Dazed, hurt, she got into the elevator; the pressure of the fast descent filled up in her ears and she looked down to see her gauze dress blowing around her knees. She hadn't seen Hammer. Her dog, the puppy she raised with Noah was cruising, bonding with Blitzie. Hammer didn't need her, either.

When she walked past Henry, she nodded, her voice quivering again. "He's *not* spoiled," she murmured, and walked out the door. Thunder and lightning had cleared from the sky; she could breathe again.

It was noon when she swung her car into her driveway. Freya got out of the car and walked up the drive to the back. The tree was gone. Sargeant had chopped the tree into firewood, earmarked for cremation. She went close to the stack of logs; the long fronds had been cut away, nothing was left but thick cylinders of wood.

"So you're back," Claire said, startling her.

"I'm back," Freya said. "Was there a letter from Noah?"

"There wasn't," Claire said. "I'm sorry, I'll bring over the rest of your mail."

Freya was standing at the wood pile looking down on it; she bent down and touched her fingertips to the smoothly cut end of the log. She had loved the tree as she loved every tree. It was a sin to cut down a tree.

"Don't take it so hard," Claire said, looking at the pile of logs. "Anyway, it will cost you one hundred bucks. I paid Sargeant." She heard Claire groan. "Sounds like you had a great time these past two weeks."

"I'm tired, that's all. I'll write you a check."

"No rush. You're not leaving town," Claire said.

"But I am. Monday, maybe."

Claire's eyes widened. "You're leaving town Monday? *Where . . . ?*"

"I'm going to Israel."

"You're kidding!"

Freya shrugged. "Going to Israel isn't joining the French Foreign Legion."

"You're strung out, my girl. Cool off in the pool. One dip in and out."

"I can't. There's barely time to get ready. And Noah took all my suitcases."

Freya unlocked her back door and stepped in to a sweltering kitchen. She pulled her dress off and opened windows, plugged in two fans, and opened the refrigerator to feel a rush of cold air. The refrigerator was mostly empty except for a bottle of ketchup and a jar of mustard. "I don't believe it!" she cried aloud. She hadn't cooked the shark for Pushkin! Dora would find it in the back of his tiny refrigerator, stinking. Why couldn't she have left her father's house with a clean slate?

The phone rang.

"I'll be over tonight," Evelyn said. "If you're not home, leave a note in your mailbox."

"Bring a suitcase," Freya said.

"Why a suitcase? Don't tell me . . . If you're moving in with Pushkin, I'll slash my wrists!"

"Don't start that, Evvy. I'm going to Israel."

"You're *what?*" There was silence and then Evelyn moaned. "Of course you are," she said, softly. "Tell me before I start Mea Maxima Culpa for shoving you out."

"I should have gone long ago, Evvy, but there was always something . . . Momma just died, before that, the divorce. Now there's a war. I have to talk to Noah face to face before he gets heroic ideas. I have to know what he's up to."

Evelyn hesitated, then she blurted out. "It cuts me up! You want to fix Noah but it's yourself you have to fix, not Noah."

"Evelyn! For Christ's sake, there's a war! And if Noah is in it, I have to get him out."

"Get him out! My God. How would you know *how* to get him out? You think you're tough enough to pressure the *Israelis?* You think you're tough enough to waltz your way through that thicket of Arabs and Jews? For Christ sake, Freya, you'll be like Little Red Riding Hood!"

"That's enough out of you, Evvy. I have a lot to do."

Evelyn's voice was breaking. "Shall I store the Audi? Listen, I'll leave supper for Norman and I'll be over to do your laundry," She hung up.

Freya pulled the heavy Chicago phone book off the shelf and thumbed through the first few pages to *Airlines Domestic and Foreign.* She called TWA. She was put on hold. She hung up and called Pan Am. She was put on hold. TWA. The same. Her eye traveled up the column: El Al. The same. She dialed TWA again. Hold.

She sat in the kitchen, the phone book across her knees. She felt sudden panic that she wouldn't get a reservation after all. Would she still go if she had to wait for days, maybe weeks? There was no time to jump off. The boxcar was at full speed.

She had been sitting there ten minutes. She had to phone Michael's parents.

"Rudy? Freya Gould. Yes, it's been a long time. Oh no, I can talk to you as well as Sarah."

54

"Yes, Mrs. Gould?" *He wasn't calling her, Freya. He was freezing her out right off.* "I wanted to call you before dinner time. I'm going to Israel, Rudy, maybe Monday if I can get a reservation. If you have a message for Michael, something you want to send him. . ."

"You're not giving us much notice."

"I'm sorry, Rudy, I only just decided. But I could pick something up tomorrow."

"I'll talk to Sarah about it," he said, abruptly.

"Rudy, I'm sorry about Michael joining the army. It must be hard for both of you."

And then she had a shock.

"What do you mean, *hard?*" Rudy said. "We're *proud* Michael joined the army. We're proud! What do you mean, *sorry?*"

"I didn't mean to offend . . . I assumed . . ." She had got what she deserved, pushing friendship where it wasn't wanted.

Rudy Haas seemed to be on some kind of kick; as though for years, he had been waiting to put her down. "You shouldn't assume anything," he said. "When Sarah and I sat in the concentration camp we couldn't fight our enemies. Whatever the dogs wanted to do to us, they did."

"I know," Freya murmured.

He bellowed back. "You don't know! We can fight back now! Like any people should fight for their survival!"

"Yes," Freya said.

But Rudy Haas wasn't finished. Oh, he remembered the sign on her lawn during the Vietnam days, all right. *War is Not Healthy for Children and Other Living Things!* "Ridiculous!" Rudy said. "They would have sent you to the cattle cars like we were sent, Missus America."

She interrupted. "If you want to send something, Mr. Haas, let me know."

"Nothing," he said. "Michael has everything he needs. And maybe you *should* understand, Mrs. Gould, we have hard feelings about Jews like you. Vietnam!" His voice was tinged with scorn. "How hard did you fight for your own? At least we tried to teach Noah . . ."

55

She hung up and sat there, feverish, frozen. How dare they! How dare they give Noah corrective lessons? The heat was rising in her face. She went to the refrigerator and took an ice cube out of a tray and sat in the chair, pressing the ice cube to her forehead, moving it around and around until it dripped onto her nose and over her chin. She had been stoned like a French traitor.

She closed her eyes, remembering how urine had splashed down her legs when Poppa showed her the pictures of Jewish corpses. And it had all come home to Beaulac. While the Haases were making it big in America, hot fire scorched beneath lucrative real estate deals and Jews like herself.

She sucked in her breath from the shock of the cold drops trickling between her breasts. Momma's words were in her ears. "Love your own, not the world." Momma had not sidestepped the Hell of Jewish *Weldshmerz* and the Haases had seen their duty to Noah. It wasn't fair; it wasn't right.

But their attack pushed her forward. She cleared her throat to make sure she could speak when she phoned El Al. She was told to wait. Yes, finally, they could ticket her for one seat through British Air, Flight 646, to Heathrow, London, Monday 21st of June, seven o'clock PM, from O'Hare, Chicago. Transfer in London to Flight 401 *El Al*, Tuesday, 22 June, 1:10 PM, London to Tel Aviv. "You're lucky on such short notice," the ticket agent said. "You are very, very lucky."

When she turned from the phone, Claire had slipped in through the back door and was coming toward her. She was carrying a suitcase and there was a sad, half smile on her face.

JERUSALEM

CHAPTER FIVE

It WAS FOUR HOURS after leaving Chicago when Freya saw the sun come up over the black ocean beneath her. Half awake, she fantasized turnings of the earth without sensation of movement. Soon, she and Noah would exist in the same light and the two year lapse wouldn't matter. They would be close again. She leaned her head against the headrest of the seat and closed her eyes imagining Noah's lanky frame, his resemblance to Momma in gentle brown eyes and agile fingers, of herself sitting on the bleachers cheering at his swimming meet triumphs. Vaguely disconnected, she watched the clouds beneath her break in a crimson wash and a fast-moving mist, weightless, the feeling of floating in midair. And here she was, descending on Noah like the Mother Demeter swooping down to make a deal with Hades. How would it be and how would it end?

She felt herself drifting off. Evelyn was saying: *Why couldn't you live a normal life? Why couldn't you shop at Bumpers and join a geography class or go beetle hunting with someone who would tickle your tits? Why do you have to go rushing off to a war to pull your kid out of a ditch?* And Evelyn was saying: *You think you're tough enough to waltz your way through a thicket of Arabs and Jews?*

Freya's eyes opened; she was shaken by the convulsing plane in heavy turbulence. She stiffened and looked away from the window where thousands of feet beneath her, she imagined the water swallowing itself and whitecaps sliding and disappearing into the foam. The huge plane was shivering, rattling, and her hands clutched her handbag. She had a sudden panic that she was travelling in a time warp, backed up in a journey she was supposed to have taken twenty years ago. Then the turbulence began to subside. It was over. She leaned back, her forehead relaxing until she was on the cusp of sleep. *Asher's deep blue eyes were looking at her with a suggestion of a smile, giving her license, allowing space for uncertainties while he pressed her to make a critical examination of every doubt in its most simple form as though it were a chemical compound. She would tease him: "I wonder why I bother with you?" And he would answer, grinning, "You know why," and he would take her in his arms.*

She felt the tension going out of her shoulders and the back of her neck. Fear, after all, was only a kind of precaution, an alarm system for survival, an awe of probability. But it was fear that had made Momma the Quantity of X. She felt herself take a deep breath and when she jolted awake and looked out her window, the chimney pots of London were beneath her.

She stood still for a moment on the terrazzo floor of the International Terminal at Heathrow Airport; she was trying to breathe evenly so her dizziness would pass. She had hardly slept or eaten for two days; her eyes blurred in the cavernous terminal where currents of people crowded at ticket counters, at the money exchanges, or moving up and down on escalators headed for departure or arrival locations. Sound was cacophony, bits of human voices hummed in a pastiche of languages moving past with a sometimes discernible broad English A or a high pitch of laughter piercing through sibilance.

She started to walk again but halfway across the polished floor she set her hand luggage down. Stroboscopic flashes of perpetual movement and pulsing light pressed against her eyes. A cluster of Sikhs with white turbans walked around her. Children turned to look at her and a young boy wearing *lederhosen* and a backpack smiled while a black-garbed and heavy-bearded cleric of some Eastern sect moved on in a long-gowned swish.

She pulled her sunglasses out of the pocket of her suit. Her eyes were wet behind the glasses and she was telling herself that it was lack of sleep, tension, turbulence on the airplane, turbulence inside the terminal.

Her sunglasses were sliding over her wet face and she lifted them to wipe her eyes with a tissue, stiffening to keep herself erect. "God!" she said aloud and caught herself. After all these grueling months why fall apart *now* at Heathrow Airport?

She was on the escalator, not certain she was going in the direction of the Buffet. The plane to Tel Aviv had been delayed four hours. Freya felt herself go weak again. A small sign lit with red neon letters said, *Buffet*. Clusters of people carrying trays were circling like flocks of birds looking for a place to squat in the large expanse of ice cream parlor tables. A young woman was sitting alone and Freya moved quickly. "Would you mind?" she asked, and the young woman nodded that she wouldn't.

The dining area of the buffet was a labyrinth of suitcases, parcels, people and smoke. Freya sank down on the fragile ice cream parlor chair and the young woman sitting opposite her, shimmered in the smoke of her cigarette.

There was a long line at the buffet. Freya got up and sank down again. The thought of coffee, English coffee, brewed strong for mass consumption, made her queasy; tea would be better, strong and black tea with milk. In a moment or two she might trust herself to get up. There were four hours to wait. When the plane circled over London she had felt such an excitement to see and mix and mingle in the city of all the writers she loved.

But now she only sat there wanting tea. The young woman across from her was finishing a sausage roll and a pasty with chopped meat creeping over the edges. A large drinking glass half full of milk was in front of her. Freya smiled. The young woman smiled back. She was probably in her late thirties, not much younger than Freya herself, and she had a thin face with soft, buff coloring, sherry colored eyes and long, tawny hair.

The woman was putting out a cigarette in an ashtray loaded with butts. "Are you all right?" she said to Freya. "You don't look very well. Can I bring you a cup of tea?" Her English was fluent; she had a multi-colored voice like a mezzo soprano.

"I'm not up to standing in line yet," Freya said. "But thanks."

The young woman studied her for a moment and got up. "Never mind . . ." she said. "Hot tea? Something else?"

Freya murmured, "Yes . . . thanks so much . . ." and watched the tall, young woman in snug blue jeans drift off while Freya waited for twenty minutes fighting back queasiness. The young woman came back with a mug of black tea, a small jug of milk, and a package of water biscuits.

"Thanks . . . thanks," Freya said, smiling. She took two dollar bills out of her purse and held them out.

"American money is always useful," the woman said, taking the money. She sat down and lit another cigarette. She blew out a long stream of smoke and rested her cigarette against the mound of butts in the ashtray. "Put three spoons of sugar as well," she suggested. "It will do you good."

The tea was bracing. Freya held out her hand. "I'm Freya Gould."

"Layla Ayoub," she said, taking Freya's hand, smiling broadly with even white teeth. "I knew you were an American the minute I saw you! I subscribe to *Ms* Magazine. American women are models for me!"

"I'm not much of a model."

"But you are! Every person feels ill from time to time. Otherwise, I can see that you are independent, competent."

Freya smiled. It occurred to her that Layla Ayoub was a Middle Eastern name, not Hebrew, possibly Arab? In her research, somewhere she had read that the Arab word Ayoub meant Job. Layla was different in a way, but not really different: her tall, slender frame, her tight blue jeans hugging small hips, a neat white shirt blouse belted over her jeans were universal, but there was something more in her walk, the tilt of her chin. The look of Layla Ayoub was out of Arabian Nights.

Freya finished her mug of tea, smiling back at Layla. "I'm fine now. Will you save my seat?" She changed ten dollars for English money and then she went for more tea and a cheese sandwich.

They talked for what seemed like hours. Layla's intelligent face animated, excited at imagining America; she poured out details like the editor of a travel magazine: The Sears Tower in Chicago, the World Trade Center in New York, Bloomingdale's, Central Park, the Golden Gate Bridge, Half Dome in the Sierra Mountains, skiing in Colorado, Harvard University in Cambridge, Massachusetts. "You see," Layla said, glowing. "My mother has a brother who lives in Detroit. America is such a great hope for us."

"I wish our children thought so. Have you just come from visiting your uncle?" Freya said.

"I wish I had! Some day, though. I've been on holiday. Here in London. I must go home now."

Layla Ayoub seemed in no rush to go anywhere or to say where home was. She wanted to go on talking about America, about American women and what she had read. The way she pronounced equality was in capital letters. No woman was chattel. A woman could be a man's boss. How good. How free! But Henry Kissinger. Was he truly even-handed in the Middle East? How could a Jew be truly even-handed whether he observed the religion or not? How could he possibly be even-handed?

Freya was taken off guard but the second time Layla said, *even-handed*, with such emphasis and emotion, she knew that Layla Ayoub was an Arab. "Kissinger is an American," Freya said, suddenly defensive.

"Of course." Layla lit another cigarette.

Layla hadn't offered any information about herself and Freya had the instinct not to ask. She had been too busy recovering. Now she was coming alive in the presence of Layla, who had read so much, who was so well-informed and so full of vitality.

They walked about in the airport, making two passes in the duty-free shops where first, Layla bought four cartons of cigarettes and ten huge bars of Cadbury milk chocolate with almonds. On their second pass they tried on silk scarves and Layla bought one that Freya said looked wonderful on her. When they stopped by the candy counter again, Freya bought five Cadbury chocolate bars too. "Let's get fat," she said to Layla. The sting was going out of anxiety: she and Layla were laughing.

"You're going home?" Freya asked, when they found another place to sit. "Home is a good place to go, I suppose. Well, most of the time," Freya said, and looked down at her watch. "Two hours before my plane leaves," she added. "Talk about time flying."

Layla looked at her watch. "I have two hours as well," she said. "I'm taking *El Al* to Tel Aviv."

"The *El Al* one-ten to Tel Aviv? That's my flight!"

Layla stared at her for a minute, not responding, her expression stony.

"I tried to get another flight," Layla said. "But *El Al* is the only plane flying to Tel Aviv today. My . . . my mother is quite ill. I had no choice but *El Al*."

"Anything wrong with *El Al*?"

"Nothing, no, no, I didn't mean . . . *El Al* is good, very secure. They take great care, especially now with the war in Lebanon."

"I'm *glad* you'll be on the plane," Freya said.

"Are you feeling better?"

"A lot better. Do you live in Tel Aviv?"

"In Nazareth," Layla said.

Where was that exactly? "Someone will meet you?"

"Yes," Layla said. "And you?"

"I'll take a taxi to Jerusalem. I booked a hotel in a hurry. The Shiloh. It was all so fast. I just came ahead! Prepared or not!"

"My God, I have a cousin who works at the Shiloh!"

Freya nodded.

"The war didn't stop you?" Layla sighed. "American women are like that," she said. "American women aren't afraid of their impulses."

"I've had tons of second thoughts."

"*Why?*" Layla said. "Isn't it better than planning every move, every step and every turn, afraid to make a mistake you'll be sorry for?"

"You have a point," Freya said, unwrapping a bar of Cadbury's, biting in.

It was an hour and a half before embarkation. Layla said the El Al security check took forever. Especially now with the war in Lebanon.

Freya felt her pulse jump. The war was moving in and out of her consciousness. She finished the candy bar while they walked to the embarkation area.

"We'll see each other in Israel?" Layla said.

Freya started to tell her about Noah and then she stopped herself. Layla was surely an Arab. "I hope there's time," she said, and corrected herself quickly. She saw Layla wince.

"Maybe you want to know more about who I am," Layla said, crushing her cigarette.

Freya looked at her, puzzled. Layla asked her again if she felt well, and finally she straightened her backpack on her shoulder. She was frowning.

There was a long line at the embarkation area for *El Al.* Four British policemen surrounded the area. Layla stood with Freya waiting their turn to present their passports and be checked through security before passing on to the boarding area. Layla smoked. They were quiet. Two American tourists were in front of them, a middle-aged couple, reviewing their itinerary in Israel and making a great point of fanning at Layla's cigarette smoke with travel folders. They never mentioned the word, *war.* She was reassured. They were looking at folders of Masada and the mud baths at Ein Gede.

Half an hour later Layla and Freya were in the inspection area. They went through the metal detector to meet two women and a man who stood there beside a long table inspecting passports. Layla was ahead of her. They looked at her passport and then they asked her to open her backpack. Clothing, underwear, blouses, blue jeans, the scarf and the chocolate bars and cigarettes she bought at the duty free shop were all removed. Toiletry and perfume bottles were opened, smelled, a nail file was removed from its case, inspected, and put aside on the table. Two books were opened, pages flipped, bindings flexed. A small gift-wrapped package was undone and the contents unfurled—a mini cassette recorder. The recorder was inspected.

"A gift for my mother!" Layla cried. "My mother loves music . . . cassette recorders are more expensive in Israel . . ." There was pleading in Layla's voice. The cassette recorder was put on the table with the nail file. "Why can't I take them?" Layla said, her tone quiet but shrill. "It's a gift. For my *mother!*" The security guard's eyes passed over her. He asked her to move out of the queue; he told her to wait.

Freya felt herself start. What was happening? Why was Layla taken out of the queue? Because of a *cassette recorder?* She looked over to see Layla's face pale, her jaw set with anger and humiliation. Layla looked away from her. What was going on? She had a cassette recorder too, and cassette *tapes*, Mahler and Mozart and Vivaldi. The recorder and the cassettes would be taken away. And she would be taken out of the queue, detained, held. Karl would get a message at his office while he was examining a patient: *Your former wife removed from El Al Flight 401. Suspicion of being a terrorist.* Was *that* it? Terrorists! *Layla?*

She showed her passport to the officer. "Tourist?" he asked her. "How long will you be in Israel?"

"I'm going to see my son. I have an open ticket," she said.

"Were you given any packages on the way to the airport? Did somebody ask you to deliver something for them in Israel? Did anyone help you pack? Is there something in your luggage you should declare?"

"No, nothing like that," she said. "But I have ten cassettes and a tape recorder. For music. The inspectors in Chicago even put the cassettes aside so they wouldn't get erased." Her mind was scanning to Evelyn's husband, Norman, the lawyer, who had connections everywhere.

The officer looked at her passport again. "Where is your son in Israel?" he asked.

He was watching her as though he could hear her thoughts. "On a kibbutz," she said. ". . . Gan Zvi."

"The woman you were talking with. You know her?"

"I met her here!"

"It's okay," the security guard said, waving his fingers. "Go on."

A female inspector moved her hands quickly over Freya's hips. "Enjoy your stay in Israel," she said and waved Freya on through double doors to the boarding area. She was on the other side from where Layla was being led to a booth by a female security officer. Other passengers were looking at her, whispering, shaking their heads with relief as though they had been rescued from a burning plane. Layla glanced at Freya sideways and went with the guard without resistance.

Baffled, Freya stood in line waiting to board the plane. What had happened to Layla? Layla had cassettes. She also had cassettes.

Freya watched the inspections proceed at the boarding area. Three more people were taken out, a heavyset woman with a black shawl around her head, carrying a child of not more than three, and a man. A child of three could hardly be a terrorist. Cassettes, tape recorder, she had been passed through with a smile: "*Enjoy your stay.*"

Armed British policemen moved about the area while she boarded the huge 747. The jolly American tourists were now in front of her looking for their seats, joking with a stewardess who was laughing back at them. Whistling in the dark. Mud baths . . . hot springs. She found her seat and strapped herself in. The two tourists had crossed to the other side of the plane.

It was twelve-fifty London time; in twenty minutes they would take off. She sat there studying the flow of passengers still boarding. Where was Layla? Ten minutes went by and then, two aisles over, she saw the man and the woman holding the child in her arms come up the aisle. The man and woman wore black looks and the child was crying. They had been cleared for boarding. Still no Layla. Another five minutes went by and suddenly Layla was there, her backpack over her shoulders; she was checking her boarding pass, looking for her seat. Freya sat on the aisle of her row; the seats to the left of her were empty. She bolted up and went quickly to the front of the plane so she could cross to the other side where Layla was searching. "Layla!" she called. Layla looked up, saw her; she smiled. Freya caught up with her. "There's a seat next to me," Freya said.

"Good."

They were settled, seat belts buckled; the doors had closed and the engines were whirring.

67

"I was worried," Freya said. "Did they give you back the tape recorder?"

"Yes," Layla said.

"And your nail file?"

"Yes," Layla said.

"What's going on?" Freya asked.

Layla didn't answer.

"They didn't react to my cassettes or my tape recorder," Freya said.

Layla shrugged; her face was tight. "That's you," she said.

"What's that supposed to mean?"

"You are an intelligent woman," Layla said.

"What's that got to do with it?

Layla's voice was flat, controlled. "You're a Jew," she said. "Right? I'm an Arab. They know who the Arabs are. Nazareth is an Arab city. We're all terrorists, a fifth column of terrorists. They even searched the baby. Can you imagine a mother wiring her baby with explosives?"

Freya felt herself stiffen. She had already guessed. "Your light eyes and fair hair," she said to Layla.

"Crusader genes, that's their joke for not lumping all Arabs into the swarthy and the black."

Layla smiled, rummaging in her flight bag and pulling out her cigarettes and a lighter. "Now that you know how dangerous I am, will you still have time to see me?" There was a haughty and bitter cast to her voice. "Well," she added gaily, "I'm an Arab, you're a Jew. Shall we be friends or enemies?"

Freya did not smile. "I don't have enemies," she said. "At least, not yet."

A steward came up the aisle and stopped where they were sitting.

He looked directly at Layla. "May I see your seat assignment, please?" Layla showed him. "Why did you change your seat?"

"Because I invited her!" Freya said.

The steward shrugged and made a notation on his master list. "As long as we know."

The seat belt light went on and the stewardess was holding up a card, about to demonstrate the equipment. She was a short, curly-haired woman. "Shalom," she said. "Monsieurs and Mesdames, Bienvenue. Ladies and Gentlemen, Welcome."

Freya waited but the curly-haired woman did not say welcome in Arabic.

Layla pulled a copy of *Ms* Magazine out of her flight bag, clutching her cigarettes and her lighter. She shuddered and took a deep breath. "Take-off," she murmured. "I freeze until the smoking sign goes on again." Then she looked up. "Oh my *God!* I'm in the no smoking section!"

CHAPTER SIX

F REYA BOLTED UP and looked at her watch. The sun was streaming in through the window of her room at the Shiloh Hotel in Jerusalem. She had slept ten hours. She was sweating; her hair was tangled and wet. Had the night actually happened? Somehow she was here in this small room with pale blue walls and casement windows and pale blue curtains and a TV set turned on and hissing with snow.

She swung her legs over the bed and went to the bathroom. A towel was on the floor from her quick shower when she was half dazed from jet lag. Her toiletries case was open on the sink and, quickly, she swept a hair brush over her hair.

She looked at her watch again. Seven AM, Wednesday, June 23rd. A flyer lay open on a small desk opposite the bed. SHALOM! ENJOY AN ISRAELI BUFFET BREAKFAST! She went to the window and pulled back the curtain and, slowly, she opened the casement window. There were hills bathed in early light all around her, sepia and white, undulating like a sea of stones; radiant buildings rose out of the hills like white mesas. Sun was flooding the dry, sand-scented air and when she breathed in she had a whiff of jasmine. On a street beyond the hotel, a huge bus was swallowing a crowd of people. The bus moved on. An emptiness settled on the street and on the silent hills under a cobalt sky, like her mother's glass beads, cloudless, hot. Shimmering outside was a picture in the book of Bible stories Momma gave her when she was twelve. This was Jerusalem. These were the hills of Judea.

She turned away from the window with a feeling she couldn't pin down, not excitement, not exaltation, but with a flood of impulses that wanted to open the window again and stare out at the stone hills and feel the sun's heat in her eyes until she understood what was glistening out there in some larger meaning. "*Stone worship,*" Poppa would say. "*Idols.*" But Poppa had never smelled the jasmine.

Downstairs, the breakfast buffet was still spread in the middle of an empty dining room. It was eight-twenty; she was a late comer. Young men in white waiter's jackets were clearing tables, lifting tablecloths, chattering to each other in guttural sounds. The young men were dark-haired, dark-eyed, slender, and when one came up to her where she stood waiting to be seated, she was surprised, pleased, that he spoke English.

"Where you would like, Madame," he said, pointing to empty tables. She went to the buffet—a panoply of platters. Fruits and vegetables, yogurt and cheeses, smoked fish and herrings, and a huge cornucopia basket of assorted rolls, breads and coffee cakes.

The white-jacketed young man who spoke English was at her side. "I will refill if you would like," he said, nodding to the basket of coffee cake. She heard herself laugh. "*Refill!*" He was smiling back at her. "I will bring coffee to you," he said, with the polish of a diplomat, intelligent, cultured; his sly smile seemed to catch the absurdity of his offer to refill. "I'm glad you speak English," Freya said.

"I speak very well."

She smiled. "Yes. You do speak very well."

Freya sat at her table alone; she had a glass of orange juice, a dish of yogurt and a slice of white toast. The young man brought her coffee and she took a sip of the strong, full-bodied brew. The young man refilled her cup. He would be about Noah's age. Noah . . . Noah . . . he was within spitting distance! Her heart began to thump. She would see him today, this very afternoon, in a few hours.

She was sipping her second cup of coffee, her confidence sinking with every swallow. At nineteen, Noah would have a shock at his mother having arrived unannounced to check on his

71

movements—to make sure he stayed on the rails. What if he didn't *want* to see her? What if he had left the kibbutz?

She looked at her watch and then she opened her purse and took out her address book, clutching it while the elevator went up.

"May I speak to Noah Gould, please?" Freya asked the kibbutz operator.

"Just one minute," she said. "I'll connect you to somebody."

Freya waited. She looked at her watch; she tapped her finger against the phone. Five minutes went by. She called, "Hello!" into the phone, tapping her finger up and down, up and down. A male voice was saying, "Shalom, Mrs. Gould. I am sorry you had to wait so long. My name is Uri."

"I came to see my son," she said. "I'll be at Gan Zvi today."

"Noah didn't write to you?"

"Write *what* to me?"

"Where are you now, Mrs. Gould?"

"In Jerusalem. At the Shiloh Hotel."

"Yes," he said. "I must be in Jerusalem today. Will it be convenient if I come to you?"

"I came to see my son."

"Yes, yes. But he's not here. I would like very much to see you in Jerusalem."

"What's happened? Where is Noah?"

"Nothing has happened to Noah. We heard only yesterday about how well he is . . . Please, it's nothing that's happened to your son. He's not here, but if he didn't contact you before, I think I should come to see you."

And then she understood. "He's in the Army!" she cried. "Why did you let him go to the war?"

"We didn't send him, we didn't let him," Uri was saying. "He went, he volunteered. I was against it." She could hear his voice pleading, while a chill went over her. But she had known and dreaded this moment.

"We're American citizens! My cousin's husband is a powerful lawyer." She was shrill, her voice was breaking in a sound she hardly recognized. She dropped the phone onto the hook.

The maid had already cleaned. A royal blue spread covered the bed. Clean towels hung in the bathroom. She splashed water on her eyes and buried her face in a towel. It was too late, too late. She was always too late; too late for the boxcar, too late for the bread. And then she sank down on the toilet seat and wept.

The sun was beginning to creep down in the western sky when Freya lifted herself off the blue bedspread. She had lain there all day in her moss green dress dozing, turning from side to side, not knowing what to do about Noah, not knowing where to begin, sitting up long enough to turn on the tape recorder with the cassettes she had unpacked. In a barely audible volume she played the Mozart Requiem over and over: *Te Deum. Sanctus. Credo.* Then the Mahler 8th, then Mozart again.

The music was background, more than that, it was a system of logic that tapped into some larger meaning, the way things came together, what came first and what followed were not accidental; the sounds and rhythms, sonorities and phonic effects were linked to a larger pattern as though she were holding a small map to a universe she could understand. She could pick through the smallest details of the landscape. She could pick through painful memories without being menaced or crushed. Like Noah's graduation party, when Evelyn first quipped: *What scares me about your kid is that he might find what he's looking for.*

When Noah came downstairs the morning after the graduation party, she could see that something was wrong. The sunlight shafting through the glass at the bay window sent a red cast over his hair and his expression was intense, searching, fishing, in his turbulent moral sea. He had a homily look, like the message on his T-shirt: Save the Oceans. She focused on the little mole near his nose. He had a good face—direct, nothing to hide, a face without guile or pretense, but his expression now puzzled her; he seemed regressed to childlike defiance.

"Something wrong, Noah?"

He shrugged. "Maybe," he said, sounding tentative, rehearsed. His eyes were directly on her. "If you want to know, there is something . . . it's hard to tell you . . . I plan to go to Israel with Michael Haas."

"And when are you going?" She hadn't known what else to say. He might have said he was planning to go to Pakistan or to Rio de Janeiro.

"In a couple of weeks," he said.

She smiled. "Would you have left a note in the mailbox?"

"This isn't a joke."

She felt her pulse quicken. "For a vacation? Why Israel? Why not upper Michigan? What's the matter with you?" she asked, getting up from the bay window and crossing the room to sit in the chair by the coffee table. There was a look of quest about him. She tried to sound detached, pragmatic. "Then what? You'll be back in time for college?" She sat down again.

He pulled a footstool over and sat down in front of her. "That's the point, well, part of it. I need to think about Pre-Med. A summer away might help. I want to be a photographer; I do not want to be a doctor," he said, enunciating DO NOT WANT like pistol shots, his hands clenched. "But I never had the guts to face Dad. In his mind, I'm already carving out tonsils."

She saw his eyes and they were like her own, too big, too deep; he was too thin for his height; he was too solemn.

She couldn't move her eyes from his face. "Why, all of a sudden? You never said . . . I thought photography was your hobby like the swimming team. You misled . . ."

He straddled the footstool. "I didn't mislead anyone!" He was leaning toward her, pleading. "When did Dad ever ask? And it was you who drummed it into me that I should be true to myself. 'Be yourself, Noah, be yourself.' Okay. I've decided to be myself if it means fighting Dad."

There was a rush of confusion inside her. Something was coming together in the conflict between herself and Karl, in Karl's arrogance that unless his standards were met you were a failure and a disappointment. Is that how Noah thought of himself? She was being tested: "What does a child owe a parent? Are children not entitled to freedom?" She had written that to Asher twenty years ago.

74

"Noah, listen a minute," she said. "All those walks we took, we talked about everything, and now I see we didn't talk about anything important. But there was more going on in your head, and now, all of a sudden . . ."

Noah was examining his fingers. "Dad enrolled me in medical school before I was born! But there are other things. For instance, Paradise in Beaulac."

She stared at him. "You have something against Beaulac?"

"Don't you, Momma? You never hung around Beaulac, did you? Neither did Dad. He was this neat-looking Jewish doctor on the most wanted list for snipping out Christian body parts. He was so proud of it! And you were the Mom they couldn't even call pushy. The only place anyone ever saw you was in the library . . . you even worked in a museum. But none of it was enough to get me invited to parties. I got left here to tough it all out!"

It was as though a stone had hit her. She hesitated, bewildered, afraid to encourage the full story for fear of how it would end. What was underneath that paradoxical look? "That's quite a speech," she murmured. "You think a summer vacation in Israel will fix it? That's what Michael believes?"

"Michael believes in a lot. He tried to get me into this organization for a long time. A few months ago I joined."

"What organization?" Her hands were going cold. She imagined Moonies and Hare Krishnas and Manson cults and her mind raced to connect them with Noah or with Beaulac or with Israel or even with Michael.

"A Zionist youth group," Noah said.

"A what?"

Noah didn't answer; he ran his hands over his head and looked away. "I'm not doing this to hurt you, Momma. So far, I'm going for a summer."

She felt her face go hot. "So far?" she cried. "This doesn't make sense."

"I've made a commitment," he said. The way he said, commitment, was as though Noah had found his true worth.

"It doesn't matter how we feel about it? How I feel about it?" She was numbed. How had it happened? What had come together?

75

"Noah . . . !" She stopped, giving herself time to absorb, cautioning herself not to blow issues out of proportion.

Noah shrugged. "I don't want to fight you, Momma. Why can't you believe that I'll have a good experience? They can use a good photographer. Especially . . ." He broke off.

"Especially what? You mean for wars? Is that what you mean? I read, Noah. I don't always like what I read. There are two sides trying to survive. Not just one."

"I'm not on both sides," Noah said. "Are you?" Noah was smiling. "Isn't Grandpa's favorite saying, You can't sit on two chairs with one ass?"

"That's right! Pushkin has one chair. You want to be like Pushkin? Is that what you want? One chair, one side, one ass."

He got off the footstool and turned away and started up the stairs, then he turned back to call out to her: 'Just be glad that the worst stuff you ever had to sniff is epoxy, not gas."

"Noah!" she called out to him. Somehow she had to set this right. She wanted to smooth his hair, to hold him. He came down to her and she put her arms around him, trying not to cry. "I . . . I'll try to help you, Noah."

He pulled himself free. "Don't cry, Momma," he said. "Just remember that I love you. Okay?"

He left her there sitting on the chair, her heart pounding. One swallow didn't make a summer. One summer in Israel didn't make a lifetime. The question she had asked herself for most of her life surfaced again: What does a child owe a parent?

Her dress was a crush of green when she switched off the Mozart Requiem and went to look out the window at the deepening light. Heat poured in to the air-conditioned room. The color of the stone hills had taken on a golden hue. How quiet the stones were. Even the buses and the passengers loading were soundless. She brushed at the creases of her dress and looked at her watch. She took a deep breath of air. In Chicago it was ten in the morning. Blitzie would be taking Hammer for a walk. Karl would be at the hospital. Dora would be making dinner for Poppa. Life was going on. Somewhere Noah was going on.

76

She picked up the phone and asked for room service: Would they send her a cheese sandwich and coffee? Half an hour? She put the phone down and started unpacking her suitcase. She had packed for a short visit. But she could buy blue jeans if she needed them. Or an extra pair of shoes.

The phone rang. Her heart bolted. Noah. She cleared her throat. "Hello," she said. It was Uri's voice on the line.

"Mrs. Gould," he said. "I'm in Jerusalem. If it is convenient for you."

"I ordered my dinner."

"I don't want to disturb you. But I'm here. In the lobby of the Shiloh."

"Please. I didn't mean to be rude. Come up."

She opened the door to a slight, slender man maybe in his sixties, maybe older, with round wire-rimmed glasses and gray hair that swept back from a high forehead; his hair was bushy like Pushkin's hair. Socialist hair. Uri wore a white shirt open at the neck with a navy blue cotton jacket on top that hung like a suit coat over beige cotton pants—an outfit, a workers going-into-town outfit. He was weathered, smiling a full mouth of gold teeth at her. She was staring. Uri was a type of person familiar from her childhood. But what did he have to do with Noah's experience at Beaulac High School or girls named Patty and Chris?

Uri was smiling gold at her. "I can come in?" he said.

"I'm sorry. Come in. Please."

"May I sit?"

"Of course."

She stood there while he sat down on the chair; she sat at the foot of the bed, facing him.

"You had a good trip?"

"Yes, fine."

"Noah was right," Uri said, smiling at her. "He said his mother was very, well, you *are* very pretty."

"Mr. . ." she interrupted him. "I don't know your last name."

"Levitsky," he said. "But don't call me that. Uri. That's our way here. And you? You have something to go with Gould?"

"Freya."

Uri's smile was delighted. "Freya! Where did you get such a name?"

"My mother heard it on the radio . . . Please. What about Noah?"

Uri Levitsky closed his eyes and scratched his head. "Yes, Noah. I feel sorry he didn't write to you. I told him . . ."

"I wish you'd just *say* it! Noah joined the Army. What else is there to tell?"

"Maybe you don't understand what happens in Israel. We accomplished too much to let anyone destroy us. We stand on a watchtower."

Impatiently, she interrupted him. "Noah is an American. Your enemies aren't his."

"American Jews don't often express such an opinion."

"Not every Jew in the world is on your tower. You turned Noah's head! You gave him a cause."

"No one can give a cause. A cause pre-exists. It's ready to be recognized."

"With help. Don't forget all the help he got. Michael Haas. And *you!*"

"Michael Haas," Uri said, shaking his head. "Michael Haas was a trigger. As for me . . ."

She waited. Her anger was spilling over.

"Listen," Uri said. "I told Noah to go home. We don't need more heroes; we need doubters, the kind who ask questions, who isn't so ready to cry Doomsday and climb on the tower. I told him to go home. Do you know what he said?"

She didn't answer. She was staring at his eyes that were peering back at her through the thick glasses; his eyes looked large, magnified.

"He must have read a lot," Uri said. "Romantic poetry. Too much. He said that a home was where your feelings and acts agree like a subject and a predicate."

"Is that what you came to tell me? That Noah ran away from a rotten home?"

"No! No! Don't misunderstand me. It's not unusual that young people from America should feel something here they never felt before. Everything here is obvious. Issues are clear and out in the open. There is no wilderness of ambiguity. There is no time here for self-conscious introspection. And at the same time there is poetry jumping out from thousands of years. The Bible. Stones that stand here with us and don't crumble. You see, we are seduced by the obvious."

"You amaze me with your English," she said.

"I was once a teacher of English . . . but my grammar isn't always so good when I talk."

"Yes," Freya said. "Of course you were a teacher."

"You think I am tricking you with well-spoken English."

"I didn't say that. You have a great flair for explaining my son."

"Listen," Uri said. "When Noah first came to the kibbutz, I used to think that if you flashed the card, *purpose*, in front of him, his brain would go into chaos. But something here touched him; the way he worked in the orchard and learned to speak Hebrew was as though he were getting acquainted with himself for the first time."

He was sitting on the edge of the chair, his hands on his knees, his body leaning forward so he could see her more clearly through his thick lenses.

"What do you know about Noah's life?" Freya cried. "He was *weaned* on purpose . . . his grandfather questioned the purpose of every thought, every act. His father . . ."

He waved his hand at her. "You prove my point. You prove my point!"

"No! I'm not going to ask; I know what you're getting at— Noah's sense of purpose belongs to him. I'm saying, no, to that. I'm saying he's a dumb kid with no sense of direction. He doesn't belong here. I want him home. And I'm not going to sit in your seminar any longer. Where is Noah? *Where?*"

Uri was looking at her. He lifted his hand and dropped it to his lap. "You talk about a sense of direction? Michael Haas is the one with a sense of direction. Something you should know.

Michael Haas left for Paris. His father wired him a ticket. You should know that," he said. "It upset Noah very much."

She heard herself gasp. "When? *When?*" she whispered.

"A month ago, before his voluntary training ended."

They *knew* Michael was in Paris when she talked to them. She could hear Rudy Haas hissing that she wasn't a proper Jew. And Noah was to be taught a lesson meant for her. Noah would be left to tough out reality.

It was as though Uri were not sitting opposite her. She covered her face and then she looked up, her hands tightly clenched. "Before I came I phoned Michael's parents . . . his father didn't tell me . . . he was angry about the Holocaust, he was proud that Michael was in the army. They didn't tell me. They didn't . . ."

Uri got up and moved toward her; he reached out and put his hand on her shoulder. "Sometimes," he said gravely, "We Jews who went through the Holocaust, punish Jews who didn't suffer."

Uri took off his jacket and showed her his arm. Just below the short-sleeved shirt was the indelible tattoo.

"We don't forget this," he said. "The problem is, not forgetting has become the rock of the kingdom. Every Jew takes their turn to free the sword from the rock."

"They wanted to save Noah from *me!*"

"No, no! I think when they came to their senses about Michael, they realized they had already paid a big price."

"You're apologizing for them!"

"I'll tell you something, Freya. When I was in the camp, the most important thing for me was to polish my shoes. The shoes were full of holes. But every day I would pick up a piece of charcoal or a stone or anything I could find to polish them. It was foolish. But it was my way of not feeling like an animal. I had no hair on my head, I had no name, no food in my stomach, I was bruised from head to toe from blows, but I could polish my shoes. I was a full human being with values and standards; it helped me look at the others without disgust."

Freya was silent, looking down at the floor.

"Don't feel sad, Freya," he said. "For some reason, I don't know why, you've been cast with us. There's no logic in it that we

80

should make claims to your life, to the life of your son. Israel is a legitimate country, not a social contract with every Jew in the world."

"I came to see my son! You said someone saw him yesterday. Where did they see him?"

"I don't know, exactly. The man who spoke to him was a soldier from our kibbutz. He saw Noah in the North. His unit is probably in South Lebanon. I'm sure of it."

"It's too late!"

"Tell me, Freya, is he your only son? Your only child?"

"There's just Noah."

"Listen to me. It's not too late. An only son does not go into combat. Especially foreign volunteers Noah's age. That's the way we do things here. Unless their parents give permission by signing a statement, they don't go to the army until they are citizens. The Haas boy got out because his parents withdrew permission. I thought you knew. That's why I didn't mention it. If Noah didn't write to you for a permission statement, he forged signatures. He would be discharged immediately."

She felt a tremor go over her. "You're sure?" she murmured.

Uri nodded, reaching out to touch her shoulder. "Yes, I'm sure. But you have to understand. The incident could embarrass him very much. We have to find a way. Also, the question is, should his mother expose him, make up his mind for him?"

A weight lifted from Freya's heart. She could get Noah out of the army! "I never dreamed it would be that easy," she murmured.

"You think that's easy? You'll expose your son, cause him disgrace and make a decision for him he couldn't live with? Noah is nineteen years old. You don't think you have to talk to him first? After all, he came here to be on Gan Zvi."

She was silent, confused. Uri was pleading as though he had a prior claim, some hold on Noah. "I came to take him home!" she cried. "Now you're saying I would disgrace my own son if I did, that I would be running his life!"

"I don't envy you," Uri said. "Whatever decision you make, I don't envy you. I only ask that you should talk to Noah first."

81

There were three polite raps on the door. She opened it to the young waiter who had given her coffee in the dining room. He was holding a tray. He must have been shocked by her ravaged face, by the man sitting at the edge of the chair, and he backed off. "I'm sorry, Madame," he said.

She signed the check. The waiter smiled and left.

She put the tray on the night stand.

"Please," Uri said. "You must eat."

She looked at the tray with the sandwich covered by a white cloth and a decanter of coffee beside it.

"Please," Uri said.

She poured herself a cup of coffee. "Let me ring down for more. Perhaps you'd like something to eat."

"No, no." He looked at his watch. "It's late. I have a long drive." He looked away for a moment and then he turned back to her. "Come to the kibbutz with me," he said. "My wife and I, Naomi, we'll take good care of you, believe me. We can talk more. We can help you think in such a serious decision."

She shook her head. She took a sip of the coffee, hesitating. "No, I'm not going anywhere."

"But you need friends to help you see Noah before anything about his circumstances is revealed," he said.

"There's an American consul here."

Uri was shaking his head. "No. That isn't a strategy. You need Israelis," he said, shrugging. "But more than just people. You need friends with influence, *protexia* we call it. When you're in Israel a while you'll know what I mean. I have some contacts. I'll try my best, believe me, but maybe you know people here to make the time go fast while you wait?"

Freya hesitated. She began to speak and stopped, then began again. "There is . . . perhaps there is one person," Freya said. "He's been here twenty years; he got a Ph.D. here. He's probably a professor. Do professors have *protexia?*"

Uri shook his head and laughed. "We have one professor for every kilometer in Israel, maybe more!" Uri said. "Unless he's well-known, of course."

She hesitated again. "His name is Asher Frank," she said.

82

Uri's eyes magnified even more behind his glasses. "There can't be two Asher Franks with a Ph.D. You mean, *General Asher Frank?*" he said. "You *know* the General?"

Freya felt her breath catch. "No, I don't know him now. I knew him twenty years ago." She was trying to take in what Uri was saying: *Asher, a general?* "He's a *general?*" she asked.

Uri's gold teeth glistened, and then, suddenly, he was somber. "Yes; he's a major general in army intelligence. He has *protexia,* more than anyone should have; but he's a genius, he's respected. I have some reservations but he probably feels the same about me. His wife, Aviva and I, she's much younger, of course . . . She was born on Gan Zvi, the kibbutz I went to live on."

She had known in her heart that Asher would have a wife. "Aviva," she repeated.

"Does General Frank know you're here? You must phone him. If anyone can help you . . . With a flick of his finger he could arrange for you to see Noah."

Something was draining inside her. Phone him? *Do you mind that I suddenly turned up in your war to ask a favor?* Freya shook her head. "I doubt he'd remember. And a general has more on his mind than one soldier."

"Phone him," Uri interrupted, "I see you have trouble being aggressive. But you *must* be aggressive. *For Noah's sake.* General Frank lives in Jerusalem. I'm sure he's away, but Aviva would know where you could write to him."

"Yes," she said, getting up quickly, facing him. He caught her signal and got up, too. "I ask you again to come to the kibbutz," he said.

She shook her head. "No. Not now."

"You plan to remain at the Shiloh? For how long?"

"I don't know. I can't think."

"Nothing moves quickly here. That's why I want you to come to the kibbutz; waiting will tax your nerves."

"No, Uri. I'm grateful, but I can't. I'll find things to do," she said.

Uri looked at her for a long moment, his eyes seeming to float behind his glasses. "Well then," he said. "Try to keep yourself calm. See this beautiful city. I'll be in touch with you about my connections for Noah."

She watched while he scratched on a scrap of paper, his nose bent close. "General Frank's address and phone number," he said. "I don't think I'm making a mistake to give you the information; otherwise, you would have to contact him through the army and it could take weeks. If someone asks how you got the number, say that Uri Levitsky is related to a friend of yours at home. Say I gave you the number. Aviva knows me very well."

Uri handed her the slip of paper, and stood back watching while she took it from his hand, avoiding his eyes.

When Uri left, Freya went to the window and opened it wide. The cast of gold on the buildings was deepening to shadows. Eucalyptus trees lined the boulevard beyond; and there—mimosa trees, surely they were mimosa. Gardenia was somewhere in the air, heady, exotic.

Her eyes were hot from the glaring light but, still, she stared, not moving, the scrap of paper with Asher's telephone number on it damp in her fist. How long had it been since she heard Asher's name? Sher she used to call him. Sher. Tears were streaming from her eyes. If she never left this spot at the window, perhaps she would wake up in another place.

On Thursday, it was six in the morning when she looked at her watch. She had slept intermittently between reading and looking through the window. She was staring out again, as though there were an answer hiding in the palm trees. Uri said she only had to tell the army that Noah had forged his parents' signatures and he would be out. What did it matter if he suffered embarrassment? You had to be alive to be embarrassed, and she was determined that Noah should stay alive. She took a deep breath, giddy with relief. Noah would be coming home.

She sat up in bed and leaned over to phone room service. A woman answered with a heavy accent. "It's not my business," the voice said. "I shouldn't interfere. There is a good Israeli breakfast downstairs. Fruit. Cheese. Tomatoes. Last night you had a turkey sandwich and a coffee. You don't feel well?"

Astonished, Freya leaned over in bed, clutching the phone. Then she began to laugh. Where, any place on earth, would room service answer like that?

"I'll be down soon," she laughed into the phone.

The shower was bracing. The cold water poured down over her hair and her face and, suddenly, the little room with blue walls and blue curtains at the casement windows was cozy, secure. Noah would be coming home. She let herself gurgle into the spray of the shower and then she dried her hair with a towel while she sat at the edge of the bed, her legs stretched out in front of her. Maybe she had only dreamed that Uri had gold teeth and Asher was a general and that the card on her desk was telling her to enjoy an Israeli breakfast. She let the towel drop to her lap. If she weren't dreaming and Asher was a general . . . her philosopher, a general? How did one speak to a general? Like one spoke to Patrick, the alderman in Albany Park?

The phone rang. Was it Momma summoning her to get herself down to breakfast?

"Layla!" she cried. "You remembered the Shiloh . . ."

"It wasn't hard to remember. I told you I have a cousin who works there. Do you feel well?" Layla said.

"Fine. I must have slept for two days."

"Have you made a heavy schedule?"

"Not yet."

"I'd like very much that we should see each other."

"Where are you?"

"In Nazareth."

"I'd love to see you, Layla. Anytime. I have no schedule, not yet, anyway."

"Are you free today?"

"I've lost count. Is this Thursday? Yes. Why not? Will you come to Jerusalem?"

"Of course I will!" Layla's multicolored voice rang out with delight.

85

"You know where the Shiloh is?"

They fixed it. Layla had a car; she would drive to Jerusalem that day and arrive at the Shiloh about noon. She wouldn't need to be back in Nazareth until late that evening.

The sun surged through the window. It was seven-thirty. The linen dress was crushed but it would have to do. Thank God she had had enough sense to bring a pair of walking shoes. She combed the tangles out of her hair, slowly, one strand at a time and tied it back with a scarf.

"Good morning," the young waiter said.

"Good morning. And it's time I knew your name."

He smiled. "Suhail."

"Suhail?"

"All the service people here are Arabs."

She thought it was odd that he should inform her so abruptly. "Fill up the basket, Suhail. I'm starving."

After breakfast she bought a guide book at the hotel gift shop and then, hesitating, she asked the shopkeeper where the pay phones were. He sold her telephone tokens. *Simonim*, he called them. One *simon* for local, two for long distance and more for both if you go longer than a minute. No, the operator won't tell you. You will be disconnected, that's all. "Best to put in three or four. Five, maybe. Do you want a Jerusalem Post?" he said. "It's in English." The man was round with a flaccid face and one hunched shoulder.

She glanced at the paper and recoiled. Beirut had been bombed again.

"You want something else?"

She had noticed a shelf of notebooks when she came into the shop. She went back to the shelf and chose two. "I'll take these," she said. "And ten letter stamps for the States. Will you take American money?"

He shrugged. "Why not?"

She felt him watching her when she walked out of the shop.

It was after eight-thirty. She stood poised at the pay phones, her heart racing. Uri was right that she had to see Noah before

86

she told the army about his forgery. And Asher could help her with a flick of one finger. How much would she do for Noah's sake? One, two, three, four tokens and then she dialed the number Uri wrote out.

"*Ken?*" (yes) a woman answered. She had the impulse to hang up but she held on.

"I'm sorry to disturb you," she said. "Mrs. Frank?"

"Aviva Frank. *Ken* . . ."

"I'm sorry . . . I don't speak Hebrew . . . ."

"English is perfectly good," the voice said.

"It's somewhat awkward to be phoning, Mrs. Frank. I'm Freya Gould. I'm a school friend of Asher's from Chicago. We haven't seen each other for many years, and well, I happen to be in Jerusalem. At the Shiloh Hotel. I'm not sure Asher would remember me, but I thought I'd call to say hello."

"I see. A school friend. Freya Gould. That's very nice," Aviva Frank said. She had a husky, smoker's voice. Was there a hesitation? "My English isn't perfect," Aviva said. "But that's very nice you should phone us. I'm interested that you knew the number to call."

A sinking feeling: strangers should not phone a general at his home. "I'm sorry, Mrs. Frank," she said. "I shouldn't have . . . A mutual friend encouraged me to phone. Uri Levitsky. He gave me your number."

"Uri Levitsky? You know Uri Levitsky?"

"I only just met him. He's a relative of friends in Chicago," she lied as instructed.

There was a curious silence for a moment. "Listen," Aviva said. "If you're a friend of Uri and an old friend of Asher's, especially from the same school . . . I do remember Asher once spoke of a Freya something or other. The Shiloh is not far from us."

Aviva hesitated for a long moment, and finally she went on in a rush as though she had made a reckless decision. "Perhaps you would like to have dinner with us? Tomorrow night? Shabbat. There will be some other friends here. I would like very much to meet you. Can you join us?"

"Dinner? Tomorrow, Friday?"

"Unless," Aviva chuckled, "you have something else to do at night. In that case, I wouldn't interfere." And then she added, "I'm making a joke."

She was imagining Aviva sitting there, waiting for an answer, the right tone of, yes, the wrong tone of, no. What had Asher told her? Why was she so eager? Freya bit the inside of her lip and in an unguarded moment, she answered: "I'd love to come."

"Now," Aviva said, as though she had resolved one issue and was taking up another. "Would you be offended if I asked for your passport number? You must forgive us that we are such slaves to security. I'm sure, of course, that everything will be fine and someone will pick you up at six o'clock."

"You shouldn't go to such trouble,"Freya said, "A phone call was all I . . ."

"No, no, no," Aviva said, gaily. "You mustn't feel offended at our precautions."

Freya reached in to her purse for her passport and read the number out to Aviva, her voice barely audible.

"Would you repeat, please? I didn't catch . . ."

Freya repeated the number.

"You don't mind, I hope you understand," Aviva said. "I look forward to meeting you. I'm quite prepared to enjoy the school reunion," Aviva said, chuckling. "I'll phone you tomorrow."

Freya stood at the phone, not moving, a telephone token pressing into her palm. She lifted her hand to phone again with an excuse to cancel. But that would be more awkward; it would be suspect. Why had she allowed Uri to pressure her? *School* reunion? She had got what she deserved. Humiliated, she put her passport back in her purse.

She looked down at her crushed green dress. There had to be dress shops in Jerusalem.

A dark, slim woman in a tight black dress with a neckline just short of her cleavage was at the reception desk. Her hair was black, cropped and spiked, and a huge silver loop dangled from one ear lobe.

88

Freya regarded her with a smile when she moved toward her. "Zafra at your service!" She smiled a mouthful of small white teeth. "That's my name. Zafra. Can I do something?"

Freya smiled back. "I wondered if you might suggest a dress shop?" she asked.

An older woman stood behind Zafra, the kind of woman she remembered from all the years growing up in Albany Park. A comfortable, stout woman with thinning salt and pepper hair and a sallow skin. She glanced over Zafra's shoulder, frowning.

"A fancy dress?" Zafra asked.

"Just something nice."

"So what is nice for you? Sheesch or plain?"

"Well, for dinner."

"Just dinner? At a house? At a restaurant?"

"At a house."

"Hmm," Zafra said. "Something plain, something nice. Special people? I mean, someone, you know what I mean, *special* people . . . Where do they live?"

Freya looked at the address. "Beit Hakerem."

"Beit Hakerem," Zafra repeated. "Near the Knesset, near the University. The brains of Jerusalem. *Snawbs*. Believe me, they look at every seam." Zafra reached over the counter and lifted the ends of Freya's long hair. "You want a haircut, too?"

"No, no, not a haircut," Freya said, astonished.

"For a nice dress you go to Max's," the older woman chimed in. "On the Jaffa Road. A walk from here. Tell Max you're going to Beit Hakerem for dinner. He'll treat you right, believe me, he has taste."

Zafra fingered her one earring and turned to the older woman with a shriek. "*Max's?*" she cried. "What for are you telling her, Max's? Listen," she said, turning to Freya. "Don't listen to Malka. Stay out of Max's. His sizes begin with *her*, not you. You want a nice dress? On King George Street there is a shop, *Boutique Rene*. A designer from Morocco. I'll show you on the map. She also gives haircuts, something wonderful. That is, if you want."

"And what about an earring?" Malka murmured sarcastically, and added through tight lips, "I don't go to such places." And then she caught herself, straining to keep up professional appearances.

Zafra was unfolding a map of Jerusalem. "I'll show you King George Street."

Malka looked as though she would snap. Freya moved back. "Look," she said. "It's not important. I think it would be fun to discover a shop for myself, don't you? Actually, I need to cash an American traveler's check for Israeli money. Two hundred dollars?"

"Why don't you go to a bank?" Zafra said. "You'll get a better rate."

Malka glowered. "What are we here for? Not to cash checks?"

After the money exchange, Zafra said gaily, "I marked King George Street. My advice is, go to Rene's. Also, only if you want, she'll give you a haircut. When I had such long hair like yours my head filled up like a bomb." Zafra gave her head a vigorous shake. "See how free I am now? You have a hat? Keep it on your head."

Freya smiled. She felt a rush of warmth toward Zafra's open, funny face, her spiked hair and one earring and her sweet, bird-like voice. Turning away from the desk, Freya called back to her. "I have a cousin like you!"

Freya stepped out of the hotel to a blazing sun. She rummaged in her bag for sunglasses and looked at her watch; it was almost nine o'clock. If she took a walk for an hour or two she would be back in plenty of time for Layla. She could even look into shops. She stood there for a moment studying the city map of Jerusalem where Zafra had marked King George Street. An easy walk if she could get across the intersection where cars seemed to be going in every direction with a set of bewildering traffic lights that gave no hint about which direction to cross.

CHAPTER SEVEN

THE TRAFFIC PATTERN was chaotic and Freya waited until one green light or another flashed on, none of them relevant to where she stood, until, finally, she made a dash for it. Cars came within inches of hitting her as they screeched toward the maze of an intersection and Freya ran to the opposite side where she caught her breath in a wash of relief. She started up Herzl Boulevard where up close, on terraces, stood the buildings she had seen through her window. Not Byzantine, not European, not massive, but a blend of simple modernism resonating in the past, solidly confident in the future, dazzling white in the sun, breathtaking in their scope and their impression of endurance. She opened her guidebook.

"*The Benyenai Ha'ooma Concert Hall. Three thousand seats.*" She scanned. "*. . . In the war of Independence, the Orchestra piled into trucks which lurched and jolted up to Jerusalem.*"

The sidewalk was virtually empty but the boulevard screeched with traffic. Open army jeeps whizzed by before she had a chance to take them in. A few yards ahead was another divided boulevard; she crossed over and read the cluster of street signs: Tel Aviv pointing west. The Biblical Zoo pointing north. The Old City, pointing east, down Jaffa Road, a normal-looking street with shops and apartment houses on both sides.

She was walking downhill. The scene changed abruptly from a sleepy, dense, and deserted boulevard into a seething labyrinth

91

of humankind. Chains of connected two-story houses with wrought iron balconies were on either side of her; they were old, lopsided. Smells of fish and fresh-killed poultry and strong coffee with a burnt chicory aroma laced the air. People were rushing everywhere. Women with their heads covered in dozens of styles and colors of shawls or turbans, carried fishnet shopping bags or leather sacks. Suddenly there were bearded old men with long black coats and fur-trimmed hats walking with a strange swing of arm and forward thrust of shoulder. Then, for the first time, she saw soldiers. Young men, young women, lounging at corners, appearing out of nowhere, milling about with guns under their arms. Short-barreled, snub-nose guns were slung over their shoulders from leather straps—Uzis.

The soldiers were in and out of shops, gathering at vendor stands in twos or fours, gesticulating at each other, laughing, talking in the high chatter of Hebrew. Khaki mini-skirts and short-sleeved shirtwaists on long-haired young women soldiers. And on the men, crushed trousers were tucked into ankle-high boots, open-necked khaki shirts, sleeves rolled to the elbow, hats fastened into shoulder loops. Nothing was military about them except their boots and their guns. Every shade of complexion, every stamp of country, marked their faces; all of them playful elbow pushers, full of nonchalant fun, meandering in the crowd. The Uzis slung over their shoulders distorted their youthfulness; the gun was another pair of eyes, an extra set of limbs, a signal of consummate strength. If Noah had gone this far, he was beyond motherly advice, out of her reach.

Jaffa Road grew more crowded the farther she walked. Army jeeps were parked at every corner, the soldiers inside casual, watchful. Every variety of shop was on either side of her; shoe shops multiplied by shoe shops, leather goods, watches, tobacconists, newspaper kiosks. The old men with beards and long black coats serpentined around her. Little boys with feathery hairlocks, great hats and black coats like the old men wore, pushed past her. The women with turbans and shopping bags sidled through the crowd undaunted. It was before nine o'clock but there was a frenzy over the morning, everyone was moving—fast.

It was Thursday, the last full shopping day before the eve of the Sabbath.

She came to a small intersection: in both directions was an open market. The smells of fish and fresh-killed chickens shot out at her. She turned right and moved on through the market, incredulous. The barrows and stalls, the rush, the callings out and yellings back, the human colors and shapes, animal and vegetable—arms, legs, balance scales in perpetual motion; chickens, oranges, cabbages, flying from barrow to baskets.

She exhaled in a gasp. She wanted to meld with the crowd and the movement. Her eyes were everywhere. Barrels of pecans, almonds, and sunflower seeds. Barrels. Barrels. Pickles and olives, marinated red peppers and eggplants. Macaroni, broad egg noodles, barley, bulghar, dried peas and lentils.

A vendor was yelling and holding out a fish; she stopped, shaking her head at him while the powerful bouquets of odors permeated her senses—she was released, liberated from Bumpers, from the corridors and tight packaging of all the grand supermarkets of her life.

She stooped down to smell each one: trays, casks of seeds, leaves, sprigs, and powders. Once she had done research for a label about the spices of the Bible: Anise used as the basis for fiery Arak. Coriander, likened in the Bible to manna. Cumin, described by Isaiah. Garlic, which the Israelites craved in the desert. Marjoram, the Biblical hyssop for purification. And saffron, the precious spice of the Song of Songs, overflowing in a cask before her eyes; bay leaf, called *oren* in Isaiah; celery seeds, exempt from tithing, chicory, the Passover bitter herb; cinnamon, the spice for the holy oil of the Temple and for praise of the bride in the Song of Songs. They were all here, brimming over, together with mace, mustard seeds, mint, and rosemary, drowning out the odor of live carp, smoked herrings, and freshly killed beef and chickens that were changing hands over the stalls of the narrow street. She was muttering aloud, "What a scene."

Between the barrows laden with mangoes and artichokes, sweet yellow peppers and persimmons, were rows of cheeses. A gray-haired and anorexic-looking woman in a blue denim apron stood next to the stall. A long sabre of a cutting knife lay on the shelf in front of her. "You want?" the woman asked her.

Freya smiled and turned back to the spices. A vendor who hadn't been there a moment ago was there now, a cocoa-colored and bushy-haired fellow, his shirt open to his navel, stood over her with a measuring scoop in his hand.

"*Ma'at rotza?*" she heard close to her ear and she turned, quickly.

"I don't understand Hebrew."

"Ooo, Ahh," he said, cooling his fingers. "From your face you are Shepardi so I make a mistake . . . I ask you in Hebrew what you want . . . But you are a guest, rich American lady! *Todah Rabah.* Thank you very much for visiting our poor country. Thank you." He was mocking her, bowing from the waist, feigning obsequiousness.

When she turned away, he called out. "Don't get insulted so fast. I'll buy you a coffee. You and me? Yes?"

She walked on, curious about the Jewish Afro and she turned back to him, smiling.

"Actually," she said. "I'd like a bag of pecans. You speak English, don't you?"

"You can't hear English when you hear English? Go there," he said, pointing to another vendor standing near the casks of nuts. "Then coffee?" he said. "You and me?"

"Okay," she said. "Coffee. You and me."

He said his name was Moshe and after she bought pecans he bought her coffee at a two table cafe right in the middle of the market, a hole-in-the-wall cafe with a beaded doorway. The menu was simple: Espresso. A sandwich made of pitah bread, ground chick peas, and spices—*houmous.* Moshe said he was Moroccan. They settled the question of whether or not she was Jewish. He knew English as good as a lord because he worked as a driver for an American in Haifa. He spoke with hands, with pointing, with facial expressions. His open shirt flaunted the hair of a gorilla. "You like *Mahane Yehuda?*" (Jewish Market) he said. "*Mahane Yehuda,*" he repeated, "*here.*"

94

There was something repelling, exaggerated, magnetic about Moshe. He was inflated with opinions of himself. He was full of Macho hunger for sex.

She answered him. Yes, she liked the *Mahane Yehuda.* "It could wake up the dead," she said.

"It could *make* dead quick," Moshe said. "Wait. Some time here a little bomb go off. Fsst . . . Fsst . . . You like?" And then, looking up, he smiled at what was outside: an old Arab woman swathed in a black robe that spilled over the great pillow of her body. She lingered at the beaded doorway of the coffee shop balancing a loaded basket over her arm. Moshe nodded at Freya.

"See? You think she has *farsomin* in her basket? How you say, per-simmons?" She looked out at the Arab woman, pondering her. Moshe's laugh was a roar.

"I see you are soft," he said. "You are pretty. I make you afraid? Don't be afraid. The soldiers in jeeps take care of everything. Don't worry. Anyway, Arabs, they are nice people; only they don't know what's lies and what isn't lies.

"Listen. A story. Two little boys were bothering an old Arab in an olive grove and the old one said, 'Go over there! They are giving away candy.' The boys went. The old man sat and he thought and he thought . . . If they are giving away candy over there, why should I stay here? So he went after the boys." Moshe howled. "That's Arabs," he said. "You don't need to know more."

He switched subjects. "Where you live in America? Are you rich?"

"I live in Chicago." She found herself laughing at him. He was straight out, no frills, no pretenses. He wanted a bed. She felt drawn to the way he pared it all away; she could almost imagine . . . "No, I'm not rich."

"If it's with *you,*" he said, lifting his brows. "I don't care," he said, looking her over again. "You very pretty. It's okay not rich."

Freya smiled. She looked at her watch. "I have to go," she said.

"To where you will go?" Moshe said. "Some night, maybe, you and me?"

The proprietor of the coffee shop was drying his hands on his pants with one long stroke. Moshe gave him a couple of coins. She made a gesture to pay for her own and he brushed her off. "You answer me? Some night, you and me?"

"I won't be here long," she said.

"Where you stay?" Moshe said.

"A long bus ride," she said. "Look, Moshe, I really liked meeting you. It was fun." She put her hand out.

Moshe grasped her hand; he massaged her knuckles in slow, deep movements. "I here all the time," he said, in his coarse, sensual voice.

She pulled away. "Good-bye, Moshe."

He stared at her left hip. "American lady!" he said. "If you will need to take some bus tomorrow? Friday, Shabbat, three o'clock, bus finished. Come back, I take you."

She looked back and waved.

It was 10:30 when she realized she should have retraced her route and should not have walked farther down Jaffa Road. She found herself in the center of Jerusalem under a large Coca Cola sign at the corner of Kikar Zion. She got her bearings. Small shops were on both sides of the street, souvenir shops, tiny cafes with a few tables and chairs out front, book stores, newspaper kiosks, tourist agencies, silver shops, bakeries, shoe stores, shoe stores, shoe stores. The guidebook in her hand said the walks they mapped out would take her through every neighborhood, past every monument, church, synagogue, and antiquity. So far, she could have bought shoes and brass menorahs.

She stood still for a moment, adjusted her sunglasses, and looked around. The sky pierced blue in a blaze of sunlight. People were crisscrossing everywhere in no pattern; the undertow of noise compelled movement. A young woman police officer stood outside the cordoned off square, directing traffic in the middle of the street—she was gorgeous in a navy blue mini-skirted uniform with a white hat and insignia. She couldn't have weighed more than 100 pounds, not older than twenty, but she was strong enough to wave her arms around as though she were conducting the Philharmonic. "Zus! Zus! Zus!" she yelled in a voice like a circus barker to the cars circling around her, and when a driver put his hand out of the car and cooled his fingers at her, she pulled a whistle out of her breast pocket and blasted him with it.

Freya heard herself laugh out loud. Costumes! Old men, bearded and tattered, sitting on the ground, begging. Moustached, turbaned younger men, sitting on the ground, waiting, gesticulating, motioning to passersby for what seemed like selling their services for a day's work.

She felt high, excited. The pungency of the brash Moshe was still with her, and suddenly it was as though she had plunged from the window of her hotel room into this pixilated pool of fur trimmed hats and fezes and kaffiyehs and turbans and gibberish, and policewomen smashing the air with regulation whistles while flaunting long sexy legs. What kind of a place was this? She felt tears of excitement rush to her eyes. All this had caught Noah and it was catching her too.

Freya had come down to the lobby of the Shiloh a few minutes before twelve o'clock to meet Layla who was there, waiting, dressed in tight yellow jeans and a sleeveless white shirt unbuttoned recklessly low. Her tawny face was aglow with pleasure and when they embraced Freya caught a whiff of musk.

"I'm so glad you're here," Freya said. "Is your mother all right?"

"Yes . . . yes, we'll talk later. I'm full of happiness to see you. You look *well!*"

"I had quite a morning! Shall we have lunch?"

"Not here," Layla said. "I have my car."

They went out to the parking lot of the hotel where a red Volkswagen bug of ten years vintage sat, waiting. "Is it all right?" she asked Freya when they were settled in the car.

"It's beautiful!"

"I'll give you a tour," Layla said. She knew her way through Jerusalem; the traffic was reckless, insane, because Thursday was marketing day. No, it wasn't necessary to put on seat belts; anyway, there weren't any; but Layla plowed ahead undaunted, holding her ground, smashing her horn and averting collisions by a hair. And all the while she called out streets and sights and narrated like the driver of a tour bus. She made a screeching turn into a great open space that was a parking lot and came to a stop.

"The Jaffa Gate," Layla said. Freya caught her breath, and then she looked up to what she couldn't quite grasp: a massive stone gate, the entrance to the citadel was rising before her eyes, gray stone walls and ramparts, floating in a tension of sun-drenched space.

They got out; Layla locked the car and motioned her on. Open jeeps of soldiers were parked, sitting. Layla's walk was swift and graceful and Freya found herself lagging behind, hurrying forward to catch up. Layla's stride, the way she called back to Freya who was dodging cars swinging into the parking lot, was a transformation, full of swish, sure-footed, almost swaggering on her own Arab turf.

In front of the great gate of the citadel, clusters of old men in keffiyehs sat on long stone benches outside, fingering their prayer beads, selling baubles or jabbering with cronies. And then she startled by a high wail that filled the air. She turned to Layla who pointed to a tower. "Call to prayer," she said. "Five times a day."

Freya stood still, compelled, trying to catch the cadence of the liturgy. The soldiers around her were alert, on guard, while the wails floated out from the Mezzuin tower.

Inside the gate, the cobblestone alleyways teemed with shoe-shine boys, pretzel sellers, vendors, and more soldiers. "Don't think about soldiers," Layla said. "There's a war. Let's have one sight before lunch!"

Freya was led through the maze of what Layla called the Sukh. Layla stopped in front of a storefront where, inside, old men in kaffiyehs were smoking huge water-pipes. She pointed to them: "Nargillas," she said. The smell of tobacco wafted out and mixed with the pungency of honeyed nuts and baklava—it was a dizzying aroma.

"Is it hashish?" Freya asked.

"You want hashish? I'll take you to Uncle Zaki. He has a dress shop. Not far. The purest hashish in the Middle East." She laughed. "But I warn you, he is seven feet tall and he weighs three hundred pounds; he is like Sinbad and to every man or woman he sells hashish, he kisses them on the mouth—hard."

They laughed and then Layla was suddenly sober. A line of bearded, black-frocked men with broad black hats trimmed in fur walked by, their bodies leaned forward, their hands locked behind their backs. "The religious Jews," Layla said. "They are going to pray at the *Kottel Ma'aravi,* the West Wall, some call it the Wailing Wall."

Layla took her down the alleyway to David Street that bordered on the main square. From there the streets led off, she said, to the heart of the Armenian Quarter. A bagel vendor wandered over and Layla bought two huge soft pretzels sprinkled with sesame seeds. The air was alive with spice. They turned into a doorway. Four wooden flights up in the Petra Hotel was the roof; Freya followed to a door leading out. They went to the railing, munching their pretzels and squinting against the sun.

Layla's finger was pointing, her arm outstretched to the desert that shimmered beyond the Old City. "Out there," she said, "My great-grandfather was a Bedouin. And his grandfather before him and before and before . . ."

The way she said before and before and before, with such pride of possession, gave Freya a feeling of having been alerted to something she could not pin down. Layla pointed to The Pool of Hezekiah, an ancient waterworks and reservoir where garbage was dumped now, then to the Dome of the Rock, to the walls of Sulemein, to The Garden Tombs.

The half-eaten pretzel was in her hand when she turned to face Freya. "I think I would die for Jerusalem," she said.

Freya could barely see her tall, slender figure in the sunlight. "*Die* for it?"

Layla reached out to put her hand on Freya's arm. A strand of her long sandy hair fell over her face and she pushed it back. "Now is a good time to tell you," she said. Layla's eyes seemed to swim in the sunlight.

"Don't be angry with me," Layla said. "I hardly know you but I *want* to tell you. It was a lie about my mother being ill. It's my brother, Fawzi . . . the police are holding him. He insists on organizing demonstrations against the war without a permit. They arrested him. They think he's maybe PLO."

Remembering the spectacle at the airport, Freya felt jolted, suddenly on guard to what she didn't know. "Is he PLO?" she asked.

"No; *no*! We're citizens of Israel! We were born here! Shouldn't we have the same rights to protest as Jews? Why are we seen as a fifth column?"

Freya had heard it somewhere. Or read it somewhere, but now it tumbled into her head: *A stone thrown from within is not the same as a stone thrown from without.*

"Is equality such a strange idea for an American?" Layla cried.

"No," Freya said. "It's not strange. I hope Fawzi will be okay." She began to feel enmeshed in a struggle she hardly had begun to grasp.

When they left the rooftop of the Petra, the narrow streets of the Sukh, the bagel vendors, the tea carriers, the religious Jews pushing forward to the West Wall, seemed to Freya like passages winding through a threatening imbroglio. Layla looked at her watch. "We're expected for lunch. We're late."

Layla's friends lived in the Armenian Quarter on a cobble-stoned mews within a courtyard and up two winding flights of stone steps that ended on a balcony. Purple bougainvillia hung over the rail, the smell of jasmine and orange from below was mixed with the aroma of coffee coming from the house. Inside, there were straight-backed chairs, a long table, and an over-stuffed plush sofa. The house belonged to Yusef and his wife, Amal, but there was another visitor there, Khalil, a man of about thirty with a huge black moustache.

The men got up when she and Layla came in. Yusef extended his hand to greet her: "Welcome," he said. Amal repeated, "Welcome." She shook Freya's hand and promptly disappeared. Layla scolded her back in Arabic but only the sound of laughter came from beyond the sitting room until she reappeared again briefly only to stutter in English, "Not speak." She went and came back and went and came back with plates full of dolma and stuffed zucchini, fruits, honeyed nuts and almond cakes and huge bunches of grapes, the size of plums. She brought coffee, and poured everyone a cup; she coaxed food onto everyone's china plate and then she was gone. Layla kept calling her back, but Amal had disappeared.

Yusef explained that they all learned English in British-run schools. They ate. They touched on trivialities. Of course, Layla had told them that Freya was a Jew. An American Jew.

"Do you consider us enemies?" Khalil smiled. "You aren't afraid of us?"

"Khalil!" Layla interrupted. "She is a guest!"

Yusef sat by munching nuts, picking grapes off the stem and eating one by one. "I always respected Jews," he said. "Their accomplishments. Their culture. Do they respect mine? Do Americans respect my culture?"

Freya cleared her throat so her voice would be audible. "I hope so," she said.

"I agree with Yusef," Khalil said. "Arab history, art, science, mean nothing in the West. Only Jewish history and Jewish science is significant."

"There is no Jewish science," Freya said.

"Stop this!" Layla cried. "In the West they know Jews because there was Freud and Einstein and more. Our civilization was snuffed out in the fifteenth century. What should they know about us?"

"See how Layla talks?" Khalil said, grinning. "She talks like a Jew. She's lived with Jews so long she's one of them!"

Yusef was shaking his head. "Never mind, Khalil. The Jews need to feel that everything they did in Palestine was done *morally*. For good and for justice and because of the Holocaust. *Not* like the British, who didn't care. The Jews are idealists. We bother their sleep. They will tear themselves apart before we do."

"You make me ashamed!" Layla cried. "My friend does not make the rules here. Have you forgotten how Arabs treat guests?"

The tension in the room was relieved by Amal coming in again. She began gathering plates. Layla got up to help, a stricken look on her face. But something more was roiling in Freya: something was going on in Layla she didn't understand, that Layla hadn't expected to spill over.

They were outside in the mews again. The soldiers milled about. A jeep was parked a few feet away and a young man in a white T-shirt and blue jeans seemed to be taking a long time getting a light from one of the soldiers in the jeep. He eyed them for a long moment while he leaned back against the jeep and took a long drag on his cigarette. Layla noticed him watching; she hurried Freya on. When they were outside the mews, Layla turned to her: "I could kill myself."

Freya smiled but she said nothing.

"I'm humiliated. I can't look at you," Layla said.

"I thought you said American women were tough?"

"You're decent, Freya. Why did I involve you with Khalil? I'm ashamed."

"Why did you, then?" Freya said.

Layla was taken off guard. "I don't know," she said, on the verge of tears. "I just don't know how to cope anymore. I keep looking for someone to understand me. Oh Freya, you don't know! My brother, Fawzi. There's too much to tell . . . there's my fiance, Toufic. And Sani, my sister . . . she's a nurse and she's gone to . . . to . ." Layla stammered and hesitated as though she had to think of a place where Sani went. "France," she said, finally. Layla stopped talking and stood there, her eyes stricken with apprehension as though she might have said too much and blundered into a costly mistake.

Freya slowly began to walk beside Layla. Why had Layla blurted *France* as though it would burn her tongue? Hadn't she just come from London? Why wouldn't her sister be in France?

Baffled, she put her arm through Layla's while they walked the cobblestone streets, out the Jaffa Gate to the parking lot.

She stepped aside for the large gatherings of bearded Jews with fur-trimmed hats and long black coats that were filing through the Gate in the direction of the West Wall. She caught the eye of a young one and she felt a wash of tenderness at his wistful eyes peering out from under the fur-trimmed hat. His long black coat hung on his skinny, hunched shoulders like a hand-me-down.

Freya startled awake on Friday morning. Aviva's dinner party. What had possessed her to play along and say, yes? Just because Uri said she should? What did Uri know about her life? Why had she listened to him pontificating about protexia? She imagined Aviva checking on her passport, sending someone to pick her up if she *was* who she said she was. Aviva hadn't mentioned whether Asher would be there. And if he were, how had she imagined facing him? Somehow, she had to wriggle out.

She was soaking wet from the shower when she went to the window to look out at the hills and the hot wash of sunlight. Every sight she had seen, every shift of feeling she had experienced was a tunnel on the way to dinner at Aviva's house. She reached for the phone.

"Aviva?"

"Aviva. Ken."

"This is Freya Gould. I hope you won't mind if I can't make it tonight. I seem to have worn myself out sight-seeing. I'm hardly over jet lag. I'm really sorry."

"But everyone is looking forward to meet you! I was about to phone you that we are delighted."

"I'm very tired. I'm not sure I'd be good company."

Aviva laughed. "From what I hear, you are very good company. I think you are *shy!* It wouldn't hurt your jet lag to come."

Freya winced. "If you don't mind," she said, "Would there be another time? A rain check?"

"But there may not *be* another time soon. Asher does not stay here long. It takes five minutes to pick you up," Aviva insisted.

Asher *would* be there. A sensation of draining went over her. How much did Aviva know? Had she made a point of phoning every guest to say Asher's childhood nymphomaniac was coming to dinner? She felt her face go hot as though she were in debt to Aviva, as though she needed to show herself a decent woman and apologize for knowing Asher. "If you won't mind that I leave

103

early? If that's all right . . . ?" She heard her voice as obsequious, apologetic. She had no right to insinuate herself. How could she face Asher? How?

Freya took the blue cotton suit she had worn on the plane off its hanger and inspected it. Spiritless, functional. She gave it a shake, hung it up again and went to look out the window. She had made a damned fool of herself.

She stood there, squinting at the sun that was still round and hot. The sidewalks were deserted. A few cars moved on empty streets. She imagined the frail young boy in the long black coat saying his prayers on the Sabbath.

At six o'clock, a man named Yonah Klar rang her room and she went down to meet him in the lobby. He was short, tubby, and he had a crop of graying hair and thick eyebrows fringed across a small forehead. There was something theatrical about him; he was dressed in a scruffy blue jeans jacket, as though he had just come from a camping trip with Fidel Castro. His bottom lip protruded when he said, in excellent English, "I won the raffle to be your chauffeur, Mrs. Gould. Shalom. Shabbat Shalom."

She tried to smile back at him. "It's nice of you to pick me up."

"I see I was lucky," he said, looking at her. "You are Asher's school friend?"

"A long time ago. Twenty years."

"That is a long time. You're lucky to catch him. He comes in and out of Jerusalem."

"The war . . ." she said, flatly.

"Ach, even before the war . . ."

Yonah was hesitating. She hesitated too. "Do you want to see my passport?" she said.

Yonah laughed. "Your *passport?*" He shook his head. "We are not so crude as you think," he said.

He took her arm and they left the hotel. In the parking lot he took a deep breath. "If nothing more," he said, "Jerusalem is the one place in the world you can smoke and not get bronchitis."

104

He had a dark blue 1980 Volvo with leather seats and a dash board that lit up like Halloween.

Traffic was thin. "Your first time in Jerusalem?" When she nodded, he glanced at her. "This, right where you are, is the heartland of Jewish culture. The Knesset is over those terraces. The University. The Israel Museum. The Shrine of the Book where they keep the Dead Sea Scrolls. You see, our culture came home to Jerusalem." He glanced at her again to catch her reaction. Freya sat there motionless, silent, her heart pounding.

The outside of Asher's house turned out to be a thicket of brambles, cactus, and banana trees. The door was open and Yonah led her inside to a reception hall tiled in black and white terrazzo. Beyond was a sunken living room from where Aviva was coming up the two carpeted steps to meet them. The blue cotton suit paled on Freya's back. Aviva was dressed in a long yellow robe that vibrated in multi-colored cross-stitching. Her face was like painted porcelain that shone under her cropped, jet black hair: she was tall, her eye makeup was subtle, elongating black eyes to an oriental slant. Gold loop earrings hung almost to her shoulders. Freya had a sudden image of her first week in Beaulac when Claire had hosted a coffee for her to meet the new neighbors: they were tall, wonderfully attired, and it seemed to Freya that she had landed in a world of porcelain sirens where she would be devoured.

When Aviva kissed her in a light brush against her ear, she exuded an exotic and expensive fragrance. Asher had married the kind of woman Karl had wanted for a wife.

"Freya, at last," she said, holding Freya at arm's length, inspecting her. "I didn't know how to think about you! Asher kept his old friend a secret!"

"We were children," she murmured.

"You met our Yonah, of course," Aviva said. "Did he brag to you? It's all true. He was a general. Now he sells water cannons, pretty things like that—he's a tycoon!"

"Don't make a drama," Yonah said.

Freya's heart jolted. Asher was suddenly there, not smiling, standing off a bit; he was older, tall and deeply tanned, a khaki shirt was open at his neck with an array of ribbons on his breast pocket and insignia on his shoulders. She felt her breath catch at the first sight of his dark luxurious hair streaked with gray, the strong curve of his jaw, matured to the full measure of strength. He was staring at her as though he could see her heart pounding; a muscle flexed in his jaw; his eyes, blue and intense were taking her in as in the first moments when they would meet after a day's separation.

She brushed a strand of hair from her face. How was he seeing her? It wasn't possible to see one's self. In Asher's eyes what was she now? Middle-aged? She thought she saw him frown and she felt herself draw inward. He was disappointed; twenty years was nothing but an embarrassment.

Asher waited for Aviva to move back and he reached for her hand. "Freya," he said, his voice hardly audible.

Their eyes met in silence. *Sher . . . Sher* echoed in her heart. She withdrew her hand quickly.

"Asher!" Aviva cried. "I didn't know how to think about this school friend of yours. Why didn't you tell me how lovely she is?"

Asher's eyes never left her face when Aviva introduced her to the others. There were handshakes. David Aaron. Ageless, maybe forty. A writer, a professor of literature. She was imagining everything, not hearing exactly what Aviva was saying. "You should have a lot in common with David Aaron, Freya. David is a writer and Asher said you were keen on poetry."

David said his wife couldn't come that night. One of the kids had a cold. David shrugged. "One of the kids always has a cold."

"So," Yonah said. "That means you see very little of your wife?"

"Don't listen to Yonah," Aviva said. "He's a Nazi."

Two other people came down into the living room. The Savirs. Savir, who seemed to have no first name, or had she missed it? He was a tall heavy-set man with glasses, and Tanya, his wife, was tiny, her skin yellow. Savir was a demographer for the government. The room seemed to expand and shrink but Freya never lost the sense of Asher's eyes. Aviva motioned her to a blue

106

velveteen sofa and she sat down, Savir on one side, David Aaron on the other.

Abruptly, suddenly, Asher excused himself. Important business. He was sorry but it couldn't be helped. Her eyes followed him as he left the room, the back of his head, his stride, his presence fading.

Aviva was smiling at her. "The army," she said. "Asher is annoyed with the war tonight, I can tell."

Aviva crossed her legs under the flowing yellow gown and slowly shook her head, meeting Freya's eyes. "You must excuse him," she said. "He is pressed, poor man. He has no time for a person in the present let alone from the past. He would ask you to forgive his bad manners."

Freya nodded, not answering. Aviva had put her in her place. Asher knew his priorities and she had no claim on his time. The room was stifling and she sat very still, something inside her, crushed.

Asher came back. He was running his hands through his hair in that way she knew so well—when he was agitated beyond speaking. "I'm terribly sorry," he said. "Something came up. I have to leave."

He came to stand over her. She had to look at him or everyone would notice how vulnerable she was.

"I'm sorry, Freya," he said. "How long will you be here?"

"I don't know . . . a month, longer, maybe. I'm not sure."

"We'll phone you at the Shiloh," Asher said. "Very soon."

Aviva got up, her earrings swinging, an amused expression on her face. She turned to Freya. "You see?" she said, laughing. "War has no respect for old school friends."

Asher was gone. Freya was telling herself: Sit still. Be calm. How many times over twenty years, bathing her mother, arguing with her father, living with Karl, had she told herself to sit still and be calm? Now, thousands of miles away, in Asher's house, she was telling herself again. Sit Still. Be Calm. Evelyn had said that she coped, hadn't she? Like Mrs. Miniver, she was.

She would leave, inconspicuously; she would disappear, before dinner. She would make excuses. Jet-lag. She had a headache

from the heat. A sudden anger was roiling inside her. Anger at
Uri Levitsky, at Noah, at her father—all those who had nudged
her to this moment in Asher's house. But anger revved her
up, displaced the humiliation and pain. She would *not* sit here
like an orphaned child in a crumpled suit. She held on to her
martini glass as though it would support her. She looked at her
watch again. An hour and seventeen minutes had gone by. Still,
no dinner.

She had finished her second martini and David refilled her
glass. Whiskey made her shiver and cough but she had controlled
herself through two martinis and this one was going down like
cool broth. She took another swallow. And another. The room was
opening up, the atmosphere was getting better. The coffee table
burgeoned with nuts and dips and wedges of pitah bread. Freya
stared out to the room; if she didn't look at anyone directly, they
wouldn't guess her relief that Asher was gone; no one would notice, or care, for that matter, not even she, that Asher had walked
out on the school reunion.

She kept her eyes on one object or another, a hodge-podge
from Occident and Orient of artifacts from exotic trips: fertility
goddesses and aboriginal rugs, hand hammered Bedouin coffee pots.

David Aaron was watching her finger a Bedouin coffee pot that
sat on the coffee table. "Home-grown roots," he said. "The truth
is, Bedouins would give a lot for electricity to plug in a kettle.
You see all this?" he said, waving his arm around the room.
"Aviva is the most expensive interior designer in Israel. The
more primitive the culture, the more money she makes."

Freya began to laugh, then she caught herself and fixed her
attention on what jumped from every corner and every shelf and
every surface: sixteenth-century icons, antique Limoge figurines,
thick-lipped wooden gods, Ibu masks, ancient glass, water jugs,
oil lamps, Chagall lithographs. All the while she had been writing
labels in the Museum, Asher and Aviva had collected the artifacts.

There were voices around her: Was she enjoying her stay?
(Aviva was attentive.) Did she have a family? Just one son? (It was
clear now that she had no husband.) How *brave* she was to come

to Israel alone when there was a war! "You see! A free woman does as she likes," Aviva said. As for herself, she had no time for children and Asher was never around to produce one. Was Aviva calling her a free woman?

Aviva urged Freya to try the eggplant salad on a square of pitah. "When I'm in America I eat hot dogs," she said. Then, smiling, she added, "Asher never takes me to Chicago. He's *hiding* you!"

"Hiding me?" she asked.

"You haven't seen Asher for twenty years?"

"Twenty years."

Yonah was telling her that she should not miss the diggings in Jerusalem. He would take her there himself. She forced herself to listen. She accepted another martini and nodded at Yonah. Such interesting people! She would stay to dinner after all. If she were to be eaten by Israelis, she might as well enjoy herself first. She focused hard on the conversation around her, on every word and every gesture.

"Diggings," David was saying. "Every foot we dig under the Temple Mount is ours. When we dig through to Rome, that will be ours, too."

Savir was looking at a huge art book. "You should see this," he said, motioning to Freya. "A rare medieval Passover Hagaddah, the story of the Exodus from Egypt." Looking over Savir's shoulder, she saw translucent colors of Hebrew calligraphy, page after page of letters like flowering vines. Savir said to her: "From slavery to redemption."

Freya sighed. "Gorgeous . . ."

"We've learned a lot since then," Savir said, shrugging and closing the book. "Maybe we've *over learned.*"

"What if we had *under learned?*" Yonah said. "If you overshoot the runway you have choices, if you miss it going down your options disappear."

"Under learn, over learn, everything comes out to disaster," Savir said.

"Asher is right," Aviva said. "He says we're obsessed. Every time we open our mouths we make self-fulfilling prophecies."

"Maybe the General learned something from the Arabs," David Aaron interrupted. "Their self-fulfilling prophecies are a pleasure! *We'll force the Israelis to their bellies. Islam will be the Main Street of the world.* And how long will that main street be? From Tripoli to Mecca by way of Ben Yehuda Street. Israel will be the nightmare that didn't last long. In the meantime we have the best pilots and fighter planes. We're slaughtering them in Lebanon and they go on, planning the Main Street of Islam."

"They shouldn't count on it," Tanya Savir said. "You say we're slaughtering *them?* You mean we're not dying, too? Why do you bleed for the Arabs and not for your own?"

David shrugged. "Let's not start," he said.

"We have to die to teach the Arabs we've beaten them," Yonah said.

"What do you mean, *beaten?*" David said. He scooped up a handful of nuts. "We've beaten them? Because they lost some land? To the Arabs that doesn't mean, beaten. They mark it down in a ledger. So much land lost. So much property taken. So many dead. They know the day will come. What does it matter how long it will take? Listen, I knew an old Arab who planted an olive tree when he was eighty-five years old. 'Why do you need to plant an olive tree at your age?' I asked him. And what did he answer? 'My son,' he said. 'From your question it's clear you don't understand Arabs or olive trees'."

"*Yofi*," Aviva said. "Wonderful . . . *Yofi.*"

Freya repeated. "Yofi."

"Never mind," Savir said. "I understand Arabs because I understand Jews. We like to pull out decayed myths and pretend they're new antibiotics. We do it and they do it, too."

"Change the subject right now," Tanya Savir demanded in a thin, little voice. "Our guest doesn't have to hear how we shout at each other." She turned to Freya. "I have a story, too," she said. "When I was a child my mother had a friend with a glass eye. I was afraid of her because when she came to tea she would sometimes take out her glass eye and lay her palm over the hollow opening on her face. I asked my mother why she did such a horrible thing. 'She keeps checking to see if something is growing back,' my mother said."

Freya gasped and put her finger to her eye.

"You're a better liar than I am, Tanya," Yonah said. "Now tell the one about Medusa and what grew back in place of her head." He turned to Freya. "She's talking about the rebirth of Jewish culture. It's supposed to be obscure."

David Aaron was looking at her.

Aviva had lighted a gold-striped cigarette and she was crushing it in an ashtray. She leaned over to Freya. "I hope you weren't offended that Asher had to leave; the war is his mistress . . ." She smiled broadly. "You must be starving," she said.

"If you don't produce dinner soon, I'll take Freya out for a sausage," Yonah said.

"Asher would want us to treat her better than that," Aviva said.

"With Asher she has *protexia*," Savir said. "A way in and a way out. The lady from Chicago doesn't know how lucky she is."

"I think she does," Aviva said, glancing at Freya and leaving the room with a turn that set her earrings swinging.

David Aaron turned to her. "I'm taking an interest in you," he said. "Are you left brain or right brain?"

Again, Freya began to laugh. "*No* brain," she said.

"Don't laugh," David said. "Take me, I am despised for my right brain. Take Asher, both sides of his brain fight each other; they don't cooperate. So don't laugh. For you, Freya, it will make a big difference in what you see here."

"Now tell her your ideas on reincarnation," Yonah said.

"Let him sneer." He looked at Freya. "You need to know one thing, nothing more. We're put on this planet to work things out."

Freya liked that idea and repeated it to herself several times until it was straight in her head. *We're put on this planet to work things out. We're put on this planet . . .*

Dinner was served by a slight, dark girl, a Jewish girl from Morocco, Aviva said, her *ozeret*, maid, who understood only French and Mograbi, an Arabic dialect. The girl was shy and ten-

111

tative; she did not look at the guests while she served them, attending only to Aviva's rapid, fluent French. Fifteen minutes after dinner, Freya got up very carefully and excused herself.

"You hardly ate!" Aviva said.

She had a headache . . . jet-lag overtaking her, but would they please excuse, forgive . . . She was unsteady on her feet as though she were sleepwalking. Aviva's eyes were amused. She thanked Aviva and, yes, she did understand about Asher's emergency, and it had been very good to see him even for a moment, and to meet her and all their good friends. She was sorry to leave so early. Aviva was kind to have invited her. She stopped herself. She was being excessive. She was drunk.

Yonah took her back to the Shiloh and insisted on walking her to the elevator. "Will you be all right?" he asked her.

"I'm not really a lush, you know."

Yonah was smiling at her. "I didn't think so," he said. He held her hand to say good-bye. "I haven't met anyone like you for a long time." Freya slid her hand free. He handed her his card. "I'd like to take you to see something of Jerusalem. Will you phone me?"

"Perhaps," Freya said. "Sometime . . . I can't think now . . . but sometime. . ."

CHAPTER EIGHT

ON SATURDAY MORNING, Freya got up with a pounding headache, a hangover after four or was it five martinis and no food. Maybe it was only the altitude of Jerusalem or jet lag, she told herself, as she lay there inert, living down Aviva's dinner party, images jumping in her head: Ibu masks, fertility goddesses, Aviva's porcelain face, a woman flipping a glass eye out of its socket, left brain, right brain, no brain. Asher staring at her . . . *Sher, Sher.* She rolled over on her stomach, put the pillow over her head and wept.

She had to get down to the shop for Bromo-Seltzer. But it was Saturday, Shabbat, the shop would be closed. She got out of bed and rummaged through her suitcase. A bottle of aspirin, thank God! Under a cold shower, she mumbled to herself: how stupid, how stupid, stupid! She had made a huge fool of herself. She had to pull herself together, have coffee, a lot of coffee.

Breakfast for Saturday morning, Shabbat, was modified to what didn't need heating. The toaster had been removed but the coffee was kept hot in thermos jugs. The dining room was full of tourists. Another woman cut in next to her at the buffet. "It's crowded," Freya said.

The woman smiled. "There's a table, there in the back," she said. "For two." She spoke with an Israeli-British accent. They sat down together.

Freya sipped coffee as though she had found an oasis in the desert. The woman was looking at her. "Oh," Freya said. "Coffee . . . that's how I am in the morning."

"That's perfectly all right," the woman said. "I'm the same."

"I'm Freya Gould."

"Lizabet," the woman said. "Lizabet Avidan."

The waiter brought more coffee. Not Suhail this time. She asked if he might have Bromo-Seltzer? He shook his head, no. Tomato juice then, a very large glass. She glanced at the woman sitting opposite her.

"The waiters are so pleasant," she said.

Lizabet smiled. "Very much so. You're on a tour?"

"No," Freya said.

The waiter brought her a huge glass of tomato juice, a glass of water, and a small bottle lettered in Hebrew. She looked questioningly at the bottle but the woman filled her in. "Bromo-Seltzer," she said. They didn't speak for a while. The woman was looking at her obliquely. "You're on your own?" she asked.

"Yes, on my own," Freya said.

"I'm a native," Lizabet said. "On Saturday mornings I take a rest from my kitchen."

Freya was puzzled, but why? There was nothing puzzling about taking a rest from the kitchen.

"And what have you seen so far?" Lizabet asked.

"I hardly remember."

"The Old City?"

"Yes . . . Yes, the Jewish Market, The Jaffa Gate, afterward, the Armenian Quarter."

"In the Old City on your own? That's not so wise at this time." Lizabet shook her head.

"A friend took me."

"Only to the Armenian Quarter? Not the best shopping. Unless you buy rugs."

"My friend has relatives . . . we had lunch at their house."

"In the Armenian Quarter?" Lizabet said. "Lucky for you to have Arab friends. Someone you just met? You knew them from before?"

"Just one friend," Freya said, puzzled that Lizabet assumed she had Arab friends. She dissolved the Bromo Seltzer in water, swallowed it, and glanced at Lizabet. She was matronly, heavyset, in her forties; she was dressed in a navy blue sack dress with small stripes of yellow and gray. She had exceedingly small eyes of a nondescript brown. But she was not unpleasant. As for the questions, she had already had a dress rehearsal from the two women at the front desk.

"Better now?" she asked. "You are Jewish?" When Freya nodded, Lizabet went on. "You're very wise to remember that Arabs live here, too."

Freya nodded again.

"You met the friend here?" Lizabet asked.

"In London," Freya said. "We took the same plane."

"London? Yes; a good place to meet friends," Lizabet said, adding a chuckle. "Arabs are twenty percent of our population. Your friend lives in the Old City?"

"In Nazareth," Freya said.

"Of course, Nazareth! If you haven't already, by all means visit her there, . . . the heart of Arab Christian Israel. I come here every Saturday. Perhaps you'll let me know what you think about it, what your friend shows you. I have very good relations with the Arabs. Perhaps I even know your friend."

Freya had a feeling of being cornered. The woman was asking for Layla's name! But was she? Lizabet was sipping coffee, waiting for something, if not Layla's name, what? Freya was silent.

"I'm glad you'll see the country," Lizabet said, "I didn't forget your name. Freya Gould. From New York, yes?"

"No," Freya said. She hesitated, but it was too late. Lizabet was waiting. "From Illinois," Freya said, looking directly at the matronly woman who wanted to get out of her kitchen. "A small town," Freya said. "You wouldn't know it."

They said good-bye. Lizabet said she would look for her again the following Saturday.

Lizabet finished her coffee, shook Freya's hand and left, turning with a smile so that Freya caught the light of her small eyes that sent another wave of puzzlement and discomfort through her. Never mind. Her headache had lifted. She was okay, alive again: she was swearing to herself never to touch a drop of alcohol as long as she lived.

When she got back to her room she opened the window to catch a breeze from the fragrant and shimmering quiet of the streets. Sabbath was wrapped around the city and the high, dry air was heady. Out in the clear distance, the mountains of Moab were supposed to be visible. Her mind kept tracing back to Lizabet Avidan.

The phone rang.

"This is Uri. Uri Levitsky."

"Uri," she said, deliberately cool.

"I have news, Freya, yes. At least I have an address you can write to Noah."

"Oh, Uri!" she cried. "Do you mean it?"

"Wait, wait . . ." Uri said. "They didn't tell me much. My connection is in the public relations department of the army. Maybe with a lot of red tape they would have told you the same without me. Noah is a photographer assigned to a tactical unit. They film front line battle scenes for tactical analysis. That's all I learned, nothing more.

"Yes, something more that is important for you. They are holding a letter he wrote you before he went to the army. Our post office sent it to Chicago instead of the place where you live and it was returned 'Address Unknown.' It hasn't gone back to Noah yet. You want it sent to the Shiloh? You see?"

"He did write!" Freya cried.

"You have a pencil?"

She rummaged in her purse, pulled out the little notebook and a pen and, in a rush of excitement, scribbled the army address Uri spoke into the phone. She was laughing and crying. Uri waited. "I'm floating, Uri. I'm in a helium balloon."

"I thought you would be. Didn't I tell you Noah would have written! Did you phone Asher?"

At Uri's last words, she hesitated and kept her eyes fixed on Noah's address. "I was there to dinner," she said, hesitating again, finally blurting out, "I told Aviva I knew you, but she asked for the number of my passport."

"Your passport?" There was a pause. "Don't take it personally," Uri said, chuckling softly. "In the General's house they would check out a fly. You saw Asher?"

"Yes," she murmured. "You should never have sent me there, he's very busy."

"You didn't tell him about Noah?"

"Uri, Asher left after ten minutes!" Her voice was catching and she swallowed against the dryness in her throat, angry at herself for breaking down.

"I see," Uri said. There was a pause. "I hear how disappointed you are, Freya. But now you have Noah's address. At least that."

She couldn't hold back. "Uri!" she cried. "Maybe I'm stupid, but I don't seem to grasp anything here! Every time I make a move, I get caught on some hook. I went to breakfast this morning and I swear I was being interrogated."

"By who interrogated?"

"A woman. Lizabet Avidan. She says she lives in Jerusalem but she comes to the Shiloh for breakfast every Saturday."

"Innocent enough . . ."

"She asked about my friend. I swear she was fishing for a name."

"Your friend? You have a friend here beside Asher? You didn't tell me."

"We took *El Al* together from the London airport. I saw her yesterday and she took me to her cousins' for lunch."

"*Where* for lunch?"

"In the Old City. In the Armenian Quarter."

"Your friend is from the Old City?"

"From Nazareth: she's an Arab."

Uri hesitated. "This Lizabet," he said. "How did she meet you for breakfast?"

"She edged herself over to me at the buffet. We found a table together."

"She was fishing for the Arab woman's name? You told her?"

"Of course not!"

"Why do you say *of course not*? What would be wrong with telling a name?"

Freya caught her breath. "Why should I tell her anything? Look," she said. "I was raised in a home where I learned not to throw names around. Anyway, I had a headache."

"But why not tell the name of this friend?"

"Uri! It's not important anymore. Now that I know where Noah is, nothing is important."

"It *is* important," Uri said. "I won't ask you about your friend. I'm going to tell you that you're right not to give names. I don't know if you were questioned officially, maybe yes, maybe no. If it was official they already have the name of your friend. It's *you* they wanted to see. This friend of yours in Nazareth may be in some trouble. If you go places with her, they'll watch you too. Especially now."

Freya was remembering the man sitting on the jeep, smoking, watching, while she talked to Layla. "But Lizabet *encouraged* me to visit my friend in Nazareth."

"That's what she told you? What did you say?"

"I didn't answer."

"Listen, Freya," Uri said. "I asked you, maybe ten times, to come to the kibbutz. At least until you know where you are. Why won't you come? It's possible you were being questioned. And maybe not. You know what Freud said? Sometimes a cigar is just a cigar? Still, I think you should be careful. Go slow. Learn what's going on here."

"Never mind, Uri. It doesn't matter. I have Noah's address. If I can't see him, I'll write and tell him that I plan to get him out."

"You think it will be so easy for Noah, and for you, I mean, in the future, when you both will have to live with it . . . that a mother informed on her son and they told him to pack up and get out?"

"Don't say that to me, Uri! It's not my fault that Noah didn't hear from me. His letter was kicked around everywhere but where it belonged. The kibbutz didn't give him my messages. It's *you*, it's your kibbutz, it's your telephone system, it's your mail system!"

Uri was silent. She heard him clucking. "Freya," he said. "It won't help you to sit and blame the post office or the switchboard of the kibbutz or even Noah. If you'll be bitter you won't grasp anything, you won't be able to think. When the country is mobilized for war nothing works the way it should. Don't ask me why we are courting suicide, or why you or Noah should be mixed up in our fate."

She heard herself stifle a moan. "I read the papers, I watch the TV. I wouldn't accept this war, not for my country, not for yours."

"Many of us don't accept it," Uri said. "We have no choice."

"I have a choice," she said. "Thank God, Noah has a choice."

Uri was clucking again. "That is not a choice. You have to talk to him. You must."

"Why? *Why?*" she cried. "You confuse me so I don't trust my own instincts! I met someone at Asher's house. Yonah Klar. He was once high up in the army and I'm going to phone him, maybe now."

"Yonah Klar? The arms dealer?" Uri repeated. "Freya, listen to me. You are impulsive. You don't know enough about this place to have instincts. Tolstoy knew the truth: *Everything settles down.* And believe me, what happens in a book depends on what characters show up. Yonah Klar is surely not the person for you. I want you to wait, be quiet until you can talk to your son. If Noah ends his stay in Israel ashamed of himself, humiliated, there will be more for you to resolve than now." Uri broke off; there was silence.

"Are you still there?" Freya said.

"I'm here," Uri said. "I have to tell you. You're not stupid. You're a foreigner. Your instincts don't work here as though you were in your own kitchen. You don't know what operates here, what or who will do you good, what or who will harm you. You don't know the complexities."

Uri began an uneasy chuckle. "Look at it this way, Freya. Did you ever see a fish nailed to a wall with its mouth closed? It's better to keep your mouth closed. If Asher would have helped you, that would be different. He's a general in military intelligence, after all. He's straightforward and he wouldn't let you or Noah get embroiled in an awkward situation with the army. Trust me, Freya. I know what I'm talking about."

Freya broke down. "Uri! Uri! You don't listen. Asher isn't *interested* in me! He was cold; he shook my hand and he left."

Freya put the phone back on the night table and sat looking out the window, trying to calm herself. Sit still. Be calm. Why should Asher be anything but cool? A general is responsible for his country, and she was an embarrassing triviality from his past. Asher was committed to an ideology, and if he remembered, he knew that Freya Pushkin wasn't right for Israel. Her only politics was conscience. How many times had she said that to him? How many times had he answered: *Conscience is a blind corner.*

She sat on the bed hugging a pillow to her chest, burying her face in it, thinking conscience overlearned, underlearned, right brain or left brain. She had never wanted an icon like Evelyn's picture of Freud on the wall of her studio, or Pushkin's picture of Lenin pinned to the wall in the kitchen, or Asher's portrait of Ben Gurion. She had nothing but the love for animals and flowers and the works of Donne to Dryden. She had no pole star, not even a box of glass beads.

She closed her eyes. Sit still. Sit still. She got up. At least she could write to Noah.

My dearest son, Noah,

It seems ironic that the letter you wrote before you volunteered for the army was the one that got lost. If that letter had reached me, I would have come here sooner. I can imagine you're shocked, and you're wondering what I'm doing here. I came because I can't quite believe that you'd volunteer to be in a war that has nothing to do with you. I can't believe that you wouldn't pack up your cameras and get out. Like Michael Haas did. Somehow the Haases' instincts seem right though it was unfair of them to convince you and then cop out. Did Michael try to persuade you to leave?

Anyway, I shouldn't go any farther without telling you how I know all this. When I got to Jerusalem, I phoned your kibbutz and talked to Uri Levitsky. I plan to visit the kibbutz one day. But not now. I had hoped an old friend from Chicago would help me see you where you are. He's actually a general here. Asher Frank. Imagine that I've been here for such a short time and already I've learned the meaning of protexia. *I want to see you and talk to you, Noah. Uri thought General Frank could help but so far the General seems unavailable. Uri is trying to do what he can so we can meet.*

What will I say to you? Because you forged our signatures, I'll have to ask you to reconsider your mission. Not because of the forgery, but because of the mission. Uri tells me you would be discharged from the army if they learned of the forgery but I wouldn't do that until I talked to you in person. You might not listen; but, Noah, do you remember that we always ended up going to Cooper's for ice cream no matter how much we argued? I know we can't end up at Cooper's, but maybe there's a better place for both of us than where we are. It's not too late. Will you think about it? Your father is still unhappy that you wouldn't go to Pre-Med. But he was glad I was coming to see you and he was generous with my trip. He's taking care of Hammer. Poor old Hammer misses you so much.

By now, I think you have the tearful letter I wrote you when your grandmother died. She loved you, Noah. Now, Dora is taking care of Poppa.

Would you believe, Noah, that I'm beginning to love Jerusalem? I've been reading a lot. I've been walking. I bought pecans in the Mahane Yehuda. I even went to the Old City and saw the religious Jews going to pray at the West Wall. There was an ecclesiastical looking boy coming through the Jaffa Gate. I don't know why, but my heart went out to him. For a moment, I changed eyes with your grandmother. And most amazing of all, I see the desert from my window. I think I see everything from my window.

Uri says that the army will send me the letter that went astray. I love you, Noah. I know I will see you again.

Outside, Freya followed her map to the post office where she dropped her letter in a slot. Then she walked back to the little park on King George Street. Rene's Boutique was on the corner and she went to look in the window. A black linen suit was on a svelte mannequin; a black mandarin dress of Chinese silk stood at its side; another mannequin dressed in a mauve dress with a black sleeveless coat draped over it watched at an angle. No one would be spared; the well-heeled, the sophisticated, the pampered, the protected. Rene's Boutique was preparing its clientele for mourning.

For the next week Freya checked every day for a message from Asher; every day she swallowed back disappointment. She scoured the shops with the receptionist, Zafra, who insisted that her lunch hour belonged to Freya and there was plenty of time to wander around Jerusalem, Zafra swinging her hips in a tight skirt, and Freya, walking alongside. They stopped to see dresses at the department store on Ben Yehuda Street, where two guards regularly stood outside the entrance to inspect handbags. "You have a bomb?" Zafra asked her. Freya smiled. Zafra pronounced bomb as usual, sounding the b at the end. "I have no bomb," Freya answered solemnly, pronouncing the last b. Once as they walked, Freya said: "Remember the dress I wanted to buy? For the party in Beit Hakerem?"

Zafra nodded. "Did you buy?"

"I went to the party. I got drunk."

Zafra stopped in her tracks and turned to Freya. "You didn't buy a new dress? You didn't have a haircut. You got drunk? What did they do with you?"

"They didn't do anything. I staggered out. I left."

Sadly, Zafra shook her head. "It's okay for me, drunk. For some other Israeli, drunk, but for you! You're an American! You don't stagger out!"

"How do Americans go?"

Zafra put her hands on her hips, tilted her chin, stuck her nose in the air, and marched ahead, two giant steps. People were looking at them. "Like that," she said. Freya burst out laughing.

In a few words, Zafra had taken the sting out of Aviva's dinner party.

She wrote to Noah every morning. She wrote Aviva a note to thank her for the dinner and she checked with Zafra just to make sure she didn't miss a phone call. Every few days she spoke with Uri on the phone. Every time he would tell her: "I'm trying to make connections for you, but in a war, the wheels of bureaucracy move as though they were square. Be patient, yes? What will you do today?"

"I'm busy being a tourist," she would answer lightly, so he wouldn't ask her to come to the kibbutz where the stir of every leaf would fan her angst about Noah.

She hung up she sat down on the bed and stared at the wall. Where would she go? The library? What was Zafra wearing today? She got up and went down to the front desk and made plans to go shopping again. "*Yofi!*" Zafra would sing back.

After breakfast, at the end of her second week in Jerusalem, Suhail fixed her a lunch to eat outdoors, a large thermos of iced coffee, a roll spread with cream cheese and an apple. No, he didn't know whether they would let her eat in the sculpture garden of the Jerusalem Museum, but, surely, she could sit in the shade of a eucalyptus tree and sip coffee.

The sun beat down relentlessly when Freya climbed the wide, graded, staircase of the museum, higher and higher and step by step, the vista of Judean Hills enlarged to a city bleached white in shimmering heat. She stood at the top of the stairway, high over Jerusalem. Where, but in a museum, was her sanctuary?

Inside, she felt the sudden blast of air-conditioning hit her face and she was inhaling deeply to take in the coolness and the atmosphere of marble and wood and glass.

A group of about ten young people, dressed in shorts and beanie hats, were chattering in English near the great door of the foyer. An aggressive male voice rose above the others: "*Natan drags his feet on every field trip!*" As soon as he said it, a man with thick arms and muscular legs dressed in shorts, a T-shirt, and a beanie hat on his head appeared from behind the group; he leered at the boy who spoke up. "Drag feet?" he said, icily. "Do you know how long it takes to piece the fragments of a vessel together? Think about it, Mike. If you don't have patience to put a vessel together, your study is in the wrong field. And maybe you came to the wrong country." There were muffled grunts while Natan motioned with his arm to the outdoors and announced that the Egged bus would take them to the site. *That* was what she wanted.

She moved closer to the group and to their leader, Natan. "Excuse me," she said. "I don't mean to push myself in, but may I join you? Just to look and listen? I work at the Natural History Museum in Chicago."

"You're an archeology student?" Natan asked, looking her over.

Freya smiled. "Not a student. I write labels. May I join you?"

Natan looked her over again and managed a small smile. He turned to the group. "Is there space in the hearts of you diggers for one more American soul?"

There was a buzz. Mike, the boy with the aggressive voice, spoke up while he looked at Freya. "You betcha!" he called out.

The ride on the bus lasted little more than ten minutes.

The site was virtually on the slope of a hilly city street between the railroad station and the Old City. When they got out, Natan led the climb up another hill to a stony knoll over a deep valley, a roped off area with police barricades around it. Mike stayed close to Freya; he held her bag after she lifted out a scarf to protect her head from the fierce sun. All the way up, Mike talked. He was from Indianapolis and he had already tried accounting, then English lit, then psychology, and now he wasn't sure if archeology was his thing, but how did you know until you *tried?*

Freya laughed at him. "You're a real American."

At the site, Freya bent down to tighten her sandal and her hand brushed the ground. She let her fingernails scratch into the hot and crumbling dirt. What century had come to the surface? She straightened up, holding a small clump of dried mud in her palm so she could feel the reality of this place where Fiats and Volvos and Volkswagens whizzed by and a gasoline filling station stood on the hill below them.

Natan gathered the group at the ropes beyond the police barricade. Four square plots back to back had been excavated to smooth hard walls of rock and earth. Inside the plots, stone benches lined the sides. She recognized the burial benches in different sizes, the largest for the patriarch, then for the mother, and the smallest ones for the children. Natan began to talk, his feet planted wide, a pointer in his hand.

"The artifacts are stored in the Museum and we will see them next time. For now, we see the layout of the site and what we found here.

"We are high above the Valley of Hinnom," Natan lectured, pointing to the east. "The site had been calculated to appear at this precise spot which separates the historical city and the new city. The valley has interesting Biblical connotations, because it was the center of foreign cults for the Jerusalemites in the seventh century as mentioned in the book of Jeremiah. The site was close to the city and yet outside the city which doesn't compete with other land uses and is an ideal location for a cemetery. The distance of the cemetery from the heart of the city reduces the contagion and magic fear of the dead. But close enough to make visits. There was probably a filling station here for donkeys. Now there's a filling station for Fiats and whatnots. The spot was also the site of the encampment of the first encounter with the Romans, Pompey in the year 63 B.C. and for the auxiliary camps of Titus."

She stood there riveted to Natan's narrative of centuries, periods, civil uprisings, change of populations. He pronounced the name Jerusalem, *Yerushalayim,* like a poem, like a prayer.

She thought of Layla, looking out to the desert on the roof of the Petra Hotel: *I would die for Jerusalem, wouldn't you?* Another time, Layla had called Jerusalem *Al Kuds,* uttering it like a poem, like a prayer.

She closed her fist over the earth between her fingers; if the hard mud had but one Jewish grain of sand, she could feel it turning moist in the sweat of her palm.

Freya ate in her room again that evening. She had been dropped off in the center of Jerusalem and she was tempted to eat dinner somewhere beside the Shiloh, perhaps at *Yossi's*, Zafra's version of Cafe Paradiso. But Natan's long lecture under the hot sun had drained her, and there was something about Jerusalem itself she needed to distance—the stage-setting with stones as props. She started walking in the direction of the Shiloh; what she needed now was her small room with its very small windows.

She came into her room to hear the phone ring. She felt a rush of excitement.

"Oh," she said, "Layla."

"I called to apologize again for taking you to Yusef's."

"Don't be silly. There's nothing to apologize for."

"Are you feeling well, Freya? You sound like someone took your breath away."

Freya laughed. "I just got in, and, listen, I'm not exactly an invalid."

"I'm sorry. I feel concerned about you."

"Why?" she asked.

"No reason. You're alone and I want you to feel well."

"Don't worry. Or *is* there some reason to worry?"

"Nothing . . . nothing. I want you to have a good time. Why else would you come here?"

Freya was silent. Now Layla was fishing. She had never said she came on a tour. She had never given Layla any reason for having come at all. "There's so much to see," she said.

"Will you come to Nazareth? It would please me so much. I could show you the Synagogue Church where Jesus studied the Torah."

"I don't know. How would I get there?"

"By bus! Very easy. A bus from the Damascus Gate. It's simple."

She had the image of Lizabet Avidan munching a roll and watching her. "Layla," she said. "I don't know. I may be busy from now on," she said.

"Freya, please come."

"I'd like to see you, Layla. Let me think about it."

"Oh my God!" Layla said. "I told you about Fawzi and now you're afraid to be with me."

"I'm *not* afraid to be with you. That's not how I am."

"I wouldn't blame you," Layla said. There was a catch in her voice. "We are Arabs. Even to you we look like trouble. But let me tell you," Layla cried, "We have a right to live like human beings. We have a right to make friends."

"Layla! We didn't show our passports before we dared speak to each other!"

"I'm sorry. I didn't mean to embarrass you. I was beginning to feel normal, worth something."

"How can you say such a thing? You're worth as much as any one. I'll phone you when I can come to Nazareth. Really, I will."

Despite all of her frolics with Zafra, or her puzzling friendship with Layla, when the sun went down, Freya settled into her little room, aware of her loneliness. She felt enclosed in a chrysalis that held her at its center while she wound round and round spinning the same thread.

After the call from Layla, she had a sandwich and coffee sent up and switched on the TV movie channel. For the next four hours she stared at the final episode of the *Forsythe Saga* with Hebrew subtitles. Then a program instructing new emigres how to use launderettes. Next *The Last Tango in Paris*, where Marlon Brando and the girl had it off in Hebrew. Freya bit in to an Elite chocolate bar. Marlon Brando and the girl were humping against the wall, panting in the holy tongue. It was after one o'clock when she switched off the TV, and lay on the bed awake, watching the greenish after-image on the TV screen slowly disappear.

On Sunday, Freya came down to breakfast just after eight o'clock. She had been at the Shiloh for three weeks and she had walked in Jerusalem until she had worn out her shoes. She and Zafra shopped, ate, and after she was off duty, they went back to the shops again. She had followed her guidebook on walking

127

tours tracing architecture to periods and she had been to the Israel Museum at least twice a week, studying the artifacts from the Hinnom excavation and other ancient sites.

Uri said he was working on another contact to help her see Noah, but by now, she was seized with restlessness, wrestling with the impulse to phone Yonah Klar. One more day and she would phone him. In the meantime, she would go to Steimetsky's to browse and buy another book.

She wore the first thing at hand, a white sleeveless blouse and a yellow skirt she bought at a tiny shop off Zafra's circuit. She cashed a traveler's check and Zafra, now alone at the desk, took her in, frowning. "I see you don't need my shops," she said.

She was about to answer but Zafra's expression changed suddenly; her eyes grew wide and she was staring as though an apparition in royal robes had appeared. Freya turned. A tall and striking man in khaki towered over her. An array of silver and brass insignia were on both shoulders of his open shirt; silver wings were pinned on the pocket. He was deeply tanned and his deep-set eyes, piercing blue, were looking down at her and seemed to be everywhere on her face at once.

"Asher?"

He didn't answer; his eyes kept searching her face. He reached for her hand and released it. "Freya," he said softly. "I'm sorry to burst in. I had to leave that night you came to dinner. I just got back . . . I have some time. Are you scheduled?"

She struggled to speak. "I was going to the book store. Steimetsky's."

His face creased in a smile. "You were always a reader. Have you had breakfast?"

"I'm late for breakfast."

"We'll have breakfast together," he said.

He took her to a small coffee house off Ben Yehuda Street. They sat in the back and he encouraged her to have an omelette but Freya ordered coffee and a raisin roll.

"You still snack," he said.

Except when he was driving, he hadn't taken his eyes off her. They sat in the cafe staring at each other, not speaking, she, squinting, forcing back tears.

"You should have let me know you were coming to Israel," Asher said. "When Aviva told me, I didn't believe it. I wasn't very hospitable. I had to leave."

"We hardly spoke," she said.

"That's why I'm here, Freya. I took special leave so I wouldn't miss you . . . I want to apologize. I didn't want you to leave the country thinking I was rude."

Freya brought her coffee cup to her lips and set it down again. She pushed a long strand of hair behind her ears. "I wouldn't have thought that," she said.

"I shouldn't have left you with strangers," he said. "I'm sorry." He had been watching while she arranged the loose strand of her hair.

She nodded at the insignia on his shoulder. "Your ancestor would have been proud of you."

He smiled. "I'm not spit and polish enough for a southern officer."

"It's been a long time," she murmured. "Twenty years."

"Yes," he said. "It's hard to imagine."

"I know. Legends aren't supposed to materialize."

He smiled at her in the way she knew so well, half-teasing, half-serious. "You're no legend," he said.

The strain of being near him, making trivial talk . . . She had to act normal, somehow. "You're not curious why I burst in on your war?"

"It can wait. I want to look at you."

She looked down at her coffee cup. "You're older, Sher," she said, and drew back. She had called him by the old name she had given him long ago, the name that belonged only to her and to him. She looked away, embarrassed, and began tracing her finger around the rim of her cup. "I used to imagine how you would look older," she said. "You're fiercely handsome."

Asher had caught what she had called him. He hesitated for a moment and when he answered his voice was low and sober. "Can a girl of fifteen imagine anyone older?"

129

She felt herself going pale. "It's embarrassing. We must have been pretty wild for those six years."

Asher made no move toward her. "Yes," he said. "It may seem that way now."

Her eyes were brimming with tears. "You've had a good life . . ."

Asher nodded, looked at his watch, and abruptly got up. "What would you like to see?" he said. "The least I can do is show you something interesting. A special tourist deserves a special guide."

"Anything," she said.

"All right, then. I'll show you a bit of our country."

The proprietor glanced at them obliquely, aware of the string of insignias on Asher's shoulders. Asher paid the check, spoke to him in Hebrew, and the proprietor, murmuring, *"Ken, ken Aluf,"* (Yes, yes general) hastened to pack two large pitah breads and a container of olives and four avocados. He dared a half-smile at Asher, and then he added a slab of cheese and two persimmons as tribute. Asher went out to the car and brought back three water bags to fill. In that instant of the proprietor's deference, his obsequiousness, Freya understood the meaning of power and *protexia.*

They left Jerusalem by way of The Mount of Olives on a barren and parched descent that Asher told her was the ancient caravan route from Jerusalem to Jericho. His expression was intense, focused, like when he would show her something he had written, something he wanted her to share with the same intensity that had fired his creation.

One could see the mountains of Moab, and far across the Jordan river was Amman. As they drove, the Judean Mountains followed them to Bethany where Lazarus had risen from the dead.

The farther they drove the more she argued with herself about whether she should speak of their past, make casual references. From the periphery of her vision, she could see the handsome contours of his face. She glanced at his arm in the khaki shirt decked with the marks of his rank; everything about him commanded authority. How strange to see him this way; yet, she told herself, Asher had always given the impression of strength. And, now, that same strength restrained him: if he could not acknowledge their past, he would be courteous.

She would rather he were rude! She could defend against rudeness. His good manners and courtesy were leaving her with nothing but hollowness and homesickness.

"You're on the West Bank now," Asher was saying. "The New York Times sells a lot of papers here."

He rambled on: Over there was the area of the Byzantine and Crusader periods. And there was the stop between Jerusalem and Jericho in Biblical times. This spot was the border marker between tribes spoken about in the Book of Joshua. Asher smiled over at her. "The book of Joshua says, we will possess this land. I'm not sure Joshua had the right map."

The windows were open in the Army staff car and the back of Asher's khaki shirt was wet; she uncorked the water bag and he took a long drink while he drove.

"Now you," he said, handing her the bag.

She was telling herself: Sit Still. Be Calm. She could feel the hills around her where sheep grazed and camels wandered in long legged struts. He suggested that she visit Wadi Kelt to walk on the Herodian aqueducts and climb the path to St. George's monastery which was suspended from sheer cliffs and where the remains of a fifth century Byzantine church still stood. Down below on the Old Jericho Road were the excavated remains of the first century Hasmonean and Herodian palaces.

The sweep of the hills beyond the curving road went past her in a dusty vision that made her turn in her seat to look back at boulders and rocks, the color of sepia, the color of gold.

Asher pointed to a sign. The Dead Sea was amber glass spreading out in front of them in hardly a ripple, holding its breath. "The lowest point on earth, below sea level," Asher said.

There was another sign on the right side of the road. "Nebi Musa," he said. "The Moslems say Allah buried the bones of Moses in this place."

"Maybe the Moslems are right," she said.

"Maybe. Here, everyone claims to be right. No one is ever wrong." He glanced at her and smiled. "We would kill over a stone."

Scattered from field to field there were more signs, markers, prohibiting entrance beyond the barbed wire fence.

"They're mined," Asher said. "Or they're firing ranges."

"It's a kind of blasphemy to the land, isn't it?"

"No," he said. "It's not blasphemy; it's survival."

He was quiet then; it was as though any reference to what either of them thought, felt, remembered, had to be avoided. Their reserve was strained and unnatural; they were awkward, clumsy, focusing on what was outside themselves: signs, road markers, fences, the quiet of the sea, sites, historical or archeological, defenses against looking back, as though to recall the past would turn them to pillars of salt.

He pointed out huge metal silo tanks that he said were desalination pools. Then, near the shore of the Dead Sea, he made a turn south. "I wanted you to see Qumran," he said. "And then we'll cool off."

Again, they didn't speak. The end of the world was a steaming desert surrounded by dry, discolored cliffs. Black Bedouin tents dotted the landscape. Young Arab women, their heads covered in black and wearing pantaloons under their flowered dresses were working in the fields.

Asher stopped the car and they got out. He moved toward where she stood and, suddenly, he seemed to be reaching for her, but just as suddenly he pulled back. Her heart began to race when he glanced at her and their eyes met. "Freya . . ." he began, his hand smoothing her hair. "You have no hat." He handed her his khaki desert hat and they walked over the dried sepia earth until they were face to face with a Bedouin standing next to a camel; he was inviting them to take his picture. Asher waved him away but she caught the eye of the old Bedouin; an old eye, the oldest eye she had ever seen; it spoke nothing, showed no emotion, no expectation, no judgement. She imagined Layla's great-grandfather and looked away from the old Bedouin to follow Asher toward the cliffs.

He shaded his eyes and pointed. "Up on the cliffs you can see the desert in all directions. A good place to spot your enemies."

"They found the Dead Sea Scrolls here!"

"Come on then." He held her hand while they climbed the rocks until they could see the very rooms where the ancient scrolls were written, where the meeting halls were carved into the rock, rooms for potter's kilns, ritual baths.

She pretended to hang on to his every word while her heart grew heavier. Asher was a special tour guide for a special tourist.

They sat down on a large protruding boulder and Asher handed her the water bag; he took a long drink after her.

They sat there in heavy silence. She scoured her mind for some casual approach to him, something neutral, light. "I never imagined you would be a general," she began. "I still have the copy of your great-great-grandmother's diary you gave me about the Siege of Vicksburg. It would make a great movie. How . . . how is your sister? Does she live in Israel too?"

"Amalia? She visits every two years to the day. She lives in Atlanta. Amalia is southern to the core."

"So was your mother. You're like her, Asher."

Asher laughed. "You think so? It was my father who was the Zionist, not my mother. She once told my father he could be all the Zionist he liked, from south of the Mason Dixon Line. I think about that sometime," he said, then, abruptly, he asked, "I remember your family very well. How is your mother?"

"My mother died."

Asher pursed his lips. "I'm sorry. *When* did she die?"

"I think it's almost two months."

He turned to look at her; then he stood up. "You were a dutiful daughter." Then, abruptly, he took her hand to help her up. "The sun isn't friendly," he said.

She couldn't tell what he was thinking about the death of her mother. Perhaps he took it as a hint that she had finally arrived as she had promised, twenty years later, and she expected him to be free and waiting for her.

"I'm divorced. Asher. I have a son. . . ." she said.

He stood looking down at her. "I'm glad you have a son. I envy you. Come now, Freya. You'll burn." Asher was calling time on the tour.

She sat there, resisting, looking out to the brown expanse of stillness and golden sand and burnished cliffs. The old Bedouin, his head swathed in a keffiyeh, stood in the blazing sun in his long black robe and a wool jacket. It seemed that he would die in the sun. The camel beside him didn't move as though he were painted in a book. There were no shadows. If she sat here time wouldn't move. But Asher was pulling her up, closing the book.

"He'll burn up," she said.

Asher smiled. "They know how to live with the desert." He was standing in front of her to shield her from the sun.

"I'd like to be like that camel," she said.

Asher laughed. "I'd dress you in bangles and beads." He took her hand and held it for a long and breathless moment, meeting her eyes; then he looked at his watch. "The sun is too hot . . . we should go now, Freya."

They drove to an oasis—*Ein Fasche*, Asher said, where people went for picnics and swimming. After a few miles more, he pointed to the cliffs of *Ein Gede*, the place where David hid from the jealous and raging Saul. "Nothing changes," Asher said, "Kings kill to stay kings."

South of the cliffs was a kibbutz. Date palms lined the road. They drove into another curving road and Asher parked the car. "Keep the hat on; it's not too far from where we'll stop to eat."

He took her hand and pulled her along until they came to a grove of tropical trees and towering palms. They found a shaded cluster of rocks and ate the lunch Asher brought, sitting close, neither of them moving from the contact of their arms and legs touching. He peeled the avocados with an army knife and left the skins where he said an animal would find them.

"*Animals?*"

"Of course! Wild goats, and ibex, and leopards. *Leopards,*" he repeated in a deep, teasing voice. "And rare birds, a jet black beauty with deep orange patches on its wings. I remember that you loved that sort of thing. Do you still?"

"I work in a natural history museum," Freya said. "I write labels."

"*Labels?* Not poetry?" Asher smiled. "Well then, that is your kind of place for both. But it's hot, even in the shade. It's a short walk to where you'll cool off."

Asher took her hand and they walked up a trail surrounded by rocks and cascading waterfalls. She was panting, but he pulled her on until a small mountain pool shaded by great palms spread out in front of them. They stood there on the wide, slate ledge of the pool looking at the pristine water.

She sat down at the edge and Asher settled beside her; he unbuckled her sandals and slipped them off her feet. The silence of the place and the shimmering heat around them was drying the breath out of anything alive. She could feel Asher's eyes on her while she dangled her legs in the icy water. "Is that better?" he said. She nodded, afraid to look at him.

They sat there at the water's edge for what seemed forever. Tears and sweat spilled over Freya's face and Asher soaked a white handkerchief in the water to stroke her eyes and cool her sunburned face. It seemed to Freya that twenty years were drifting by as they sat, hands entwined, surrounded by silence in the shaded, mottled sunlight. How could she explain the feeling inside her, heightened, after so long a time and so much unrealized struggle? He was there beside her. She was *with* him, gripped by a terrible weariness and a draining sense of disbelief, but she *was* with him. Even if the world around them had changed and if they had changed with it, what remained was a still point—they themselves, their need for each other, their common language, her mind and her body imprinted so early in life that no subsequent experience could prevail. She leaned against him and he held her quietly while she wept with the frustration and anxiety of years of longing, so long suppressed.

Asher found his voice as she trembled against him. "Don't cry, sweetheart, don't cry."

"At your house, you hardly spoke to me."

He turned her face toward him. "Freya, Freya . . . couldn't you see? When Aviva told me you were coming . . . I couldn't be in a room full of people, acting, pretending."

"It's too late for us," she began. "Aviva . . ."

Asher covered her mouth with the palm of his hand. "I married her. I waited for eleven years before I married her."

In an abrupt, impulsive outburst, her heart racing, she was telling him: "I *wasn't* a dutiful daughter. Every day I prayed that Momma would die. Nothing was honest in my life. I let myself get pregnant and I married Karl when I didn't love him. It was only my son I loved. I have to tell you. You have to know."

Asher's voice was low, patient, curious, when he asked her, "Why, Freya? Why did you come now?"

"Noah is nineteen. And now . . ." She stopped; she couldn't say it.

Asher took her hand and pulled her up to hold her in his arms and murmured against her hair. "Freya, you said you'd be here a month, but don't you know, Freya; I won't let you go."

Asher did not leave her that night. He bathed her sunburn with face cream. He brushed the tangles out of her hair. She sobbed in his arms and they talked and made love and dozed and woke to talk again and make love. When lazily she mentioned Aviva, he covered her mouth with the palm of his hand, murmuring, Freya, Freya, *sheli*, my love. His arms were tightening around her, dimming out everything to where the events of her life hadn't happened: she had never been married to Karl, she was not a mother, and Asher was not an image in a dark room. Tomorrow, she would talk to him about Noah. Tomorrow . . . Tomorrow . . . No one would understand. No one would grasp the meaning of Asher and Freya or the claim they had on each other.

CHAPTER NINE

F REYA PHONED TO have breakfast sent up and Suhail wheeled in the cart; he seemed wonder-struck at Asher's uniform and insignia. Asher spoke to him in Hebrew when he gave him a tip, and Suhail bowed a little, fawning obsequiously, "*Todah, Aluf, todah rabah.*" (Thank you, general, thank you very much.)

Freya hardly ate though Asher coaxed food on her. He matched her elation and recklessness as though they were on one of their clandestine camping trips when she was sixteen. How could anything ugly exist? How was anything threatening to life connected to the swings of joy she and Asher had given each other last night?

"What are you thinking?"

"I want to tell you about Noah."

"There's something we need to talk about first," he said.

The phone rang.

"Layla?"

"Will you come to Nazareth, Freya"?

"I don't know. I can't talk now."

"Oh Freya, don't be afraid of me. I want to show you everything."

"I'm *not* afraid of you. I just can't plan anything now."

"I hope soon, Freya."

"I'm not sure. I'll let you know. I really have to go."

"Oh Freya, try. Will you try?"

"I will. I promise."

137

Asher looked curious. "Who are you not afraid of?"

"It's nothing . . . a woman I met on the plane," she said. "She came to Jerusalem to see me and I promised to visit her."

"Visit her? Where?"

"Nazareth," Freya said.

"Nazareth?"

Asher picked up his coffee cup and for a moment he was silent. "What do you know about Arabs?" he said.

"Arabs in general, or this particular Arab?"

"In general, in particular. . ."

"Aren't they people like people are usually?"

He smiled. "People are not so usual here. Our situation is far from usual."

Suddenly, it was clear that a discussion about Arabs was the wrong overture to Noah. She put her hand on his. "My friend doesn't matter. I've tried since yesterday to tell you about Noah."

He looked curious, waiting.

"My son is in your army, Asher. That's why I came."

He look stunned.

"I found out where he is from Uri Levitsky."

"Uri Levitsky! What does Uri Levitsky have to do with you or your son?"

"Noah was on Uri's kibbutz. I didn't hear from him. There was the war. When I got to Jerusalem I phoned the kibbutz and Uri came to see me. When I mentioned your name, he was surprised. He told me you're a general."

She saw him draw back. "I understand," he said. "Is that why you came to see me?"

"Asher! You *don't* understand. If you think. . ." She stopped herself. He was right. Why else had she phoned Aviva, had she gone to Asher's house?

She looked up at him. "What has my life been without you?" she murmured. "Last night, I lived twenty years."

"You didn't come in time of peace."

"There was never peace for me!"

"Only for you, Freya?"

138

"Asher, you have to listen. Oh, Sher, *please* . . .Someone I met at your house, he said we were on this planet to work things out. We can help each other. I love you, Asher."

He sat looking at her, his face grave. "Yes we have to work things out. We need time alone together; we need a chance to know each other again, Freya."

"We're here together now."

He shook his head. "If I visit you here, well, it's not easy to explain. We're both exposed but I wouldn't care about that so much. You'd be exposed to possible danger, so would I. It's not as secure as I'd like."

"You'd be embarrassed?"

Asher reached for her hand. "Not that, believe me. Can you understand your connection to me is like catching the measles? If I come to be with you here, any waiter, any bus boy. . ."

"Because they're Arabs?"

"Because getting to you is getting to me. The probabilities are, nothing would happen. But I wouldn't want security here and I can't risk you."

"I'll get me to a nunnery. . ." Freya interrupted.

Asher grinned but he was serious. "How about a place where I wouldn't have to petition the Mother Superior? I have friends with an apartment. I hold the key while they're out of the country. You can stay there. As long as you like."

"That doesn't make sense! Why would I be safer *alone?*"

"You have to trust that I know," Asher said. "There's no risk at the apartment. And I'll be with you as often as possible. Can you manage it?"

"What about Aviva? What about that risk?"

He sat looking at her for a long moment, and then, barely audible, he repeated, "Aviva."

"Freya, listen, Aviva is my problem, not yours. When Aviva isn't in Paris on business, she's in Haifa arranging shows, managing her shops. Between her business and the army, we're separated half of the time. She's in Haifa now; I may not see her again for weeks. It doesn't pain her, it doesn't pain me. That's how we manage."

139

She lowered her eyes. "I'm sorry."

"There's nothing to be sorry about. Freya, *look* at me, don't you think we have to face what we did to our lives?"

She looked away. "What we did to our lives is already done."

"Maybe it is, but you're *here*. What shall we do about it?" He leaned over and turned her face to meet his eyes. "What did you *think*, Freya? That we could see each other again, say hello and good-bye and go back where we came from?"

"It's too late." She leaned against him, her face wet with tears.

"Freya, listen to me. I said I wouldn't let you leave, but if you don't feel the same, I won't press you. Don't you think I need honesty as much as you? If we're doing the wrong thing, we can change it. We'll know."

"How can we talk about honesty when there's Aviva?"

"Freya, who will suffer if we say good-bye now? Who will be victimized? Don't you *know*? But if that's what you want, we can call what happened last night an accident."

Freya had covered her face with her hands.

Tenderly, he lifted her hands away from her face. "Freya, my beautiful Freya, don't we deserve time together to sort out our lives?"

Asher said he had meetings until late afternoon; he would pick her up then. He helped her pack and took her luggage down to the lobby. When he was gone, she went to the dining room for more coffee. Lizabet Avidan was standing at the buffet, filling her plate in clear view of the lobby, watching her obliquely. But Lizabet Avidan ate breakfast at the Shiloh on Saturdays, not Mondays. To get out of her kitchen? Freya turned away from the dining room and went to the elevator. Had she seen Asher too?

Zafra's eyes were glazed with curiosity. "You're going to another hotel?" she said, picking at her earring. "You're going with, I mean, you're going somewhere else? Listen," she said, "I don't put my feet in your business. I wasn't here last night. This morning they told me you came back with the general. I knew him from his picture in the papers. Who doesn't? I almost fainted." Zafra smiled coyly.

She handed Zafra the telephone number Asher had given her. Zafra examined it. "It's a hotel?" she asked.

"No. Just forward the calls."

"Forward calls is one thing. Not to see you again is something else," Zafra said flatly.

"Zafra. Don't be silly. You're my only friend in Jerusalem. You'll get *tired* seeing me!"

"For sure? You promise me?"

In her room again, Freya switched on the cassette recorder and flipped through guidebooks. She hardly heard the music or understood what she read. Lying back on the pillow, she closed her eyes and relived the night. A miracle had happened. As she lay there, her excitement faded to apprehension. She was trading the safety of her little room to be with Asher. Somewhere— where? It was rash. Their hunger for each other had always excited their risks. But she had *known* the risks then—they hadn't mattered. Where would this risk lead? What if their being alone didn't work for him? He had Aviva; they could go on living, not caring for each other. But how would she live the rest of her life? She closed her eyes listening to the heavy thump of her heart. She couldn't leave him. No matter what the risks, even if he never helped her with Noah, she couldn't leave him.

He drove to Ramat Eshkol, high on the southeastern slopes of the Judean hills, a literal ascent, and if it was romantic, Asher said, it was also an uphill, exhausting walk from town. But there was a bus, the Number Four, that stopped directly in front of the apartment house.

"Choice property," Asher said, his arm out of the window, pointing, while he drove through the winding streets of rose-beige stone apartment blocks. Acacia trees lined the narrow parkways. There was a laconic look on his face. "A real killing. We took it from the Jordan Legion in '67. Our visionaries bought land here for peanuts. It was empty, a chain of brown Judean hills. A lot of people died for this real estate."

"The way you talk, am I supposed to think you like it or you don't like it?"

141

"Let's say, I'm aware of the cost."

The three story apartment house turned out to be on a non-street with strictly limited access, a tiny square entered from a main street into a courtyard flanked by a small supermarket on one end, a one room stone synagogue on the other. The street that went by the courtyard was called Mitle Pass Street, and the adjoining street was Six Days War Street; the empty stretch of sand across the wide Eshkol Boulevard that flanked Ramat Eshkol was Ammunition Hill. The key to the Old City from the north, Asher told her.

"That hill is a landmark, a war monument?"

"Everything in Jerusalem is a war monument."

"There were no Arabs here before '67?"

"There were, of course. The Arab sector is a few blocks east on the Nablus road. Sheikh Jarrah. Upper class. They look on us as occupiers. But never mind what they think. The city is united, no Jewish Jerusalem or Arab Jerusalem, only Jerusalem."

"You make it sound as though Jerusalem were part of the Jewish lymph system."

"Something like that if that's how you want to put it."

"What if it's part of *their* lymph system?"

He glanced at her. "It's not my job to be political. I try to be fair."

"Does *fair* go with survival?"

"Listen, Freya, You haven't been here a month. It might be a good idea if you saw the whole movie before you write the review."

"Asher? Are you still a socialist?"

"Let's go in," Asher said.

The three room apartment was on the second floor of an apartment house built of thick Jerusalem stone. The shutters fit tightly against the heat and there were two balconies, one in back, one in front off the small living room. The furnishings were sedate and pleasant.

"It's nice," Freya said. "Whose apartment is it?"

"If I told you a name would it help?" He looked at her. "Freya," he said. "I wouldn't do anything to hurt you."

She smiled at him. "I like that sofa," she said. "Everything is just the right size."

"The air-conditioning is free," Asher said. "When you're hot, lie on the floor."

She slipped off her sandals and felt the cool terrazzo under her feet.

He opened the doors of the front balcony and they stepped out. The Judean hills were clear, close, and when Asher told her to look East, light was falling over the hills and in the direction of the desert—colors shifting from white to rose, coral to magenta. The stones and the air were saturated with color and in less than an hour, Asher said, Jerusalem would turn to gold.

He put his arm around her while they looked out to the hills. "Shall we give ourselves a chance to know how to go on? Or we could finish it now. I won't press you."

A squadron of jets smashed the sound barrier and disappeared like a streak. She was startled.

"Don't be frightened," he said softly. "They're ours."

When they turned back into the apartment, the whistle of the jets was still in her ear. She turned toward him. "If it's honesty we want, you have to know that I came to see Noah."

He put his hands on her shoulders, searching her face. "Nothing else?" he said.

"I don't want to use you."

"When you thought of coming to see Noah, you didn't think about us?"

She looked away. "It was Noah who brought me here."

Asher's jaw was tightening. "Shall I take you back to the Shiloh? I said I wouldn't press. . ."

She was hardly breathing. She shouldn't stay; she couldn't leave him. She searched his face. "Even if I wanted to leave. . ." she murmured. "I couldn't, Asher, I couldn't."

He took her to see the neighborhood and to stock the apartment with food. "Everything you need," he said. "Complete with a babel of tongues from New York, Moscow and Montevideo."

Two short blocks toward Eshkol Boulevard was a shopping square with a supermarket, a beauty parlor, *Marit of Paris*, a launderette, a coffee shop, a small post office, a branch library. In front of the Bank Le'umi, Freya persuaded Asher that she didn't need money.

He took her to meet Haim, the green-grocer, and introduced her as a new *Olah*, an immigrant to Israel. "Teach her some Hebrew," Asher said to Haim, in English. "Don't speak English. If this lady speaks English to you, pretend you're deaf. *Beseder?* Okay?"

Haim was leaning against a large sack, his hands on his hips. His face was deeply flushed, his eyes glassy, the whole look of him, on the verge of eruption. He spoke English carefully, deliberately, as though she needed to hear and understand. "Tell me, General," he said. "For what is she a new *olah?* You sent her an invitation to sacrifice her sons for the Galilee?" His voice began to break. "Last week my son came back from Lebanon with a bullet in his head . . . he won't die, he won't live. . . ."

Asher put his hand on Haim's shoulder and spoke to him in Hebrew, mollifying, comforting sounds, and Haim pulled a red cowboy bandana from his pocket and blew his nose. Asher waited until Haim stopped crying. *"Beseder?"*

Haim nodded. Asher waited a moment more, watching, until Haim raised both hands and dropped them again. *"Ze lo ha-'baaya shelha!"* (It's not your problem!) he said. "What shall I give you to buy?"

"It *is* my problem, Haim. Oranges can wait."

"You want oranges?" He began throwing oranges into a sack one after another, not counting, each orange splitting another rent in the sack so that oranges scattered and rolled over the floor. Asher stood by, making no move to stop him, his face grave, until he called out in a strained and hoarse voice, "Haim! *maspic!"* (enough!) And then he replaced the sack and bent over to pick up oranges. He handed Haim a bill with Ben Gurion's picture on it.

144

Haim looked at the bill and with a great sob he reached up with his free hand, held his fist in the air and smashed it against Asher's chest.

Asher flinched, and Haim, his face ashen, shrank back, as though his hand were clutching hot coal. "My God! My God!" he cried out. "I strike a general of Israel?" Haim cried. "I don't know what I'm doing! Arrest me! Take me away!" he sobbed, waving the money away. "Aluf! Aluf! May my hand wither! Forgive me!"

Asher took Haim's hand and pressed the bill into it. "*Lo, lo,* No, no, Haim. Maybe I'm the one who has to be forgiven."

For a full block, they were silent. Asher carried the sack of oranges and Freya, shaken by Haim's outburst, followed him. Asher broke the silence: "Did Uri Levitsky tell you he lost a son on a commando raid?"

Already stunned by the way Haim's fist had come down on Asher's chest, his words didn't register until she stopped where she stood. "He lost his son?" She stared at Asher. "Uri lost a son? And Haim's son. . ." Asher put his arm around her shoulders. "We pay a high price, Freya. . ." he said.

"But Uri was in a camp!" she cried. "He lost his son. How many times will he pay?" Asher's face was set, inscrutable.

They stopped by the supermarket and Asher bought an assortment of staples and foods and a cooked chicken for their dinner.

When they got back to the apartment, they filled the kitchen's empty cabinets in silence. She went to sit in the living room and Asher followed, settling himself in the chair opposite her, his legs crossed, his elbows leaning on his knees, his fingers interlocked. His eyes were fixed on her: he waited while she shifted her position, her hands holding both sides of her chair. "Noah is very much on your mind," he said.

"When I brought it up before, it seemed clear," she said. "Now I don't know how to begin."

"At the beginning."

145

She told him about Noah joining the kibbutz and volunteering for the army. But she was not ready to tell him about signatures and forgeries.

Asher listened. There were no signals on his face although his eyes were tender, but from time to time she saw a muscle tighten in his jaw and his forehead crease into a frown. When she was finished, he said nothing.

"Asher, I didn't mean to burden you! It's too late to help Uri's son, or Haim, but Noah can be helped! Uri says you can do anything."

They were facing each other; the twilight had deepened and the last streaks of sunset were shafting through the living room door.

"Why did Noah come to Israel?" Asher asked.

"I don't know. His friend, Michael, convinced him and then Michael left and Noah stayed."

"Noah stayed but why did he volunteer for the army? What made him do it?"

"I don't *know!*"

Asher interrupted. "You didn't talk to him before he came?"

"I did! I tried to reason with him."

"Against coming to us," Asher said, flatly.

"The way you put it makes me sound *against* you. I'm not against you. But Noah's life is in America. His future is there."

"Do you know exactly where a person's future is?"

Freya looked out to the glow outside. What had been so obvious to Uri was turning out rejection for Asher.

"I didn't mean that," she said. "We used to talk, Sher, remember? You said my attitude was indifferent, as though Zionism were just another flawed novel. I tried to tell you that I couldn't get excited about grand ideas anymore." She looked away, pondering. "I was drowned in grand ideas."

He had thrust his hands into his pockets and suddenly she felt like an obstinate child, exposing her ignorance and selfishness. Asher sat back, not speaking, before he took his hands out of his pockets and leaned toward her.

"New information never changed an attitude . . . it's predispo-

146

sition that counts, how you feel about the need for a Jewish Homeland. How much you're willing to pay for it. And, maybe, a compromise for what you call *fair.*"

"Reason has nothing to do with it?"

She felt the tension between them mounting.

"Maybe it was in Noah's heart to fight for a Jewish Homeland. Is that possible?" he asked.

"Uri said he tried to tell Noah the realities!"

Asher was brusque. He did not want to talk about Uri.

She sat there searching for an answer. What had been poised and waiting in Noah? "I don't *know* what Noah felt," she said. "Karl pressured him about Medical School. Noah didn't want it. He's a talented photographer. And there were other reasons."

"More positive reasons?" Asher said.

"I don't know! He felt some anti-Semitism where we lived, in his high school, in the people who lived around us."

"And you didn't feel it because Americans aren't expected to feel anything but universal acceptance?"

"I don't *know* what I felt. I was preoccupied with my mother and I was trying to make my marriage work. When Noah was older, I went to work at the Museum and wiped Beaulac out of my mind. But I had friends, I liked my neighbors."

He looked at her for a long, searching, moment. "What do you want me to do?"

She hesitated. "Asher," she said, looking down at her hands. "Until Uri told me, I didn't know you were a general."

"That doesn't matter," he said.

She hesitated, then she forced the words out: "Noah forged our names for permission, and Uri says the army won't keep him unless we consent. I *can't* consent."

She could tell by his look of shock that she had thrown him off. Asher leaned back in his chair, not speaking. "Well then, Freya," he said finally, "If you can't consent you have to file a complaint with the Ministry of Defense. They would discharge him in a flash."

"But I should talk to Noah before I do *anything*. Uri says he's a photographer in a special unit. If you could arrange for me to see

147

him, Asher. Noah has to understand that I would never hurt him. He has to hear my side of it, why I want to him to come home."

"So," he said, running his fingers through his hair. "You came here for my help to take your son home." He was hurt, roused to anger, but it was hard to tell in the half light. "How long do you give me to get the job done?" he said.

She stopped talking, silenced by how he must have heard her, of what seemed obvious. Not that she wanted Noah out of the army but that she wanted to take him home, that she had come for his help and she would leave again after she got her son back.

"Oh, Sher! It's not like that!" She bolted up and went to kneel beside him. "I was so afraid you'd misunderstand, that you'd think I only wanted to use you. I didn't want to tell you. . ." She put her head in his lap. "Do you think I want to leave you? I couldn't leave you, Sher. Is this how we find out about honesty or where we are in our lives?"

He stroked her hair. "Yes," he said. "It's a way to know."

After dinner, Asher explained carefully: when it was occupied, the apartment was always under surveillance, she would never be alone in any emergency or even if she found anything suspicious. A David Barr would be assigned to her security. He wrote down a strange number—0001.

She laughed. "Who answers? James Bond?"

Asher smiled while he lit a match and burned the paper in an ashtray. "Not James Bond," he said. "David Barr." The number was easy enough to memorize but she was not to repeat it to anyone. She should dial the number—he repeated it, "0001" only if she needed anything, if she wanted to be driven somewhere, or if she received suspicious mail, or unexpected packages.

David Barr carried a long distance beeper and he would be at her side within minutes.

"You look as though you're about to be eaten," Asher said, grinning.

Asher had more to say about Noah. False credentials in the Army was sensitive and awkward and had to be handled carefully.

"We're not in the business of shanghaing foreign conscripts," he said. "We take these matters very seriously." She listened, feeling enmeshed in what she couldn't imagine, a situation that needed careful handling. It wasn't going to be easy after all. "Sher, I don't know what to imagine."

"Don't imagine anything, Freya. David will do the groundwork on Noah's case, and when I get back, he'll know the easiest way to arrange a meeting. Special leave, only if his commander can spare him. I don't like to interfere."

Alarmed, she interrupted. "I don't want Noah to be hurt! I wouldn't even want to embarrass him."

"I didn't say any of that, Freya. Don't dwell on the down side until I know more."

That night they opened all the casement windows of the apartment as wide as they would go. The air was cooled by the mountain heights and though Asher held her close to him, the image of Haim striking him and of Uri's dead son was with her.

She sat up suddenly. "Do you believe in this war, Asher?"

Asher seemed taken off guard and he sat up to lean against the frame of the bed.

"The Defense Minister doesn't query my personal feelings," he said. "The survival of the country is what I believe in."

She pondered it. "Can this country survive when it's surrounded . . . millions of Arabs?"

There was a cast of bitterness in his voice when he answered, "Good question."

"If you were in politics. . ."

He laughed. "Make shifty deals? Pander to the public? No thanks. Where I am now, the goal is clear cut. Protect the child until it's grown up enough to make the right deal for itself."

She lay her head on his chest. "You were so studious, Sher, you had an omnivorous mind."

"The trouble with books is survival doesn't have a text. It's your instincts, what you can let yourself feel."

She was quiet as she lay against him. "Sher?" she murmured. "You told me how it is with you and Aviva. Why did you marry her?"

He wound his fingers through her hair. "It's not very interesting," he murmured. "She was married before. Aviva marries men who can help her. I was willing. Maybe I helped her become a star in her world. And I have my work. That's all there is to it, what we call facts."

He was reaching for her, pulling her close. "You're the only fact that ever counted," he whispered.

The cool air from the mountains blew in and she murmured against his chest, "Noah can go back to Gan Zvi. I love you, Sher. I couldn't live. . ."

"We have the chance to live now, my Freya, my sweetheart."

At five in the morning, Asher woke her. He was dressed, ready to leave. He had to stop at home for clean uniforms and then he had another meeting. Aviva was in Haifa preparing for her Paris exhibition.

"I'll get back to Jerusalem as soon as I can. How will you fill your time? What will you do?"

She sat up, half asleep. "I'll find another dig somewhere."

"Are you awake, Freya?" He was rubbing her cheek. "I want you to hear me. You're close to the Arab sector, a short walk to the Old City. Try to remember we're in a war and from their point of view, you're an occupier. Right or wrong, what we did to them, they won't forget. No one forgets anything here. David Barr will take you anywhere you want to go. Did you hear, Freya? Promise me."

She nodded. He nudged her gently. "Promise," he said. Suddenly, she was awake. She had seen a pistol when she opened the glove compartment of Asher's car to look for kleenex.

"David carries a gun?"

"Every soldier carries a weapon."

"Is that what he is? He's in uniform?"

150

"No, he won't be in uniform. Yes, he's armed."

She pulled away and looked up at him. "You make me feel like a private in your army."

Asher grinned. "I wish you were. I'd take you with me. Now listen to me. If I didn't feel you'd be safe when I was gone, I couldn't leave you."

She put her arms around him. "It's you who has to be safe, Sher. Noah has to be safe."

"I haven't forgotten Noah," Asher said.

At six in the morning, the sun was still low and the air was cool on the front balcony when Freya took out a cup of coffee and her guidebook. A group of children were playing on the stone steps in front of the building, jabbering in Hebrew, gesturing to each other, making noises in a teasing rhythm recognizable in any language. A voice from somewhere yelled out to them. "*Sheket! Sheket!*" (Shut-up!) The children looked at each other and laughed uproariously. "*Sheh-ket!*" came down again in a shrill warning. A lean, red-haired boy cooled his fingers in the direction of the voice and shoved one of the other boys. Before they moved on they spotted her, and cooled their fingers at her too.

There were prams on the street, wheeled by young mothers with one or two toddlers embroidered at their sides. The bus stop across the street was crowded. Men in cotton wash pants and short sleeve shirts, women carrying net shopping bags, a group of soldiers with their Uzi guns slung over their shoulders.

In the hum of sound she could not make out anything clearly except that she recognized Russian words, Russian rhythms. A bus stopped with a great blue sign on its side in Hebrew, Arabic, and English: EGGED. It screeched to a stop and the crowd at the doors melded into a heap thrusting forward.

Bougainvillia and morning glories hung from the balcony, open, alive; she leaned over and touched the petals that were wet from the mountain dew. She snipped a petal of bougainvillia and held it to her nose. The phone inside was ringing and she went in quickly. Noah, maybe.

151

"Poppa! You returned my call! How are you feeling, Poppa?"

He was shouting and she moved the ear-piece away from her ear. "Don't shout, Poppa. I hear you."

"To make a call to you is worse than to call to a space ship."

"You don't need to shout, Poppa. I can hear you as though I were in Beaulac."

"I never heard you in Boo-luk," Pushkin said.

"How are you, Poppa?"

"That mixed-up girl at your hotel said you moved. She wouldn't give me your number. 'Security,' she said, like I was Arafat."

Freya laughed. "That must have been Zafra. I told her to forward my calls. How did you get me then?"

"I told her, 'This is the lady's *father.* I've got papers. I'm connected to Kissinger.' "

"Poppa. I was going to call again."

"Don't Poppa me. I read the newspapers plenty. You saw Noah?"

"Not yet. But soon. A friend of mine here is helping me arrange it."

"Ahh, a *friend! The Socialist-Zionist with the two behinds?*"

"Poppa! Why are you shouting? Why do you have to throw insults?"

"The Socialist-Zionist, he poisons the vegetables Arabs eat?"

"Poppa! Stop it!" There was a silence. "How's Dora?" she said.

"Her hands are peeling from scrubbing potatoes. Why are they throwing bombs on Beirut?"

"You want a political discussion at three dollars a minute, Poppa?"

"I'm going to tell you something, Freya. I knew a Socialist-Zionist. He went crazy. Someone asked him will you save the Zionists first? He said, sure. Someone else asked him, will you save the Socialists first? Sure. So, he thought to himself, if I save the Zionists first and the Socialists first what will happen, in the meantime, to the Zionists who aren't Socialists and the Socialists who aren't Zionists? Then he went crazy."

"Is that worth three dollars a minute, Poppa?"

"You wrote me Noah is in the army. He enjoys himself?"

"Don't say things like that! You know how I feel."

"You *feel?* Last time you told me, you *think.* I'll tell you something, if you think and you feel, ask yourself how many people fell on their knees because someone put a war on them. In 1948, did the Jews fall on their knees?"

"We're talking about terrorists, not people."

"Listen; in the jungle every animal is a terrorist if he don't have a place to live. Take away their place and you'll see big teeth in your face. I don't have to work in no museum. . ."

"I'll write you a letter, Poppa."

"Explain to Noah the story. Every animal needs a place where to live before it will know how to live. It won't help Noah to throw bombs."

"Noah isn't throwing bombs!" she shouted back.

She sat on the bench next to the phone in the foyer, shaking her little finger in her ear. One more conversation with Poppa and she would go deaf.

She glanced at the phone. Zafra wasn't giving out her number unless you knew Kissinger and if Uri had news he wouldn't know where to reach her. She picked up the phone.

"Levitsky is not here," the kibbutz operator said. "He's in Tel Aviv. He'll be there for some days."

She left her name and number. Yes, she would like him to call her.

Half an hour later, Lieutenant David Barr called to introduce himself. Did she want to go anywhere? Sight-seeing? To the Old City? The General left orders that she should not go places alone while he was away; he was responsible for her security. He had tickets. She liked music? Theatre? Very nice.

Freya listened; she thanked him but, no, she did not like music or theater and he didn't need to buy tickets. She did not mention the trip she planned to Nazareth the following Saturday.

"You make me feel like Catherine the Great," she said.

"Maybe you are," David said. "Listen," he added, "I have a long range beeper. You can reach me day or night."

She dressed quickly. It was nine o'clock. The library would be open. She left the apartment and walked across Eshkol Boulevard to Ammunition Hill and stood there at the rocky rise, staring at the gouged-out bunkers. A squadron of jets was suddenly above her, deafening, splitting the sky.

CHAPTER TEN

"No," Zafra sang out, "No buses on Shabbat." Then she began racing her words, hardly taking a breath. No, Freya should not take the Arab bus to Nazareth. An Arab bus was like going in a storm without an umbrella, God forbid. An Arab *sherut,* a taxi that takes many people at once, okay. But not to talk to anyone in the sherut. Sherut at the Damascus Gate, seven in the morning.

Zafra stopped to breathe. "Why are you going to Nazareth?" she said. "You're becoming a Christian?"

Freya laughed. "I'm a Jewish tourist."

"Okay," Zafra said. "*Kadimah*! (Forward!) You should see Mary's Well. The churches are *Yofi!* You don't have to be a Christian. I cried my eyes out in Rome when I saw in St. Peter's the marble toe of the baby Jesus worn away from people kissing. Why shouldn't you go on a tour? I'll arrange it."

"I want to go alone, Zafra."

"Okay, okay," Zafra said. "Anyway, I gave out your phone number. He said he was a relative. I asked him. 'A father,' he shouts back. 'What you think, a lover?' "

"I *told* you to give my phone number. *Please.*"

"So. You had another call this morning! With you, I don't know what's what. Uri Levitsky. He said he'd call back next week. Shall I give the new number?"

"Zafra! For God's sake!"

"And you think that's the end of the headache?" Zafra said. "A lady asked for you this morning. I saw her here for breakfast once. She's from Jerusalem. She wanted to know if you checked out. So, I said no. I mean, considering the general and everything like that, maybe it's something delicate. I did right?"

"I don't know her," Freya said. Her heart had given a jump. "I told you to give my number to people I *know.*"

"Am I your address book?" Zafra said. "How I should know who you know?"

"You did right, Zafra. I love you."

"Then come over and I'll take you to my special place, where we didn't go before. My cafe, *Yossi's,* in Kikar Zion. Special. The coffee is so-so but the people who go there—*Yofi,* don't ask! The war didn't clean out Jerusalem altogether."

"*Yofi!*" Freya repeated. "How about when I get back from Nazareth? Let's see. How about Monday after I get back? We'll go to Yossi's."

"Monday? Not the best for *Yossi's.* Saturday night, after Shabbat is best, but Monday is okay. We have a date, yes? You say in English, date, like what you eat? Yofi. In the meantime, let's go to Rene's. There's a black lace nightgown. I almost fainted. I would spend my last penny."

On Wednesday of the following week Freya went into the center of Jerusalem to buy books at Steimetsky's. She thought of stopping at the Shiloh to pick up Zafra, but the drenching heat had exhausted her and Zafra had probably gone home early.

It was almost five and Freya decided to stop for a cold drink at *Yossi's Cafe* before she went home. Yossi's was on Kikar Zion opposite her Number Four Bus on Jaffa Road and she was curious about what shenanigans, exactly, went on inside Zafra's Singles Heaven.

When she went inside, clearly, Zafra was right: Yossi's was the setting for intrigue, a huge semi-dark room with a wood plank floor, a pastry counter, and small tables and ice cream parlor chairs scattered in total disorder as though every customer rotated on cue to strike up new conversations. The atmosphere reeked with tobacco smoke; young men sat drinking coffee and reading newspapers with their feet propped on chairs that were swung over from other tables; clusters of standing coffee drinkers stood over coffee drinkers that sat. There was a carefree cacophony of raised voices, laughter, crockery rattling. Every table was full, half-full, missing a chair, or there were three chairs added. Where did she sit? Everyone seemed to know everyone else, or it didn't matter who knew whom. Zafra wasn't there.

A waiter came over and pointed a fat thumb at her in the direction of a table with a stack of books on top and a woman with short, blond hair and tortoise shell glasses sitting behind them. Freya sat down and the waiter came over, his weight shifted to one side: he was glaring.

"Ken?"

"Coffee," Freya said.

"*Turki? Espress? Mah?* What?"

She shrugged. "Coffee. Cafe kar . . . ice coffee. Yes?"

The blond woman laughed. "Best of luck," she said.

A writing pad was in front of her and she was twiddling a pencil between her fingers, engrossed in what she had been writing. She took off her glasses, her eyes half-teasing. "Want to take a bet on the ice?" she said, biting on the stem of her glasses. There were not-so-young shadows beneath her eyes and she was not pretty but she had dash and an adventurous look, with subtle mauve lipstick and a khaki journalist's vest and jeans to match.

"They'll bring you exactly what they want you to have," she said, before she poised her pencil again. She glanced up at Freya. "From the States?"

"Yes. Chicago. And you?"

"Canadian," the woman said. They shook hands and exchanged names. She was Carole MacKinley from Canadian News assigned to Jerusalem and Beirut.

"You've got a war on your hands," Freya said.

Carole MacKinley nodded while she moved the books to a chair. "You bet we do," she said. "I just limped back." She sat for a moment looking at Freya over the frame of her glasses. She asked finally, in a curious tone, "You don't strike me as a lady on a tour. But you seem new at the local chaos. Ice doesn't live here anymore."

157

For almost an hour, Carole offered opinions and Freya took her in, amazed at the easy way she lectured. If there was a spirit of bloody mindedness during a war, she said, this war was especially bloody minded—a *dirty* war. She had been covering the beat for years. Morale was never lower than now.

"As for women in Israel," Carole said. "It's one step forward, two steps back. Hard times for men put women in their place. Fetch, carry, and plenty of you know what. I should know. I've coddled three husbands through one crunch or another."

The books were on the chair next to Freya. A title was written across one in large red letters: *The Lebanon Network*.

"Do you get time to read books?" Freya asked.

"My homework," Carole said. "It's like wearing a flak jacket."

"It takes guts to work in Lebanon."

"Not really, not if you stay half drunk."

And finally, Freya couldn't hold back. "My son is in Lebanon," she said. She was not sure why she blurted it out except that Carole MacKinley was a stranger and strangers in foreign countries seemed to tell each other everything as though they had been friends for years; strangers existed in the sphere of the peripheral and unaccountable; they had no past, no future.

"He's not the only American there," Carole said, studying her. "I'm getting the picture," she said. "I'm sorry. You're not having a ball, are you?"

"No. I want to go see him."

Carole MacKinley's pale, blue eyes widened. "Go *see* him? You mean like popping in to a side show?" She shook her head. "This is one side show you can miss. Anyway, they won't let you."

"But you go to Lebanon," Freya said. "I'm an American citizen. Why couldn't I?"

"You wouldn't get past the first check point. Look. I have credentials. As much as the Israelis would adore to, they can't keep the press out. Anyway, you could be a spy," she said.

"I'm not."

"Possibly not. But who knows these days? A shy little nun could be a spy. Doesn't your son get leave? A pass? The Defense

Ministry might even find it amusing, good PR, to accommodate a worried Jewish mother from Chicago."

It took almost the full hour before the waiter brought Freya coffee, steaming hot. "See?" Carole said.

The room was stifling. Freya tried to smile. Carole was watching her. "Pay no attention," Freya said, trying to cover up.

Carole lit a cigarette. "Listen," she said. "I never had kids, but I can imagine how tough it is. It may not do you much good, but I have a friend at the American Embassy. You'd have to go to Tel Aviv." She opened her bag and took a card out imprinted with her name and the address and phone number of her office. She wrote on the back: "Lloyd Baxter, U.S. Embassy, Tel Aviv."

"It can't hurt. Tell him I sent you."

Freya took the card and looked up at Carole. "How can I thank you?"

"Don't. I get back here now and then. Call me, tell me what happened. I like you, Freya Gould."

Freya reached for Carole's hand.

Carole smiled. "You like feasts? I'll take you to the American Colony Hotel for a buffet you'll never forget. Strictly a pig-out. And ice, plenty of ice." She crushed her cigarette in a saucer and collected her books. "Good luck," she said.

All that evening Freya toyed with the idea of contacting Lloyd Baxter at the American Embassy. She decided against it, uncertain of American law on citizens in a foreign army.

She wrote in her notebook that night: *No matter how I try, I can't seem to put things together here. One day I understand, the next day, I don't. Is there a way to know that, yes, this is the problem and this is the solution. What are the consequences of making war and what are the consequences of not making war? I worry about my vacillation. I seem to agree with the last person I talk to. As though my mind were a file cabinet where the last thing filed is the first thing remembered. I never thought of myself as not having a mind of my own. Maybe if you were born here. But Asher wasn't born here; neither was Noah. I keep thinking about the Hebrew language and what I read long ago,*

that the Golem was made up of combinations of Hebrew letters.
That seems to be the cultural link, the magic of the language.

I'm afraid to draw conclusions, afraid of what they might be. I
keep thinking about what my father said, that there must be a
place WHERE to live before a person can know HOW to live.
Not merely to LIVE, but to live without humiliation. I don't think
I'll ever forget Layla's face when she was taken out of the queue
at Heathrow airport. It wasn't only anger and humiliation on her
face, it was FEAR. What comes next? HATRED? Like the hatred
in Khalil's and Yusef's eyes? And when I think of how Haim
struck out at Asher, I forgive him because of his grief over his
son, but I'm angry all the same. If Evelyn were here, she would
call Haim's lapse some Freudian slip.

For a week, Freya did nothing but study Hebrew from books
she got at the little library on the square. She visited Haim and
asked in Hebrew for fruits and vegetables. She asked about his
son every day and he looked at her as though *she* were the one to
be pitied, a naive girl, a concubine imported for the general's
pleasure. "*Yofi,*" he would say to her every time she pronounced
a Hebrew word properly.

The television obsessed her. The dark-haired young woman
with the silken voice broadcast in English from Amman, reciting
calamity, flashing pictures, voicing over comments from enraged
Arab spokesmen.

The television news in Hebrew was even worse. Language
wasn't necessary. Moslem West Beirut lay in ruins. In East
Beirut, Christian Lebanese women, gorgeous, dark, model types
posed in bikini bathing suits, one of them waving a tiny Israeli
flag. Israel was liberating them from the yoke of the PLO. The
faces of the inhabitants of Tyre and Sidon, were haunted, destroyed, staring with hopelessness, with emptiness at the rubble.

The pock-marked cities of southern Lebanon screamed from
the TV screen like toppling cedars while the Israeli fallen rolled
back on trucks to Tel Aviv and Jerusalem. Israeli fighting units in
khaki and berets moved up the coast past Nahariya, past Rosh
Hanikrah, where girls in scant shorts and halters waved to them

and blew kisses. Children on the beaches of Tel Aviv cooled off from the hellish heat, securing their sand castles with mud.

The English language papers and magazines were riddled with editorializing *firsts:* For the first time Israel had attacked a country without direct provocation; for the first time Israel brought destruction on living cities and on an Arab capital; for the first time there were official lies, contradictions, posturing and much talk of the Holocaust. For the first time, soldiers replied to reporters that they didn't know what they were doing in Lebanon. For the first time there were lectures on moral credibility and Jewish experience. The word, *terrorist,* was on everyone's lips. Poppa's words rang in her ears: *If an animal has nowhere to live, you'll see big teeth in your face.*

On Israeli television, the war was a photographer's paradise. Hebrew wasn't necessary to start her heart racing. It was Asher's duty to do what he was doing. But this was *not* Noah's duty! She was facing him as she did when he was ten years old and he shot a bird down from her tree with a sling shot. She was telling him what a vicious thing it was to prove his power over the bird. When he did it again, she tried to frighten him: the bird and the tree could turn monster and the stone could ricochet back to him.

But it was SHE who had suggested that Noah take pictures of birds instead of shooting them down.

The neighbors in the apartment house smiled at her a lot; they were bursting with curiosity but the fact that no one knew more than a few words of English protected her. Tami, her first floor neighbor, full busted and sensuous, without a shred of inhibition, was on the cutting edge. She imagined, probably, that through teaching Freya Hebrew, she would find out everything about the general. Every day Tami intensified Freya's Hebrew lessons. They took walks in the neighborhood while Tami stuffed her with Hebrew vocabulary, gesturing to Freya to repeat after her. Once, in a brave and defiant tone, Tami threw out the trial balloon. "General?" she said in English, with a thick guttural accent. When Freya didn't answer, she tried Hebrew. "General in Hebrew—it is, 'Aluf!'. Your Aluf, I see him. He is—(she fanned her fingers) *yofi, beautiful.*" Poor Tami, her husband was in Lebanon and she was captive to her mother, Batia, a Valkyrie of a woman who had nine children and left eight of them and a husband in Morocco when she snatched up Tami and made off for Israel.

At night Freya had trouble sleeping; she went through spells of imagining herself signing permission for Noah to stay in the army; she imagined his explosion when she refused to sign. And the mystery of awkward consequences. When the spells were intense she got out of bed and went to the kitchen table to write in her notebooks—making up letters to Noah that she would not send.

What is in your heart when you take pictures of the kind that are beamed back to where I sit in Jerusalem? I want you to know that I'm one of the horrified viewers looking at pictures taken by objective journalists. But you AREN'T an objective journalist. You work for the army, Noah; there are huge demonstrations in Tel Aviv. The Jerusalem Post gives them front page coverage. I've never seen such angry faces, arms raised in fists, anti-war slogans . . .

Zafra was her only relief. They talked on the phone. They ate standing up at pitah stands. They took sandwiches to the little park. Twice, Zafra took Freya to see the black lace nightgown. "Just to see," she said, and when the clerk produced it, Zafra rolled her eyes up. "I'm fainting!" she cried. When they were out of the store, the nightgown left behind, Zafra confessed, while they strolled up Jaffa Road, that she had a new lover, with a technique unimaginable, a first prize winner for sex. A hot blooded Russian, just come from Odessa. Of *course*, she met him at Yossi's!

Twice, Zafra spent the night on the sofa at Freya's apartment and they spent half the night howling over Zafra's bawdy descriptions of the techniques of her boy friends. Like Sheherezade, she reeled off stories: An American girl, on a tour at the Shiloh, once told her about erogenous zones. Zafra would twiddle her earring. If the American girl were to try Zafra's new lover, she'd find zones she never dreamed of. Her new man from Odessa never laid a finger on her until she danced the Seven Veils. "Then, don't ask!" He was a first prize winner.

At six-thirty in the morning on the following Friday, Uri phoned her. No, there was no news of Noah. "I've been away quite a lot. And you're full of mystery," he said. "You took an apartment!"

"Yes," she said. "Asher Frank helped me find it."

He was silent. "That's nice, yes, very nice," he murmured. "At this moment you're alone?"

"Yes."

"You're busy?"

"Not busy at all."

"Would you like to hear a lecture? At Mount Scopus, the University, not far from you. I'm lecturing to political science students in English."

"I'd love to."

"It's an early one, ten o'clock. On Friday everything has to be early or Shabbat will catch us by our necks. Can you be ready by half-past nine? Mt. Scopus is a few minutes from you. Shall I pick you up?"

"I'll be ready!" She tried to explain her address across from Ammunition Hill to Mitle Pass Street and Six Days War Street, and then she was laughing. "I can hardly get it out of my mouth."

She sat quietly next to Uri in the car while he talked and talked; they were winding higher and higher up the hills to Mount Scopus until they could see the Old City. Open army jeeps passed them on the narrow road one after another and Uri slowed down for the curves at the edge of the steep ascent. Jerusalem and the Judean wilderness were stretched out in hot, gold light under an unbroken sky. Domes, steeples, arches, merged with the desert in synchrony and rhythm like drums and tambourines.

"Breathtaking . . . it's unimaginable," Freya said.

"Yes," Uri said. "Unimaginable. Since I am here there are five wars. Not such dirty wars like this one."

"*Dirty* war. Someone else said that to me. Why is it a dirty war? What is a clean war?"

Uri glanced at her while he drove. "In our war of independence, seven Arab armies attacked us. Self-defense is not dirty. We win or we die. But when self-defense becomes a pretext . . . it gets dirty."

Uri seemed to be in rare form; she had never seen him so animated.

He pointed across the horizon "You see?" he said. "In our war of independence we took it back. It was a miracle."

"Maybe you'll have another miracle," Freya said.

Uri laughed. "I'll tell you about miracles," he said. "A crippled and blind man prayed that God would give him back only one thing—his legs, so he could walk. He didn't ask for sight because it wasn't as important to him as moving, walking, feeling like a person. So God made the miracle for him and he walked. Not only did he walk, but blind as he was, he learned to dance and he won contests and everyone praised him for his courage and his wit.

"One time, though, when he was walking alone, he knew he was walking close to a precipice but he was very confident. Even reckless, arrogant. God had made a miracle especially for him, hadn't He? How can you give back what God gave? Would God make such miracle only to walk him over a precipice? He had legs; he could walk away. What the poor blind man didn't see was the slippery gravel in front of him; well. . ."

"You don't have to tell me!" Freya said, laughing. For the rest of the drive Freya was silent; thinking about Uri, squelching a morbid impulse to ask about Auschwitz, what it was like, more than what she had read and seen on film—the calamities of world's end, a mire of striped suits and shaven heads and huge eyes aflame in their sockets. She found herself glancing at Uri, who spoke of the camp casually, who gave himself away only when it seemed that he was deciphering every word, noticing the slightest shift of tone or cast of eye. She could not resist asking: "How did you survive the unsurvivable, Uri?"

She had taken him off guard. He looked over at her and went on driving. He seemed to be thinking. Finally he said: "If you're lucky to have a high number so they don't get around to killing you . . . if you know exactly how to balance a boulder on the tip of your nose, you survive."

He stopped, drove on, and after a long pause, he began again: "Of course, if they convince you that you're not human and that everyone around you isn't human either, you don't survive. I told you how I used to polish my shoes? As long as I polished them, I was a human being who wants what every human being wants, a little dignity to remember when they had it. A human being who remembers dignity, doesn't tear a crust of bread from someone else's mouth."

It was one thing to hear accounts of Uri's survival as an heroic ode; it was another thing, to think of serving the legend lunch. They drove on in silence.

The lecture hall was already full of students, at least a hundred young men and women. She sat down on a seat next to a young man in army uniform on the aisle, his Uzi gun resting on his lap, the barrel pointing toward her. She sat looking at it. "Excuse me," she said, "Would you mind getting your gun out of my ribs?" He shrugged, turned the gun around so the barrel pointed toward the aisle. The young soldier shrugged again, crossed his legs, and shifted the Uzi to a precarious angle.

Freya leaned forward in her chair, her elbow on her knee, her hand supporting her chin while Uri's slight frame moved to the lectern. He looked up at his audience and smiled. "*Shalom,*" he said. "*Shalom.*"

He glanced at a few pages of paper in his hand and then he looked up at the audience, adjusted his wire glasses and began to speak.

His voice was low and Freya had to lean forward to hear him. The audience had come to life; students were pulling notebooks and pencils out of their bags, poised for notes.

"The ideology of Jewish nationalism and universalist ideologies cannot be reconciled. You must confront

the question of whether or not a Jewish nationalist state, unlike other nationalist states such as France, Britain, et cetera, can be fully democratic. That is the post-Zionist problem and it will exist as a problem long after the Arab-Israeli conflict is resolved, long after the PLO and any Arab nations recognize Israel. Do we advocate a society where the game is open for all of its citizens, where all ethnic groups are encouraged or, because of our uniqueness as a Jewish society, will we write the rules in the way they were written for Jews in pre-Napoleonic Europe?"

Freya's throat was dry in the stuffy room and she wanted to get out, breathe air, but Uri wasn't finished. There was whispering in the audience—they didn't like what Uri was saying, accusing Israel of not being democratic, of not being universalistic in the Jewish tradition. She shifted in her seat. Uri was about to begin again when a young soldier was on his feet, agitated. "One minute!" he cried out. "You are trying to connect the war with the proposal that we are fighting for the wrong ideas? If I accept what you say, shouldn't I convince every Israeli soldier in Lebanon to go home? According to you, our soldiers are there to protect a theory, not to isolate those who want to butcher us."

Another young man was on his feet, his voice shrill: "You think they are not a real enemy? We made them up? Our army is a people's army and if peaceniks like you want to influence the people, join the army. That's where the people are!"

Uri's head was thrust forward, peering at him intensely through his thick glasses, while the young man spoke. He looked as though he had gone flat, that the question had pierced his spirit.

"Wait!" Uri shouted. He pulled more paper out of his jacket pocket.

Freya glanced at the empty seat next to her. The soldier and his Uzi gun had left half-way through the lecture. She sat tensely in her seat. A student in front of her was muttering *hara* (shit). Uri straightened his shoulders and began again:

"You don't like polemics or dialectics? All right then. I have a true story for you, a story of action that might interest you more." Uri's face had paled and Freya saw his shoulders slightly slumped as he began to speak again.

"I will read an entry from a brigade commander's diary: he sent a soldier on his final action. Sending soldiers to their death was a great burden on his conscience, from which he never recovered. He was killed a few months later. I translate from Hebrew into English what he wrote: this part of the diary is written as though he were speaking to the father of the dead soldier. He wrote:

'I had to pick the people from the group for the Beirut operation one by one. The choice had to be a real personal choice because of the special character of this task. I decided that your son would be in the first unit that had to act and implement this complicated raid on a PLO headquarters. From the point of view of the action, the opening had to be absolutely decisive for the success of the whole mission.

The demand from this unit of two people was to undertake a very unconventional movement, namely to shoot the guards while walking as though nothing special was happening. It wasn't just infiltration; it was the exact opposite. Ordinary walking, non military and in this way to reach a short distance from the guardsmen, to identify them, to stop, to light a cigarette and to go on walking. I think that such a thing demands to play it cool, with a lot of composure and tranquility. And then you have to shoot point blank, before the guardsmen understand what's really happening. And if it really succeeds in such form, then the chances for the whole raid succeeding from site to site, will be tremendously increased. We knew there was a big force there, inside the building itself, so the whole operation had to be started in a special way.

I didn't have many misgivings about choosing your son for the raid. I didn't choose him because he was an officer of his unit. I didn't have a shortage of officers, not in his unit and not in the brigade. I thought that your son would be good for this job. I saw in him exactly what we were looking for in courage and in cool. He would do what he had to do despite misgivings about the mission, to shoot while moving inside Beirut. It's not so simple. And there is certainly here a moral problem, the actual act of shooting first, this meeting between two people, one standing up and shooting at the other one, shooting fast with deadly accurate aim.

I didn't want to get a professional killer. I don't want murderers, but exactly the opposite. It looks absurd, because actually I could hire people; I could hire a gangster, a murderer. But even if I could, I wouldn't have done it. I thought about this from the point of view of your son, how does he look upon these things? Namely, if he had any misgivings on this subject or not, I don't know. But I put myself in his place, and after all, both of us were educated, if not in the same house, but in the same way; we were educated to be humane. And if I were in his place, I would have misgivings, the matter of conscience is of tremendous importance. You can say that about any matter relating to war, but this action was special and unique in character. It is different from an ordinary act of war, and after all, your son was not involved personally in any terrorist act: no terrorist had attacked his kibbutz or his family or his friends; no one he knew was kidnapped in a plane or something similar. There is nothing here of any personal revenge feeling or any other primitive emotion of this kind. This is very important for a person who is educated all his life in a certain way, to say it in beautiful words, to love people, and to regard human life as holy and sacred and to all the things we call humanism.

*Your son was educated in a kibbutz and enlightened
in all those things so he must have been put in a big
dilemma. I wanted to speak to him about this before
the raid. Would this be a disturbing factor? But I
didn't get the chance to have that talk and your son
answered his call to the duty that was demanded
of him.*

*The day after the raid when your son was dead al-
ready, I had a depressed and terrible feeling: After all,
it was my duty, sometimes I go and sometimes I have to
send others. As long as you go up the scale of military
hierarchy, you have to send others. Not every day there
is a war that you have to send someone to personally
shoot another human being point blank. My feelings,
my emotions, after this raid, after your son was dead, I
cannot describe to you. I imagined myself standing before his father, telling him that it was I who picked his
son for this task. It is I who sent his son to his death.
And what do I ask of his father and mother? And of all
the fathers and mothers of the enemy guardsmen and
all the sons blown up in the raid? Forgiveness? But
that is not the way of war. Missions and operations
have no faces, thoughts or misgivings; missions and operations require hitting the target. To speak of enlight-
enment and humanism is absurd. Not to speak of it may
tell more of what we and they have become.' "*

Uri's eyes seemed lost behind his thick glasses as he stared out
to the audience while they stirred in an undercurrent of voices.
Then Uri spoke again: "I only want to say, only to finish, that the
soldier the brigade commander was talking about was my son".

There was a stunned silence.

Freya convinced Uri to let her fix him lunch. While he sat on
the sofa in the little living room, she served a salad of greens and
avocados and olives with pitah bread to eat on their laps. She sat
on the chair opposite him while his eyes scanned the room. "The

apartment is comfortable. I don't suppose Asher is here very often." It was his signal that he knew she was living with Asher.

Freya felt herself flush, but she answered, "No, he isn't. He disappears until he comes back. He never says where he goes. I never know where to imagine him."

"Of course he doesn't tell you. We're at war. You aren't supposed to know, no one is, not even Aviva, so don't try to imagine."

They ate in silence, the story of his son still fresh in her mind. She reached for his hand. "I'm so very sorry, Uri."

He seemed oddly calm, stoical. But he held her hand. "I told the story for you, Freya, I wanted you to know, not to dwell on it, only to know."

Freya drew back, a sense of confusion coming over her. "I don't understand. Now I know this terrible thing in your life and . . . and still, you want Noah to make his own choice?"

"Yes," Uri said. "If Ephie had had his choice, apart from the social pressure of the moment to be heroic, if the brigade commander had taken the time to talk to him about the morality of the task before the raid, as he said he wanted to do, I know that Ephie would have chosen not to go. It was against his grain. The commander said he wanted soldiers, not murderers, but what did he make out of Ephie? Not a murderer? It had to be done . . . many things have to be done in a war, but my son did not have to do it. He didn't have a choice but to die, not on the battlefield in his uniform, but he died wearing a keffiyeh, dressed like an Arab, the clothes of the enemy. He died like a terrorist. I said we shouldn't dwell on it. There's nothing more to say."

Freya was biting the inside of her lip. Uri had closed the subject of his son. There was bitterness in his restraint, in the quiet of his voice. Now, he was glancing around the apartment. "When I told you about Asher's *protexia* I never dreamed you would come this far."

Uneasy, she tried to sound nonchalant, "I've known him most of my life. I was very young." She looked back to meet his eyes. "I was wondering, when you told me to keep my mouth shut, were you thinking of Asher, too?"

"No, no," Uri said. "But, it's reasonable, isn't it, that the Asher you knew when you were very young in America may not be the Asher of now?"

"The first part of your lecture, Uri, about democracy in Israel, I don't think Asher would like it."

"No, he wouldn't," Uri said.

His staring eyes had the distortion of a squint through the thick glasses. He smiled gold teeth at her.

"You're in a . . . a *mood*, Uri!"

"Maybe that's how you affect me."

Their eyes met for an instant. "I'm not exactly an Asher but I'm not a talking head, either," he said. "You affect me. I find myself thinking about you and Noah, too much perhaps."

Freya was taken aback, but Uri guessed that he had jarred her; he folded his hands on his lap and after a moment of readjusting himself on the sofa, he picked up his plate once more. "What have you decided?" he said. "Will you sign permission for Noah?"

Freya sat there looking down at her hands. What did he mean . . . too much perhaps? He sounded as though he were responsible for her and Noah; he sounded possessive. She sat there, searching for the right tone of voice that would set things right again. "I can't give permission," she said, finally. "Maybe I see too much on television."

Uri was himself again. "What you decide has nothing to do with the television. It has to do with whether you should make Noah's choice for him without trying to make him see another point of view . . . what the commander did not do for my son."

"Noah is full of runaway impulses." She stopped herself and glanced at him, "He doesn't think he has a choice."

"Whose choice?" Uri asked. "You could force dishonor on him and maybe he would be grateful to you when he's older; but maybe he would never recover. Such a dishonor could influence his whole life. Because you're more powerful doesn't entitle you."

"Noah doesn't know what he's doing!"

"By now, he knows," Uri answered her, shaking his head. "If he had doubts he also knows that he could be discharged honorably by confessing his forgery. A mix-up in communication, let-

171

ters that got lost, and so forth. But if you exposed him, you would take him home broken. He wouldn't come back to Gan Zvi." She was silent for a moment, confused. Was that his point? That Noah had to be discharged honorably and go back to Gan Zvi? With her persuasion and with Asher's *protexia?*

"Is that it, Uri? You want Noah?"

She saw him wince; the eyes behind the glasses closed for a moment. "My son was a commando, yes. Special forces was his choice of service. I tried to talk him out of it. I tried. I wasn't successful. You will do better, not with force but persuasion."

Uri's face had gone pale; as he spoke, he was squinting at her. "There are influences on our children we can't do anything about. We think we know them until those other influences show up. Models, heroes. When they show up, there's nothing we can do."

Freya was silent. "I don't understand," she said. "Weren't you a model for Noah?" Uri's enigmatic personality had shown up once more, entangled and dense. She had to change the subject. "I'm going to Nazareth tomorrow."

He looked at her sharply. "Listen, Freya," he said, after a long and uncomfortable moment. "Nothing is normal now. Don't plunge into water where people won't even go fishing. Lizabet Avidan didn't bother you again?"

Surprised at his question, she hesitated before she answered. "She showed up at the hotel when Asher was there. She didn't speak."

"Asher saw her?"

"No. Would he have known who she is?"

"Probably not. But she knows who he is. She'll be careful about her report. How are you going to Nazareth?" Uri said. "I feel I should take you."

"No one needs to take me! Asher had a person assigned to take me everywhere. I don't want to be taken."

He scratched the corner of his eye behind his glasses. "If you think you can hide, you should understand," he said. "This person Asher assigned to you is Asher's aide, Intelligence. Asher will find out anyway. So better tell him."

172

"No, Uri. I was brought up in America."

Uri made clucking sounds. "Why do you insist on applying your American values?" he said. "Nothing works here in the way you were brought up."

"I don't know how else to live," Freya said.

"You think throwing yourself into a sensitive situation is independence? It's arrogance! Anyway, Asher wants a close watch on you. Living with him could be dangerous."

"Like getting the measles?" she asked. When Uri didn't respond she sat quietly, studying him before she spoke again. "Are you and Asher friends, Uri?"

He wiped his mouth with the napkin. "Well," he said, "You have to understand how we in Israel feel about Army Intelligence, about all of the secret services. We're not like the people of other countries who call their secret services bad names. Israel's Intelligence are royalty. We wouldn't have a country without them, we would have been slaughtered.

"When Asher went into the army, he was already a prince. His superiors saw how capable he was. Mission after mission proved them right. He rose in the ranks very fast. I suppose Asher knows how important he is, but he's reasonable. Most of the time."

She turned away from his questioning smile. "It surprises me quite a lot that he brought you here," Uri went on. "I always found him aloof. There is a certain detachment about Asher that gives an immediate impression of power. Aviva was swept off her feet when he wanted to marry her."

Freya was listening, her eyes fixed on him, puzzled that he would bring up Aviva now, when he knew she was living with Asher. Instead of answering the question about whether he and Asher were friends, he had given her a lecture on the Israeli secret service. There was a certain deviousness she couldn't pin down in Uri's mixed messages.

But Uri went on talking. "Aviva doesn't make demands on Asher's time. She's satisfied with his connections."

Uri took his glasses off and wiped them with a tissue from his pocket. Then he looked at her in one of his strained glances.

"Did Asher tell you that Aviva was married to Asher's best friend, Danny? Actually, Danny was from our own kibbutz. When Danny's plane was shot down and he was killed in the Yom Kippur War, I think Asher felt guilty."

"Why should Asher feel guilty? What did he have to do with Danny's getting killed?"

"Asher can influence people just by the way he looks, a tall and handsome hero."

Something wasn't making sense. "By the way he *looks*, Uri?" she said. "Everyone in this country goes to the army! Is Asher to blame for every one that gets killed? My God, Uri, you're not making sense."

Uri shrugged, then he was silent. She felt he was struggling with an irrational part of himself; something was dangling in front of her like a fishhook.

She began to speak again but Uri cut her off. "Yes, I thought you would want to know about Asher and Aviva. It's very simple. When Danny was killed, like a brother, Asher married the widow. And it served Aviva very well to be married to a young Major on the rise. Doors opened. Maybe that eased Asher's guilt."

She was growing angry, bewildered; her hands were sweating. "Okay, Uri, that's enough. I'll make coffee."

Later, Uri seemed reluctant to leave her but he said nothing more. She promised to phone him when she got back from Nazareth. Then, standing by the door, he held her hand for a warm handshake. "Freya," he said. "We're a small country but everyone here, Jews, Arabs, come from mixed up lives. We go on living mixed up lives. Sometimes I think we'll be buried under a wall of complexity. There's no peace in ourselves. I doubt we could recognize or accept peace if it came."

The sun was ablaze. She stood there on the balcony watching Uri get into his car. The story of his son was still in her heart. Had she misunderstood? Had she offended him? She needed Uri! Without Uri she was vulnerable in a country where she was foreign, where she didn't know where it was safe to go fishing.

She stood on the balcony feeling baffled and alone, isolated, as though she were lost in the Judean wilderness that loomed in front of her, fighting the impulse to call Uri back. But Uri was revving the engine and pulling away.

Later that morning, Freya phoned two taxi companies in Jerusalem and both times she was told that the few taxis running on Shabbat were already spoken for. They assured her that there was no sherut to Nazareth on Shabbat. She could take an Arab sherut at The Damascus Gate. Seven o'clock in the morning. Next time, if she wanted a Jewish taxi or sherut she should call a week in advance. Both companies gave the information curtly and hung up.

She would have to walk to the Damascus Gate in the Old City in time for a seven o'clock sherut. She gathered her map of Jerusalem and her walking tour guidebooks to organize her route from Ramat Eshkol. If she asked Tami for directions she would phone her mother right off that her American neighbor was headed for the Damascus Gate at an ungodly hour.

She ate an orange at her kitchen table and considered phoning Zafra again to check on dinner plans for their date on Monday. She was smiling as she imagined Zafra's gasp when she presented her with the black lace nightgown, gift- wrapped in silver paper and tied with a huge black bow. Which of Zafra's boy friends would get the sneak preview and win the first prize? Zafra's face would beam like a spotlight.

It was four-thirty in the morning, dark, but the eastern sky glowed in faint tints of rose-gray, a kind of bleak beauty. Nothing stirred in Jerusalem's dark, outer husk—the quiet of the Sabbath was like dead air and the old stone buildings with their iron balconies and grill-work hung there behind a curtain of haze as in a dream state. Freya, with a scarf tied around her head and with maps and her walking route written in a notebook secured in her handbag, crossed over into Arab East Jerusalem at the Nablus Road.

The street was deserted. The large apartment blocks had disappeared; small houses were set on terraces or slopes of hills, white-beige stone with grill-work around small porticos and doors hung with blue or green beads in settings shaped like the fingers of a hand or an eye. She peered into a courtyard where she saw poinsettias growing. The smell of jasmine was overpowering. The little garden breathed a wisp of Alhambra.

Freya stopped and rearranged her long braid under the scarf; she was sweating. The call of the Mezzuin wailing across the city sent a chill through her, a sense of displacement and unpredictability. She walked on, grateful for her olive coloring, her dark hair, her anonymity. *They look at you as an occupier; they don't love you, believe me.* On the way home, she would get a taxi.

At the Damascus Gate, she was early with an hour to spare. The sky was growing hotter, whiter, streaked with mauve. The old men were there, wearing red or black checkered keffiyehs, sitting on the stone ledge outside the great arched entryway to the Old City, fingering prayer beads, gesturing to one another, lifting out bits of jewelry from packing boxes, wares to sell for the day.

She looked at her watch. There was just enough time. It was only six-ten and she ran up the stairs that led to the ramparts. She stopped to catch her breath and stood there, not moving. Behind the walls of the Moslem Quarter a world was hiding. She stared. Patchy gardens were hidden behind the labyrinth of shop lined alleyways, a miracle of carob and acacia, oleander and fig trees that stood amidst the junk of the outer streets.

The plotted courtyards with small domed houses were like honeycombs, layer upon layer, century upon century, the domes undulating everywhere, tightly packed in, leaving land to grow on. Vineyards and peach trees, artichokes and gooseberries. Huge walls of cactus boundaries. The pendulum inside her was swinging again.

The Valley of Kidron lay below and Freya stood there, transfixed. The hills in the distance sat there like an ancient rock garden, eroded, striated with ledges, catching the sun and the olive trees in a wash of gold. Where had David wept for his son, Absalom?

Freya stood still, shading her eyes. She could see Mount Moriah where Abraham took Isaac for sacrifice and the First Temple

176

was built by Solomon. And the Temple Mount where the silver dome of El Aqsa glistened next to the golden Dome of the Rock. The churches of the stations of the cross were down below in the labyrinth of streets and cobblestones, in David's City . . . the Third Commonwealth. She could hear the Hebrew litany, *Ariel, Ariel* . . . the Third Commonwealth, Israel. She startled at the wail coming from the minaret; and then it was as though, or was she only imagining, that the dirge of the *miserere* was settling down below over the Via Dolorosa; it was something she once had heard in a requiem mass that was now a dream sound in her head: *Stabat Mater dolorosa, Juxta crucem lacrimosa.*

Creeping down the rampart steps, she sensed what it might be like to go mad. The great circular staircase to the upper level and the plaza was now crowded but she had the feeling of silence, of vacancy.

Taxis were standing in uneven patterns with no indication of what was a sherut and what wasn't, which one went to Nazareth and which one went to Amman or Damascus. She went up to a lanky driver standing beside his taxi, waiting. The driver nodded at the word, Nazareth, vexed. It was five minutes to seven when he peered inside his taxi. There was one place vacant and he was waiting for his full quorum of passengers. He nodded at her impatiently and she maneuvered herself over the occupied jump seats to the back where the middle seat was empty, reserved for latecomers. The passengers were all Arabs, two wearing keffiyehs sitting on the jump seat ahead, another with a keffiyeh at her right, and a feather bed of a woman in a black embroidered dress and white shawl hunched up close at her left. The passenger next to the driver wore a tidy business suit and was the only one to respond with a blink when she murmured apologies for being late—the woman had averted her eyes to look out the window. She was huge in bulk and Freya felt a heat that would stifle her on a long ride. She strained not to touch the knee of the man on her right and when she wriggled as far from him as she could without leaning against the Arab woman, she noticed that the man up front had turned around to signal another blink. The

177

blink was not a defective eye or a nerve; it was some sort of signal that shouldn't be spoken about.

The driver revved up the engine and they started out with a jolt. When she asked how long it would take to get to Nazareth the heavy jowled man who had the face of a disgruntled civil servant blinked again and answered. "Much more than two hours." How long could she sit frozen into one position? Zafra was right: why was she going to Nazareth?

"You are an American?" the heavy jowled man asked in French accented English. When Freya nodded, he went on. "You're on a pilgrimage to Nazareth? Christian?" She started to say no, but she gave up and merely nodded.

Blink! It was hard not to blink back for now it was clear that he wanted to establish the fact of their being two western Christians in this pocket of Arabs. He was from Paris, he said, a contractor on business. He came here often but with the war in Lebanon and conditions as they were in Israel, the whole damn place was bound to blow up sooner or later. And good riddance!

He spoke English over the drooping heads of the Arabs in the jump seats. The keffiyeh of the Arab at her right fell over his face, his hands lay limp in his lap, a string of amber prayer beads wound through his fingers. If the man from Paris would stop shouting back at her maybe they would go on sleeping, but Claude Biton, as he had introduced himself, had already taken a dynamic interest in her. Was she touring alone? Just for the day?

She answered his question with nods, then turned her attention to see what lay outside: eroded hills and a sea of stones, olive trees rising in tiers and terraces while the saplings sat, waiting.

She was cramped, hot, and this trip wasn't making sense anymore. Why was it so important to Layla?

Claude Biton would not be quiet. He wanted to know what an American thought about this cassoulet of Jews and Arabs. He gave a stifled laugh.

The Arab woman was very much awake but she made no sign of understanding. Freya kept her eyes peeled to the window. They were deep into occupied territory. It was as though stones had rained down over the land with wild flowers growing between the cracks like hidden messages.

Close into the town of Ramallah, men were huddled around small tables playing a board game, smoking from the great vases and decorated piping of nargillas. The only signs of occupation were military jeeps here and there, and sometimes, on an isolated hilltop, the blue and white flag of Israel shimmered in the sun.

For what seemed like hours she stiffened in her slender middle place. The driver was in no hurry, but the taxi rocked over the snaking inclines and descents; he was making full turns of the wheel, round and round, dizzily, dizzily, while the Arabs on either side of her pushed and fell away.

And then the descent to the valley of Shiloh was gentle, the small hills settling. She opened her guidebook. David had brought the Ark of the Law to Shiloh to rest on his way to Jerusalem where he would unite the northern kingdom. The heights of the Emek Valley, the Plain of Esdraelon, with its rich black soil that once was the settlement of the Canaanites and where the prophet, Deborah, fought the battle for the sea. Gilboa . . . where Saul fell upon his sword when his son Jonathan was killed by the Philistines.

A voice from behind the keffiyeh startled her when he began speaking in English: "You see stones on the hills?" he said. "We are the stones." He had understood every word Biton said! His title for the movie, *We are the Stones*, was sure to have some deeper meaning, some threat beyond the obvious. She turned to him but he said nothing more; he was counting a string of amber prayer beads, shifting his weight.

"Are you from the West Bank?" Freya asked him.

"The occupied territories, you mean?" he asked in clear English. "No; I am from Nazareth. From Israel."

The air was stifling, soporific. The scenery outside repeated and repeated: brown field and brown field and hill after hill dotted with stones and more stones—incessant, relentless.

For close to an hour there was no more conversation until a clock must have chimed inside Claude Biton's head and he woke

179

up to announce joyfully, "Afula!" Actually, the map showed that Afula was at least another thirty miles from Nazareth, but the relief for Claude Biton must have been the map itself—the occupied territories were behind them and now they were within the Green Line where there would be Jewish towns and at least he wouldn't get kidnapped or terrorized. The terrain would rise and fall again to the banks of the fresh water Sea of Galilee—the *Kinneret*. Freya, too, began to breathe easier—they had arrived safely in Israel.

Every one of the Arabs were still with them.

CHAPTER ELEVEN

LAYLA WAS NOT at their meeting place when the sherut arrived. Claude Biton asked if he could take Freya somewhere, show her the city or, perhaps, just a coffee? It was not easy to shake him off. It occurred to her that if a shy little nun could be a spy, why not this contractor? She said a polite, no. Cheerfully, he went his own way.

At their appointed place, Freya stood waiting for Layla. What a *Yofi* idea to come here! Nazareth was a bustling city! Full of history, full of legend. She was standing at the Greek Orthodox Church of Saint Gabriel with its enclosure of Mary's Well below, set in an arched stone wall. She opened the guidebook again. She was virtually on the same spot where the archangel had appeared to Mary. For the first time since she came to Israel, she wanted to buy a postcard.

The street beyond the Church teemed with men wearing white keffiyehs, women in long dresses and shawls over their heads, young mothers in western dress holding little children by the hand—it was market day. Two priests came out of the church and went by her. A nun nodded and swept up the stairs in her white habit.

A red Volkswagen bug was honking and Layla was waving from the window. Freya got into the car.

"I'm not late, am I?" Layla cried. "Did you have breakfast? I'll take you."

Layla parked on the sidewalk of a side street and they walked to a hole in the wall cafe on a cobbled alleyway planted between gray stone buildings that might have been hundreds of years old. Arab music thumped, smoke rose in a grey haze; the smell of meat frying in olive oil and onion spiced with cinnamon and clove and fresh cucumbers shot out at her.

"I'm *starving*," Freya said.

They sat down but it was half an hour before a young dark haired boy, not more than twelve, sidled over. Layla seemed to be scolding him in Arabic; then she ordered breakfast. "Is it all right? Pitah, *leben*, yogurt, and *babagenush*, egg plant salad? And rice in grape leaves? Then espresso? Yes?"

Freya chuckled with delight. "Even for a starving American!"

"Enjoy it," Layla said.

Layla's manner was strangely subdued. Her sandy hair was knotted and pulled to the top of her head, adding grace to her long, Modigliani neck but her face was drained of color. She was solidly downcast, chain smoking from a pack of Farid cigarettes, chattering with forced and pressured nonchalance about the distance from Jerusalem to Nazareth, the heat, the rise in prices, heavier case loads in her agency. Too much stress, too much frustration. Freya watched her puff her cigarette, crush it, light another.

"I'm not making much sense, am I?" Layla said.

"Is something wrong?" Freya asked.

"Everything is wrong. But you didn't come to hear my troubles. I want to show you Nazareth. I want you to meet Toufic, my fiance."

Freya tried to measure Layla's mood and why she had lowered her head when she said, Toufic. "What's wrong?"

Layla fingered the paper napkin on the table. "I didn't tell you that I was supposed to marry Toufic this month. I don't think it will happen." Layla said nothing more. She was tearing off edges of the paper napkin into small strips. Her box of matches was empty and she rummaged in her purse. She looked at Freya. "I forgot you don't smoke," she said, getting up. She came back with a box of matches.

There was silence again. Layla took a deep drag from her Farid.

"Let me try one," Freya said. Layla gave her a light and she

182

took a drag. "Woo!" Her eyes were flooding. "Why won't your marriage happen?" she said, crushing the cigarette in the ashtray.

"You wouldn't understand," Layla said.

"I'm not exactly dull."

"But you don't know our conditions here!"

"No. I don't. We decided that on the plane."

"I don't want you to think I brought you here to pump you full of our heartaches. I want you for a friend!"

"I don't want to pump you, either, Layla. I'm sorry. Let's leave it then," Freya said.

Their eyes met and Freya guessed that Layla wouldn't hold back. Layla hesitated and lit another cigarette. "Toufic is from the West Bank," she said. She stopped, waiting for a reaction. But Freya didn't grasp the import. "You *see*?" Layla said. "You're an American. You don't understand. You're a free person."

Freya shifted in her chair, impatient at Layla's hedging. "You don't have to tell me, Layla!"

"I'm sorry, I'm sorry," Layla said. "I want to tell you. It's a matter of where to live. If Toufic and I marry, I must go to live on the West Bank with him."

Freya was silent, then she thought she understood. "*Ms* Magazine," she said, smiling. "You don't follow the man. I get it."

"You *don't* get it!" Layla cried. "I have no choice! Toufic wouldn't be *allowed* to live in Israel! He's from the occupied territories so, legally, he's a citizen of Jordan."

"How do you see him at all, then?" Freya asked.

"To visit in Israel is no problem. He comes every Saturday. But he can't *live* here. And if I go there to live, I could come back to visit, too, but not to make a home! Especially that we don't know what will happen with Sani. You don't believe me! I see by your face you don't believe me!"

Sani again. "Give me another cigarette," Freya said. She lit up and inhaled. Her eyes flooded again. "Why couldn't you come back to make a home?" she said, flatly.

Layla shook her head. "You aren't dull, Freya. Why *would* you understand? Toufic can't live in Israel with his identity card from the West Bank or his de-facto citizenship from Jordan."

"Isn't there some sort of appeal?" Freya said.

Layla shrugged. "Oh, yes. There's always an appeal to Israel's Courts. And maybe with *protexia* . . . Some cases like ours were worked out in the Courts. But we don't have *protexia*. Far from it. Especially now, with Sani."

"Wait a minute," Freya said. "Let me digest this."

Layla shook her head. She glanced around her and lowered her voice. "You don't understand what limits we have. In principle, I could come back if I left. But in practice, I could be old before it happened."

Freya leaned closer. "Layla," she said. "Don't hold a dumb question against me. It's not ideal, but even so, why wouldn't you live on the West Bank with Toufic?"

"My God!" Layla cried, muffling the sound. " To be occupied! What happens to our children? Should they live a life of identity cards, to be searched and suspected and stamped, always with a hundred papers and questions. Always watched and humiliated! Always punished! I will *not* be occupied! You want me to live like Yusef and Khalil? Full of hatred? Every day, every minute?"

Freya turned away.

"I know what you must be thinking! You don't think I love Toufic enough to live with him anywhere? I love Toufic. I love him! But I can't do that for him. I can't!"

Layla lowered her eyes. "I love him, Freya. But I can't live on the West Bank. I would die of shame and anger. We should never have fallen in love."

Freya reached out for Layla's hand. "But you did fall in love," she whispered, stunned, confused. "There has to be a place you can live."

Layla raised her head and pulled a handkerchief out of her purse. "I'm so sorry, Freya," she said. "The pressure is too much, sometimes. I don't know what's worse, my situation with Toufic or with Sani or with Fawzi."

She waited, impatient with hearing the name, Sani, then nothing. "What is it with Sani?" Freya said. "You said she was in France."

"Did I tell you that?" Layla said, looking around her. "I shouldn't talk about her now. Oh, Freya," she garbled, "Do you know what it's like, when you try, you can't think of anything positive, not one thing?"

Freya pressed Layla's hand. "You have to level with me, Layla. Has Fawzi done something serious?"

"No! Maybe unlawful demonstrations." Layla's anger was kindling.

"Nothing more?" Freya asked, remembering the demonstrations during the war in Vietnam when they were chased by police.

"What reason would I have to lie? Fawzi was expressing frustration! Nothing more. I swear it."

"They'll let him go," Freya said.

"Yes, they did let him go. They taught him a lesson and now he's a target. When an Arab citizen is disobedient, they're enemies, a fifth column."

Freya met Layla's eyes. "What about Sani?"

Layla was furtive. "Sani is different. She's foolish. My mother will die of it. I might die of it, too."

Even with Freya's eye on her, probing, Layla said nothing more about Sani. Instead, she spat out an epithet in Arabic. "We've been waiting almost an hour," she said to Freya.

Fifteen minutes later, the food came. Layla scolded the boy again. He looked at her, shrugged, and walked off.

Freya said nothing more, determined that the next time she asked about Sani she would get an answer. But Layla didn't owe her an answer; it was none of her business. She was not the Grand Inquisitor. She bit into the pitah bread. Something in Layla was about to spin out of control.

Layla was stirring a fork through the yogurt in her plate. "Sometimes I think I won't be able to stand it," she murmured. "I'll go to live in London. Or Chicago, maybe. I'm thirty-eight years old and, for me, life is too long."

Freya was struck by Layla's face, her waxen color, her haunted, trapped eyes. She found herself searching for answers, for something to say that might give Layla hope. She waited, wondering if she dared mention, then, tentatively, she said, "I know someone,

Layla. I can't be sure, but it's just possible I could tell him about your situation with Toufic."

Layla's face was tear stained but she was looking at Freya with awe. "Why are you in Israel, Freya?" she said. "I promised myself I wouldn't ask, but you're hinting that the someone you know has influence."

"He does. And I shouldn't have said that. I'm not trying to impress you. The truth is, I do know someone."

"You're not a tourist, you don't act like a tourist."

"I came because of my son, Layla. He's in Lebanon. I want to get him out."

"Freya!"

"Yes," Freya said. "I won't go any farther, Layla. I mentioned it because I couldn't stand the look on your face. You think *protexia* would work?"

"My God!" Layla said. "*Protexia* works miracles! Toufic could get a permit to live here from the Civil Court! I don't want to think about it. I didn't want to hope. But I knew! I knew you would help me from the moment. . ." She broke off speaking, her eyes opened wide, as though something awful had slipped; she looked down at her plate.

Freya bit into an olive. How did Layla know that she could help her? Because she was a Jew, any Jew?

She stared at Layla. "Layla," she said, finally. "How did you know I could help you?"

Layla's face colored. She looked up at Freya helplessly and didn't answer.

"Layla, how did you *know*?"

"You have to believe me that it happened after I first phoned you . . . *after* we went to lunch at Yusef's, *after* I phoned you that morning at the Shiloh. I wanted you to come to Nazareth before I knew, Freya! You have to believe me!"

"What was there about me in particular that made you think I could help you?" Freya insisted.

Tears were filling Layla's eyes when she looked at Freya. "I told you right away that I had a cousin who works at the Shiloh. It's Suhail, he's my cousin."

She hadn't remembered anything about a cousin; their talk in the airport was all of a piece. Now she remembered Suhail gawking at Asher when he brought breakfast to the room. Layla knew more than she was willing to say.

Silently, they finished the meal with Arab coffee from tiny cups. "I wanted to show you Nazareth," Layla said, sadly. I wanted to show you the Church. I wanted you to meet Toufic. I've spoiled it now, haven't I?"

Freya sat, preoccupied, thinking about murky waters where she shouldn't have gone fishing. Yusef and Khalil and Lizabet Avidan. Her heart was beating rapidly and suddenly, she had the impulse to leave Nazareth immediately. Nothing had actually happened but everything had happened too fast; every door she opened had a dark interior. But if she left, her imagination would leap beyond reason: every person she met would be spying on her and every move she made was meant to lure her into a trap. Suhail's telling Layla was gossip, nothing more. She and Evelyn gossiped constantly. Did they end up murderers and kidnappers?

"Layla," she said. "I *mean* this. I never want to hear the word, *protexia* again." Layla shook her head. "Never, Freya. Never." She began to explain that she had only mentioned to Suhail . . and Suhail had only mentioned to her . . .

"Don't, Layla. Please. Do you still want to show me Nazareth?"

Layla's face brightened. She wiped her face with a tissue. Then she was glaring in the direction of the proprietor and his child waiter, shaking her head in vexation. "I'm so *angry* at them," she said. "Arab time. Two hours!"

Because they liked each other, they somehow found a way to refocus. Before long, Layla was ebullient and Freya found her heart softening. She understood. In a few hours Layla would see Toufic.

From the town center, they walked west to a cobbled lane next to St. Joseph's Church where Joseph's carpentry shop was enshrined. If she wanted to, Layla said, they would go to the Church of the Annunciation and see the little grotto that was Mary's kitchen. Freya had the image of a billboard along the highway in Illinois: JOSEPH'S CARPENTRY SHOP. MARY'S KITCHEN.

The Greek Catholic Synagogue Church was farther on, built on the site of the ancient synagogue where Jesus studied the Torah on the Sabbath. "But who knows for sure?" Layla said, smiling; her spirits had soared. "It's a matter of belief."

The synagogue church was dark and cool inside; above the altar a statue of Jesus was nailed to a huge cross. Layla led the way to a corner of the church where a long stone bench stood against the wall.

Did Freya know that Jesus studied his Hebrew lessons on this spot? And (she was laughing) the story went that only Christians were able to move the bench, Jews couldn't budge it, neither could Muslims, not half a centimeter. If anyone wanted to hide their Jewishness, they should never, never, try to move that bench.

"Let's try!" Layla said. "Then we'll know the truth about ourselves! Am I still a Moslem in my heart? Are you a Jew in your heart? If so, no use trying to move the bench."

"A theological polygraph test!" Freya laughed.

Layla looked at her watch. "It's late," she cried. "They're waiting for us!"

"Who?" Freya asked, alarmed.

"My relatives in the village. And Toufic!"

Kfar Salah was an hour's drive west of Nazareth toward the Carmel mountains, an old and prosperous Arab village of four thousand, the showcase of villages, Layla said. Layla turned sharply to the right, off the main road winding higher and higher while the wheels kicked up dust and the tires bumped over the parched cracked earth where the sun had roasted every blade of grass. Small houses hung over with bougainvillia began to appear on rocky terraces and a pattern emerged in a kind of concentric circle fanning out from what Layla said was the meeting hall, small shops, a school, the clinic. Telephone wires were strung from poles and most of the houses had television antennas perched on their slate rooftops.

Layla parked the car on a precarious slope and then for what seemed like an hour to Freya, they walked uphill and uphill and uphill toward a house set inside a shaded courtyard. The wide rim of wall around the courtyard was lined with huge clay flower pots overflowing with geraniums and the panorama in the distance was the humped ridges of the Carmel Range. Layla pointed out where the mountains and the sea met and where Acre and Haifa lay beyond.

They stopped for a moment to catch their breath from the heat and the climb. Freya's cotton blouse was stuck to her body, drenched. The air stirred vaguely in the courtyard where a line of dazzling white sheets were hung to dry in the hot wind. From the house the beat and whine of an Arab chanteuse filtered out. Layla took her inside where a huge woman dressed in a green polyester dress and matching jacket was waiting for them. Her face ran with sweat and from looking at her jacket, Freya was certain the heat would swallow them all before a word was spoken. The huge woman kissed Layla; she shook Freya's hand when they were introduced. "Welcome," she said. "Welcome."

Freya was to call Layla's aunt, Violet, nothing more. The room they had entered from the courtyard was a sitting room lined with plush sofas against the wall, a long table in the center of the room and straight-backed chairs facing out rather then set under the table; a phonograph beating out Oriental rhythms came from the back of the house.

Freya shook hands with Violet and tried to smile but a wave of nausea and weakness was coming over her; apprehension, the climb, the heat, had left her heart racing and a steady throb had begun in her head. "May I have some water?" she asked Violet.

"My word, my word," Violet answered in her Arab accented, husky, British English. "A moment, just a moment." She sailed out of the room in all her graceful bulk and came back with a tray of purple liquid. "Grape and papaya," she said, "a mixture of this and that—but very cold."

"This heat. . ." Freya said. The sight of Violet in her polyester dress and jacket was making her head spin.

"It's not only the heat," Violet said. "It's the wind that blows from the desert and soaks up the air this time of the year. The Hamsin. It makes many people quite ill."

Violet insisted that they sit down, that Freya roll up the sleeves of her cotton blouse and that she drink a second glass of the purple juice. When Layla lit a cigarette, she stamped her foot. "No!" Violet was tyrannical.

A young man came into the room, a dark-haired and dapper fellow who looked like a clerk at Tiffany's. He shook hands with Layla; his name was Nadir and Freya was introduced as a good *Jewish* friend from America. He was Violet's son, Layla's cousin. How many cousins did Layla have? A network of cousins stretching from one border of Israel to the other?

Violet served the purple juice; she disappeared again and came back with a tray of dates and sugared fruits and fresh grapes and round sesame cakes.

Layla's face was lighting up, and she rushed to open the door to another young man in a loosely knit cotton sweater, a large man, stooping a bit, straightening his shoulders when he came in and saw Freya's unfamiliar face. He had a high forehead and great dark eyes; when he brushed Layla's cheek, she seemed to disappear behind him.

"Welcome," Toufic said, when Layla introduced him.

Violet interrupted. "You must get her something . . . she is ill from the Hamsin." She inclined her head toward Freya. "He is a doctor," she said.

"I'm fine," Freya said. "I really am."

"Never mind," Toufic said, and he left the room and returned with a large glass of water and two pills which he held out to Freya.

"For goodness sake, don't drink the sweet juice," he said. "Only water. As much as you can." He had a lovely baritone voice with soft English inflections.

"Take the salt tablets," Toufic encouraged her. He brought a chair and insisted she stretch her legs across; he sent Layla for a cold compress. "You shouldn't exert yourself, Freya. The Hamsin cuts the oxygen."

She looked at the tablets, argued with herself, looked up at

Toufic, and held the tablet out to him. "The water and the cold compresses are more than enough," she said, smiling.

He took the tablets back, a puzzled look on his face. "They are only salt," he said. Embarrassed, she reached out for the tablets. "Don't be afraid. They're only salt."

Toufic sat close to Layla but their hands did not touch. Every now and then they glanced at each other; she saw their eyes meet furtively, as though their relationship was a tender secret, not to be revealed, never to be expressed. A rush of warmth was in Freya's heart. There had to be more for them on Saturdays than sitting on straight-backed chairs! The glow on Layla's face. The longing in Toufic's eyes. Layla was so lovely, so full of grace. Freya looked away with a heaviness in her heart. How could she ask Asher to help them when she had hardly dared ask help for herself?

Violet was asking about America, about Detroit where she had cousins. She asked about Disneyland and California and the state of Colorado where her cousins went to the mountains. Desperately, desperately, Violet tried to keep the conversation casual.

Nadir sat staring at Freya. "Chicago," he said, pointing two fingers like pistols. "Al Capone. So what can we teach you about our problems in Nazareth?"

The casual conversation had petered out.

"Don't start that, Nadir!" Layla said. "Your lectures won't solve our problems in Nazareth or in the villages!"

"You think they intend to solve our problems?" Nadir laughed. "And do you think we know how to solve our own?" He was sneering.

Freya avoided his eyes, her reaction floating in the suspicion that she was viewing a performance. Kfar Salah was the morality play. But something inside her knew that this outburst was no play; something was spilling across the room like the oranges Haim had spilled across the floor of his shop.

Violet had caught Freya's eyes and she looked at Nadir with alarm. "Everything is not the fault of the Israelis!" she cried. "How do we treat each other? I was not born in this village. But we lived here thirty-five years and we are still looked on as refugees!"

191

Then she turned to Freya. "You must understand," she said. "The place an Arab is born is his home forever. Send an Arab five kilometers from his home and he is a refugee. If Israel made us refugees, our own people make certain we remain refugees. After thirty-five years should I still be a refugee in this village?"

Freya sat there, listening. Violet was telling the truth: she was remember the thrashing Rudy Haas had given her on the phone. In one way or another, refugees, displaced people, lived out their lives in a logical system of madness.

Freya sat back, searching for words. "Why do you stay here?" she murmured. "My family left Russia for something they thought would be better."

There was a snort from Nadir. "Of course! If we don't like it, we should get out. The world is big. The world is beautiful!" He turned to Violet. "You see? The Jewish mentality asks the right questions."

Toufic had been sitting quietly, glancing at Layla, glancing back at Violet, then angrily at Nadir. "Nadir!" he cried. "One more word will be too much."

"Forgive me, Mrs. Freya," Nadir said, ignoring Toufic but lowering his voice. "*Why* do I stay here? Because this is *my* country, Mrs. Freya!" There was a serene smile on his face.

Violet got up. Beads of sweat ran at her temples. "Enough!" she said. "We have a sick woman here. Let her rest."

"I'm *not* sick," Freya said.

"If not, they'll make you sick in a hurry," Violet said. "Don't listen to them. They want to shock you just like the Jews want to frighten us." She looked at her watch. "You know it is half past six o'clock?"

"Oh my God!" Layla cried. "Freya will never make the seven o'clock sherut to Jerusalem. The next one is nine."

"We will take her to Jerusalem then," Toufic said. He looked at Layla and she stared back at him. "Yes," she murmured, barely masking her disappointment.

Freya could tell by their faces. Layla and Toufic did not sit on straight-backed chairs every Saturday. They had a better place to spend the night than driving to Jerusalem and back. "*No one* will take me," she said. "I'll catch the nine o'clock sherut."

"No!" Layla insisted. "You will *not*. We will drive you."

"Please," Freya said. "Let me do it my way." By now, she was dissolving from heat, curiosity, suspicion, and stirred by Layla and Toufic who sat on the edges of their chairs, waiting for this evening to end so they could make love.

Freya looked at them while they discussed how she would get home. She caught Nadir's eyes . . . such anger, such anger. She felt chilled; if she didn't leave soon . . .

Toufic got up. "We must let her do as she likes," he said. "I won't stay for dinner." He turned to Layla. "I must go get supplies for my office in Acre. I'll meet you in Nazareth with Freya. At the sherut before nine o'clock."

After dinner, before they were ready to leave for Nazareth, Violet brought out a package to give to Freya. Two velveteen squares of cross-stitched embroidery. "Antiques," she said, embracing Freya. "We gave you a heartache. I want you to have something nice. You are welcome here, always." She hesitated before she spoke again. "I think you should know," she said. "One of us *didn't* remain here. Sani. Layla's sister. We pray for her."

And Freya, though she wanted to, knew better than to ask.

On the drive back to Nazareth shades of crimson light streaked the sky. Twilight was coming down on the hills and the fields like a setting for revelations and divine apparitions.

"It's too beautiful here for so much struggle," Freya said.

"The beauty is dangerous, it's a magic potion that makes you fall in love with the land so you would kill to possess it."

Freya didn't answer. Layla had said it again — she would kill for the land, her land. They drove on in the twilight without speaking.

CHAPTER TWELVE

THOUGH TOUFIC AND Layla drank one cup after another of black, Arab coffee while they waited for the nine o'clock sherut, Toufic allowed Freya to drink only water. "When you get home, make yourself a cup of hot tea if you like," he said. "But keep drinking water. A lot of water."

The sherut turned out to be virtually empty except for a young Arab couple traveling back to Jerusalem. Toufic spoke to the driver in Arabic and Freya saw him hand the driver two bills while she stood with Layla, saying good-bye. Toufic had already shaken her hand with a broad good-natured smile.

"Take care of yourself in the heat," he said. "You're not used to it."

Layla's face was sad, downcast, when Freya held her hand. "When I see you next time," Layla said, "I'll tell you about Sani. I promise."

"I don't need to know," she answered Layla, looking at her directly, hoping that she had heard the last of anyone's secrets, or of any proposition that would swing back and forth in her head, or any debate on the pros and cons of what was happening here.

But Layla wouldn't stop. "I want to tell you. I trust you. The authorities know about Sani, that's why they watch us." Layla had clasped her hands; her face was pained and in the semi-darkness, Freya could see that her eyes, looking into her own, were full of tears.

"Sani is a nurse in Lebanon. She went to help the Palestinians.

There isn't much more to tell." Layla's voice was faltering as though she had revealed something against her will. She added quickly, pleading, "You remember when we met, Freya? I didn't know you were a Jew, you didn't know I was an Arab. We were just two women. Can you understand how rare that is in my life? To make friends without showing passports?"

The driver of the *sherut* was motioning that Freya should get into the front seat next to him. "Toufic tipped him to take you directly to the hotel," Layla said. "By the time you get to Jerusalem there won't be taxis at the Damascus Gate. He'll take you to the Shiloh."

"But I'm not at the Shiloh!" Freya said. "Does he speak English?"

Layla look puzzled. "You moved somewhere? Never mind." She spoke to the driver again. "Yes; he speaks English enough. And he says you should tell him where to take you. But if you're not at the Shiloh, where shall I phone you?"

The driver was starting the car. Freya called out, "*Zafra at the Shiloh will tell you!*" Layla and Toufic were holding hands as the sherut pulled away.

The impact of Layla's revelation about Sani was beginning to sink in. Sani, an Israeli Arab, was a nurse with the Palestinians in Lebanon. Treason or something like it. But if the authorities knew and they were being watched, why hadn't Layla told her about Sani before coaxing her to Nazareth? Something was going on. She remembered Evelyn's warning that Freya Gould would become the Middle East version of *Little Red Riding Hood*.

She moved as far from the driver as she could, but his frequent glances made her edgy and she looked away from him to stare through the window at the blacked-out hills, rolling in stone, reflecting in the headlights of the car. The young couple in the back were whispering, dissolved into themselves.

Exhausted, jangled from thick, Arab coffee, she closed her eyes and leaned her head against the window, thinking of Layla and Toufic and their quiet restraint. Sitting on straight-backed chairs. Not touching.

She opened her eyes with a start. To hint that she could help them! And Suhail, so intelligent and charming, had tipped off Layla about Freya and the general. Her head had begun to ache again and she shut her eyes. Sit Still. Think of Monday. Think of Zafra and her first prize lover. On Monday, she would cook a real Mulligan's stew; afterward, she would whisk out the lace nightgown. Ta . . . Da!

Suddenly, the driver was waking her.

The sherut had stopped in the deserted and moonlit plaza of the Damascus Gate. Freya sat up, dazed, forcing herself awake. The young couple seemed to have disappeared. She handed the driver her fare.

"My friend paid you extra to take me home," she said.

"The Shiloh," he said.

"No. Not the Shiloh. Ramat Eshkol."

"*Where* you say?"

"Ramat Eshkol. I'll tell you how to get to the street."

He looked at her for a long moment and then he shook his head. "I don't take there; no, not to Ramat Eshkol. I don't take who lives on my brother's bones."

"My friend paid you! It's after midnight. Then get me another taxi!"

"Lady," he said. "You find it hard? We find it hard." He put his hand in his pocket and held out the bills Toufic gave him.

She tore the money out of his hands. "Get me another taxi!"

He looked around him and shrugged. "I don't see taxis," he said, and then he pointed across the street to the empty bus station where a light was still burning. "Maybe there," he said, leaning over; the taxi door swung open.

She got out on the verge of hysteria. It was after midnight. She had been dumped in the middle of nowhere. "What's the number of your sherut?" she hollered into the cab window. He glanced at her and revved up the engine, starting off; his license plate number was elusive and shimmering in the moonlight.

The tiny bus station was about to close when she went in. An elderly man in a black checkered keffiyeh stood behind a counter.

He spoke to her in Arabic, then in Hebrew, and when she shook her head, he said, "No more bus. Finished."

"Taxi," she said.

"Finished. I . . . to home now."

She stood there staring at him, clutching Violet's package. A slow panic was fanning out inside her. Her head was throbbing; the downhill walk had been an hour and a half, uphill would be even longer. She would never make it home.

She opened her purse and took out her address book where she had written the numbers of the Jewish taxis.

She pointed to the telephone. "Please?" she asked. He nodded. Both lines were busy. She made a motion to the man to wait another moment. He was shaking his head. She tried the taxi numbers again. Busy. And then, looking up at the man in the keffiyeh who was still shaking his head and pointing to the door, she felt her panic erupt and, quickly, she thought of David Barr. What was the number? Why hadn't she paid attention? It was something easy, James Bondish. Not 007. She was immobilized. The attendant growled. "You go now." As soon as he said it, she remembered. David Barr. Jerusalem 0001.

"You are *where*?" Barr asked.

"At the Damascus Gate. I'll have to wait outside the bus station. The attendant is going home."

There was silence. "Okay," he said. "Wait there. You're by yourself?"

"Yes."

"Very foolish, " he said. "Fifteen minutes. Not more. Can you take cover somewhere?"

"NO!"

"Don't move from there."

She went outside with the bus station attendant and while he walked he glanced back at her standing alone outside the door. He shook his head, tossed the ends of his keffiyeh around his neck and moved on.

She stood there in the plaza staring out to the Citadel rising beyond the great gate in an expanse of moonlit space. There was hardly a breeze and she could hear her own breath in the silence. She was telling herself that the night was mellow, the air full of perfume. There was nothing to fear.

She looked at her watch and took a deep breath. The deserted plaza had a threatening vacancy, something in its emptiness waiting to erupt. She had to fill her head with something . . . *The Hunter of the East has caught, the Sultan's Turret in a noose of light. Think, in the battered Caravanserai, Whose Doorways are alternate Night and Day.*

She had stuffed Violet's package into her handbag and she stood there, shifting it from one shoulder to the other, scanning the plaza, trying to remember more lines, when suddenly, the feeling of isolation and exposure began to dry her throat and set her heart pounding.

She strained to see if David's car was in sight. Against all of her will, she was caving in, proving Asher's point. On the dark route through the Arab sector, a massacre waited for any Jew who trespassed.

A blue Peugeot pulled up to the bus station and David Barr got out and came toward her—she felt giddy with relief. She had the impression he was smiling, which irritated her like a scratch. She had caved in and set a precedent. Now her options were gone.

He was medium height, young, fair haired and good looking. Uri said he was *Shin Bet*. A spy for a baby sitter.

His footsteps were light and quick as he moved toward her. "Shalom, Mrs. Gould."

When she was settled beside him, he started the engine and swung out of the bus station lot. She sat there while he glanced at her, looked back at the road, and glanced again. She could tell him a story about why she had phoned him—a sprained ankle, (she wasn't limping) a lost wallet, (maybe).

"To be here after midnight, Mrs. Gould? You had an adventure?" David Barr spoke before she had a chance to state her alibi.

"Your municipal services aren't exactly convenient," she answered brusquely.

"You don't want to tell me?"

198

"*Tell* you? Why should I tell you anything?"

"Ooo-ahh," David Barr said. "Why so angry?"

"I'm sorry, I'm very tired."

"I can imagine," he said. "So where did you go?"

She took a deep breath. "I went to Nazareth to see the churches and I missed the seven o'clock sherut. The driver was paid to take me home but when we got back, I was dumped. Is that enough?"

"Ooo-ahh," he said. "I see you more angry than I thought. I don't blame you. To be *dumped*?"

"He wouldn't take me to Ramat Eshkol. The crown jewels wouldn't tempt him, and since he said I was living there on his brother's bones, I don't blame him. We have it hard; they have it hard."

"His brother's bones? That's what he said? He has it hard and that's why he dumped you in the middle of the night? You got his sherut number? They get their license from us. The General won't like it."

"Well then, you have a gun. If the General doesn't like it, you'll just have to shoot him, won't you?"

David Barr glanced at her and was silent. "General Frank should have warned me," he said after a moment. "I don't think we get along very well."

"It's nothing personal."

When they pulled up to Six Days War Street, David parked the car. "I'll come up with you," he said.

"Why will you come up with me? You want a drink? You want something else?"

He held his ground. "You were gone all day. I want to see if the apartment checks out."

"Why wouldn't it?"

"Mrs. Gould. Don't give me aggravation. Anywhere the General might go needs to be checked out, especially tonight. I'm not going to fight with you so don't make me crazy." He thrust his hand out. "Give me your key or I'll use the one I have."

He came up with her and walked through the rooms. He tapped a few drawers, opened the kitchen cupboards, walked out to the two balconies. He opened the windows and inspected the window sills while she watched, astonished.

"Okay," he said. "Shalom. Have a good night."

"Why do you have a key to my apartment?"

"Security," he said. "Don't worry about it."

"Security," she repeated after him. "Right. Thanks for picking me up. I'm sorry I was rude. The heat. . ."

"It's okay," he said. "It's *okay*. Next time you want to see churches, call me. You'll see. We'll get used to each other."

When he was gone, she went out to the front balcony and took off her blouse so the cool mountain air would soak into her. The dew of Jerusalem was beginning to gather and she sank down on the lawn chair, trying to put the day together, asking herself if it were possible to live here without shouting the golden predictions and singing the fiery songs? Could she live in ambivalence and vacillation without taking a stand? And if she lived with Asher what stand could she take?

On Sunday morning before eight o'clock, she phoned Uri at Gan Zvi as she had promised.

"Who is calling him?"

"Freya Gould."

"You'll wait?"

"I'll wait."

After ten minutes of waiting at the telephone table next to the kitchen, she dragged the long telephone cord with the ear-piece under her chin to put up the kettle and drop a tea bag into a mug. Last night, she hadn't had the cup of hot tea Toufic said she could drink. The morning was scorchingly hot and she remembered Momma telling her that drinking hot tea was cooling, not cold, hot. It was worth a try. The kettle whistled before Uri finally answered. "Freya!" he said. "Why didn't you call me last night when you came back from Nazareth? I was afraid you were angry with me. I rang your phone until midnight. What happened?"

"I got home late."

"You're all right?"

"Why should I be angry at you, Uri? I'm snug in my safe house." She poured the boiling water into the mug and balanced it in the air while she sat down on the telephone bench.

"I'm glad you're safe. I'm relieved," Uri said. His voice was strangely unnatural.

"Excuse the racket," Freya said. "I'm sipping hot tea. After gallons of black Arab coffee. Anyway, I want to ask you something: I'm curious about Arab marriages. I'm not sure if the story is true or not."

Uri interrupted her. "I'm glad you're back safe," he repeated. "You'll ask me everything when I see you."

"Uri? Is something wrong? You sound so faint. Are you coming to Jerusalem?" she said.

"I have a meeting there on Wednesday. But you sound very good. Nazareth was interesting?"

"I think so; I'm not sure."

"Freya," he said. "From the way you sound, I assume you didn't hear the news."

Her heart leapt. *"What news?"*

"There was an incident in Jerusalem last night—Kikar Zion. A bomb in the toilet of a cafe. A woman was killed."

"My God! So that's why he said *especially tonight* and checked out the apartment."

"Who checked out the apartment?"

"David Barr. He brought me home from the Damascus Gate."

"Freya," Uri said. "The woman, I hate to tell you, I think you knew her . . . a young woman who works at the Shiloh. On Saturday nights she was a regular customer at the cafe; they said she used to spend hours fooling around and drinking coffee. She went to the toilet at the wrong time."

"Who at the Shiloh?" She stopped; her breath caught in her throat. "Not Zafra," she whispered.

"Yes," he said softly. "Zafra Bernstein."

She gasped. She tried to catch her breath but her voice was gone. A few monotone words came out. "Uri, *Please*. Not Zafra."

"Freya, I didn't know how to tell you. There's never a good way. You're alone. Should I come to Jerusalem now?"

She was silent while Uri called out to her. Finally, with a start she heard herself shout, "No! Don't come! I want to get out of here!" Her hands were trembling and suddenly, her full mug of boiling hot tea was spilling over her bare knee and leg, scalding. She shrieked and buckled in pain. The mug and the phone were on the floor.

Through the ear-piece, she could hear Uri shouting, *"Freya! Freya!"*

She limped to the kitchen and soaked a towel in cold water. Wailing, she pressed it to her leg, bending over to inspect the burn. The phone was clicking.

She reached to replace the phone to its hook and moments later, the phone rang again.

She picked up. "I can't talk to you, Uri," she muttered. "I spilled my tea. My leg is burned.

"Don't hang up, Uri! Please don't hang up." Freya was sobbing into the phone. "Don't tell me it was Zafra, Uri. Not Zafra . . . I have to get out of here. I have to get Noah out of here."

"Freya, listen to me. I can't fly to Jerusalem. You must take a taxi to the clinic or ask a neighbor to take you. Getting out of here isn't the point for you now."

"I have to get out . . . I have to get Noah out."

"Stop it, Freya! That won't help Zafra. If you won't go to the clinic, I'll call the Jerusalem police!"

"No," she sobbed. "There's no help for anyone here, not Zafra, not anyone."

"Freya, you are a stubborn woman," Uri said. "Go to the clinic immediately!"

"Don't call the police. Don't call. I'll go."

Whimpering, she hung up and limped to the refrigerator for ice to wrap in the towel, and then she sank down at the kitchen table, her leg stretched across another chair. The sun streamed in from the open door of the kitchen balcony and a hot gust of dry wind swept past and caught in her throat. The Hamsin. The air soaked up. Her leg was beginning to swell in huge blisters.

Zafra's murder wasn't sinking in. They had a date! Monday, not the best night for *Yossi's*, but better than never. Saturday night was the best. But Monday . . . How could she give her the black lace nightgown that was so beautifully wrapped? What about Zafra's first-prize winner? The Russian who knew where the zones were? Was Zafra murdered on a toilet seat wearing pretty clothes from Rene?

The icy towel was hot against her leg. *Zafra was planting her hands on her sexy hips, slinking toward the door of the toilet. . . . I know death hath ten thousand several doors . . .*

She sat there, bent over her leg, then she stood up and peered across the square to another balcony where laundry was flapping. Baby sheets. Diapers . . . *An infant crying in the night/An infant crying for the light/And with no language but a cry* . . . The heat would soon squeeze the breath out of her, like that day in Beaulac when Hammer sat under her feet at the kitchen table and her mother was dead and it was weeks since she had heard from Noah.

A squadron of jets smashed the sky and Freya covered her ears while she limped out of her kitchen. In the shower, she peeled off her clothes and let the cold, hard, spray beat down on her while she wound her fingers through her hair and let out what was accumulating in her throat.

Malka, at the Shiloh, told her that the funeral had been delayed until ten o'clock on Monday morning. Freya got into a taxi that would take her to the cemetery. The day before, a young woman doctor at the clinic had treated and wrapped her burn and given her antibiotics and tablets for pain. Dr. Tamar had picked up the phone to make arrangements for Freya to go into the hospital when Freya cried, "No! I'm not going anywhere!" Dr. Tamar put down the phone and spoke to her sternly: She was to take her temperature twice a day and if she had fever she would go into hospital whether she liked it or not. She was to stay in bed, off her feet entirely, and to report back to the clinic every day—by taxi—for examination and fresh dressings. She was not to come alone, but with someone to help her. Was there someone at home to help? Did she live on the ground floor? "Yes, oh yes," Freya lied to both questions.

203

When Freya showed up at the Shiloh at nine o'clock the next morning with a taxi, two soldiers with Uzis milled about in the lobby. Behind the desk, a tear-streaked Malka reached out to clasp Freya's hands.

No, Malka wouldn't go the funeral for fear she would not sleep for the rest of her life. She leaned over the desk, closer to Freya, her voice hushed. She was sorry to put Freya to the trouble of coming to the Shiloh but she wasn't supposed to give information on the phone. She told her too much already, but she knew how Zafra treasured Freya, her new friend. Malka wiped her eyes and Freya limped back to the waiting taxi with directions to the cemetery. Suhail was coming out of the dining room and when he saw her, he shook his head sadly. She lifted her hand as though to greet him, then she caught herself, turned, and barely smiled.

The cemetery was on the Tel Aviv Road. Two military jeeps were at the entrance and the taxi driver stopped and helped Freya out. He sighed deeply, clucking his tongue, shaking his head, cursing *Aravim* (Arabs), in low guttural syllables. Two soldiers stood at the gate and the taxi driver called back to Freya, "*Geveret*, Madame, you must on foot." he said, looking at her bandaged leg. "You can walk? You speak Hebrew? Someone must help you."

She shook her head. "*Lo Ivrit*, no Hebrew," she said. "I can walk. *Todah Rabah*, Thank you very much."

"Goot-by . . . Shalom, *Geveret*," the driver said. Freya turned and went back to reach into her purse. But the taxi driver shook his head. "*Lo, Lo*," he said. "*Lo, Lo*". He swept his hand in the direction of the gravestones. "*Anachnu besira ahat!* (We are in the same boat)."

The taxi driver left and a soldier examined her purse, looked her over in a long sleeved cotton dress and regarded her heavily bandaged leg curiously. He began to speak Hebrew and she interrupted him. "English!" she said.

"What happened to your leg?"

She was taken off guard. "A burn. . ."

He nodded. "Wait here." He lifted his walkie-talkie and mumbled in Hebrew. "You can stand by the tree," he said, when he was finished talking. "Someone will come to you."

He pointed to a small rise a few feet away. The second soldier walked with her slowly as she limped up. The pain tablets were wearing off and her leg was throbbing as she stood on the rise. She thought of sinking down to the grass. The soldier next to her was watching closely and she remained standing.

A huge assembly was gathered a good distance away. Military men were everywhere. Journalists. Cameras. She could not see the grave site and in the crystal clear air wild shrieks tore at her ears so that she drew back to lean against the yew tree, bewildered at the Uzi guns and cameras and the shrieks of a mob. A large TV truck was parked between where she was standing and the grave site. She felt her heart sink. This was Zafra's eulogy before the earth swallowed her up.

Shifting her weight, she leaned into the shade of the tree while she pulled several tissues from her bag to wipe the sweat from her face. Several young men, civilians, walked past her in the direction of the grave. She wanted to call out to them. Were they Zafra's lovers? Was the tall one the Russian from Odessa? Oh, how Zafra would want him here!

A buzz of excited sounds were layered against the steaming sky, radiating like the inside of a kiln. The young soldier reached out to steady her. He spoke to her in Hebrew and she shook her head, "*Lo Ivrit*," she said, shriveled from sweat and humiliation. "I *told* you. No Hebrew!"

Every now and then, shrieks pierced the air as though someone were in the last stage of childbirth. A woman's cries in Hebrew were amplified over a loudspeaker, carrying above the others, and Freya felt herself shrink as she imagined the cries of Zafra's mother, her arms raised over a pine coffin, and Zafra inside, blown to bits. No! That was *not* Zafra! She was lying there inside, quiet, decked out with one earring and clothes from Rene, lying there smiling, sweet as honey. She exhaled a stifled cry. The soldier was watching her, and she felt, suddenly, that she was like her tree that had fallen over in her yard, a transplant with no roots. She was a suspect stranger like Gulliver or Alice with no justification whatsoever for existing in this place. Abruptly, she turned to the soldier.

"I don't want to stay. Please let me pass." Inside her, Zafra's voice teased: *Stay, Freya! We had a date on Monday! You promised!*

She turned away from the tree to move up the hill toward the grave, but an army officer and a policewoman with bright, red hair were coming toward her, blocking her way.

"*Geveret,* Madame," the officer said.

"*Lo Ivrit!*" she interrupted, brusquely.

"We already know that," the officer said, in English. "Your name, please?"

Astonished, she stated her name.

"You live in Jerusalem?"

"I'm an American. I'm visiting here. And I intend to report this."

"What was your connection to Zafra Bernstein, please?" the officer interrupted her.

Something in his eyes told her she should answer his questions. The pain in her leg was building, her heart, pounding. "Do you think I'm stupid?" she cried. "Everyone in this mob isn't connected to Zafra! I met her at the Shiloh Hotel . . . when I was staying there."

"You don't stay there now?"

"No. Why are you questioning me?"

"May we see your passport, please?"

She took her passport out of her bag and handed it to the officer. He inspected it and handed it back while the policewoman stood by. "What is your address here?"

"I live in an apartment."

"Where is this apartment? The full address, including the apartment number."

She hesitated, glancing from the officer to the red haired policewoman whose eyes had never left her bandaged leg. "Number four, Mitle Pass Street, apartment three," she said.

"Why did you decide to rent an apartment? You plan to renew your visa?"

"I thought I would. I don't know. Why are you questioning me?"

The officer spoke to the policewoman and turned back to her. "I hope you realize, *Geveret,* this is not a tourist event. We have many officials here. The policewoman must inspect your bandage."

Freya heard herself cry out but the policewoman already had squatted, her fingers bearing down with heavy pressure over every inch of her bandage. A wave of stabbing pain and nausea went through her and she staggered. A recollection flashed of Layla standing near the boarding gate at the London Airport, the way Layla's face had paled, the way she had turned away, ashamed that curious faces were watching.

The officer was steadying her, but Freya drew back, gasping, "How dare you! How dare you! I have a severe burn on my leg. I'm an American citizen!" And then, trembling, she blurted out. "Phone him. . ." she sobbed. "Phone David Barr. 0001. I need to go home. I don't feel well."

A look of astonishment came over the officer's face. The policewoman straightened up and reached out to steady Freya.

"David Barr?" the officer repeated. "What is your connection to Barr?"

She glared back at him. "I told you to phone him!"

He seemed to recognize the strange number of 0001. His eyes were squinting at her against the sun.

"Young woman, *Geveret,* there are many officials here and anything could be hidden in a bandage."

He lifted his field phone and dialed the number.

After a moment of conversation, his expression changed. "You should have told us," he said, disconnecting the phone, inclining his head to take her in. The surprise on his face faded to comprehension. "*Geveret,*" he said. "We meant no harm. If we hurt you, *Geveret,* we'll have a doctor here in a moment."

She cried out while she stumbled, her body giving way. "I want to go home!"

The officer grasped her arm, supporting her. "My God, we didn't know. . ." He beckoned to two soldiers standing nearby.

"We'll take you home immediately. We are very sorry, *Geveret.* You must appreciate the logic of our tight security."

She cast him a scornful glance. *Logic.* There was even logic in madness.

That night, Freya had a dream about Sani, hiding in Violet's great flower pot where the gardenia plants were withering from the dry, dusty winds. Suddenly, Zafra and Noah were hiding in the flower pot, too, Zafra grinning at him seductively and Noah coiling the ends of her hair.

She woke with a start, bolting up and sinking back again to stare into the darkness, her eyes stinging. Zafra's death had been elevated to martyrdom, to a graveside soapbox, to a media event. What did it matter that Zafra was always at the cusp of a wink, that she balanced spike heels to give her hips a better swing, or that she spread her warmth around and gave prizes for sex? Who cared? Terrorists had paid for their headlines with Zafra's blood. Israelis kept score. For whom were the mourners howling? For the sweet person of Zafra? Freya buried her face in the pillow.

During the next week, Uri phoned every morning. And every morning, Freya put him off. No, he didn't have to come to Jerusalem; she had been to the clinic and was discharged with a minor burn. She had exaggerated in panic and she was sorry. "Forgive me?" she asked each time. "I'm resting and eating and reading . . . Okay, Uri? Don't come." She did not mention the funeral.

"I'm relieved," Uri said. "The kibbutz has business in Cyprus. Can I go in peace?"

"Yes, Uri, go in peace."

David Barr phoned, pretending not to know anything about the incident at the cemetery. Oh, but he knew all right. He accented the word, *health.* Was she enjoying good *health?* "General Frank told me," David mentioned casually, "if something is wrong with your *health,* we should phone his friend, Colonel Ben Ami, an army surgeon. If you were hurt . . . or something unpleasant

208

happened that someone was supposed to tell me . . . Of course, we don't want to make trouble with a report that would upset the General . . . You sure you don't need Colonel Ben Ami? He's a very busy man, but he could arrange. . ."

She got the drift. We don't want to make trouble? Trouble for himself, he meant. He was informed on the field phone and made things worse when he didn't show up to get her. If she didn't understand Hebrew she understood now that Barr had decided to hush things up. For that, he needed her cooperation. She was not supposed to see the Colonel; she was supposed to keep her mouth shut. If Asher knew how she had been frisked and brutally jostled, he would be furious. Instinctively, she knew it was better to play their game, not to make enemies of the *Shin Bet,* David, in particular, whom she had already crossed with the cab driver.

"Why would I need a doctor?" she said to David. "There's nothing to report!" When they said good-bye, David Barr's tone was cagey and relieved.

But Freya was far from well. The police woman had handled her leg so roughly, it was inflamed and swollen under the bandage. Tami was hysterical when she heard Freya's cry as she crept up the steps. Within the hour, Tami's mother, Batia, showed up in Freya's bedroom with a supply of exotic herbs that she mixed with an aromatic liquid she kept in a thermos. Without a word, she undressed Freya and lifted her into a sitting position, held a glass to her lips and barked in English: "DRINK!" Freya, astonished, tried to break free but Batia wouldn't back off. Freya drank while Batia held her in an iron grip. But Batia wasn't finished. She unwound Freya's bandage while Freya sank back to the pillow and wailed helplessly. Batia applied a gummy mixture and wound yards of linen loosely around her leg. Then, she pulled a brush off the dresser and returned to Freya, lifting her up again to brush and braid her hair with long, fierce strokes while Freya shrieked.

Every day afterward, Batia arrived to change the linen on her leg and apply gooey stuff; she brought huge containers of food from Katamon, the other side of Jerusalem, a section that swarmed with Moroccan Jews. When Freya balked at eating, Batia forced a fork between her fingers and commanded, *"EAT."* Judging by the quantities of squash stuffed with lamb, whole fish, rice, ground chick peas, eggplant salad, it seemed that all the families in her three story *shikun* had organized to cook for Freya.

Finally, at the end of the second week, Freya decided to turn Tami and Batia off and show that she was perfectly fit. She hobbled downstairs to have tea. Tami had laid the table as though it were an official reception, and after they drank tea and stuffed themselves with cakes and nuts, Tami took out a photograph of a brown-skinned young man, in olive drab army fatigues smiling on the rim of a tank.

"My husband," Tami said, smiling proudly. "He marry me with a big bride price." Then she began a story about a boy's parents who had dishonored a girl by offering a low bride price—which meant they didn't think much of the prospective bride. Her brothers were sent to settle the score of dishonor. The brothers found the bridegroom in the Mahane Yehuda buying walnuts; they kidnapped him, cut off his ear and stuffed, blood and flesh, into the pocket of his shirt.

Freya gasped. "That's disgusting!" she cried in shock. "Did the police get them?" Tami didn't answer; instead, she looked hurt and bewildered.

"Our people handle their own business!" Tami cried. "You look like you think we're not civilized!" Freya was taken aback by Tami's stricken face. "I'm sorry. I'm sorry," Freya murmured. She asked Tami to forgive her, and Tami, brightening in a great smile, put her arms around Freya and embraced her.

"My mother says you don't go to doctors. She fixes you. With herbs she brought from Morocco. Much better."

"Oh yes," Freya said, cheerfully. "Oh my, yes." How would she have managed without Tami and Batia and an apartment house full of Oriental Jews? Overkill for the mistress of a powerful general. What did it matter! The truth was, since she began taking Batia's herbs, she slept like a log.

Instructed by Batia on how to limp properly with an ebony cane Batia brought her, and encouraged to use her leg so she wouldn't forget how, Freya limped to the supermarket once a day and to the branch library for books in Hebrew and for Hebrew language tapes. In one of the books the epigraph read: *The legend goes, that the cosmos is made up of Hebrew letters.*

When she finally returned to the clinic, Dr. Tamar made disgusted noises. She gasped at Freya's homemade dressing and the goo underneath. "I told you to come back every day! What crazy person did this to you?" she demanded, outraged.

"She's not crazy. She's from Morocco."

"You let some Moroccan witch-woman treat your leg? We'll have to start over. From the very beginning. I told you to go into hospital!" Dr. Tamar turned away for supplies and when she turned back she was muttering: "Despite that uncivilized woman, it's healing . . . you're just lucky."

Freya didn't answer. Dr. Tamar said that if she allowed that Moroccan woman to touch her leg once more, she, Dr. Tamar, would wash her hands of any responsibility.

Freya put her arms around Batia. All those years she had taken care of Momma, there had been no one to take care of her. There had been no one to lean against as she leaned against Batia's great bosom now, as though she were a child, while Batia sang simple tunes in Mogravi. Freya explained as gently as she could that she had to go back to the clinic.

Every day, Dr. Tamar removed a bulky dressing from Freya's leg and replaced it with an equally bulky one. Dr. Tamar warned her to stay out of the sun, rest as much as possible, and continue the antibiotics for another two weeks. They would see her the following day. Freya refused sleeping pills. With Batia's herbs she slept soundly, fresh the next morning. The nightly slide show about Violet's flower pot had disappeared. Batia's folk medicine, her scolding and soothing had healing powers far beyond Dr. Tamar's condescending professionalism.

Almost three weeks had gone by since Zafra's funeral but Freya's mood would not lift. On Thursday, Uri phoned in the morn-

ing to say that whether she wanted to see him or not he would
see her in Jerusalem at seven that evening. He had been in Cyprus all that time. She waited for him at the kitchen table where
she sat writing in her notebook, cooling her feet on the tiled
floor. At five o'clock the sun was still hot; the apartment was stifling but the strange letter Uri had given her in translation, a
letter from his son before he died, had been pressing in her mind
since the first time she read it. It was not the kind of letter that a
soldier at the front should hear. If soldiers needed morale, Noah
also had to think about what he was doing. She would not let him
believe that only in death, he would understand his life.

Dearest Noah,

*At the museum, I once worked on a label about mummy wrap-
pers in Egypt: the mummy wrappers stored all the human organs
in separate finely wrought vessels, the heart, the kidneys, all the
organs that made a human being function. They threw away the
brain because they didn't know what it was for.*

*Are you asking my point? I'm not sure I know. But I keep
thinking about the logic of myths.*

*When I worked in the museum the atmosphere of creatures and
totems and myths and taboos were all around me, and I would
often believe that particular myths were invented to fill in something crucial to life but had no model in reality.*

*So, what I'm getting at, is that Uri gave me a translation of
the letter, Ephraim, his son, wrote and posted the very morning
he was killed. I've read the letter many times and I understand it
now as a strange, waking, dream. That's what myths are made
of. Ephraim's dream state is a gentle myth about the magical
properties of war and death written by a Shaman.*

Dear parents,

*I have such an urge to tell you what has sprung into my heart,
the idea of erecting, in some lonely and isolated place, a tombstone, which in its quiet way, will hold all my thoughts. The
thoughts will be so clear, everyone will know what they mean
because they are a soldier's thoughts and they are aged like wine*

with struggle and tears. *No one will mistake their meaning and everyone will know what pain and doubt lives inside a soldier's heart. In the evening, before sunset, is the best time to visit the tombstone because then is the time when my thoughts will look backwards. Backwards in thinking and remembering.*

To look backwards and actually SEE is a very magical power and without a tombstone like mine it can't be done. Good moments stored inside will line up with black moments and each will lead to an act that my tombstone will judge. I myself will often climb up to my tombstone in this special, isolated place and listen to the judgment. Then I will know how each thought brought me to where I am now.

CHAPTER THIRTEEN

W<small>HEN THE DOORBELL</small> rang at seven in the evening, she pressed the buzzer and limped down a few steps in her bare feet to meet Uri.

When she went to fix him a glass of iced tea, she called, "I'm glad you're here, Uri."

"I should have come sooner." He was standing at the door of the balcony, staring out, not speaking. She carried the drink to him and he turned to look at her bandaged leg. "You're all right, Freya? I thought I would find you with your bags at the airport."

"You think I would leave without Noah?"

"Freya," he said, reaching for her hand. "You've been with us such a short time and you're already bitter."

When she withdrew her hand quickly, he drew back.

"Aren't you bitter?" she cried. "You lost a son! Don't you look at the television? Don't you read the news? I went to Zafra's funeral to cry; instead, I got frisked."

"Freya, I think you should stop. There's nowhere to go with it."

"Not for you. I know where to go."

"Listen," he said, his face flushed. He came to where she sat on the sofa.

"If you like it or not, I'm going to tell you something. War or no war, most people here live a normal life. I warned you not to expose yourself to your friends in Nazareth, not to go fishing."

"Would you say that Zafra lived a normal life?"

"Zafra was a victim. All countries have victims. Even America."

"Layla doesn't live a normal life! Every last Jew on the street has a tragedy. Haim was ready to kill."

"I see," he said. "You're a real American. You come from a comfortable, fat, country. Good for you. We don't have such luxury. There is death here; death grows with the flowers."

"That's fine for you; I can't live like that."

Uri shrugged. "You act like you made discovery of grief. Zafra isn't the first one who died. As for Asher, you think because your handsome cavalier belongs to someone else you can sit here and suck on it? Let me tell you! Lovers are lost; husbands, brothers, children, are lost."

"That doesn't make it better for me! I don't glory in loss. I don't feel honor in loss."

Uri was breathing heavily, his hands clenched, "What are you saying?" he said. "You *insult* me with your tantrums. *My* child is lost, not yours!"

"I didn't mean to insult you. I want Noah out before he's lost! I want to take him home alive."

"Don't ruin his chances to remain here with honor. Gan Zvi was Noah's first choice. Remember that!"

She felt herself going pale; Uri's tone flashed on how Rudy Haas had spat through the phone at her. Uri wasn't able to contain his resentment of the safe American Jew, either.

Uri was sitting on the Danish chair opposite her and they were staring at each other; she tried to speak and then, looking at his thin face with the wire glasses askew, his white hair flying in wisps, his slight frame stiffened, she could not hold back a slow smile.

"You have a nerve," she said.

"Yes, I do. Without nerve, we don't survive here."

She was studying her fingernails. "I have nightmares. Every night gets worse."

"I didn't come here to give you another nightmare. But, nightmares can be enlightening."

Silently, she sat looking at him. No matter how she fought it, she could not help feeling awed by his survival in the camp. *Why,* not *how,* had he survived? If not dumb luck, wasn't survival an assessment of probability, instantaneous judgment, as though the calculating mind continued to function by rote? Outside the ravaged body, two plus two still equalled four. She found him daunting, formidable, as though she had to follow his lead if she and Noah were to survive.

She told him everything she dared tell: from the walk on the ramparts, to Violet and Nadir in Kfar Salah. And Layla and Toufic and the ride back to Jerusalem when she was dumped and had to phone David Barr.

"He checked out the apartment. I understand it now," she said. "The bombing."

"What did you learn in Nazareth?"

"I didn't learn anything! I can't tell a sales pitch from the truth." She was leaning toward him. "I'm afraid of everything here, Uri," she said. "The Mezzuin frightens me. The look in Layla's eyes frightens me. It frightens me that Barr has a key to my apartment. It's as though everyone knows something I'm not supposed to find out until I step on a mine."

Uri took her hand. "What about your notebook? Are you writing everything down?"

She nodded. "Four notebooks by now."

"Come to the kibbutz, Freya; you won't be frightened anymore . . . of him or anyone. I promise you."

"No, Uri." She said it emphatically, in a way that was new for her, definite. What did he mean, afraid of him or anyone? Innuendoes again. She had to assert her independence. The shock of Zafra had jolted something into place inside her . . . either she took responsibility for herself or depended on offers of protection which guaranteed nothing.

"Freya, listen," Uri said. "All right, then. You won't face reality, will you?"

"What are you talking about? At the moment, the only reality I have to face is telling Asher about Nazareth."

Uri took off his glasses and wiped them with a tissue. "You *still* don't understand," he said. "Or don't you *want* to understand?

The *Shin Bet* will tell Asher everything. Just remember that they can't control you. You're an American citizen."

"I care about Asher."

"I see," he said. "Your existence here will depend on General Frank? Whether you upset him or you don't upset him? Then you have to decide if you want to express your own thoughts or if you want to please the General. You think Asher would tolerate the two of you working on opposite sides of the street?"

She sat in silence, not knowing how to answer. If Uri had appointed himself monitor to her relationship with Noah, he was now monitoring her relationship with Asher. It occurred to her that the reason her feelings see-sawed perpetually was because of the controls on her; her impulses were restrained by any player that happened to show up. She was an easy target, a foreigner who didn't know the rules of the game or even what the goals were. But not only that. Others had always ordered her life. She was flushed with the impulse to say it aloud but she was silent. Issuing a proclamation of freedom would be arch and juvenile. Anyway, Uri's motives were always generous, always in her best interest; in a way, she had learned to love what he was . . . for all his suffering, he had not turned mean or vengeful.

With forced ease, she turned back to their conversation. "I can't accept that business about working on opposite sides of the street," she said.

"Then you need a few lessons on how to master a maze. Tell me, you drive a car in America?"

"Everyone drives a car."

"You can drive on your American license. Rent a car. You won't have to rely on Asher's watch dogs. Get off the general's leash. If you want to know where it's safe, ask me. Come," he said. "I'll take you to dinner at the King David where it's cool. You can rent a car there."

"Uri," she said, puzzled. "What *is* it between you and Asher?"

Uri startled. "I'll tell you at dinner."

The inside of the King David Hotel was elegant, cool, with high, beamed ceilings and spacious lobbies, luxurious sofas, arm chairs and long tables. The bell captain pointed them in the direction of the Rent-A-Car agent.

"Dinner first," Uri said, and they went out to a glass-enclosed and air-conditioned veranda where tables were set with white cloths and gleaming silver. The room was only sparsely filled and suddenly Uri's hand was on her arm.

"Aviva and Asher are here," he said. "It's not the end of the world."

She winced at the sight of them sitting at a table and half turned to leave but Uri held her back, clutching her arm.

"It's better to face reality," he said.

He virtually pulled her along to where Aviva and Asher were sitting. Asher looked at her, stunned. He got up. "Freya," he murmured. He shook Uri's hand, "Shalom, Uri."

"Freya!" Aviva said, gaily. "And Uri! *Shalom. Shalom. Ma Shlomha.* How are you?" She began to speak again in Hebrew, but she caught herself. "A surprise! Sit down. Sit down."

Freya glanced at Asher's astonished eyes. She looked away and tried to protest sitting down—no; no, she didn't want to interrupt their meal. But Uri was pulling a chair out for her at the empty seat next to Asher.

"Of course! Freya said she knew you," Aviva said to Uri. "I think she knows *everyone* in Israel by now." And then she turned to Freya, glancing down at her leg. "For goodness sake! What have you done to yourself?"

Asher startled; his eyes shifted from her leg to her face.

"It's nothing," Freya said. "I spilled a mug of tea. It's nothing."

"I'm so sorry," Aviva said. "Are you all right now? We're not taking very good care of you."

"It's really nothing," Freya repeated. Asher's eyes were on her.

"Your note was very kind," Aviva said. "Yonah Klar was quite taken with Asher's lovely school friend from Chicago."

"Freya's son is on my kibbutz," Uri interrupted. "She came to see him."

"Why didn't you tell us that at the party?" Aviva said. "Anyway, I was going to phone you at the Shiloh."

Freya nodded. The atmosphere was thick with tension but

Aviva held on to her gold-striped cigarette and turned to Uri. She seemed highly amused. "Can you imagine? Asher dropped out of the sky this morning. After all, it wasn't the first bombing in Jerusalem. I think something went to his head."

Aviva looked cool in her white linen suit; her black, clipped hair was smooth and silky above dangling ivory earrings. Not only was she imperious, the kind of woman Karl wanted, elegant and expensive, but she had a black pearl at her throat. She was like the house in Beit Hakerem with all of its artifacts solidly implanted, the sunken living room, the house she wouldn't have recognized as belonging to Asher. The apartment in Ramat Eshkol was suddenly makeshift, their wide bed, ephemeral. Rented furniture for a temporary arrangement. Freya felt herself draw inward; she was under-dressed; she was childlike with a long braid and a huge bandage on her leg; she was absurd.

A waiter appeared with menus. "Just coffee," she murmured.

Uri looked at her: "Just coffee, " he said, and then he turned to Asher. "We came for Freya to rent a car."

Startled, Asher turned to her. "You're taking a trip?"

" No. No," Freya said.

"It was my advice," Uri said. "A car will be convenient for her."

Aviva smiled. "It's no pleasure to drive in Israel. You need six eyes," she said, glancing at Freya's leg. "And a steady hand. After all, we must feel responsible for you."

"She's perfectly all right to drive," Uri said.

The waiter brought the coffee. She had not taken a sip when Aviva asked if she was still at the Shiloh. Did she know the poor desk clerk?

"Freya has an apartment," Uri interrupted. "She'll be here a while. Maybe a couple more months. At hotel rates, an apartment is the better solution."

Asher was looking at Uri, an openly-vexed expression on his face.

Aviva took out her address book and Freya gave her the telephone number while Asher watched, his jaw tightening.

"It's a pity," Aviva said. "I should have taken you to see some sights. Unfortunately, now, I have business in Paris."

When they finished their coffee, Uri got up and Freya pushed herself back from the table.

"I'd invite you to come home with us," Aviva said. "But our house is a fright . . . Keep in touch, Uri, I'll be back from Haifa for a few days before I leave for Paris. Perhaps you'll bring Freya for a drink?"

Freya sat there. Asher's eyes had hardly left her face. Abruptly, he got up too, towering over Uri. "Good-bye, Freya." He made a gesture to help her out of the chair, but Uri had already taken her arm, a look of triumph on his face as Asher glanced at him, his eyes narrowed.

Aviva called out: "Take better care of yourself, Freya!"

By the time they got to the Rent-a-Car Agency, Freya was drawing back.

"Come to Gan Zvi with me," Uri said.

She shook her head. "Drive me to the apartment. I can't manage driving."

"I'm telling you to drive, Freya. You're stronger than you think. What kind of car do you want? A Subaru with a manual shift? It's better for the hills, but with your leg, better an automatic. No?" When she nodded, he said, "So it's settled. Let's get the car. I know a Moroccan place across the street; we'll catch a bite of kebab. You must begin to eat properly. After supper, you'll follow me."

After supper, she followed Uri all the way to Eshkol Boulevard past the Mea Shearim until the turnoff to Ramat Eshkol. Traffic was heavy in the twilight. With a sharp intake of breath she felt a stab of pain in her leg; her hands were slippery from sweat and she kept drying them one at a time on her skirt until she switched on the air-conditioning while the windows were still open. On Six Days War Street, Uri signaled for her to pull over.

He parked in front of her rented car and got out while she sat behind the wheel, trying to even her breath. "There's nothing to it. Next time you'll drive to Gan Zvi."

She was relieved that Uri could not come in. She wanted to

sit by herself or lie on the floor, not speaking, not thinking, not feeling.

Freya waved when he drove off and before she turned in to the apartment house, she saw the Army staff car parked several spaces down; she turned into the building and limped as fast as she could up the stairs.

The door to the apartment was opening and Asher's arms were tightening around her; he was breathing against her hair. "Freya, sweetheart, I was afraid you went off in your car. Where would I look for you? What have you done to yourself?"

He lifted her in his arms and carried her to the living room sofa to examine her leg and brush his fingers over the bandage. "How did it happen?"

"It's nothing. I was drinking hot tea when I heard about Zafra. It's all right now. I don't even use the cane anymore."

Asher's eyes widened. "The *cane*? Who treated your burn? Why wasn't I told? Did David know?"

"I was very well taken care of," she said. Then she lied. "No, David knew nothing about it."

He was holding her but she felt herself resisting, still bruised from the farce at the King David Hotel.

She moved back. "I feel so wrong! How could it feel *right* to you!"

He reached for her hand. "Do you have an idea that might feel better? Shall I take you back to the Shiloh, buy you a ticket home, say good-bye and we both develop amnesia? We tried that once. You married someone else, I married someone else."

Their eyes met. She unbuttoned his shirt and pressed her lips to his chest. "I wouldn't have lasted, Sher. Not another week. When I saw you at the King David. . ."

"I wanted to shut Uri up."

Amazed, she sat up. "He was trying to protect me!"

"Don't be naive, Freya."

Asher got up from the sofa and went to stand at the balcony door. He seemed preoccupied, depressed. "I promised I wouldn't leave you alone too much, I hadn't realized that you would spend so much time with Uri."

She sat there, remembering that Uri promised to tell her at dinner what there was between himself and Asher. There had to be something.

"I haven't spent much time with Uri, not with anyone but Zafra. We used to go shopping." Her voice was breaking.

Asher came back to take her hand. "I'm sorry about Zafra," he murmured. "For us, it's something we live with; for you it was a terrible shock. I couldn't rest."

"I wanted to go to the funeral," she said, and broke off, remembering her unspoken deal with David that nothing be said about the funeral. She wasn't there. She had not been grilled. Her bandages had not been fingered. And there had been no bowing and scraping after, obviously, David told the officer about her relationship to Asher. But what if she had been just some suspicious anyone, some ordinary suspect? What then? How much longer would she have been detained, grilled?

She blurted out: "What happens in this country when you're just an ordinary guy without *protexia?*"

Asher, mystified, quipped back. "You know some ordinary guy who needs *protexia?*" he said, grinning. "Well, everyone can find a scrap of *protexia* here and there. It's like buying wholesale."

"What about Arabs?"

Asher hesitated. "What about them?" he said.

"I mean, what about if you need to get stamped in and stamped out and still, you're always suspect."

"Let's go out to the balcony," Asher said. "It's a beautiful night."

She brought oranges and the Turkish delights she bought in Nazareth out to the back balcony. Asher set up two lawn chairs. The twilight had deepened, and his bronzed face, his deep blue eyes, faded into shadows: he leaned back and stretched his hands behind his head while they sat there quietly. Every now and then she reached out to touch him and the familiar excitement began to build inside her.

"You didn't answer my question," she said. "Can Arabs find *protexia?*"

He leaned over and brushed a strand of hair out of her eyes. "You see? I go away for a few weeks, I come back, you're politicized."

"Politicized is getting in touch with your conscience."

Abruptly, he got up. "I'll get coffee. Would you like a drink, Freya?" She shook her head, no.

He came out again with coffee and a large drink he had poured for himself and set the tray down on the little table between the chairs.

"Okay. There's no reason why Arabs shouldn't have *protexia*," he said, lifting the glass again.

"In principle or in practice?"

He took a swallow, held his glass for a moment before he set it down to hand her a cup of coffee, making sure it was stable in her hands, eyeing her obliquely. "You're worked up about something, Freya. What is it you want?"

"No one gets what they really want."

"Not if they don't really want it," he said, mocking her. He took another swallow and put the glass down on the table. "Okay. Let's have it."

"I'm trying to get hold of reality."

Asher looked away. They sat there, not speaking, and then he reached for her hand. "Reality," he repeated, flatly. "I know all I want to know about reality. For now, I prefer Scotch."

His voice was constricted in a way she had never heard before. She reproached herself. Instead of trying to comfort him after his long, grueling weeks, she had triggered an argument.

She sat back, uncomfortable in the silence, searching for something light and amusing. She reminded him of their comic excesses when they were young lovers. "Remember when you wrote me a love poem and I made corrections all over the page? You were so mad you tore the paper to shreds in my face." They were leaning toward each other. Asher barely smiled. She was suddenly solemn.

She reached for his hand and brought it to her lips. "It's your fault, Sher. I was never *with* anyone after you. I thought I gave Noah everything he needed but I must have hurt him."

"People don't become what someone else makes them. Was I meant to be a general? I used to fight with dead philosophers. I didn't kill women and children."

"I never killed anyone. I didn't make anyone happy, either," Freya said.

"Not even now?"

"Now isn't a word with a future."

He cupped her face in his hands. "Now is everything we need for our lives. We were derailed once. It won't happen again."

The mountain air had turned cool in purple blackness. Asher had moved from the chair and sat leaning against the stone wall of the balcony: he had helped her up and she sat down next to him and stretched her bandaged leg out in front of her.

She was not sure why she didn't tell him about the ride home from Nazareth. She considered it for a moment, then dismissed it.

He reached out and drew her into his arms, settling her against him. "My God, if you had gone to that cafe with Zafra . . . It makes me anxious as hell."

"You want to know what makes *me* anxious?" She sat up, facing him. Despite her resolve, an argument began again; something flew out of her without thinking, images from all the television she had watched were surfacing . . . a corpulent Defense Minister with pointers and maps. Human rights violations in Gaza. "If the country wants to survive, its policies have to change! Why can't you see that?"

Asher ran his fingers through his hair, restless and agitated. "I don't make the policies," he said. "If I did, I wouldn't make the ones we have."

He got up to lean on the ledge of the balcony. He was staring out at the night. She managed to stand beside him while he spoke into the darkness. "You have all the bright questions tonight, Freya. Arabs and *protexia*. Gaza and the future. Human rights."

"Didn't you tell me to learn?" She watched his tall silhouette and rested her head against his arm. "I love you, Sher. I see how troubled you feel."

"What I *feel*," he said. "Such a luxury. It's only when I think of you that I feel."

"Let's not argue anymore, Sher. Not tonight. Smell the air. It's so sweet."

"I'd be satisfied not to smell anything," he said. He put his arm around her shoulder. "If the TV could stink, the country would retch on their carpets.

"You want to learn? I'll teach you. We've been playing Monopoly with the Phalange, with every Christian faction, every renegade militia, one after another, every self-proclaimed saviour in Lebanon. From '76 on we gave them civil war tokens. They were supposed to have cleaned the terrorists out. For *themselves*, not for us. The PLO runs Beirut like a state within a state with plenty of caviar thrown in. They think they run the whole show and every one else is out of tokens. They've learned that *we* run the show."

He turned back to stare out at the night. "I didn't come back for a symposium," he said. "We shouldn't waste our time."

"No," she said. "You're here now."

But Asher couldn't stop: he took off his shirt, mopped his neck with it and went back to finish his Scotch on the lawn chair. She followed him and when she sat down he began talking haltingly, as though he were measuring what he could say, the amount of bitterness that could creep into his tone without losing credibility. He was tentative, constrained.

"Asher?" she said. "Maybe we should drive into the desert. I've never seen so many stars."

He wasn't listening. "I'll tell you something," he said. "You won't see it on sterile TV. How much vengeance and hate is cooking. We're not talking soldiers and gentlemen. There are women guerrillas who carry plastic bags full of amputated fingers and earlobes. Trophies. Did *we* tell them to do that? We only helped boil their poison."

"But how long can people live with identity cards? Refugees won't stay in their place. They have rights. Why shouldn't they settle their score?"

He turned to face her, frowning. "You've been listening to Uri," he said.

"I have my own opinions. Does that astonish you?"

"No; of course not. That's why Uri impresses you. The simple and the humane and the bottom line—American apple pie."

"Asher! That's not fair. You're more an American than I am. Your great-great-grandfather was American history!"

"That's enough!" Asher cried. "Maybe you should face a few facts about me. I speak another language now."

"That doesn't put you in another galaxy. You're in the world! For God's sake, can't you see what these wars are doing to you?"

"If I believe in our state, I believe in the wars."

"This war, Asher?"

"No matter what Uri drummed up in your head, I don't slaughter civilians and walk away satisfied. I almost strangled a Lebanese colonel when he fired on a civilian bus."

She leaned over to smooth his hair. "You look so tired. I don't think I've seen you frightened before."

He looked out to the blackened hills of Judea. "When will some stupid kid in the territories throw the first stone? When will the Moslem fanatics try to knife us?"

She couldn't reach him: as soon as she came close, he pulled back. If Asher went on fighting himself, dividing himself, defending what he couldn't defend, it would kill something inside him.

"You used to say that philosophers question everything. Where are your questions about what your army is doing?"

She hadn't stopped there. While Asher sat, finishing his Scotch, she was striking out once more at the corpulent general, the Defense Minister, with his pointers and maps, his articles of faith. *"Is that your faith too?"* she cried.

Asher slammed his glass to the table, bolted out of his chair and stood looking down at her. In the darkness she could see the glint of his eyes, a hot glare, exploding.

"My *what?*" he said. "Who in bloody hell are you talking to? Are you forgetting who I am? I'm not writing a philosophy paper. You may not like it, but your lover doesn't run a tractor on a kibbutz. He's a general in the army. Israel's army."

He stopped, turning away from her, then he turned back, his voice terse and controlled. "Your moral splendor is making me sick, Freya. If you knew anything about us, you'd get off your star spangled soap box."

"You're right," she answered, hotly. "Morals are an endangered commodity here."

"I've heard enough from you, for Christ's sake! What do you know about our struggle? What do you care?"

She felt herself flinch while he walked away from her to the stone rail of the balcony. When he turned back something was roiling inside him.

"Get this very clear," he said, his voice shaken. "Get it straight and we'll both be better off. I believe in our army and I intend to keep it strong and superior. I don't have a list of choices for you and Uri to flap over. This is good, that is bad, this you accept, that you don't."

"Uri doesn't flap! He was in a concentration camp. He lost a son in your army!"

"It gives me pain; it won't make me run away." And then he came out with it. "Listen," he said. "I begged you to marry me, to come here with me. When did you come or take an interest in us? When your son made the blunder of getting mixed up in our troubles? How much did you moralize when we died since we first stepped foot on this land? Did you blame anyone? Did you take the trouble to *know* enough to blame anyone? So get it clear in your head, Freya. I fight for this country to win, not to lose. And I'll go on doing it. The political policies are for you and Uri to chew on."

"You're not fair!" she cried.

He turned into the apartment while she sat there in the darkness, smarting from his outburst. But there was another truth to face: she may never have seen Asher again if not for Noah. Her life was surreal, as though she existed inside a still life by Braque or Villon with objects falling off the perimeters of tables.

Asher was right. In all these years of struggle and dying, Israel mattered to her only because of Asher, only because of the Holocaust and the doomed generation. Israel was a Jewish love song.

227

She picked up the shirt Asher left on the lawn chair and limped into the apartment, panicked. But he was there, in the dark bedroom, propped against the pillow, his eyes closed. When she sat down at the edge of the bed and bent over him, she could feel his exhaustion, the tension in his body, his shallow breathing. Why did she have to make him miserable before she made him happy?

She pulled off his boots and his socks and leaned over to press her lips to his shoulder. "Asher?" she murmured. "I didn't mean to hurt you."

"Asher? Are you asleep? I have a surprise for you! In the morning, when I go to Haim's for vegetables, I say: *Boker tov, Haim, Ma shlomha? Ma nishma?* (Good morning, Haim. How are you? What's new?) *Ani rotza tapuhim* (I want apples), *agvaniot,* (tomatoes), *ve'oolay ktsat gezer.* (and maybe some carrots). Then I go to the supermarket, in the evenings usually, and since Mrs. Ziv is already my friend, I say, *Erev tov, shalom, Geveret Ziv (Good evening, Mrs. Ziv) . . . Eize hom ha'yom!* (What a hot day!) *Ani lo re'eva.* (I'm not hungry.) *Ad matai yamshich ha'hom ha'ze?* (How long will the heat last?) *Matai yered geshem?* (When will it rain?)"

Asher opened his eyes and in the slender rays of moonlight from the window she could see the lines around his eyes creasing with amusement and pleasure.

"And this is what I say to you!" she said. "*Aluf Asher Frank. Ani isha Amerikayit po levadi.* (General Asher Frank. I'm a lonely American woman here.) *Ani meoevet meod baheyal.* (I'm so much in love with a soldier.) *Efshar l'azor li?* (Can you help me?) *B'vakashah!* (Please!) *Ech ha'ivrit sheli, Aluf?* (How is my Hebrew, General?)"

He pulled her down to him and drew her close; he was laughing. "I don't believe it."

"I am an excellent student. Mrs. Ziv says that I already speak better Hebrew than Golda. I'm a genius."

"Lonely American genius . . . just you wait."

"I can't; I'm too lonely; I'm love sick."

He held her close. *"Freya, sheli,'* he murmured. *"Sheli* means mine. Freya is mine."

"She is, Sher. She always was."

"What I said about Noah was a cheap shot," Asher said. "I didn't mean it. Noah brought us back."

"He shouldn't have been the reason."

"There would have been something else. There had to be something."

"When I rage against the war, Sher, it's not only because I hate it for what it is. It's Noah! He's my son!"

"I know, I know, sweetheart. If we had a son, don't you think I'd look after him? I'll arrange for his leave. David should know where he is."

She wound her arms around his neck and she could feel her heartbeat against his chest while he unbraided her hair.

"Asher," she was murmuring, "Did you know that the Golem was made out of Hebrew letters?"

"Is that a fact?" he said. "And did you know that there are female Golems made for love?"

Later, when he was almost asleep, she murmured to him again. "I don't think David Barr likes me. I gave him a very hard time."

"I thought you would," Asher said.

CHAPTER FOURTEEN

By seven in the morning Asher had gone to a meeting and she awoke to see a note on the bedside table: *When I get back, I'll know more about Noah. And, what do you think? Can we use a three day leave? I said we had to catch up and we've hardly begun. Think about a house my friend owns on the beach. We deserve it. I'll be back before lunch.*

There was a knock at the door. Still half asleep, she opened it to Tami who had a letter in her hand. "*Boker Tov, motek,* Good morning, sweetie," Tami said, peering into the apartment. She held the letter out, still peering in.

"*Todah rabah.* Thanks very much," Freya said. With a last look inside, Tami turned away.

Freya's hands were trembling; she tore the letter across the army censor stamp and the forwarding mark from the Shiloh.

Dear Momma,

After I got over the shock of your letter—I haven't gotten any others since that one—I was able to think about how glad I am you're here. It hurts me more to remember how grandmother suffered for so many years than hearing that she died. I won't ever forget how good you were to her, Momma. Did I ever tell you that Grandma used to teach me things? She used to tell me, Love your own, Noah.

I don't know what to say about your not getting the letter I wrote about permission for the army. When I never heard back

*from you or Dad—I knew things weren't good between you—I
figured you had a big fight and you both gave up on me. I also
figured you wrote me off, Momma, and I was pretty mad and
hurt. I should have told Uri Levitsky to get in touch with you. I
never did anything that dishonest before, but, I tell you,
Momma, there are times when there isn't much choice.*

*You could make it hot for me, but I know you would never do
that. I promise you, someday you'll feel good about my work
here. Dad might even be proud if he saw my photographs. After
this, I could write my own ticket as a press photographer. And
there's nothing much dangerous about what I'm doing. No matter
what you see on TV, the atmosphere around here isn't that much
worse than a crowded night on the beach on the Fourth of July.*

*Anyway, leaves, even passes, have been cancelled for our unit.
But sooner or later I'll get one. Wait for me in Jerusalem. I want
you to meet my girlfriend, Carmella. She's in the Nahal, one of
our military agricultural posts outside Jerusalem.*

*You say you know General Frank? I'm impressed, but I don't
want his protexia. I'm no different from anyone else around here.
Everyone in this army has a mother. We'll have to wait until I get
leave in the usual way. Unless you intend to report my forgery. I
know you wouldn't do that, but I'm nervous.*

*I'll write again as soon as I can. Don't worry about me. I'm in
great shape. I even get to practice my swimming stroke. And I've
taken up the guitar. Anyway, Momma, it does my morale a lot of
good to know you're around.*

She sank down on the chair, looking out to the room with the
letter in her hand. Momma had issued the same aphorism to
Noah that she had heard so often in her childhood. Love your
own. What had not responded in her had skipped a generation.

She glanced back at the line where Noah said he thought she
had written him off. Noah *thought. . . Was this Noah dissem-
bling? How did she begin to understand him? And if they met
when he was on leave, how much more would he dissemble?
Taken up the guitar? Practice his swimming stroke!* What about
the 700 Israeli dead? What about death stinking around him?

When Asher came in, he was not himself somehow; there was a deadly composure about him while they sat in the kitchen drinking coffee.

She wanted to ask him about the beach house, but something in his expression held her back. "You look so tired."

He didn't answer.

She felt her heart jump. "Is there news of Noah?"

He sipped his coffee. "Noah is fine. We could arrange for a leave. But not now. There's a lot going on."

She was quiet; he was brusque, preoccupied, withholding information she had waited so long to hear. "I had a letter," she said, and got up from the table. She held it out to him and he sat there sipping coffee, reading.

He handed it back. "You've hardly given up on him," he said.

"What hurt me was the cozy atmosphere he's trying to pull off. As though I live on another planet."

He shrugged. "In some ways you do."

"Noah is afraid."

"Everyone is afraid. I wouldn't trust a soldier who wasn't."

"Are you, Asher?"

He looked at her, his eyes direct. "Shall I say no? It wouldn't be true. Sure, I'm afraid. It has to do with how you face death, what you've got to lose, whether or not you're in the middle of something so important, not finishing it would make death an unthinkable failure."

His deep blue eyes were studying her. "Noah is all right," he said. "At the moment, there's something more pressing. . ." His smile did not cover his irritation. "You have a mulish streak, Freya. If the IDF were as hard to handle as you are, I'd resign my commission."

"What's that supposed to mean?"

His long, questioning stare was not amused. "You can't guess or you want to play dumb? I had a talk with David Barr."

"Oh Christ." She got up and turned away from the table.

"Damnit, Freya! Don't play games with me. Come back here. Sit down!" He was flushed, worked up, holding back. She sat down again.

"Sit there and listen for a change! David told me about the cab driver who dumped you at the Damascus Gate. Past midnight. Never mind that you kept Nazareth a secret. There are two *Shin Bet* reports on you, Freya. But forget those. One thing at a time."

"Lizabet Avidan, right? Your spy, the worn out housewife. Right?"

"It makes no difference who. . ."

"I went to see Churches! I'm thinking of becoming a Catholic. And if this interrogation is official, I should call my ambassador."

He bit his lip. "What in bloody hell does it do for you to cross me every chance you get?"

"I'm a grown woman and I don't like being humiliated! I'm not a bad child that needs to be watched, given permission, stamped in and stamped out."

"You're on shaky ground, Freya."

"I don't care! I don't like some strange man with a gun owning a key to my apartment. The driver who dumped me was a creep. But he's not the only creep in town. If that Moshe in the Mahane Yehuda could have gotten away with it, he would have tossed me in bed like a salad."

Asher pulled his chair closer to her; he reached for her hand. "Jesus Christ! What are you trying to tell me?"

"Not *that*, for God's sake! The driver wouldn't take me to Ramat Eshkol; he gagged on his brother's bones. I don't blame him."

"His brother's bones? You don't recognize hatred when it's breathing down your neck?"

"Why *shouldn't* he have hatred?"

"Let him have it! We want to know what he does with it."

"He dumps Jewish women out of his taxi at midnight. I doubt that stupid Arab will bring down the State."

"I see. Freya Pushkin is the good guy with the conscience. When it comes to responsibility, well, that's our problem, not hers."

"Is there a criminal category for mean bastard? That's all he was, a mean bastard."

"Those are the profiles we watch, the mean bastards. We don't tolerate mean bastards, especially when they're drivers. There's a war on; we watch where they go."

"My God! Haven't you ever done a dumb thing in your life?"

"We're talking dangerous, not dumb."

"Being nasty to me is dangerous?"

"Freya! Who are you trying to out-smart? We have a security problem here. Did you like what happened to Zafra?"

He was running his hand through his hair, his eyes narrowed. "What's wrong with you? And why in hell am I quarreling with you? Procedure is none of your business."

She was searching his face: Asher was no fool; he knew the difference between a security risk and a bully. He was reacting because of *her*.

She was staring back at him. "You're personally out to get him, aren't you? Because it was me. It was the dimwit's lousy luck to harass the wrong woman. The poor bastard didn't know he was hassling the . . . whatever I am to the general."

He leaned over to her; their eyes met. "You tell me what you are to the General."

"Don't bother with this, Asher. Let it go."

"Let it go? Even if I wanted to, it's too late to let it go."

"You don't *want* to let it go. But you're out of luck. I didn't get his cab number or his name."

He shook his head as though she were pitiful. "You want to play schoolgirl on a field trip? Now where do you think drivers get their licenses? From the state of Illinois? They get them from *us*. We know every vehicle and every driver and every route and every schedule. David had your friend checked out. It was routine. He brought the man in. You have to identify him. That finishes it."

She was taken aback. "*Finishes* it! Maybe for you, for me, not for him! What will you do to him?"

"We'll shoot him, tear off his fingernails." Asher stopped to watch her face. "Don't worry your conscience," he said. "Ordinarily, we'd warn him. In this case, we'll lift his license."

"You'll cut off his living? You, the *socialist*, say, not to worry?

But this case is different, isn't it? Because it was me. You're going to teach him that every last Jewish woman he picks up in his cab might be some general's popsie?"

Asher was glaring at her. "We're asking you to identify him, nothing more."

"Nothing *more?* You go after one cheeky Arab to show where the power is? You want me to help take away his livelihood?"

"That's enough, Freya! David is waiting."

"He can wait then. If that's what I have to do to see Noah, I'll go to the American embassy. Mr. Baxter will fix it."

It was as though she had struck him; angrily, he moved back from the table. "I see," he said. "The American flag to the rescue. In that case, I'll tell you something else for your field trip. The American embassy won't lift one finger to accommodate you."

He sat there not speaking, and then he folded his hands behind his head and leaned back in the chair, his legs stretched out in front of him. She watched him; she knew his way of backing off, of knowing when to stop pushing her. She was quiet; whatever the outcome, Asher would believe she was not on his side.

When he sat up and spoke again his voice was under control. "Why are we quarreling over things you know nothing about?" he said. "Cooperate, Freya. For my sake."

His eyes were pleading and suddenly she felt a huge sadness welling up, remembering how Karl always asked her only to cooperate, nothing more.

"You know something, Asher?" she said. "Karl used to accuse me of having the Pushkin mentality. He made it sound like a fatal disease. I would get so mad I thought I would burst, but I never defended myself. I know who I am now. I *do* have the Pushkin mentality. And I don't mind having it. For the first time in my life, I know what I think."

Asher got up and stormed out to the balcony. She sat there, for how long she couldn't tell, fuming, fighting tears. Moment by moment her bravado sank while she sat there watching him look out to the hills, distanced from her, his back turned, his hands thrust into his pockets. Only hours ago she had been so close to him; the earlier quarrel had dissolved into their long night where nothing existed but their bodies together, his breath on her face. And now, she had turned snide, sarcastic. Aviva didn't treat him that way. Aviva was on his side and didn't forget who he was; she didn't fight him, she didn't confront.

The tears she had tried to control were falling over her face when she went out to stand behind him and clasp her arms around his waist, her head resting against his shoulder.

"I'm sorry," she said softly, "I'm sorry."

He turned around to face her; he wiped her eyes with his fingers.

"Don't cry. It's over. It's all right," he said.

"I didn't mean to hurt you. I don't even know Mr. Baxter."

Asher fanned his fingers. "What a relief! Imagine being aced by the red, white, and blue."

"Don't tease, Asher. I said I was sorry. I know I'm an embarrassment. What will David think?"

"David is paid for muscle, not opinions. Anyway, you brightened his life. How often does he rescue a beautiful woman?"

"I meant what I said, Asher. . ."

"I *understand!* You won't identify him; we can't charge him. But don't get smug. Your favorite cab driver is marked whether we charge him or not. David doesn't give up so easily. If that fellow bats an eye in the wrong direction, David will get him. It has nothing to do with me."

"Only with *me*," she said. "Because the stupid son-of-a bitch didn't understand the power he was up against. Maybe there's something I could do. Get him off the hook, somehow."

"Holy God! For a genius, you just don't get it, do you? It's out of your *hands*. It's security business, *Shin Bet* business. As far as they're concerned, a potential sore just happened to surface."

"It's a sting! It's because of Zafra! Every stupid Arab will have to cover their ass or they'll scorch him!"

Asher's eyes widened; he shook his head in feigned disbelief, laughing. "You just don't know when to stop, do you? Let's say we did you a favor. If we didn't show you reality here, how else would you needle the system? Is it over now? Is it finished?"

"I didn't start it."

"Of *course* not. Anyway, I'm jealous of your grit. You remind me of when I could afford the Pushkin mentality." He put his arms around her. "Do you know how much time we've wasted slugging it out?" he said. "We could have gone to a movie. We could have played Scrabble."

It was not the right time to pin him down about when she could see Noah. "We could certainly have had more constructive discussions," she said, gravely.

Asher laughed. "Now why would I be taking a holiday with you, if not for more constructive discussions!"

But Asher was not as playful as he pretended. His hands were on her shoulders. "Listen to me, Freya." His eyes were searching her face. "Every step we take is uphill, full of rocks. Will we make it?"

She looked away. "I don't know. What else can we do?"

"Nothing. Before you came we might have muddled through with half a life. Then, suddenly, the other half shows up and we're stunned. How do the two halves fit?"

"I don't know! Nothing seems to fit."

"Are you willing to tough it out? You won't give up?"

"Never, never," she murmured.

"Is that a pact? No matter what?" Asher said.

"It's a pact. No matter what."

"Let's have our holiday, sweetheart. We'll talk about Noah on the way. I haven't forgotten Noah."

On the last day of Asher's leave at the beach house, they carried their lunch of fruit and cheese and pitah bread to the beach as they had each day. Because the doctor said she should stay out of the sun and not get her leg wet, Asher set up a beach umbrella at the water's edge so she could watch him swim before the sun turned to fire every morning. When the sun went down, they went back to the beach to cook chicken or hamburgers on a grill and twice they brought blankets and lay on the beach to listen to Freya's music tapes while they watched the sunset and went to sleep on the sand until sunrise. He bandaged Freya's burned leg every day and wrapped it in towels when they sat outdoors.

Asher's handsome face had relaxed to it's familiar ease, sometimes teasing, sometimes watching her tenderly, his eyes alert and responsive to every move she made and every word she spoke. When she checkmated Asher in one chess game and stalemated two more, Asher was more jubilant than she. Freya felt her strength and confidence soar.

She toyed with telling him about her humiliation at Zafra's funeral and quickly dismissed it. She was beginning to understand the officer's regard for security; she was beginning to understand Asher's burdens. Still, she couldn't resist checking out the facts about Layla's marriage to Toufic.

Tentatively, while they strolled slowly on the beach, Freya described a hypothetical case. If one of a couple was an Israeli Arab citizen, and the other was from the West Bank, and if they were in love and wanted to marry, where would they live? What citizenship would they have? What passport?

Asher stopped her on the sand where they walked, his hands on her shoulders. "Well, well, well," he said, grinning. "You've plugged yourself into our problems."

"Oh, Asher, why is there barbed wire in every question I ask? The hypothetical case breaks my heart. If it were happening to us, I don't think I could stand it. It's not fair, it's not human."

Asher hesitated, regarding her curiously. "Hypothetical, is it?" he said. "Not on your life. I don't provoke on this holiday. I don't take bait."

"You said you'd answer my questions!"

He sighed. "Right. Constructive discussions. So, the answer to your hypothetical case is, we're not in a normal situation. Our struggle is to normalize life for everyone including the folks in your hypothetical case. Will that do?"

"No," she said. "What happens in the meantime?"

He looked at her obliquely. "Everyone manages the best they can. Some people even manage to have a good life. Some people even help with the solution."

He had cut her off with innuendo and she returned his smile. She could be part of the problem or part of the solution. Two sides of the street. "No more questions," she said.

They had settled on the beach that last morning with fruit and a thermos of coffee and while Asher wound fresh bandages around her leg, Freya looked up to him, her eyes brimming with tears, her heart sinking. They would have to leave the beach house before noon so Asher could meet his helicopter in Jerusalem.

"If you cry," he said. "Your soldier will have a hard time going back to the war." He made clucking sounds. "Not good for morale."

"How much longer? When will it end!" she cried.

"I could give you a time . . . it wouldn't be true."

Until now, they had spoken of Noah only as to his general location and his work as a photographer. No, Asher had assured her, Noah's work was not considered high risk.

"What have you decided about Noah?" he said. "He's right, you know, about opportunities when he's out of the army."

"I don't want Noah in this war." She stopped herself, remembering their bitter argument at Ramat Eshkol.

Asher pursed his lips but he said nothing. Instead, he took her hand. "Noah is asking you not to interfere."

"Funny," she said. "Uri says the same thing . . . it's the first time I heard you two agree. What *is* it between you and Uri?"

Asher startled, then he was silent for a long moment as though he had to consider his answer carefully, then, finally, he said, "It goes back a long way . . . since I left the kibbutz."

She heard herself gasp. "You lived on Gan Zvi? Why didn't you tell me? Why didn't Uri tell me?"

"I suppose both of us would rather forget."

She was tracing a circle on her bandage with her fingernail. She was bottled up inside with unanswered questions. Why hadn't Asher or Uri told her? She was certain now that, somehow, Noah had gone to the symbolic center of a quarrel between Uri and Asher.

"Why, Asher, why?" she asked him.

Asher didn't answer for a moment, hesitating, until he said finally, as though he were beginning a painful confession: "Do you want the long or the short version?"

"I just want to know."

Asher shrugged. "Well," he said, "I was as green as Noah when I lived at Gan Zvi. Uri had Ephraim, his own son . . . I was fifteen years older. I was close to Ephie, I taught him to play tennis and soccer and I worked with him on his math and his languages."

Asher stopped talking for a moment and looked out to the sea, then he began again. "I came back to the kibbutz often after I went to the army. But Uri resented the way Ephie followed me around, the way he wanted to imitate me. Ephie was only 10 years old when I got my first medal. I suppose I became a model, a hero, for him. My first big quarrel with Uri had been coming for a long time and when Ephie took the medal off my uniform and pinned it to his T-shirt, Uri blew up, as though I had put it on him. He accused me of hyping medals and heroism to an innocent kid."

"But everyone goes to the army!" Freya said.

"Not to become heroes. When Ephie was sixteen, he asked me about the special forces, commandos, slated for the most dangerous assignments. He asked if guys in the special forces got more medals than in other branches of the services. I told him that he had a reckless idea, exactly what the commandos would not tolerate. Uri was furious with me when Ephie volunteered and was accepted for the special forces, while I was away. He knew I would have tried to stop him. When he was killed in a raid in Lebanon, Uri cut me off. . ." Asher hesitated before he went on: "At Ephie's funeral, Uri made it clear that I was to blame."

Their eyes met; Asher was quiet. "There's nothing more to tell," he said.

She sat there in silence. She was remembering how Uri's son died and that Uri had talked about Asher's guilt over Aviva's first husband, and the way Asher *looked*, a model and hero, was the outside influence that showed up in Ephie's life. And through her, Asher had shown up in Noah's life too. Why then had he urged her to get help from Asher? Because he wanted Noah out of the army, honorably, with no complications? So he could return to Gan Zvi unscathed? *Gan Zvi was Noah's first choice!* She remembered Uri's face when he said it. Asher would help get him back. Asher owed him.

"Poor Uri," she muttered, aloud. "Poor Uri." Something was draining inside her; this quarrel and competition over Ephraim between Uri and Asher, the question of sons, latent, unconscious. Asher had no son, Uri's son was dead and now Noah, her son, had appeared in the frame.

"You weren't to blame, Asher. Uri wants his son back. He wants Noah on the kibbutz. He thinks you owe him."

Asher looked at her baffled. "You asked me questions, Freya, I answered them."

They sat facing each other. "Asher," she said. "I need you to help me. I need you so badly to help me."

"How, Freya?"

"I need to see Noah."

"But I told you, his commander can't spare him now. I can't interfere."

"But for a few hours . . . couldn't he be spared for a few hours?"

"A leave for a few hours? It would take longer than that to get here."

"I could see him in Beirut!"

He was looking at her, amazed. *"Where?"*

"It's best if I see him in Beirut. Otherwise, I'll never know. I'll end up torn in half, never knowing what I did to him or what I should have done."

Asher sat looking at the sand, for how long she couldn't tell; then he looked up at her. "No stray civilian goes to Beirut. You're asking for ultimate privilege from someone you know can provide it. Is it worth taking that much advantage of me?"

"I want it! I know I'm asking a lot, but you said that you really needed to want something before you got it."

"You're asking me to risk your life in a war zone so you can feel morally correct, make one of your grandiose judgments?"

"That's *not* what I'm asking! Journalists take risks. You said even Noah wasn't in danger."

"Give me one reason in Hell to accommodate you. You want me to pass some test?"

"Asher! *Please.* Listen to me!"

"I'll listen; that's *all* I'll do, nothing more."

"Sher . . . You're the only one who can put me in Noah's place. Your place, too, Asher."

"I've tried to protect you," he said.

"Protecting me is shutting me out!"

"You want to give up your illusions? You think you're strong enough?"

"What good are illusions? So you can sneer and say I'm a self righteous prig? You've got to let me cross that boundary! I have a heavy decision to make! I don't want to decide from the sidelines."

He was quiet. She took his hand and brought it to her lips. "Sher," she said, softly. "How can anyone tell what's right or wrong if they've never gone over the brink?"

He pulled away. "You're asking for a long, hard fall."

"Maybe that's where reality hits. Maybe it's the only way to be there for each other, to understand the dimensions of *there.*"

"Do you know what you'll see in Beirut? You're asking me, for Christ's sake, to push you over the brink? You think I'm crazy enough to take that kind of risk?"

"We're all at risk," she said. "I could end up like Zafra. Why are the risks you and Noah take more justified than mine?"

"We're soldiers," he said. "We're trained to stay alive if we can."

"Think of our risks then," she cried. "We make marathon love and later we tear each other apart over what's right and what's wrong. I don't know how deep your wounds are, or you don't understand what I've become. Our pact wouldn't last; it wouldn't mean anything. Asher! Are you listening?"

"I said I would listen."

"I'm learning Hebrew," she said. "That should say something."

"I want to swim while there's time," he said.

He walked away from her and swam out beyond where the waves were breaking. There was a catch in her heart and, in panic, she went to the edge of the shore and heard herself call his name as though he would disappear into the sea. He swam back to where she was shivering in the heat.

He stepped out of the water and lifted her close to him. "You're a brave mother," he said. "It may take a while, but I'll arrange for you to see Noah in Beirut."

"Asher! Can you do it? I'll never forget. . ."

He was laughing. "Who was it that said, what good is power if you don't abuse it? Don't be afraid, sweetheart. You'll be safe every step of the way."

Tears were falling over her face. "Sher, oh Sher, no one has ever been so good to me."

He turned her face toward him and smiled into her eyes. "But listen to me, Mrs. Know It All Yankee. You have to promise. No games, no tricks. You do exactly what David tells you, every step of the way, one foot after the other. If you don't, I'll give him orders to tie you up with a gag in your mouth."

She clung to him, kissing his eyes, his mouth. "You won't be sorry, Sher! I'll do whatever David says. He can shave my head if he wants to."

"Well now," Asher said, winding his fingers through her hair. "Let's not get carried away."

Freya set her alarm the next morning at five AM. It would be nine PM in Chicago. She had a heavy foreboding inside her, an urgency to say a final good-bye before she went to Beirut. When Asher told her that David would be in touch with her soon, she felt a rush of excitement. Now, sitting alone in the living room of the apartment, her thoughts were stuck on Zafra, on the silver package with the black bow that now lay in the bottom drawer of the bureau. For Zafra, there had been no time to imagine death. It wasn't possible to imagine one's own death except in shapeless, opaque images . . . A shot, a blast, a fall, before a last truth about yourself could be reconciled. She had prayed her mother would die. She had left the shark steak to rot in Poppa's refrigerator.

She sat there, fixed on the furniture of the living room, the open door of the balcony, the Judean hills where the sunrise was shafting in purple and rose. She glanced around her. Who owned this place? Who paid for it? Was it set aside for the mistresses of generals? Was the beach house also a safe house? Was she calling this *home*? But Asher had a home in Beit Hakerem with artifacts on every surface as proof.

A weight of penance and foreboding was settling on her. Even with Batia's herbs she had not been able to sleep the night before and she had waited for the alarm at five AM, so she could phone Poppa in Chicago.

Pushkin's growl of a voice answered.

"You already won the war?" he said. "You're coming home with Noah?"

"Poppa? Are you all right?"

"Ask Dora," he said. "She's the boss of the bulletins."

"Is it hot? Are you sitting in the park?"

"You called me up for a weather report?"

"Poppa! I was feeling guilty about you, but you just won't let me!"

"Don't waste *shmertz* on me. Spend it better on your son. You got him out or not?"

"Not yet, Poppa. I'm taking a trip to see him."

"A trip to a war? That's where you'll see him?"

She was verging on tears. How close to death would she have to come before Poppa would respond. "I called you, Poppa," she cried. "I'm miserable and all you can do is bite."

"Where are you going?" he said.

"Beirut."

She heard a gasp and a wheeze. "You already made up your mind. You didn't ask my permission, so, what shall I tell you? Don't drink water there."

"Poppa! You're the one who pushed me to come here! You said I should rescue. And now. . ."

244

"I pushed you to go. To rescue Noah from what he could become."

"There are two sides to this story, Poppa!"

"Two sides. Sure! Two sides of victims, both of them killers. You hollered at me, remember, when would I stop hating history? When there are no more victims to become killers, that's when I'll stop."

"What if he won't listen! What if he finds another way?"

There was a silence, and finally a sigh. "What if . . . What if. You want to be Noah's victim? If you sit and blow your nose and don't know what you think or what you believe, that's what you'll be. Later, when you come to your senses, you'll take revenge on him."

"Poppa, I know what I believe now. You weren't wrong about animals."

He grunted. "You'll go in a space ship?"

"It will be perfectly safe." She heard him groan. Then he was coughing, catching his breath. "Perfectly safe," he wheezed into the phone.

"Poppa! You're smoking a cigarette!"

"You come back. You bring Noah back from there in one piece!" he hollered.

"Poppa, I don't want you to worry. I don't know why I told you. You have to take care of yourself."

"*Why* you told me?" he said, his voice steady again. "Why not? You was my victim, no? So you take your revenge."

"Poppa! That's rotten!"

He was laughing in a hoarse grunt. "Don't work yourself up. What will it do me to worry? I'm the Executive Director? And tell Noah if he wants to save the world he should come home, I'll make him an alderman." There was silence. Poppa was wheezing again, trying to talk but his voice was breaking.

"Freya, do you hear me? You and Noah shouldn't dare to become victims over there. You'll make me a killer!"

She lay on the sofa hardly breathing, drifting into a kind of hollowness, of loss, that she would not talk to Poppa again, that Poppa would die believing she wanted to punish him. Or she would die and Poppa would still believe it. A strange sensation to talk, like hunger, came over her. She bolted up from the sofa to the phone again. Trembling, she dialed Uri's number at Gan Zvi.

After ten minutes, Uri came to the phone. "Freya!" he said, alarmed. "It's six o'clock in the morning! What's wrong? How is your leg?"

"I had to hear a real human voice. Both my legs are stuck in cement. I can't lift my feet."

"What happened?"

"I'm afraid to watch the TV. I'm afraid to read the paper. Asher is arranging for me to see Noah. In Beirut."

There was silence and finally, she heard Uri murmur, "*Beirut?*" He cleared his throat. "Asher would never let you go to Beirut," he said.

"I made him understand how much it means to me. Noah wrote that he was swimming and playing guitar!"

There was silence again. "What can I say?" Uri said. "Journalists and medical volunteers go to Beirut without half the protection Asher will give you. Why do you want to go there? Noah must come here."

"If I don't see Noah there, I'll never convince him."

When Uri answered so resolutely, "We *must* convince him," she didn't answer. He seemed to have changed his tone when he said, *we*, as though they were on a team bidding for the same prize. Uri added quickly as though he realized. "When do you go to Beirut?"

"David Barr is supposed to phone me."

"Before you go to Beirut, I have some advice, I want you to listen!" Uri said. "You had better put away your conscience and put on your armor. When the water looks deep, it's shallow. When it looks shallow, it's deep. Don't believe anything or anyone. A Christian could be an enemy; a Moslem could be a friend. You won't know what kind of a world you're in; you won't know how sly and vicious, how false. Do you hear me, Freya? The ice is very thin so leave your skates at home."

That night as she lay awake, while she rehearsed Uri's advice, a line she had once memorized from one of George Eliot's novels surfaced suddenly: *Better the foe you know than the Brutus you don't.* Uri knew how to side step Brutus, how to survive the deep and the shallow. He knew what kind of a world he was in.

Late the next morning, Freya heard Layla's voice answer the buzzer to the apartment. "Come up!" she called into the microphone, opening the door, waiting until she saw Layla's face appear over the last few steps. "I'm so glad you're here!" Freya said. "I need company."

"How is your leg?" Layla asked, before she was up the stairs. When Freya smiled, "I'm fine," Layla winked and mimicked; "Mitle Pass Street. Ammunition Hill. I had to climb over battlefields to find you." Her eyes were drifting around the apartment.

"I'll make coffee," Freya said.

The door to the balcony was open and Layla remarked that Freya shouldn't let the heat in. She settled herself at the kitchen table. "On the phone, you sounded so nervous. I hope nothing's wrong."

Freya stirred her coffee. "Nervous? I pace. I chew the ends of my hair. Since Zafra was murdered. . ."

Layla looked at her obliquely. "Are you going to blame all of us?" she said.

Freya looked up from her coffee. "Why would I do that?"

"If you're honest, you can't help it."

"If I'm honest," Freya said. "I couldn't blame all of you."

"That's what you'd like to believe. I hear myself saying, the Jews do this, the Jews do that . . . and now I hear myself saying, the *Jews* are destroying my life because they won't let Toufic live in Nazareth."

"Do you blame all of us?" Freya asked.

Layla took a tissue out of her bag and wiped her forehead. "It's not only Toufic," she said. "It's Sani. It's always Sani."

Freya cut her off; she got up and went to the cabinet and took out a bar of Cadbury's chocolate. "For chocolate freaks," she said, determined to deflect confidences, secrets, or controversy. She had phoned Layla because she needed someone to pass the hour by hour and moment by moment wait until she went to Beirut.

Layla looked downcast while Freya unwrapped the chocolate

bar and broke off two large pieces. "I guess you don't want to hear anymore about Sani," she said.

"It's not that. . ." Layla waited but Freya did not finish what she had begun to say. They drank coffee and ate pastry and chewed chocolate in uneasy chatter, in trivia: the pastry, the weather, the choking heat of the Hamsin, prices going up, the deafening din on the Egged bus.

Freya finally came out with it: "You said the authorities know about Sani and that's why they watch you, then why didn't you tell me before you invited me to Nazareth?"

Layla flushed; she sat there staring out the window before she spoke: "Don't you see what this conflict between Arabs and Jews has done to us? We examine every Jew we meet to see if we could use this one or that one, if this one will give us protection, or that one will help us survive. And then some hateful terrorist shows up and we are all punished. Can you understand how humiliated we are?"

Freya was silent, asking herself why conscience shouldn't work both ways, for the side you were on and the side you weren't.

Freya, her voice under control, leaned her elbow on the table and looked directly at Layla: "Suhail told you about me and my friend. You both decided to use me?"

Layla had begun to falter. "Freya! No, oh *no*. It wasn't like that. Suhail only mentioned . . . it was me, my fault. It was my idea that you could help us with Sani. I hoped, I wanted it so bad, I thought I would die. My mother is fasting. You're our only hope, Freya."

Freya didn't answer. They sat there in silence, looking at each other, glancing away, turning back without words. "Just what did you think I could do?" Freya asked.

Layla was studying the inside of her coffee cup; when she met Freya's eyes, her voice was slow, deliberate, as though it were time to stop rehearsing. "Sani went to Paris," she said, "She joined *Medecin sans Frontiers* so she could work for the Red Crescent Society on a French passport.

"The hospitals are bombed out. She may be dead and my mother swears she won't eat until she knows if Sani is alive."

248

Layla hid her face in her hands, then she looked up. "I thought maybe your friend, the general, could find out whether she was dead or alive . . . it was a crazy idea."

Freya sat there, trying to make sense of it. "Why would he help?" she asked Layla. "Sani is working with the PLO, the enemy."

Layla's eyes streamed with tears. "Sani doesn't *think*, enemy! She's a nurse and she had to follow her conscience. She had to help those wretched people in the camps! You've seen the television!" Layla cried.

"My friend wouldn't care what she thinks. It's what she does."

"I thought he would do it for you!"

Freya was silent until it slowly fell into place. Layla had hoped to embroil her so that Sani would become her own cause celebre and Asher would use his influence. She felt her resentment rise to anger; her voice was brusque: "Why don't you phone the French agency there?"

Layla laughed bitterly. "Can you phone a tombstone?"

"But from Beirut itself?"

Layla's eyes met hers. Freya took another bite of chocolate and said nothing more.

Layla looked up at her. "Freya, forgive me. You have a son there. I'm selfish."

"My son is all right," Freya said. Then she was silent again; it had occurred to her that if she went to Beirut, she could find out about Sani's organization, maybe Sani herself, at least make some inquiries. She was ready to speak but she stopped herself. Asher was not sending her to Beirut to chase down a PLO nurse.

"What does that do to Sani's status in Israel?" Freya asked.

Layla shook her head. "She'll be an exile forever. She won't be allowed back, not even for a visit. Toufic will never be allowed to live here. But Freya, I swear, we only want to know if Sani is alive!"

Freya thought for a moment. "I met a journalist here who goes to Beirut," she said. "Carole MacKinley, from Canada. I could ask her."

"Freya, you don't *understand*. The whole place is a wasp's

nest. Sani is nursing with the Red Crescent Society, the PLO! My God! How could we trust a journalist who writes stories?"

Freya got up and filled their cups with coffee. Layla stood at the balcony door, her back turned. "This Israeli-Arab curse destroys everything, it will destroy our friendship," she murmured.

"It doesn't have to," Freya answered, knowing that her voice lacked conviction. Chasing around the airport together, buying scarves, buying chocolate. Anywhere else she and Layla would have become best friends. But here . . . here . . . especially that she was connected to Asher, friendship wasn't possible. Not only with Layla but with any Arab. They would all want to use her. No, they would *need* to use her. In the position they were in, how could they help it?

Layla turned to Freya again, her face the color of clay. "We're not equals!" she cried. "I'm dying of shame at how much I want to use your *protexia*."

"Layla, don't. Please. I believe you. I've had a taste of humiliation. I use people, too. I wish we weren't desperate. I wish we were luckier."

Layla looked at her watch. "I should go home," she said. "I came here to cheer you up and look what I've done to both of us."

BEIRUT

CHAPTER FIFTEEN

It was mid-August and David Barr had not phoned. The bombings of Beirut had escalated; the invasion was apparently completed.

The Hamsin wind sucked up air. Batia had ordered Freya to take cold showers and lie down wet on the terrazzo floor. The terrazzo was cool and she lay on her back, skimming one book, then another, only to shove them aside.

On a suffocating afternoon just before the sun set, a feeling of dread began to gnaw at her. The war was winding down but Israeli casualty reports were mounting.

She had settled on the floor, a small fan blowing down on her, a pillow under her head and an anthology of poetry beside her. She began reading *The Ancient Mariner*, a poem she loved as a girl when, suddenly, she fixed on a verse and could not pull free: *Like one, that on a lonesome road / Doth walk in fear and dread, / And having once turned round walks on, / And no more moves his head; / Because he knows, a frightful fiend / Doth close behind him tread.*

She snapped the book shut and put it aside while she lay looking up at the moulding around the ceiling; she was gripped by an image of Zafra, chic and swish, swinging her hips toward the toilet at Yossi's.

A decent meal might revive her; her body was hungry from the erratic way she was eating, snacking on nuts and chocolate bars or on a cucumber she would peel and sprinkle with salt.

She skimmed the Jerusalem Post. Casualties. Casualties. An Israeli general had been killed two days ago.

Karl had phoned twice, once to check on new developments on Noah, the next time to tell her that Blitzie had remarked that Freya and Noah sounded like Tweedledee and Tweedledum, totally irresponsible. Karl, Blitzie said, should put his foot down.

"Where, exactly, do you intend to put your foot?" Freya asked him.

But Karl had withdrawn into his posture of victim. His wife and son were disappointments. Blitzie was healing the wounds. For Freya, the dividend was that Karl had put the decision about Noah in her hands. If the outcome was bad, she would pay the heavy tax.

Every day she had thought of phoning Evelyn but Evelyn's last letter only fired her *angst.*

At least your house is thriving with TLC. You'd be amazed. Karl had the driveway paved; last time I went past, someone was working on the roof. As for Pushkin, you ought to come back to see the transformation since Dora took over. You keep telling me you can get your kid out of the army on a week's notice. So? What in the hell are you hanging around there for?

She imagined Karl spiffing up the house, coddling the investment. For Karl, Noah was a drop out, a butterfly chaser, and she was a property that failed. Her resources would dwindle; if they didn't sell the house, she'd be stuck. Or could she find a job here? Flashes of long arms were pulling her back to Beaulac.

The phone was ringing. She cleared her throat when David Barr's voice came over the wire.

"Your health is all right?" he asked.

She felt herself stammer. "I'm alright. I'm fine."

"We go on Monday, two days from now, August sixteenth. Six o'clock in the morning. Okay?"

"Monday," she repeated. "Six o'clock in the morning."

"You should take an overnight bag, a few changes. For East Beirut I would advise everything chic."

Freya hesitated. *"Chic?"*

"Chic, chic," David repeated. "No war in East Beirut. Only chic."

"Okay," she said, "In East Beirut, chic."

"You making fun?" Barr said.

"I'm trying to cooperate!"

"I can imagine," Barr said. "Six o'clock in the morning on Monday, the sixteenth."

"I'll be waiting out front."

"That's not necessary. I'll come up. I have keys."

She was about to protest but she said nothing. "I'm looking forward. . ." she began.

She heard his snort. "Looking *forward?* You crazy?"

They hung up.

She stood by the phone digesting David's instructions. A few changes. Everything *chic.* Halfway to the sofa, she stopped, frozen. Not more than sixty hours from now she would see Noah. What could happen to Noah in sixty hours? Anything could happen.

She turned off the light; a shaft of moonlight came through the glass doors of the balcony and she sat down on the floor in the center of the living room and stared out to the Judean hills. How would it be? That moment of meeting? Holding him close. The flood inside her was already beginning.

She got up and turned on the light. It was half past ten. Newspapers and books were everywhere and she began tidying up. Evelyn said that if she died she wouldn't want anyone to see the mess she left in the living room. She swept up a pile of Jerusalem Posts and stared down at the one on top, skimming. *Thursday, 12 August, 1982. Ten hour bombardment of Sabra-Chatilla and Bourj al-Branjneh camps . . . British express concern at level of civilian deaths . . . Arab world unresponsive.*

She lifted out another paper and skimmed: *Monday, 9 August, 1982. Seven hour air and artillery bombardment of West Beirut refugee camps after Israelis claim Palestinians have been firing from the mountains on Israeli armor in East Beirut. Paris: Six die in attack on Jewish restaurant.*

She set the papers down and looked out the glass doors to the balcony. A squadron of jets flashed over the building.

The white dress? Chic? She had bought too many clothes on shopping sprees with Zafra just so they would have something Zafra loved to do. She would have to cash more traveler's checks. She had spent entirely too much on a solid gold brooch for Batia, a bracelet for Tami, and the black lace nightgown for Zafra. Everything here was three times the price than at home. If she stayed in Israel, she would need to transfer money from her account in Beaulac. Perhaps she could contact Natan from the Israel Museum for a job.

She took the traveler's checks out of her suitcase and lifted out an envelope she kept there, the five-hundred-dollar check Asher insisted she take, no matter how much she argued. Only a loan, he insisted. Asher would not tolerate her scrimping. He had bought her two ivory combs for her hair and a slender gold bracelet engraved with both their names.

She took his check out of the envelope and tore it up and then she replaced the scraps and put the envelope back in the pocket of the suitcase. Her mother's beads were safe in her wallet and she lifted them out and put each one up to the light to catch the prism. She would take the cobalt blue beads and the combs with her.

The white dress, a skirt, a couple of cool shirts would do; she laid the clothes out on the wide bed and stared down at them lying inert against the smooth covers, touching the pillows that lay side by side. Had she lain there with Asher?

The unhooked phone was clicking, making a sound like going out of order. When she went back to replace the ear-piece, she looked at her watch. Karl might phone again and she would turn shrill and biting, less than what she wanted to be, less than what she was.

It was two weeks since she and Layla sat in the kitchen, eating chocolate. She had been cold to Layla's pleading, to her crying out that they were not equals. It was an irony that Noah felt powerless in Beaulac because he was a Jew, while here, it was she, the Jew who bestowed favors or withheld them . . . the advantage of living in a Jewish state. Uri's lecture came back to her as she dialed Layla's number.

256

"Did I get you out of bed?"

"It's early!" Layla said.

"I wanted to wish you the best, Layla . . . I'm so sorry about not being able to help you."

There was silence, then Freya heard a muffled moan.

"Layla, I wish I could. I don't have that much influence."

Layla's voice was quavering. "I didn't expect you to do what you can't. It's only . . . you're our last hope!"

"Oh, Layla!"

"No. It's *I* who should be sorry. I had no right to expect . . . It's only our desperate situation. Doors get shut in our faces. It's not your fault," Layla said. "I had no right." Layla was choking on words and then Freya heard a click of disconnection.

She stood there, her hand resting on top of the phone. Asher would *not* understand. And who would she ask in Beirut about an Israeli Arab nurse working for the enemy? It would be like asking the crocodile what he wanted for dinner.

She left the phone and stretched out on the sofa, her hand dangling to the floor, her body soaked in sweat. She sat up and pulled her dress off and lay back. By now, thank God, Karl wouldn't phone. Maybe something was wrong with Hammer. She lay there in a kind of dead stillness. Whatever was wrong with the world, she was part of the wrong. She was Layla's last hope. How deftly she slipped off high moral ground when she was put to the test.

On Sunday morning, Freya sat in the kitchen drinking tea while she calculated how much money she should cash. Something substantial for Noah in cash, something for Noah in gifts, something for emergencies, anything could happen . . . what if and what if and what if. She cashed five hundred dollars in traveler's checks at the bank and then she took the Number Four bus to Ben Yehuda Street to go shopping for Noah.

At the department store, *Hamishbar,* the security guard who inspected people before they entered the store, opened her bag, glanced in, then nodded her through. She felt a catch in her throat and looked back at the guard. "Be careful!" she said. What

257

if someone slipped through with a bomb? She could be standing at a counter, smiling, never suspecting.

At a quarter to six on Monday morning, Freya went downstairs to wait for David Barr on the steps of the apartment building. She did not want him using his key. She had an overnight case and a huge package for Noah.

She sat there munching an apple and then she poured a cup of coffee into the cap of a thermos. Why had she supplied Noah with all sorts of gifts and delicacies when soon he would be out of the army? The street was dead quiet; the sun had begun its glare, but she could see a bright light shining out from a window across the street. Who lived there? The woman with the baby? The poor thing didn't look healthy the last time she saw it, so scrawny and the little face was sallow.

In the image of the baby's pallor, she began wrestling with images and premonitions. But Lebanon, after all, was the Switzerland of the East, the jewel of the Mediterranean! There was nothing to be afraid of. Everything was normal. According to Tweedledee . . . *the sea was wet as wet could be / The sands were dry as dry. / You could not see a cloud, because / No cloud was in the sky: / No birds were flying overhead / There were no birds to fly.* She replaced the cap of the thermos.

A car was pulling up, a dark blue Mercedes. David Barr, dressed in sweat clothes, got out without shutting the door and came toward her.

He glanced at her flowered skirt and he looked down at her lightly bandaged thigh as though to assess damage. "Your health is good? Why you sitting here?" he said.

"Why shouldn't I sit here? It's supposed to be where I live."

"Your health is good?" he asked again.

"Perfect," she said. "And you?"

"We'll see."

There was another man in the front seat. David opened the back door of the car and when she got in, he introduced Beni. "Our insurance policy," David said. "Okay; okay." He started the car. "Finish your sleep; it's almost three hours to the border."

258

Beni, also in sweat clothes, turned around to smile at her as though he had never seen a general's mistress. In the lamplight, he seemed twice David's size; his face was round and dark skinned like Moshe's from the Mahane Yehuda.

David called back to her. "It's like you won a vacation! The trip is free! At the hotel, you charge everything to the General."

"I do *not* charge everything to the General!" she said.

"Okay, okay. Take a sleep. Why you all the time so angry?"

Beni smiled at her and turned around and David started down Eshkol Boulevard. She shrank into the seat and closed her eyes. She was exhausted from two nights of moving from the bed to the floor to the sofa and back again. Beni turned on the radio; an Arab beat was thumping.

When she woke up, the Arab music was still thumping and the sun was beating against the windows of the car. She looked at her watch; she had slept for almost three hours. She sat up and looked out her window to a road winding through gorges and bouldered hills. White cliffs fell to the sea; far below on the beach a runner was kicking up sand, a blue striped umbrella leaned at the water's edge.

"It's good you slept," David said.

"*Tov Meod*," Beni called back. "Very good."

"We go to the Rosh HaNikra Hotel," David said. "Eat breakfast. All you want."

In the powder room of the hotel she refreshed herself; she washed her face and rebraided her hair. A look in the mirror flashed the passing thought that Rosa Luxemburg must have turned her mind off before she got shot.

They had fruit, croissants, coffee for breakfast. Beni was silent; he chain smoked and ogled her with every snap of his lighter; between cigarettes he managed to drink four cups of chicory laced coffee, black.

"You slept through our beautiful country!" David said. He looked at his watch.

"We're at the border?" she asked.

"A few kilometers. Don't get excited and don't worry."

259

"I'm not excited."

"Bored is okay. But don't get excited; don't worry," he said. "South Lebanon is clean."

"Clean?" she repeated. "Cleaner than Israel?"

"No; I mean clean. *Clean*," he said, getting up. He was fair-haired and pleasant; he moved quickly, precisely; he couldn't have been more than thirty-five and even in his sweats she could see he had a sprinter's lean, strong, body. He left bills and change on the table before they walked out and Freya followed behind Beni, his gait wide and heavy-footed.

The drive from the hotel to the border was dazzling; below the cliffs a pounding surf rose in huge bursts of spray. Rosh HaNikra. She had read it all when she was a girl studying Latin; the rest she read in the volumes piled on the floor. Rosh HaNikra, *The Ladder of Tyre*. The ancient Kingdom of Phoenicia. And Dido of Tyre, Queen of Carthage; poor willowy Dido strong as a cedar . . . it was her love for Aeneas that killed her.

They were driving toward a tall wire fence with several huts clustered together, a larger hut was flying the Israeli flag. Suddenly she did not want to see Lebanon; she did not want to see Tyre. Breathing its air would not be the same as seeing its ruin on TV. She rolled her window down to breathe the last air of Israel. Clean. Clean. The waves below the white cliffs were beating against the rocks.

David slowed the car and pulled up to the larger hut flying the Israeli flag; Israeli soldiers with Uzi guns were on guard. Two UN guards in blue helmets stood a few feet away. David told her to wait inside the car. He opened the glove compartment and took out a blue license plate, a bolt remover, a screw driver. Then he and Beni got out and went into the hut. There were raucous exchanges in Hebrew. David was presenting papers with her passport and a soldier wearing a red beret stamped them. When David and Beni disappeared, the soldier sidled out to the car to take a look.

"*Shalom*," he said, smiling. The Arab music was still thumping on the radio. "Nice music, yes?" She smiled and looked away. The boy was young, not older than Noah.

When David and Beni reappeared, she stared, astonished. They were decked out in charcoal pin-striped suits with vests, striped ties, shoes shined to a high polish. They had identical brief cases. Where did they get the change of clothes, the brief cases? They must have been stored inside the hut!

David swung the brief cases into the front seat and Beni began unscrewing their Israeli license plates and replacing them with the blue Lebanese plates. She sat there riveted to the action. When they got back in the car, David, scrubbed like a rep from middle management, turned to tell her: "We're Lebanese salesmen; you're our secretary born in America. Okay?" He turned off the Arab music and plugged in another wire.

She smiled. "What are we supposed to be selling?"

"Diamonds. Why not?"

She shook her head. "I don't know a carat from a turnip. I'll blow your cover."

"Don't get excited. No one will believe you're the secretary anyway, so forget about it."

There was a UN checkpoint at the Lebanese border: they slowed down, opened the windows and Beni puffed fast on a cigarette. A young soldier came out, looked inside the car, inspected her passport, handled papers. With the windows down, Beni kept puffing his cigarette while he exchanged something hilarious in pigeon Swedish ". . . Ya . . . Ya . . . Ya. . ." The UN soldier took off his blue helmet and scratched behind his ear. He waved them on and David remarked that the UN guards spent their time looking for bargains in stolen goods and buying hashish in Sidon.

The two lane coast road narrowed; potholes jostled the car while David serpentined around them.

"*Katushas,*" David called back to her. "The ones that missed us."

"There were plenty that didn't," Beni said.

In the distance she could see silhouettes of mountains but where they drove nothing existed but stands of naked trees deepening into what must have been acres of orchards. They were charred, burnt out. Weed stubbled sand dunes lay at the other side of the road and a fortress of white-veined rocks bordered

261

them: below the dunes the beach was deserted but farther out two sleek boats dotted the water, gunboats or maybe torpedo boats. She had seen them on TV.

The sun blanched road was ghostly; if people once lived on this highway to Tyre, their villages were hiding somewhere or dug underground, or demolished. A few miles north, several yards beyond the west side of the road, two adobe huts appeared flying the UN flag; later, a tight cluster of tents and a ring of open army jeeps stood outside; a bearded soldier, bare from the waist up, sat on the hood of a jeep drinking from a can of Coca-Cola.

Closer to Tyre, David switched on a transmitter she hadn't noticed before; he was checking his watch and speaking Hebrew.

They drove over a deserted road where stands of dead splintered trees yawned in front of her. The car was kicking up dust and pebbles. Out of nowhere, suddenly, from over the dunes, a group of children were running, barefoot children clad only in underpants. The children were at war, sweeping the air with sticks that flew white rags, fencing, pretending to stab one another. David slowed to a stop. He opened his window and yelled in Arabic. A fuzzy haired boy took a fall on the road and lay there with his legs stiff in the air, waving his white flag madly while another boy poised his stick and went for his throat. David blasted the horn. One of the boys turned, pulled down his underpants and exposed his bare ass.

David checked his watch. "One day they'll get killed on the road and who will be blamed?" he said.

"Who do you think?" Beni answered.

They were descending to the valley, to a pocked narrow highway lined with charcoal stumps, a forest of charcoal moving past as though her own dead tree had come back to haunt the road. It was a town; it was Tyre. It was dead. Carcasses of buildings with outer walls blasted away stretched in front of her. Huge piles of blackened shrubbery were piled high at the side of the road with the rubble of fallen buildings. On top of a pile, hanging on, suspended from the slope of the debris as though still connected to some nerve, were the remains of a baby's crib. The slats of the crib were broken, the sides, gaping; the baby might have rolled away. She closed her eyes.

There was static from David's radio.

Tyre went by in spoils of buildings that leaned like lopsided jigsaws; shattered glass glistened in the street and gray stone foundations were exposed with mud-like dung, smeared, glazed with smoke. A church with a tall cross on the apex of its roof stood intact. David didn't have to tell her; the Christian Phalange had breathed dragon-fire revenge for the PLO blitz on south Lebanon. Israeli gunboats were in the harbor. She shrunk back in her seat. Clean. Clean. Death was closing in.

A line of Israeli army jeeps was moving several yards in front of them. A Peugeot with Lebanese blue plates had pulled to the side of the road, the driver inside staring straight ahead, rigid, impassive.

"Why is he sitting there?" she asked.

"Civilian plates. He doesn't stop for a convoy, they shoot."

She didn't answer.

"You like better a bomb?" Beni said.

But *they* had Lebanese plates too. David passed the convoy; she unclenched her hands and sat back. She had figured it out. Their Lebanese plates were in code. Army intelligence knew their every movement. They would not get shot. When she rolled down her window the air outside reeked. Burnt gardenia mixed with noxious pesticide, formaldehyde, like taxidermist fluid.

They were in Sidon. She left her window open. To turn away was to deny what she had begged Asher to let her see. Slowly, David rolled in. Her dead tree with its shroud of blackened needles followed like lava flow. Soldiers in Lebanese uniform were sitting in a makeshift cafe outside a bombed-out building, drinking from Coke bottles; a cassette player was belting out Edith Piaf. There was laughter. Uri was telling her: *Nothing ever ends; something goes on.*

She stared, paralyzed. Her hand flew to her mouth. At the curb, a clothes basket stood on a table and a man was lifting out wares to sell: one by one, huge, huge, spotted snakes coiled out of the basket and circled the man's body while he hawked them in sing-song. Two toughs wearing T-shirts, shorts, and baseball caps tossed coins into his box and grabbed at the snakes, clutching them by their tails, hands clamped over their heads, stretching them out like taffy. They were lunging at each other with the heads of the writhing reptiles; she could hear them, hollering commands something like Sesse. . . . Sesse . . . ; they were screaming with laughter while the snakes twisted and arched. Sesse . . . Sesse.

Undigested croissants from breakfast were collecting in her throat; she forced them down, swallowing, swallowing, while she fought to fix her attention at a battered sign: *Cafe le Poisson. Cafe . . . Poisson . . . Cafe.*

Sandbags were everywhere; waterfalls of destruction encircled the street; yet women in black were crowding at vendor stands filling huge jugs of water. People were gesticulating. Old men were smoking nargillas. A carpenter was pounding a cabinet out in the open, pounding, pounding, while the boys ran at them with the writhing snakes. A retch came out of her mouth.

The main street had petered out to gravel and David turned to look at her; she had her hand clasped to her mouth. He pulled over.

"You want to go somewhere?" he said, looking at his watch.

She nodded.

"We have a place here," he said. "Don't be ashamed. When you must stop, we stop."

He drove on for a few yards and pulled in to what he said was General Hadad's Militia checkpoint; he presented her passport and papers to Lebanese soldiers. Holding her mouth, she stumbled in to the one-story brick structure where a soldier pointed to a toilet at the back. When she came out again, David handed her a water canteen.

"I don't blame you," he said. "Take the water. Keep it. Listen; after Sidon there is Damur. Then we take another road. In the hills. Not such a nice one. No snakes; only Moslem heroes. They enjoy to snipe. You're okay?"

"I'm okay," she whispered.

264

Beni laughed. "She's good! She's *okay*," he said, handing her a flask. "Take a drink. Maybe two drinks. *Shnapps*. Then you sleep through Damur and the hills. Nothing nice to see in Damur. Don't look. In the hills, Moslem villages. They don't love us."

She took the flask and drank one swallow, then another, and another, coughing between breaths, the liquid sliding down like fire. She took another swallow and caught her breath before she handed the flask back to Beni.

"Better?" Beni asked, chuckling.

She didn't answer.

David turned. "Don't be afraid. Our people know where we are. Nothing will happen."

David, then Beni, pulled two pistols out of nowhere and they started up from the valley on a steep and winding gravel road. The mountains were rising.

"Don't open windows," David called back to her. He looked at his watch and made radio contact again, talking, chuckling to the other ear of his connection.

The shnapps was tingling, floating, through her veins. Snakes were coiling. *If seven maids with seven mops / Swept for half a year / Do you suppose, the Walrus said, / That they could get it clear?* Violet was giving her an antique tapestry and the women in Kfar Salah were strolling in the village wearing embroidered dresses over their great pillow bosoms; young girls with round, black eyes wore pantaloons and tunics colored with the purple dye of Lebanon. Layla's mother was fasting unto death, and she, their last hope, had closed the door in their faces.

The sea had disappeared and the forests grew on the winding road. The shnapps was heavenly lightness. She was a tourist in a beautiful place walking a cool, shaded trail of cedar trees and columbine. Queens with tiaras were feeding on Turkish Delights.

She would have to learn to speak Moslem. Was Moslem a language? She would learn to speak Moslem and study their history wherever it would lead. Where would it lead?

A helicopter was circling low, making one, then two, passes over their car. Star of David markings were clear; she leaned back. The helicopter turned and flew ahead. Asher had organized every moment for safety. She was a princess in a sedan chair being carried to the King. She had coolies. Layla had a claim on her as though she had been called to the bedside of a dying friend. Layla and Sani and Layla's mother, fasting unto death, deserved her royal protection.

Beyond the road squat houses sat on a slope. The helicopter circled; Beni gripped his pistol and something clicked. His head was moving in a dizzying arc. The Queen and the Duchess were playing croquet.

It might have been twenty minutes or maybe it was an hour or maybe a day, when she awoke and they were descending the slopes, bumping down, round and around, down and down over hairpin curves. She was awake. How long had she slept? How much shnapps had she swallowed? David was on the radio; the helicopter was rising. She lifted her braid and wiped the back of her neck with a tissue.

"You slept!" David called back to her. "No problem."

Beni snorted. "Be glad I have twelve eyes."

The shnapps was plugged into the back of her head. One thud. Then another.

The helicopter disappeared and the road smoothed out across a plain where olive terraces lay shelved on the slope. Women were on the road carrying baskets while children jumped around them; one little girl caught Freya staring through the window and she held up a huge crucifix on a chain around her neck. Freya waved and the little girl fanned the crucifix again. The Moslem villages were far behind them but the little girl wasn't taking any chances. Up ahead was an Israeli checkpoint. Freya reached into her purse and took out three aspirins; she swigged them down with water from David's canteen while she stared out the window. It struck her that she had not seen a live bird since they left Rosh HaNikra.

Less than a mile, suddenly, suddenly, the burnt-out graveyards vanished; green, green, cedars and pine, aspens, palms—life giving green. Where had she read that aspirin and whiskey induced hallucinations? The road was solidly road; pockmarks and craters were gone and snakes lay hidden in the wild.

They were approaching a golf course; men and women were carrying golf clubs. On another street, a boulevard, where traffic was heavier, Lebanese automobiles and army trucks, jeeps and taxis, were honking. They stopped at a Lebanese checkpoint and then David turned onto another palm-lined boulevard. *Where were they?* The aspirin and whiskey had sizzled her brain. Apartment houses faced with marble gleamed in the sun. On the bottom floors of the buildings were entrances and signs: *Boutique Charmaine. Les Bijoux des Florence.* Women were in and out of *Les Bijoux.* A long Mercedes pulled up to the curb and an exotic model stepped out of a magazine, sleek in white kid jeans. They were in Beverly Hills! On Rodeo Drive!

"Nice, yes?" David said.

"Chic," she murmured.

Farther down the avenue, standing amidst giant palms, was a graceful three-story building, all marble and glass with patios on every floor and a driveway curving behind a glistening, granite staircase. *Beirut Apollo* carved into a marble marquee stood luring her like a rhinestone-studded coffin.

David opened her door and she got out of the car. She staggered and he reached for her arm. Their eyes met.

"I'm sorry," she said. "I'm drunk from your shnapps."

"It's okay; it's okay. You're a strong woman; you have courage. Listen, I understand that you weren't able to, how you say, snitch, on the cab driver. You don't snitch, do you? I like you. Some others wouldn't be able to face not a thing. I understand now how the General feels."

She was smiling. "Did we sell the diamonds?"

David laughed. "Inside, maybe. You could sell a cockroach if you said it came from Paris."

It was after two o'clock. The lobby was crowded and David told Beni to take a rest on the patio. They went to the reception desk and David said in English to the Lebanese clerk: "We're expected. General Frank."

"Everything is ready," the clerk said to David, eyeing her. "There is a message for you, Lieutenant Barr." He handed David an envelope and Freya turned to scan the huge lobby while he stood reading.

Opulence. Opulence. The place was jammed with pin-striped business suits. Spit and polish Lebanese army officers. Perfume hung in the air and women in spiked heels wore clothes to dream about. Crystal chandeliers hung overhead. Sofas and easy chairs were arranged in conversation corners. At one end of the lobby was a bar with people crowding around, their voices raised, laughing. Oriental rugs softened the din. No snake hawking. No smell of formaldehyde or gunpowder. No war in East Beirut. Only journalists in khaki jeans and vests with cameras slung over their shoulders.

David tapped her arm. "You must eat," David said. Then he smiled at her. "Black coffee, maybe?"

"I'll wait for Noah . . . something cold to drink. . ."

"About Noah," David said. "I had a message. Noah will be delayed." She felt herself pale. "For two days," David went on. "Who can tell? His commander can't let him go right now. You have a bathing suit? There's a pool here, how you say, Olympic size?"

"Something is wrong!" she cried.

"Nothing; nothing. His commander wouldn't let him go this minute. Nothing wrong. Have a good time. Make like a vacation in Florida."

"Something's wrong. Why won't you tell me?"

"*Nothing!* Look; you're dealing with army, not a beauty parlor appointment. Nothing's wrong, believe me. To tell the truth, I have to leave too. Unexpected. We have business in the north. I have to go."

She didn't believe him. She didn't know what to believe. "What should I do?" she cried.

"I told you. Have a vacation. Eat, swim, sit in the lobby, enjoy. Beni will take care." He winked. "Don't let him drink too much."

268

"When do I see Noah? *When?*"

"You're dealing with army so expect the unexpected. Two days at least. All right? Go rest now. Pick up the phone and room service will send you food, and coffee, cold, like you drink in New York. Take out a bathing suit. Anything you want you get here."

He handed her a key to a room on the second floor. "A nice patio," he said. "Only one thing, Freya . . . Mrs. Gould." His face colored. "Don't leave the hotel. Not one foot outside the door." Then he extended his hand. "I see you in two days. Maybe less, Beni will know."

"I see," she said to him. "This time, I'm supposed to be Bluebeard's wife?"

David's face was blank. Beni was coming toward them. They spoke in fast Hebrew and Beni nodded, first to David, then to her. "Okay," Beni said.

"*Beseder?*" David said to Freya, extending his hand. "Okay? Everything under control. Shalom. Shalom. I see you."

She looked at her key, then at Beni whose round, dark face was alight with authority. She frowned and Beni shrugged. "You go upstairs?" he said. She nodded. "Okay. Upstairs is all right. I go over there." He walked away from her to the far end of the lobby where people were crowding at the bar.

Room 206 was large and elegant with creamy Empire furniture. A gilt-framed mirror graced a vanity laden with crystal perfume bottles engraved with the crest, *BA*. A marble coffee table supported by gilt claws was in front of a blue plush sofa. There were two matching chairs. A blue oriental rug lay in the center of the tiled floor. The king size bed had a gilt bedside table with a vase of long stemmed roses sitting on top. She bent over to smell them and then she saw the card. *Freya, sheli, I love you.* She sat there inert, the card in her hand, and then she bent over and pressed her lips to a rose.

She showered. Barely dry, she lay down on the bed and breathed the gardenia scented pillow slips. Why was Noah delayed? Why and why and why? She was like the African parrot, Subango, in her office at the Museum. Lush colored Subango,

269

whose feathers puffed up and his throat swelled while he squawked and burped back at her: *What do you say, Subango, what do you say?* Subango always recognized her. Would Noah recognize her? It had been two years.

The dab of cologne she had touched to her wrist was not refreshing. Nothing refreshed her, not the cool shower or the lavender soap or the scented shampoo. The smell of formaldehyde had seeped into her skin; the image of huge, coiling, snakes made her gag. She lay there, her eyes closing and when she awoke, she bolted up, disoriented. It was after six o'clock. Her head thumping, she went to splash water on her face, catching herself in the mirror when she leaned against the basin. *What do you say, Subango? What do you say?*

How would Noah look? How? Like himself in a sweatshirt in Beaulac? The long arms and legs covering distance in the breast stroke, or the rollicking yell and laugh, romping with Hammer at the beach? *Go get it boy! Good dog!* But Noah was a boy in a jeep, dressed in olive drab and boots. A soldier. A Halloween costume and a mask he wanted to wear and she wouldn't let him. Not a soldier's costume! Soldiers *kill* people. They *get* killed. *No, No, No.* And Noah let her dress him in Merlin's wizard suit she made out of black and purple poplin, with a cardboard pointer for a magic index finger. Noah wouldn't wear the finger. He had the look of *I won't forgive you* when he trudged forward with Claire's two sons and came home again fifteen minutes later with a mini box of raisins and a red apple in his sack.

The phone was buzzing. Buzzing again. She reached across the bed to answer, struggling to locate the mouthpiece that rested in whiteness and gilt on a solid gilt pedestal, buzzing, before she finally yanked it off, turning first one end then the other.

"It's Beni," he was shouting. "You don't answer. Something is the matter with you?"

"There's nothing the matter with me," she muttered, and dropped the phone to its pedestal.

The white dress was not exactly chic in its present state. She gave it a shake and put it on before she sank down on the sofa. What now? Her eyes moved around the room. There was a mural over the bed. How had she missed it? Galleys in full sail sitting in an ancient harbor. Romans? Crusaders? Dragoons in doublets and hose and flowing capes were blowing trumpets on the beach while fleshy nude nymphs with huge breasts were at their feet, grinning seductively.

Shnapps and aspirin were swimming in her head and she sank down on the sofa where she had dropped a volume of Anna Akmatova's poems she brought with her . . . leafing through them, hardly attentive . . . *like a white stone in the depths of a well...* She leafed to another: *I learned how faces fall apart / How terror peeks from under eyelids / How sorrow cuts an alphabet / Of suffering in people's cheeks.* She snapped the book shut. *Why had Noah been delayed?* If she could describe the stench, the snakes, the baby's crib with the sides torn away . . . if she could name her feelings . . .

She dozed and awoke with a start. It was after seven o'clock. She had better get something to eat or she'd stay smashed for good. But to eat in the room would be to obsess on images of charcoal tombstones. And snakes. And a little girl flashing her crucifix to a vampire peering from a car window. To stay in the room was to imagine Noah delayed in a battle or Noah delayed in a hospital or Noah dead in a morgue. She bent over the roses; they were soft, fragrant. Would she see Asher again? *Freya, sheli, I love you.*

In her white dress and her long, dark hair held in place by the ivory combs, she was out in the hallway, in the elevator, and out into the buzzing lobby where she scanned for Beni. A sign with an arrow to the coffee shop was posted on a pillar. She ate a club sandwich and drank two cups of black coffee, forcing back the image of snakes with every bite and swallow. She signed the check and went for Perrier at the bar. A stocky young man with a crew cut and cameras around his neck was standing in the queue at the bar, looking at her.

"Allo, Allo. What have we here?" he said, smiling at her. She smiled back. "You speak Queen's tongue or Frog?" he asked. "Bonjour then. Want a drink?"

"That's why I'm waiting," she said.

271

"The lady is a Yank! Allow me to introduce myself. John Czerko from Saskatoon." He extended his hand.

"Freya Gould," she said.

"I'm about to ambush this crowd, Freya Gould. This is where the real war is. Hold on!" He took her hand and pulled until she found herself at the edge of the bar, John Czerko leering while he adjusted his elbow to the ledge of the bar. "What's your habit?" he said.

"Perrier, please. With ice."

"*Perrier?*"

"I have American money. Will they take it?"

John Czerko let out a snort. "Madame, mademoiselle, beauteous lady, they will *dismember* you for American bucks. But this is on me."

He ordered scotch and a Perrier and Freya felt a hand on her shoulder. She turned and looked into the face of Carole MacKinley, the journalist from Canada she met at Yossi's.

"I'll be damned," Carole said. "You actually made it!."

"I don't believe it's you!" Freya exclaimed. "Carole MacKinley! How wonderful, *wonderful.*"

"Ted Baxter took pity?"

"No; no. And *you.* . . You're *supposed* to be here."

"Right," Carole said. "Hate City."

They made it to a sofa at the other end of the room. "Okay," Carole said. "Tell me. How in the hell did you get here?"

Beni was coming toward them. "You're all right?"

"Yes; I'm all right." Beni turned and walked off.

Carole's eyes were wide. "Who's *that?*"

"It's too complicated."

Carole was eyeing her. "You're either somebody's spy or somebody's girlfriend. Which one?"

Freya was taken off guard. "Neither. I have a friend in the army."

"That was your friend in the army?"

"No; no. He's a sort of bodyguard to my friend."

"Lawd! You're top brass *protexia!* A Colonel?"

272

Freya broke off and got up. "You're on the wrong track," she said. "I think I'll go upstairs. Come up if you like."

When they were inside the room, Carole gave a low whistle; her eyes were on the roses. Freya sank down on the chair while Carole sat on the sofa, scanning the room. "Now why don't I have roses in *my* life?" Freya smiled and Carole eyed her, "You look like you've been through the trenches," she said. "Is your son all right?"

"I don't know. I was supposed to meet him here today but they say he's delayed. I don't *know* if he's all right. I don't know what to think."

"Listen. There's a lot going on they're not telling you. The PLO is about to pull out. The Israelis are bombing the hell out of the camps. There's plenty of business in the north."

"If Noah is in the north, he's in danger!"

"You're not listening. Doesn't your Colonel tell you anything?"

She searched Carole's face for reassurance. The war was ending. *Noah was taking pictures of seven maids with seven mops.*

"Look, Freya, for whatever it's worth, I'll be around in the evenings. A cease-fire is scheduled tomorrow. John and I, the joker you met at the bar, we go over the Green Line. But we'll be back. We'll play Gin Rummy."

Freya wasn't listening. The need for relief was stifling her as it had the first time she and Carole met at Yossi's. An indifferent ear. No consequences. "They had to let me out to vomit," she blurted out. "I saw little kids playing war . . . they weren't a human species anymore. There were snakes. The forests were burnt out. There's no future for anyone here."

"That's the idea," Carole said. "I forgot what it's like to be shocked."

And then Freya was unable to stop herself. She was telling about Layla and Sani, the vow Layla's mother took to fast: Layla had begged for her help to inquire, only to *inquire* about where Sani might be.

Carole sat there listening; she was staring at the roses. "You're in pretty deep, aren't you?"

"The roses?" Freya tried to appear casual. "He's an old school friend from Chicago. There's nothing to it but friendship. Before I got here, I didn't know he was a general."

Carole laughed. "I've been married three times and I told you I don't shock anymore. *What* did you say? A *general*? Friendship can't get better than that. Which one?"

Freya hesitated. "Asher Frank," she said.

Carole's eyes glistened. *"Frank?* That gorgeous hunk in Intelligence? You're kidding! I went to a press briefing to hear him and then I tried to get past three lieutenants for an interview. I tried for a week, no luck." Carole winked. "Most of me wanted a scoop. Some of me wanted a long interview in bed. Lawd, Freya! Do you know what it means to a journalist to have access to someone like Frank?"

The look on Carole's face was as though a sack of gold had landed at her feet. "Jesus," she said. "Introduce me."

Freya didn't answer. Carole laughed again. "Sorry," she said. "Forget it. Tell me more about Sani."

Carole's eyes were narrowed when Freya finished her story. *"Medecin sans Frontiers,"* she said, thoughtfully. "The holy streakers."

"How would I do it, Carole? Who would I ask about an Israeli Arab nurse working for the enemy?"

Carole was silent for a long moment, studying her fingernails. Then she looked up. "I'll help you," she said. "Why not? Listen; can you stand another shaking up?"

"What do you mean?"

"Tomorrow there's a cease-fire and I go over the Green Line for a PLO press briefing. It should be amusing; they're wonderful theatre. Medical people connected to *Medecin sans Frontiers* will be there and you can ask about Sani. You have your passport?"

Freya nodded.

"It won't be a hassle. Your passport and my old press card gets you through the check points. No one will know. It's a risk, but for all the sugary press I give the Israelis, they won't expel me. As for the others, if they checked, by the time it reached the right desk, the war would be history."

Their eyes met. "Why?" Freya asked. "Why would you take risks for me?"

"I'm like that," Carole said, smiling brilliantly. "When I like someone . . . I like you, Freya. Anyway, who knows, you might do something for me some day."

"I promised not to leave the hotel. If anything happened to me. . ." Freya was studying Carole's face, puzzled. "You seem so eager," she said. "Why?"

Carole shrugged. "If I thought you were at risk, I wouldn't ask you, believe me. I told you there was a cease-fire. Look, Freya, it's nothing to me if you go or you don't. I'm trying to help. But if you don't go, stop kicking your ass about your Arab friends. You can't have it both ways."

"Carole! *You* could ask about Sani!"

"No; I couldn't and even if I could, I wouldn't. It's your story, not mine."

Freya was silent. Carole got up. "I'll be at the bar. Let me know."

Freya ate a small bunch of grapes and drank the extra bottle of Perrier brought up from the bar. With every bite, every chew, every swallow, she argued Carole's proposal, remembering how she had thrown her arms around Asher's neck, kissing his eyes, promising, promising to follow David's explicit orders.

She got up and went out to the balcony. The mountains rose and fell in the heavy smoke of the night sky; the cease-fire had not yet begun and artillery in the distance thumped with regularity; every couple of minutes there was a whistle, a kind of screech beyond the mountains. *Where was Noah? Why wasn't he on schedule?*

There were flashes in the west, staccato bursts of flame. The summer heat was an inferno and smoke was moving across the face of the moon. Layla's mother was starving herself to death. And here she stood, watching the fiesta of dying, people burning like matchsticks while she bathed in gardenia and lay on scented sheets, stroking her lover. She hadn't asked for this. She was in the catbird seat. Something arcane, some twist in her history needed testing beyond her will to resist. She was obsessed with the idea that she was Layla's last hope. Asher would understand that only something that compelling could override her promise to him. She would let him judge her. As for herself, her conscience would have to carry the freight.

CHAPTER SIXTEEN

AT SIX O'CLOCK in the morning, Freya twisted her hair into one long braid and dressed in clothes Carole loaned her, blue jeans, a khaki shirt, and a journalist's vest. She pulled the blue jeans over her hips, relieved that they slipped easily over her lightly bandaged leg.

The lobby was crowded; the bar was in full swing. Beni was nowhere in sight. Carole greeted her with an extravagant hug, her blue eyes sparkling. "You look absolutely *Banana Republic*. Let's eat."

In the coffee shop, Carole said that she should have hot tea and soda crackers. They would eat more when they got to the Admiral Hotel in West Beirut.

"Soda crackers?"

Carole smiled. "There's gruesome stuff out there. I don't want you getting sick on me."

She gave Freya an outdated press card and told her to put it inside the cover of her passport where they couldn't read the name.

"The checkpoint guards won't bother you," Carole said. "Here's the way it goes. From here we take a taxi to the Museum on this side of the Green Line. We get out of the taxi, get through the checkpoint and walk across a field; not far. Don't worry. There will be a crowd in both directions. They make a hell of a racket, but pay no attention. The poor bastards are trying to get across to the east before the checkpoint is closed. There's a rented car waiting for us at the West Beirut checkpoint. You have plenty of American cash in twenties?"

Freya smiled. "Twenties, fifties. Enough for ransom."

"Okay, then. When we get to West Beirut flash your passport and my press card. Make sure the twenty is on the page with your passport picture. The son-of-a-bitch won't believe you'd notice how the twenty disappeared."

Freya checked her wallet.

"Just flash the passport and press card. Like this," Carole said, with a twist of her wrist. "In West Beirut the checkpoints are like chicken pox. You might get a Shiite or some crank from a maverick militia. If they go perverse, slip them another twenty."

Freya nodded.

John Czerko, the man she met at the bar, turned out to be Carole's photographer. His head was inside a taxi window, bargaining with the driver. "He's sensational," Carole said, while they moved toward the taxi. "Even when he's drunk. Are you drunk, John? We have a guest." John Czerko's face widened in a grin; he was sober.

The crossing to West Beirut stood opposite the great marble Museum with its wide stairs flanked by ancient Roman columns. They got out of the taxi and Freya's eyes fixed on the magnificent building surrounded by Roman ruins. She felt a lift of spirit, reassured in the existence of art and civilization.

She walked on and looked back, once, then again, debating whether or not to tell Carole she had changed her mind; she would spend the day at the Museum. When would there be another opportunity to study Lebanese antiquities from the glory of Baalbek? The building itself was a treasure. She stared at the magnificent facade with its wall of sandbags. If she went to the Museum her broken promise to Asher would be a minor violation, a triviality.

"Come on, Freya!" She hesitated, looked back, and caught up to Carole, looking back once more at the Museum and at the green field surrounding it. Carole had her by the arm, John went ahead. Crowds pushed to get through the checkpoint to safety in East Beirut before the gates closed. Unruly clusters of people were moving forward, as with one pair of arms and legs. Children were howling. Women balanced bundles on their heads. She felt herself draw back at the heat and desperation.

"Come *on*, Freya!" Carole called, tugging at her.

At the Green Line, Freya showed her passport with the press card tucked inside, holding her breath, trying to imitate Carole's breeze. There were grunts. They passed.

They were crossing a field, smaller than a football field with more crowds moving across to the east. For the last few yards Carole pulled on her sleeve; they ran toward a boy in jeans, jingling keys, waiting for them in front of a black Fiat. He stood in front of another checkpoint flying a Palestinian flag. John pulled several bills from his pocket and a fifth of Scotch whiskey. Then he took the car keys from the boy, tapping his watch, negotiating time.

At the checkpoint, Carole tugged on Freya's arm. "Get your money ready," she said. But the Palestinian sentry at the checkpoint ignored the twenty dollar bill. He smiled; he was downright friendly.

"Oh *Ho*," Carole said when the boy passed them on. "Something's up. They're waltzing the press!"

John glowered at her. "That's shit, Car!"

Across the Green Line into West Beirut, muffled sounds of artillery came from the direction of the mountains. Here an unearthly quiet had fallen. There was a cease-fire.

The neighborhood had changed: long lines of men wearing keffiyehs, veiled women in black with bundles on their heads, hippie looking boys with long hair, children in scruffy shorts and middy shirts were creeping up a two lane street. A pastiche of baskets, jugs, satchels, carrier bags, swung from a crisscross of arms. A heavy smell of burnt matches seeped into the car.

"Hold your nose, Freya. I'll take a few pictures." John stopped the car. He leaned out the window and aimed his camera at a black clad woman leading a small boy, his face swathed in bandages.

They started rolling again, toward the Hamra district, through checkered neighborhoods, posh to shanty, keffiyehs were everywhere; groups of black clad women with black shawls over their heads were picking through piles of debris, searching, searching, as though there, in the very next pile, they would find treasure, in potsherds of life blown out from half collapsed craters of buildings and buried in the avalanche of stone and rubble spilling out into the street. Metal strips entwined with hanging wires swayed from apartment buildings. Bombed out dwellings gaped with open rooms of flowered wallpaper like the rooms of her ruined dollhouse Momma threw into the rubbish.

In front of a ruined building, an old woman, her head and face covered in black, sat on a rocking chair, rocking, rocking, rocking.

It happened fast. "Look out!" Carole yelled. They had crossed to a wider, virtually empty street with only a few cars and a group of children playing near what seemed like a fallen balcony and a huge, clay flowerpot splattered to fragments and surrounded with sprays of blackened geraniums. Transfixed, her eyes would not move from the geraniums. *Black, black, geraniums.*

John swerved. A tiny boy, not more than a toddler, was scrambling up a hill of rubble and then, suddenly, she saw him tumble and fall into a crater gouged from the street. Their car was fully stopped. Freya gasped and held her breath.

Screams pierced the air. Children were hollering; black clad women, waving their arms, screeching in a deafening pitch, rushed from the caves of exposed basements toward the crater where the toddler had fallen.

John held the car in park while he leaned out: his camera was clicking, clicking, clicking. Freya felt her fingernails in her palm while John went on clicking, clicking, clicking.

"The baby!" she cried, and Carole glared back at her. "Shut-up, Freya!"

Two pick-up trucks were parked on the street up ahead. "Get the hell out of here," Carole hollered. "The trucks could be wired!"

John put the car in gear and bolted forward. None of them spoke. Freya's eyes were fixed at the back of Carole's head.

The Admiral Hotel was no less elegant than the Apollo, another dissonance except for its mountains of sandbags. The lobby was teeming with journalists, photographers, men in keffiyehs. Off to the right of the main lobby was an alcove with a long line of telex machines. The bar, at the other end of the lobby, was jammed with drinkers buzzing in English, French, Italian, German, Arabic. John pushed his way in and Carole followed, pulling Freya by her sleeve.

It was not yet nine o'clock when John handed them each a huge drink. "Vodka," he said, as though it were the sole palliative for Good Morning, Beirut. Freya hesitated at the double, maybe triple shot, straight. Carole smiled. "Don't fight it," she said. "The worst that can happen, you stay sane."

Freya took a swallow. Better, stronger, than Pushkin's vodka. She took another, then another. It was the way to stay sane.

"Look," Carole said. "You thought we were monsters not to help with the kid. You better learn in a hurry. Rules of the Game. One clan's Good Samaritan is another clan's corpse. We get out, we help the kid, we never get back in the car because our throats have been cut for helping the enemy."

"A *baby!* A baby is an enemy?"

"Enemies, enemies, enemies," Carole said. "Hate City. There are factions, clans, Christians, Moslems, Druze, Shiites, from building to building, street to street, all with their own militias armed to the teeth. Enemies, enemies, enemies. Reading about it isn't like hanging your ass out of a Beirut window."

The crowd at the bar was thinning out. John and Carole turned and motioned her to follow.

The conference room was huge with rows of chairs. Up front there was a podium with a microphone. At one end of the room a table set with a white cloth and silver coffee service was laden with croissants, cheeses, smoked salmon, marmalades, sweets, coffee. A huge silver bowl was filled to the brim with caviar. Freya stared. Was this huge spread laid on by the PLO? Where was starvation?

Vodka was reeling in her head when they sat down. Quiet came over the room. Journalists were poised with their pencils and notebooks. Carole gave her a tiny pad and a sharpened pencil.

A man of great bulk wearing a baggy brown suit and a black and white checkered keffiyeh came to the podium. "A brief statement," he said. "No questions. No pictures."

He tossed one end of his keffiyeh over his shoulder in nervous, jerking movements, then he looked over the room, a smile coming over his face while he marked the silence. "I am confirming," he said. "According to agreement with Mr. Habib of the United States, the Palestine Liberation Organization will presently withdraw from this city. We were not beaten; we leave because we seek peace. Our enemies are blind to the road ahead. We will deal with them."

He lifted his arms, defying heaven to fall on him; his eyes were narrowed with searing hatred; she could feel heat on her skin. Was that the hatred that had murdered Zafra?

"You can be sure," he was shouting. "We will return to claim our honor."

"That's what they all say," Carole muttered. She leaned over to Freya, smiling: "Notice he isn't telling, not a word about when they're leaving, how, where . . . I told you. Sensational theatre. Nothing to write home about."

But Freya was not amused. She was riveted to the spokesman's mania, to his rage. Poppa knew what happened to victims that survive to retaliate. The Israelis considered themselves victims. Oh, most certainly, they were victims. The Christians, the Moslems, the Druze, the Shiites, were victims. It was an opera of victims. It was *Gotterdammerung*.

Her pencil was poised; she was doodling, observing herself writing random words, *honor and hate, hate and honor*. Then she wrote a verse from Othello: *An honorable murderer, if you will, for naught did I in hate, but all in honor.*

She saw the flushed spokesman fix his eyes on the audience while he swept his arm out toward the room.

She tried to convince herself that she was still sober. Carole was tapping her arm, leaning over, looking down at the scribble. "You want I should rush that out on a Telex?"

From the back of the room, someone was shouting a question but the spokesman threw the ends of his keffiyeh back over his shoulders; he was flushed, agitated, high. He had started to leave the podium when he turned back. "We are *not* leaving them!" he shouted. "We have assurances from the highest levels! The camps will be protected! No questions. No more questions."

The conference was over. "Get something to eat, Freya," Carole said, getting up. "I think you're a little tight. We'll be right back." The journalists were rushing to the Telex machines.

When Carole came back, she had a young man with her. She introduced him: "Dr. Tony Poudrakis from Athens. He's liaison with Palestine Red Crescent Society and *Medecin sans Frontiers.* If anyone can tell you about Sani, Tony can."

Tony Poudrakis began heaping marmalade on a croissant. He was a slender, coffee colored man, probably in his forties; there were deep grooves in his forehead and his suit coat was threadbare, the sleeves too long. He was smiling, a kindly, practiced smile.

"I hope you won't mind if I ask," Freya began easily, words rolling off her tongue, "My friend in Israel has a sister here. She's a nurse with *Medecin sans Frontiers.* . . she's an Israeli Arab with a French passport."

Dr. Poudrakis's face colored; he shook his head, frowning. "Not possible! There are no Israeli Arabs here. You're a journalist. I think you should know better than to say things like that aloud. Who is this so-called friend?"

"I'm sorry, I'm so sorry." Freya lowered her voice to a whisper. "Sani Ayoub."

It was as though she had struck a match. "She's *French. French.* Do you understand?" he said. "She volunteered to go to Sabra Camp with a group of Moslem nurses. The PLO is evacuating and the camp is horrible. Sani wouldn't hear of not going."

"She's alive!"

Then he grew sober. "A few days ago, yes. But at this moment, I don't know." Dr. Poudrakis looked down, smiled, looked up again. "We have this dreadful habit of saying, *at this moment.* But we are grateful for every moment. Sani is one of our saints."

"Dr. Poudrakis, you have to understand how desperate her family is . . . not knowing, imagining the worst!" Freya cried.

Dr. Poudrakis was somber. "Desperation is very much in season," he murmured.

"Is she at Sabra now?" Carole asked.

He shrugged. "I don't know. I have no information for you."

"If we go there, will we find her?" Carole persisted.

"You can but try. Try everything if you like," Dr. Poudrakis said, turning away.

Freya wrenched free of the firm hold Carole had on her arm. "I'm not going. I've done enough."

"Only half the job," Carole said. "For Christ's sake, Freya, you've come this far, so finish it before you develop your ass kicking complex." Carole was smearing marmalade on a croissant, then she piled cheese on another and held them out to Freya. "Take your time. Drink a couple cups of black coffee. Then we'll go."

The drive to the Sabra refugee camp was a descent into shantytown. Sabra was more than slum, more than *barrios.* Sabra was a human meltdown. Survivors crept about, their heads lowered as though they were looking for something. Everyone seemed to be searching. She did a double take at the incongruous, the unexpected: a boy on a shiny silver bike was circling around and around on the street. He might have been in Beaulac.

There were sounds of anti-aircraft popping from somewhere: *Pop, Pop, Pop.* Far away? *Pop. Pop-pop pop-pop.* Panic was throbbing inside her throat. Where was Noah? Why hadn't she gone to the Defense Department the moment she arrived in Israel? What was he *doing* here? Why hadn't he arrived on schedule? *Pop. Pop-pop-pop.*

John had written directions Dr. Poudrakis gave him. They turned onto a street of rubble, a promenade of garbage dumps. Nothing moved. The ravaged tenement buildings had no numbers; the corrugated tin huts were clones.

There was a barricade; John stopped the car and pointed to the

Press sign on the car. A skinny kid reading a book while he stood at the barricade, glanced at the three press cards. "Hospital," John said. "*Medecin sans Frontiers. Red Crescent.* We make the world very mad for you." John held out a five dollar bill. "Okay?" he said.

The boy looked at the press cards again, took the bill, looked in the car. He crooked his finger, pointed around a corner, and went back to reading his book.

Freya stared at the boy. Was he reading a book?

They inched past the bombed-out tenements and turned a corner. A mutilated dog lay at the side of the road. When anyone passed, John would yell out the window. "Hospital? Medecin?"

More pointing. More crooking of fingers. More corners. "Christ!" John yelled at Carole. "You're the one who knows the language!"

A large barracks with a Red Crescent banner draped over the entrance loomed ahead. Half the population of Sabra was collected there, inside, spilling out. Outside, spilling in. Kids, women, old men. A high-pitched shriek. An undertow of excitement threatened to escalate and explode in the consuming heat. Bargaining, haggling over old suitcases, leather bags. The Red Crescent Society was sanctuary. An outdoor market. A rations dispensary. A news agency. A place to scheme.

"Who goes in?" John said.

"*You* go," Carole said.

John got out, muttering. He disappeared inside the hut; they waited. Carole looked at her watch. Kids were gathering around the car, peering inside, grinning.

Ten minutes later, John came out with a woman; she looked Chinese, half his size but she was leading him by the arm, pushing him through the crowd. Whoever she was, she had clout. She came up to the car.

Carole flashed her press card. Freya did the same.

"Miss Wehn," she introduced herself. "I'm a nurse. I try to help here." She waved her hand at the children. "If you come out, I'll find a place more quiet, more possible to talk." Her English was heavily French-accented.

Miss Wehn took them to a corner, away from the crush of people and noise. Kids followed behind; they ignored her when she waved them away.

"Quite a situation here," Carole said.

"Quite."

There was a long pause and Freya realized that Carole was waiting for her to ask the question. It was her story, not Carole's.

"Please . . . I'm sorry to add to your troubles," Freya said, showing her press card again.

"Yes, yes," Miss Wehn interrupted.

"I'd like to inquire about a friend."

Miss Wehn was stunned. "A *friend?* You have a friend here?"

"She's the sister of my friend in Israel, in Nazareth. Her name is Sani Ayoub."

Miss Wehn's eyes flashed with terror. "You are from *Israel?*" A look of hysteria came over her; she half turned and wheeled back. "I did nothing wrong! You are *Mossad!*" Miss Wehn opened her mouth, on the verge of a scream.

Freya caught her arm. "Please, Miss Wehn, please. I'm Sani's friend," she said. "I'm an American." Freya took out her passport and showed it.

Miss Wehn stood back, she was breathing heavily, her eyes moving over Freya. "What do you want with me?"

"Not with you, Miss Wehn. Sani. Sani's sister is my friend."

Miss Wehn's forehead wrinkled. She put her finger over her mouth as though to hold back panic. She was thin, anorexic; her jet black hair was short, choppy. To overcome a lisp, she spoke slowly, deliberately.

"What do you want with me?"

"We don't want anything with you, Miss Wehn. Sani's family are sick with worry. As I was coming here anyway, well, I'm a journalist, you see."

"I already know that!" Miss Wehn cried. She seemed half-starved and exhausted, at the edge; a torrent of words was spilling over. "I *told* Sani," she said. "I tried to impress her that Sabra is a doomed place. There was no use going *on!* But I grew so fond of her. I could have left but I didn't want to leave her alone. Sani was everyone's conscience!" She circled her hand at the devastation around her.

Miss Wehn's shoulders hunched; she seemed to shrivel; her voice hardly carried. "I am an independent volunteer. I don't belong to any one but . . . *They*. . . They could have gotten me out. They had to get me out! They *promised*. Something terrible is going to happen."

Freya reached out to clasp Miss Wehn's shoulder. "What are you afraid of? What can I do to help you?"

"I don't know," Miss Wehn cried. "I have a horrible fear."

Carole interrupted, her pencil poised over her notepad. "Who are *they*, Miss Wehn? The ones who promised to get you out?"

Terrified, Miss Wehn glared at Carole. "I didn't say that. I didn't say anything!" She turned to run but Carole caught her hand. "Forget I asked," Carole said, while Miss Wehn stood there, limp.

John was taking pictures, click, click, click. The smell of human waste and sewage was overpowering.

Miss Wehn's face contorted in a smile when Freya covered her mouth. "I don't smell anything anymore," she said. "These people smell everything; they scrub and scrub with whatever they can find. If not for them, there would be plagues. Worse than the Middle Ages." She frowned at John who was jumping from place to place, clicking his camera.

"Write what I said under the pictures!" she shouted at John. "In their place, would you be so heroic?"

"When will Sani be here?" Freya asked gently.

Miss Wehn brought her hands together and folded them over her heart, her eyes on Freya. "Sani. What can I tell you? I see in your eyes . . . no, you don't want to harm. You must forgive me!"

"Forgive you, Miss Wehn?"

Miss Wehn paled and reached out for Freya's hand. "I console you, Madame. Sani was hurt, very hurt. Very serious."

"How?" Freya murmured.

"It is so terrible that for one such as Sani, her injury was stupid. She was overworked; she couldn't think anymore. A woman fell; the woman was looking in the hole where we dumped gar-

287

bage while she was holding her baby. Every one was warned against the garbage dumps! I don't know what the fool was doing there! Everyone is so hungry they lost their minds. They *knew* ammunition was dumped in the garbage. So it wouldn't be found in the houses, so no one would be taken away."

Freya was thinking of hunger and the hill of caviar at the press conference when the little Chinese woman stopped talking, her face ashen. She raised her hands in helplessness and moaned. "There must have been a grenade. The garbage was exploding like fireworks. The woman fell down like a stone. Sani went after her while the garbage was exploding."

Freya stood still, staring at Miss Wehn's stricken eyes. "But Sani will recover?"

Miss Wehn's eyes overflowed with tears. "Sani tore the baby from the mother's arms and ran back to us with her. Thank God, thank God, the baby was hardly touched. Sani went back for the mother but the explosion came from the garbage. Like the tire of an airplane blowing out. Everyone was running. Only later the paramedics found Sani still alive. The mother was gone like she was swallowed in the earth."

Freya felt her voice fading. "Where is Sani?"

Carole was balancing her notepad, interrupting as though there were more newsworthy items to question than Sani. "Miss Wehn? What kind of rumors did you hear? You say you're afraid of what will happen. Where are the rumors coming from? *Who* are you afraid of?"

"Anyone! Everyone! We won't escape! We will be massacred." Her shoulders began to shake; she was covering her face.

Freya put her arms around the quivering Miss Wehn. "You won't be massacred. We just heard from one of your spokesmen. The peacekeeping force will make sure that nothing bad happens."

Still trembling, Miss Wehn looked up at her, mocking. "You say you are a journalist and you console me with the peacekeeping force?"

"You have to trust someone!"

Miss Wehn's smile was bitter. "You want to see Sani?" she lisped.

"Yes," Freya said.

Miss Wehn pointed. "Mukhassed . . . the hospital. Maybe one kilometer, maybe less. I pray you find her alive."

And suddenly Miss Wehn was hysterical, clutching Carole's arm. "No! You must take me with you! I mustn't be left here. I have no way to get myself out. Take me with you. Anywhere!"

Carole loosened Miss Wehn's grip. She shook her head. "I'm sorry. I'm sorry. We'll check up on you, somehow. I promise."

Freya took Miss Wehn's hand. The stench was cloying, choking. "Miss Wehn . . . Miss Wehn. . ." she said. "You have to find some of your people to help you. An escort, maybe, the peace-keeping force. . ."

Miss Wehn's hand was clutched over her heart. "I want to go with you anywhere. *Please.* I beg you to take me."

Freya could not answer. Flies were devouring her. The high heat of the day was drenching her with sweat. She was encircled by ragged children. She wanted to lift them up, put food in their mouths. She had nothing to give the children. A few yards away a very old woman, maybe crazy, maybe senile, lifted her huge black dress to squat; the children were laughing, pointing, and two younger women rushed over to scream at her.

Aghast, Freya opened her bag and lifted out three fifty dollar bills. "I don't know how much this will help," she said to Miss Wehn. "They say you can bribe your way out. Maybe this will help you get away." She opened her purse again and added another fifty. "You can make a connection . . . you can try. Please try, Miss Wehn, please try."

Freya caught Carole's eyes, signalling, with slight shakes of her head. When Freya paid no attention, Carole moved closer and pulled at her arm. "Don't be a jackass," she whispered. "Do you know what you're doing!."

Miss Wehn was sobbing. Gripping Freya's hand, she sobbed over and over. "I don't forget you. Sani never forgets you."

Before getting into the car, Freya glanced back at the clusters of children and dazed old people and young wives shrieking at each other.

John revved the engine and started away from the camp in the direction of the *Mukhassed* Hospital, a few blocks away. Garbage dumps were everywhere. Seconds went by in silence; then Carole turned around to Freya. "You don't learn very fast, do you? Keep your mouth shut about passing out money."

Before Freya could answer, Carole let out a nervous screech. John was swerving the car to avoid a crater, one side, then the other, banking at a precarious angle, struggling to keep control. Artillery from the nearby mountains vibrated the car while a zigzag of people crossed the street, dazed, ambling aimlessly, paying no attention. Veiled women carried babies, dragged older children after them. The children were quiet, moving close to their mothers as though they were embroidered onto their long black dresses.

Artillery was popping. "Another bullshit cease-fire," Carole moaned.

John slammed the brake to avoid a lone kid in the path of the car, thumbing his nose. There was a thud. Carole fell against the dashboard.

Freya struggled up from where she had crashed against the frame of the side window, hitting hard at the side of her neck, her shoulder. "Oh God! Did we hit him?" Freya cried.

"No," John said. "Just missed. You okay, Car? Freya?"

Carole leaned back. "It's nothing," she muttered, holding her forehead. "Just a bump. A little ice . . . Are you hurt, Freya?"

Dazed, Freya shook her head.

"Hold on, Car! We'll get ice at the hospital." His foot was on the brake; he was leaning out his window, clicking his camera. Click. Click. Click.

"You stupid hick! Get the hell out of here!" Carole shrieked. "You see those dumpsters? Get out of here!"

"Hold on, Car, hold on." Click, click, click. "We're practically out of here, Car!" They were clearing the neighborhood, turning the corner where the partly bombed Mukhassed Hospital, flying the Red Cross and Crescent, leaned in the sun.

Carole was wailing, holding her forehead, every shred of her cool gone. "Hate City. Hate City," she moaned.

Freya rubbed her neck and shoulder; a deep inner silence had fallen over her like the time she heard that Zafra was dead. The entanglement was too deep, larger than anything she could measure in her conscious mind; there was a numbness in her throat. She sat back, staring straight ahead.

Carole, holding her forehead, turned around. "You're sure you're not hurt?"

Freya shook her head.

"My God, you can't talk! You're getting sick, aren't you? Christ! You're getting sick."

"I'm not," Freya murmured, staring. "I'm not. You don't have to come in to the hospital with me."

"Are you out of your friggin mind? We should wait here for a sniper or a car bomb?"

They were met by two young boys, of not more than thirteen, who were guarding the entrance; they wore black berets, snub-nose carbines slung over their shoulders, and white arm bands with red crescents. Both boys stood there reading books, engrossed, as though the world outside wasn't real. At a glance she saw a page of geometry diagrams. Her heart melted.

The boys grew alert when Freya, Carole, and John, elbowed in through the mass of people crowded into the remnant of a lobby.

They showed their press cards. The tall boy showed off his English with a grin. "What you want here?" he said.

Carole smiled. "We want to write the story about this hospital. Everyone in the world will get mad. You want us to do that?"

The boy shook his head, frowning.

"You're a nice looking kid," she said. "You like baseball?"

A slow grin appeared on his face. "Dodgers," he said, with a long guttural *er*. Carole inclined her head to Freya. "Her husband pitches for the Dodgers. Want her autograph?"

Awed, the boy nodded.

Freya stared. Except for the Cubs, she hadn't paid attention to big names in baseball since she was ten. She tore out a sheet from the note pad and wrote: *Best wishes to my young friend in Beirut. Some day you will be a great mathematician. Sincerely, Joe Dimaggio's wife.*

"You like that?" Carole said. "Right. Where do we find the director?"

The boy poked the smaller boy standing beside him and spoke in Arabic. They followed the boy down a corridor swarming with keffiyehs and veils and children and long black dresses. A thousand miles of bandage were swathed around the crouching, the lying down, the half standing people in the corridor. A man in a dark suit appeared, looked at the boy, then at the three of them, questioningly; he half turned in alarm. "Hey!" John called out. "We're journalists! From Canada. I thought you wanted the world to get mad! Talk to us!"

The dark suit came back. He was young with thinning hair; his eyes bulged behind black horn rimmed glasses, his nose was flat like he was once a prizefighter.

"What must I do for you?" he said in nasal English.

Carole introduced Freya. "She has a special favor to ask. Maybe you could help her first; then we'll get your story."

The man looked at Freya, waiting, adjusting his glasses, a puzzled expression on his face.

"I'm looking for my friend," she said. "At Sabra they said she was here. She's hurt."

The man laughed. "Excuse me," he said. "Your friend is hurt?" He waved his arm at the sea of wounded on the floor. "Take your pick, Madame. Any friend you like."

"Her name is Sani Ayoub," Freya said. "She's a volunteer with *Medecin sans Frontiers*. She's . . . French."

"I see," the man said. "French. What do you want with her?"

"I want to know that she's all right. That's all."

Their eyes met; he adjusted his glasses again; his eyes were skeptical, his face grave.

"Who told you Sani was here?"

Freya thought for a minute, remembering Miss Wehn's hysteria, and something instinctive said she should not mention names.

"A man at the Red Crescent at Sabra," she said. "He saw what happened."

"What is your relationship to Sani Ayoub?" he asked.

"She's my friend's sister. I promised to look for her when I came here. Her family . . . they're desperate; they don't know where she is or what happened to her."

"Let me take you to somewhere quiet," he said. He reached out to shake Freya's hand. "I'm Dr. Khatoubi."

They followed him to a cubicle that might once have been an office. Papers, folders, were piled high on the floor. A network of wires hung loose overhead.

He looked at her softly; his hands were trembling when he adjusted his glasses again. She could see that he was worn thin, at the edge, like Miss Wehn.

"A hundred times a day, I have to do it," Dr. Khatoubi said. "It never helps the next time. Sani is dead. A few hours ago."

Freya looked at him in silence.

"I'm sorry," Dr. Khatoubi put his hand on her shoulder.

Carole shook her head. John shifted his weight. "Have you got any ice?" he said.

Dr. Khatoubi, astonished, asked, "*Ice?* Are you making jokes of us?"

"*For Christ's sake, John!*" Carole turned to Dr. Khatoubi. "He didn't mean that! I bumped the dashboard." She pointed to her forehead. "Here."

"Yes, I see," Dr. Khatoubi said. "We'll try to help you."

Freya stood there, frozen. *Why hadn't Noah arrived as scheduled? What if he didn't come the next day? Sani was dead. She was alive one moment, then, at the next moment . . .*

"Would you like some water, Madame?"

"No; no water," Freya said.

"Maybe the doctor should look at your shoulder and neck," Carole said.

Freya interrupted. "Will they send Sani home to be buried?"

Dr. Khatoubi, astonished, uttered, "I'm surprised, Madame; you're a journalist. There will be no sending home for Sani. We bury the dead respectfully. But sending bodies home for funerals? You ought to know Sani's situation. You're a journalist."

"Human beings bury their dead! We're not in the wild!" *Where was Noah? Why hadn't he met his schedule?*

293

"You are not talking like a journalist, Madame. Like a poet, maybe, not a journalist. Yes; we're in the wild. I'm sorry. There are realities."

"But Sani," Freya began. "Please, Dr. Khatoubi, I want to see her. I want to tell Layla that I saw her sister. At least they can mourn." Freya met his eyes.

Dr. Khatoubi shook his head. "You're distraught, Madame. You must sit down. You must have water. But I cannot let you see Sani. If I'm kind to you as I would be to her family, you must not press me to see her."

She sat down and drank the water Dr. Khatoubi gave her but she would not let him examine her neck or her shoulder. Dr. Khatoubi swabbed Carole's bump with alcohol and pressed an oversize band-aid to it. Freya watched, her heart sinking—there was no electricity, no ice. She was remembering Karl's office, the whiteness, the abundance. How well had they treated Sani's wounds? With what?

It was after three o'clock when they got back to the Beirut Apollo. Carole was composed, stony; she had scowled her way through the check points.

"Let's go in the back way," John said. "Closer to the bar." He parked in the lot and when they got out he touched Carole's eye. "You're ripe for a shiner, Car. You stay cozy tomorrow. Orders. R and R."

"If you wake me up in less than twenty-four hours, Hayseed, you won't live to tell . . . !"

Before they went in, Carole put her arm around Freya and walked her out of earshot while John looked on. Carole's face was pale against her rapidly discoloring eye. "That Wehn woman was right about you," she said.

Carole's voice was thick with remorse. "I put you through Hell so I could get to your general. I wanted you to think you owed me. I'm sorry, Freya. I'm *sorry*. Let yourself hate me."

Freya began to spill over in nervous laughter—they were back safely; her neck and shoulder throbbed and she couldn't stop laughing. She was garbling, hardly coherent: "You should see your shiner, Car! *Kharma*. Serves you right! I'll get mine too; you'll see." Carole was looking at her as though she had gone mad but Freya couldn't stop laughing.

Freya refused a drink; the thought of whiskey made her gag but Carole insisted that she should stay close to them, not be alone. The bar had its usual crush but Carole managed to get a drink as well as a glass of ice for her head. John ordered vodka straight. Freya asked for a glass of Perrier and when the bartender handed it to her and she touched the rim to her lips, she turned back toward the lobby to scan for Beni. She heard herself cry out; there was a shattering of glass when she ran across the lobby to the sofas.

"*Noah! Noah!*"

She was holding him, clutching the strap of his Uzi gun. "*Noah!*" His name was the last word left inside her. She held him at arm's length; she could barely see him through her tears. He pulled free. "Momma," he murmured. "Where were you? I almost went nuts!"

His hair was cropped but the Noah she knew was in soft sherry-colored eyes looking at her with alarm. She took both his hands, telling herself that she did not see an olive drab uniform, an Uzi gun, a helmet and a backpack with a camera slung over his shoulders.

And then she did see: a raised flaming scar, stitched along the left cheekbone from the eye to the chin. She cried out in a stifled scream.

Noah shook his head. "Don't, Momma, I'm all right. Please don't. Let's go somewhere. I'll tell you."

She was clutching his hand, pulling him into the mercifully empty elevator. "What happened? Tell me what happened!"

They were in the room and Noah was staring from one object to another, his gaze coming to rest on the roses at her bedside. He rested his gun and his gear against the gilt vanity. She pulled him to the sofa where she sat down next to him, wiping her eyes, trying not to spill over when she touched her finger to his face. "It's my fault . . . I should have prevented. . . ."

"It's not your fault! It's only a flesh wound; it wasn't even serious enough to keep me off duty for more than a week. After the war, plastic surgery will fix it. Why are we spending our time this way? I haven't seen you for two years."

Her hands and feet were icy, there was a steady throb in her shoulder and neck. She stood over Noah and sat down next to him, stifling her impulses, not knowing what to do, how to act. She gripped his hand. "You're wounded! Two years and I find you like this! How, how did it happen?"

Noah pulled free and shrugged. "We had to get pictures of the ways in and out of a PLO nest. There was an explosion, not too close to me, thank God . . . a lot of glass."

"When did it happen? *When?*"

"I don't remember exactly. A month, maybe more. I thought about writing to you from the hospital in Rosh HaNikra. I didn't know you were in Jerusalem."

She heard herself gasp. "Rosh HaNikra? You were in a hospital in Israel when I was in Jerusalem?"

Noah's face blanched.

"No one told me! Why? Why wasn't I told?"

"Momma! No one knew you were in Jerusalem. Anyway, I decided not to tell you. It was only a flesh wound and my commander agreed that I shouldn't worry you. I was sent back to work within a week."

"You shouldn't worry me? I'm your mother!" she cried. "I'm not an enemy!" She was remembering the Lebanese baby who was an enemy. Her hands were icy, trembling. she had been in Jerusalem, waiting, waiting, while Noah was in Rosh HaNikra.

"Momma! Don't. It's over. I survived. I'm whole. I'm as good as I was before."

The silence was coming down on her. Had David known? He was supposed to have found out everything about Noah. Had he told Asher? Had Asher known that Noah was in Rosh HaNikra? She was clutching Noah's hand and Noah was pulling away. "I'm fine, Momma. It looks worse than it is. The doctor said it was a flesh wound. A plastic surgeon can fix it."

"A plastic surgeon? You think a plastic surgeon will cover it up?

Like your swimming? Like your guitar playing? What part of the scar won't show?"

"Stop, Momma! It's not my fault you didn't know about my wound. Momma, please don't cry. I'm so glad to see you. Let's not talk about the war. There are other things . . . I didn't even have your picture to show."

She sat looking at him while Noah fidgeted, eyeing his gear propped against the gilt vanity.

"You need something cold to drink. You look so warm; it's so hot," she murmured.

"Momma, I'm not warm in the least. I actually had a swim downstairs."

"I'm glad . . . I'm glad." She smiled at him, remembering his letter.

"You look terrific, Momma." They sat in silence for minutes.

"Noah? How would you say this pool compares to the one at the front?"

Meeting her eyes, "Great," he said.

"And the guitar. What is it I'm supposed to swallow this time?"

"I'm not asking you to swallow anything."

"You're not?" She knew it was not going well, that she had messed up. Why was she harsh? Had she come this far only to vent her anger that he had left their home and the life she thought she had made for him, his future? She felt herself startle. Had she gone crazy? Are mothers supposed to punish their children when they're wounded? But when their wounded children won't let them react?

"Why are you in Lebanon, Noah?" she said, adjusting her tone.

"You won't let yourself believe it, anyway. We're here for security."

"*Security!* You see security here? This killing won't stop for the next hundred years!" She had to stop herself or she'd push Noah deeper. But when she heard Noah say, "Uri talks like that. . ." she felt herself going out of control. "Uri is an *enemy?*"

"I didn't mean that. I meant that his line gets to sounding like theirs."

"His line? Has Uri been so sheltered, so protected? He lost a son and came through a civilized man. He survived the camps a civilized man. Should he be gnashing his teeth? Should he be blowing up buildings?"

"Uri never made connections between what he went through in Europe and what we need to prevent!"

"Uri is an enemy because he won't justify your excesses? How about me?"

"You tune out what you should see," he muttered, turning away.

"I'll tell you what I see! *Black geraniums* is what I see. There isn't a bird left alive. The orchards and the forests are burnt out so you can plant black geraniums. But I won't let you; thank God, I can stop you."

They were sitting on the sofa, their bodies turned toward each other and Freya was pressing Noah's hand, her grip growing tighter. He wrenched free; their eyes met and he got up to sit in the chair.

"My God, Momma! Can't you be glad to see me?"

"I came thousands of miles to see you. Why are you still here, Noah? Michael Haas left. Why not you?"

"Michael had his head in the clouds. When he faced up, down beneath he's a coward."

"Why so many answers, Noah? Why no questions?"

Noah was shrugging again, running his hand over his cropped head. There was a knock on the door. A bellboy delivered Cokes with ice. She gave him a dollar tip and went to fill a glass for Noah.

Noah took a long swallow and looked up. "Momma, forget it," he said. "You don't understand what we're dealing with. You're an American. You don't worry about your neighbors or about the next bomb."

"*I'm* an American? And *you?* You find it so easy to take this on? Why are you fighting those people? Why? What did they do to you?"

Noah was downcast but she couldn't stop. "You used to bring kids from the inner city to swim at the beach. You *cared* about

298

people. What happened to you, Noah? What did we do to you!" She stopped herself. Poppa had asked her the very same question about the Schule.

"I'm not in Beaulac anymore," Noah said. "You talk about helping, yeah, everyone but my own. I'm different now. When they try to exterminate Israel, they exterminate me. I'm a real Israeli. You don't have to like it."

"A *real* Israeli? Every real Israeli believes in this war? There's Uri. Thousands went on demonstrations! None of them are your people? All of them are enemies?"

Noah's hands were clenched. "I can't stand this."

She went over to kneel beside the chair, catching his cold hands and lifting them to her face. "Noah. I love you so much, I can't let you kill. I can't let you be killed."

Noah wrenched his hands free. "I don't kill," he cried. "I never killed anyone! I take pictures. Some day those pictures will be history. I don't do the killing."

"What if they kill you . . . they might kill *you!* And how do you know they don't have so much truth as you do. Their truth says they should kill you."

She went back to sit on the sofa. Truth. Truth. He was giving her no choice. When she got back to Jerusalem she would go straight to the Defense Department. She looked at him: his eyes were on the Uzi gun and his gear, eyeing them for a quick getaway. Her heart sank. She was losing her son.

She sat there, shriveled inside. Surrender was too easy. She had to retrace her path, fight harder before she gave up; she had to find another way to persuade him to get out. But there was no use talking politics, geographies, demographies. There were fiery songs in his head, songs of heroes.

She tried again, fighting to keep her voice steady. "Noah . . . listen, Noah. You don't have to be a hero. You don't have to carry the banner of truth. Everyone's truth has a bullet hidden inside."

"Don't, Momma," he said. "You're not helping me. Do you know what it means to be in the army with false credentials? If you tell them, I'll be disgraced. You wouldn't do that to me! How could I face Carmella!"

He turned away, his face flushed. "Okay, then!" he cried. "Go ahead! But don't think you'll get me back to Beaulac!"

His words were staggering. Noah was proving his manhood to please Carmella. He was ready to die for it. But if she shrieked her frustration at him, she would push his fiery songs deeper.

Noah's body stiffened. "If I'm a killer, what are you? An innocent bystander?" he said. "Where were you when I got here? I've been here for hours, waiting. They told me you'd be here, that you wouldn't budge an inch. So? Were you looking the place over? Taking a whiff?"

She hesitated, her mind racing. "I went to the Museum! I was trying to help. . ."

"In all this mess, you went to help the *museum*? I'm supposed to believe you?"

"I don't owe you explanations, Noah."

"That's right. You're a mother. You have a son to push around." Noah looked away. "We're wasting time. I'll have to go soon."

She was at his side. "When?"

He was tapping his fingers on the arm of the chair. His right leg was bouncing up and down, up and down. She knew his way of holding back, gathering restraint.

"They're supposed to pick me up at eight sharp."

"I won't let them!"

Noah snorted. "Why can't we be glad we saw each other? Why can't you let it go at that? I wanted to see you. I wanted to talk about . . . Hell, what's the use? You're having me bounced. I can see it in your face. If that's the way it is, say so."

"There's more on your mind, isn't there? Something about me."

"If you don't want to see it, I can't make you," he said, tapping, tapping.

"Try. When did I ever not listen?"

"Okay then. But don't forget how you wrung it out of me! General Frank. He's a hell of a lot more than a friend, isn't he? No casual civilian comes here. But *you* did. Everyone know's the General is married to a classy designer. But he's got iron tight security around *you*."

Noah's eyes were not gentle any longer; he stopped tapping, his right leg was pressed to the floor. He was rigid, worked up.

She moved back to the bed and sat there, stunned, her eyes moving over the long scar gouged from his perfect face, his cropped hair, his boots. "Noah, do you know what you're saying?"

"I know what I'm saying, all right. You're here with General Frank! How do you think that makes me feel? I heard your whiskey glass smash at the bar!"

A hot surge went through her. What had her pleading accomplished but a counterattack that discredited her? It was her fault; she had challenged him to where anything was fair game. He was sitting there unmoved, callow, puffed up with rectitude. The only values worth talking about lay in appearances. In Puritan prohibitions. In Middle America.

She was not able to speak but when she looked at his ravaged face, she wanted to rush to his side, reclaim him as her son. She didn't move.

"I see," she said, finally. "It was my *whiskey* glass you heard. You saw a general in my bed. Those are your American values? Everything but bourgeois Beaulac got dumped into the Middle East swamp?"

"You're my mother!" Noah cried.

"Oh! It's the purity of mothers you care about. Mothers buried under bombs don't bother you? Of course not! They're not human! Real, human, mothers come from Beaulac: they have short yellow hair and tennis whites and gourmet cookbooks."

"Let up on me, Momma! I never asked you to be anything for me, not even when you nailed yourself to that museum. I was glad you found something . . . I don't understand you!" Noah had begun to cry.

His tears fell over his face, salt tears that would sting his wound. She came to sit beside him and reached out to brush his eyes with a tissue. "It's all right," she said. "It's all right." She was thinking that everything was fiction. Invention covered up the terrible truth of not knowing anyone, not your mother, your father, your son.

She touched her finger to his crimson wound. "I love you,

Noah. Maybe there's nothing more left than to leave you alone. It wouldn't help to take you home."

For a moment he looked astonished, then his face was bleak with remorse. "I shouldn't have said those lousy things. You goaded me." He clasped her hand. "What will you do about me?"

She smiled at him. "So that's all it comes to? All that talk about morals and sin. You need your chance to slay the dragon."

"Lighten up, Momma!" Freya leaned back against the sofa. "Remember?" she said. "You used to run out of the house at night to search the sky. You badgered me that you had to look for something amazing, something no one ever saw before. I was a pushover, a sucker for stars. So tell me what you saw?"

"Maybe it was Carmella," Noah said.

She was not ready to talk about Carmella. She wasn't ready to leap.

"Momma, I'm sorry. I didn't want to hurt you."

"You did hurt me . . . you hurt me now." They sat in silence. The day was coming down on her. The pain in her shoulder throbbed and the fear of losing Noah lay like a stone in her heart. She couldn't live with a break between herself and Noah. But a break could be avoided if she stayed off the war, off Asher. Their silence only fired tension. She had to shift before more scars were carved.

"Your Dad got a gorgeous apartment on Lake Shore Drive," she said. "I stopped in before I came here. We're friends; we have joint custody of Hammer. He's been so generous, Noah. He cares about us. And remember Dora? I think Grandpa will marry her!"

Noah ran his hand over his cropped head. "*Marry* her! Why would he marry Dora, of all people, *Dora!*"

"Who can say?"

Noah lowered his eyes. "I don't think I ever knew Dad," he said. "He told me once that the best thing about my swimming practice was that it kept my fingernails clean. No matter what I said or did he never let me forget I was being prepped for medical school."

"Your father had to invent you."

Noah was quiet and then he glanced at her, skeptically.

"I know that look," she said.

Noah hesitated. "You're sure General Frank understands what you're about, Momma? He knows how you are? I couldn't stand it if anyone hurt you again."

She was about to ask why he had the exclusive privilege of hurting her, but there had been enough muscle flexing. What Noah really wanted to know, was the effect the general would have on her decision.

"Do you mean he might wear me down to let you stay in the army?"

"I didn't mean that," Noah said, studying her. "You're having me bounced, aren't you?"

She was twisting the end of her long braid, her hands icy, sweating, her eyes on his face. She didn't speak for a long moment, and then she murmured, "No, I'm not." She saw his face light up as though the privilege of seeing this war through was the greatest gift she could give him.

"Don't thank me," she said. "The war is almost over and the PLO is getting out. Thank them."

Room service brought them everything: she ordered a feast of steaks and potatoes and a spinach souffle and marinated mushrooms and pastries and coffee. The waiter opened a bottle of French champagne and set it in a bucket of ice. Noah filled her glass.

"To what?" Noah said, lifting his glass. "To you, Momma." His voice was tinged with love and gratitude. He held his glass high: "How about a toast to getting out of this mess we're in?" he said. "The war makes no sense. If we got rid of the PLO here, there are another million more on the West Bank. We can't kill them all."

She understood. Now that she was no longer a threat, Noah could say what she hadn't dared mention. It was not the time to give up. "Noah? Your grandfather says that every person needs a place where to live before they can know how to live. Is Grandpa right?"

"I don't know," he said. "I've thought about it. We have a tiny country. The Arabs have thousands of miles."

"I met a Palestinian family in Jerusalem," Freya said. "They don't seem to like being occupied."

Noah shook his head. "No more politics, Momma. How about a toast to Carmella?" he said, his face brightening.

She watched him drink champagne. Noah was no longer her child, but he was not one of the lost children. He was a man who called war a mess, who knew it made no sense. He was a man in love. In two years he had traveled the distance from boy to man without her.

After dinner Noah took a portfolio out of his sack. A series of photographs he took for fun, he said. One after another. Soldiers in slovenly T-shirts sitting on the ground, playing cards, one of them frowning, swatting at an insect on his bare foot. A soldier was giving a boy a haircut; a line of children were waiting. There was a soccer game with close ups of kids watching, not cheering, their arms lifted in clenched fists. Children not more than eleven or twelve were swigging from beer bottles, one tongue was stretched out to the camera. Black comedy. The pictures were more of the record Noah kept on the world.

"Your pictures say a lot," she said.

"Are you convinced about my future?"

She didn't answer. The thought of Noah's future sent a chill through her.

"I can't show what I shoot for the army." Then he pulled out a picture of a slim young girl with long black hair and Cleopatra eyes, a wistful girl looking with love at the photographer. "But I can show you this," he said. "Carmella."

Freya stared at the picture. A dark Sephardic girl that could have been Batia's granddaughter. The bride price story rose up and she shrugged it off with a sudden pride that Noah loved such a girl. Was she poor? Did she feel discriminated against? Her son, from Beaulac, had chosen her. Poppa had left his mark.

"She's beautiful, Noah."

"She is beautiful." He was quiet while he stared at the photograph of Carmella and then he put it aside, looking up at her.

"Promise me you'll stay in Israel," he said, startling her. The words had tumbled out of his mouth. "It would help me a lot, my morale. . ." He broke off, began again. "You could stay at Gan Zvi."

Seconds went by before she answered, "Promise you? Like when you were six? I'm not sure I could keep my promise now. I haven't decided where I'll go."

Noah's face was solidly downcast. "It's makes me feel sick for you, Momma! There's nothing for you to go back to anywhere, is there? But don't think I don't need you! I *need* you here, Momma."

Noah looked at her, stricken. "When I wrote you that letter about swimming and the guitar . . . My buddy had just been killed in the same skirmish where I was wounded. He was standing closer in." Noah's eyes filled with tears again. "I couldn't stand it. What did he die for?"

She felt her pulse leap. *One moment, then the next.* "Noah. Noah." She was pleading. "If you're asking questions, it's a first step. You don't have to be here. You can tell the Defense Department yourself. I won't interfere. It will come from you, not me."

Noah shook his head. "I wouldn't do that, Momma. A soldier doesn't split when his buddy is killed. If he did, what would it mean that his buddy died? I would be saving my ass, not asking questions."

Freya looked away. Camaraderie in war had already been deeply imprinted.

"I promise you, Noah," she said, softly. "I'll be at Gan Zvi when you get your next leave."

She watched him open his box of delicacies, examine his Walkman, try the earphones on, inspect the batteries, enthuse at the tapes. He was a child and she had washed out his sorrow. His eyes danced at her and his body sprung alive. *"Wow!"* He was swallowing everything whole the way he did when she gave him his first camera.

Panic was building in her. It was growing late. Dr. Poudrakis's song was resonating: *Alive . . . at this moment.* At the next moment Sani was dead. Sani's mother would never see her alive or dead. In a carefree, happy, moment just as the one now with Noah, Zafra excused herself to go to the toilet. In the next moment, she was dead.

She looked at her watch and then at Noah. His scar was ablaze, screaming, filling her with the dread that began in Jerusalem before she came. As though she knew, as though she didn't have to be told. He was alive at this moment, he was alive and she could change her mind, snatch him before the moment passed. The stink of the hospital was washing over her. It was too late; the moment had passed for Sani. "Noah! It's not too late!"

"Momma! The war is almost over! You said it yourself."

"*Almost!* The casualties are still high and 'almost' could make the difference. What if this is the exact moment to get out?"

"It doesn't matter. I belong here."

"You don't belong here! I won't let our fear of each other destroy both of us! It was my fear of losing you that landed you here in the first place. And now you *shamed* me into consent because of the General. You took advantage. You manipulated me like you did when you borrowed money from Uncle Saul to come here. I paid it all back. I could never say no to you!"

"Jees, I don't believe this. You're hard core Pushkin, aren't you, Momma? You're like Grandpa! Grandpa always said *no,* didn't he? *Ooooh,* how you hated it. Dad said *no, no, no,* every time he saw you take a breath. I was the one who always caught you crying in the living room. Maybe General Frank is the only one who ever says *yes* to you. You hated the way Grandpa and Dad bossed you around. You pushed me away from both of them."

He was wrong! If it were true that she had pushed him away from them, she had not wanted Noah to turn out a fanatic, bloated with no sense of proportion.

"Momma; listen," Noah was saying. "All the time I was growing up, you said I should be myself. What were you talking about? Wasn't I supposed to take you seriously? Were you reading out of some poetry book?"

306

She was frozen. Pushkin was riding his boxcars, only this time the fable had a twist, some logic she couldn't pin down that was squeezing breath out of her.

Her hands were trembling when she gave Noah two hundred dollars. He put his arms around her. "I don't want to hurt you, Momma. I love you; I'm sorry for the forgery. I'm *sorry*, Momma. I want you to believe me!"

She nodded, holding back tears.

Noah folded the money into his wallet. He tried to smile. "It must have been terrible driving up the coast. I couldn't sleep last night thinking about you on that ride. Some day will you tell me?"

Some day? Tell him what she had tried to tell him today? "Some day," she said.

"Is it okay to buy something special for Carmella?"

She stood there, staring at Noah's face. The scar was deep. The moment wasn't over; she could still snatch him away. Why didn't she? Why did she feel helpless when she could use her power to save him? If Noah became her victim what would she become? It was in that moment she understood. Noah would not be coming home.

"A ring for Carmella?" she said.

He nodded, beaming. "Yes."

Gently, she brushed the long scar with her lips. "My son is on his way to becoming a bridegroom."

Noah lifted his camera off the gilt table. "I want to take your picture," he said. "I want Carmella to see you."

She shook her head, flinching at the camera, at the thought of a permanent record of this moment. Her anger. Her face flushed with fear. "Not now," she whispered. "Not now. I'll wait for you in Israel. Then we'll see. We'll see."

She did not go down with him when it was time for him to meet his driver. She saw him go through the door to the elevator while the words burned inside her. She could still snatch him away; the moment hadn't passed.

A waiter came to take the service cart away with the leavings of

307

the feast. She had hardly eaten a bite. She asked him to leave the coffee. It was still hot when she poured a cup to warm herself against the chill inside her.

On the patio, the heat was like a furnace, the air was smoky, heavy with heat and unnatural mist. The red glow of flames shone in the west; she imagined charcoal tombstones decked with black geraniums.

Shivering, she went in again and sank down on the bed. Noah was alive. For how long? She had sent him back to the flames and the dying. Karl would not take a plane out. He would organize a plastic surgery team in Jerusalem on the telephone. Karl would boast; he would be redeemed that his son wasn't the kind of boy who chased after butterflies and prophets.

Something had come together in all their lives to account for Noah. Where was the fault line?

Noah, Asher, Miss Wehn, Zafra, Sani, herself? What had accumulated in nature that accounted for them and for Sabra and Hate City? She lay on her back composing a label about Sani's death. It would help her distance the images of snakes and the enemy baby who fell into a crater of black geraniums.

In history nothing is ever finished; the fault line goes on grinding at the fissure. In Sabra, while Sani was fishing in a garbage dump, tectonic plates crashed together until the earth opened up and swallowed her. The earth had a score to settle. Zafra first, Sani next.

CHAPTER SEVENTEEN

The telephone rang. Beni was shouting: "Where did you go? Who gave you permission? We let the General know how you walked out. He's on his way here. David will break my head for losing you! You satisfied?"

She hung up. The little gilt clock on the night stand said nine-o'clock. She steadied herself and stretched out across the bed again. She managed to get her shoes off. The blue jeans had hidden her bandage from Noah but now, the jeans, the vest, and blouse were too much of an effort to pull off. Her thoughts were in turmoil from the events of the day. After-images were floating behind her closed eyelids like sunspots. *Tears glistening on Noah's face. Noah searching stars like the ring he would buy Carmella with the money she gave him. Noah standing at the door with his gear where he couldn't be turned back.*

She lay there, breathing heavily, another image racing through her head, how she had stood over Poppa, prodding, prodding. His body hurt and every breath he took was a torture but she wouldn't stop: *Get up! Get up this minute, Poppa!*

Her eyes were dry and burning as she lay there, thinking of Momma and Poppa and their connection to some great compulsion in her that had been resolved in Beirut. All of her life, one thing or another had dislodged her from herself. But here, she had seen what she had seen and for the first time she had reached out in no one's image but her own. It had nothing to do with friendship for Layla; the roots of their friendship had not

309

gone so deep that they couldn't be pulled up. But Layla was a particular kind of friend, based on the impulse of a chance encounter, a stranger, who had revived her. When strangers stoop to help the wounded they are bonded—they give, they need, they use.

Her eyes were closed. Asher was on his way here. Her neck and shoulder were throbbing. It was very well to congratulate herself, but what had she done to him? Why would he trust her again? She and Asher had to begin again as though they had no history. Had something been triggered when she refused to identify the cab driver? And that time with Uri, when she stood on the balcony watching him drive away, feeling isolated and lost; he was pulling away and she was afraid that she wouldn't survive without him.

But Zafra's death had taught her a deadly lesson. Uri could not give survival lessons, teach them how to throw the dice and come up with a lucky number or teach them how to manipulate the world. Uri had not been able to save his own son and he had unwittingly used her to save Noah for himself. Poor Uri. She felt a swell of love for him imagining how he had polished his shoes with a nub of charcoal so as not to be stripped of being human. But in the end, it was only chance that saved him.

She needed more aspirin but she could not get up. There was nowhere to go. Finally, she must have slept.

It was still dark when she heard the door click. Startling, she saw Asher's silhouette coming toward her and she reached out to him, he was holding her, smoothing her hair, his lips everywhere on her face. She clung to him against every impulse to push him away so she would not have to face him. She had betrayed his trust and she was not sorry for what she had done.

Asher's voice was hoarse and low. "I'm sick with relief," he murmured, "I could strangle you. Where did you go?"

She didn't answer until she felt his hands on her shoulders, lifting her to him. She winced, trying to pull free, sucking her breath in so she wouldn't cry out from the pain in her shoulder and the pain of her maverick truth.

Struggling out of his arms, she murmured: "I went to a press conference at the Commodore Hotel." She hesitated while he stared at her. "I went to Sabra."

Through the shadows, she could see his expression turn to shock. "*Sabra!* You're dreaming. It's one of your stories. How did you get there? Who took you?"

"Two journalists."

"Who? What journalists?"

"I can't tell you, Asher. I know you have ways of finding out. The journalists are my. . ." She stopped herself from saying friends. "They're good people."

"Your *good people* are finished here," Asher said.

"Finished? Who's full of rectitude now? Noah was in Rosh HaNikra wounded! Noah was in Israel!"

She saw his astonished stare. He sat at the side of the bed while she rubbed her shoulder. He looked at her, curiously, as though he saw what she had seen in the mirror when olive skin turns to pallor. "Freya, are you ill, are you hurt? What happened?" She resisted but Asher wouldn't let her go. He slipped off her vest. "Tell me what happened."

She sat up. "Noah was in Rosh HaNikra wounded."

He sat back. She saw him take a breath. "David mentioned it only this morning. He said it was nothing, a minor scratch."

"A scratch? Noah's scar is a scratch?"

Asher took her hand. "To David it was a scratch, to Noah's commander, it wouldn't keep him off the job. There was no report to David because Noah's commander rushed him back."

"They had no right! I'm Noah's mother!"

He moved closer to her. "I know. I know. You're Noah's mother." He was hesitating, his lips pursed as though he didn't know what to do about her anymore, as though she and her son had used him up. "His commander should have reported it sooner," he said. "Barr should have known, he should have told me."

"There's a lot of things Barr doesn't tell you," she said, stopping herself, dismayed that she would deflect what she had done to Asher by mounting an attack on David.

311

"What's that supposed to mean?"

"Nothing. It doesn't mean anything. I'm upset about Noah . . . when I saw his scar. . ."

"Freya, sweetheart, David assured me, it looks worse than it is. He'll get a citation."

"You think it helps to make him a hero?"

Asher didn't answer; he lifted her to him and held her in silence, while he brushed his lips over her face and she breathed in the smell of him. Who was she apart from Asher? She was forty-one years old and she was fifteen, a child arrested in Asher's arms.

"Why did you go to Sabra?" he asked her, his voice soft and tight with control.

"I went to find out about Sani . . . my friend's sister from Nazareth."

"Sani," he said flatly. "I know all about your Sani."

Her face was hot from the rough edges of his uniform when she moved back to face him. "Oh I know you do, Sher. That's your job, isn't it? To know about people like Sani. Sani was killed! I went to the hospital where she died. Her family will never know where she's buried. What kind of a ghoulish job do you have, anyway? Your world is full of road blocks and refugees."

"I see," he said. "You'd like a world made to order. No more enemies. A global buddy system, sharing and caring."

"Sani wasn't born an enemy! If she was an enemy, it took a lot of work to make her one. She was born a citizen of Israel!"

"*Citizen?*" he said. "A PLO nurse? What in the hell do you think this war is about? My God, I begged you not to get involved. You went over the line, Freya. You wouldn't listen. Now you want another world where your conscience can be comfortable."

She watched while he went for the chair across the room and pulled it close to the bed to sit opposite her. Light was only beginning to shaft through the shutters.

"You're exhausted," he said, getting up again. "How were you hurt? I want our doctor to see you tonight."

"No, Asher, no."

312

He came back with a cold washcloth, lifted her braid and pressed the cloth to her neck. "My God," he said. "Sabra. No matter how I try to protect you, you fall through the cracks. You've been *there* now, haven't you? Over the brink."

She couldn't speak when he asked, "Did it help you decide about Noah?" She only nodded. Finally, she was able to say it: "I'm letting him stay in the army."

Asher regarded her quizzically.

"I don't want Noah in this war," she began, hesitating, searching her mind for reasons she hadn't known until this moment. "I was afraid to interfere! I didn't want to be like my mother with a child who prayed his mother would die so he could live the life he wanted."

Asher's blue eyes were tender. "Freya," he said, lifting her hands to his face. "No one could wish you dead no matter what you did. Is it my turn to tell you what kind of a world I want? I'm responsible for my country's safety. Every war against us was a war of extermination. We won't let them do it. Is that a world you can live in? Are you with me in that?"

"Not if there's only one way to be with you."

"I don't ask for one way. I don't have David's mentality. But we need feeling in common. We have to work at the same puzzle. Are you willing?"

"Another pact?"

"Maybe one we understand better. We grew up fast since the last one."

"You know where the pieces fit before you move them, Asher."

"You forgot what I taught you about chess?" He smiled at her. "You have to *know* what might happen if you make the wrong move, like answering David's questions. Did you tell Noah where you went?"

She swung her legs over the side of the bed, facing him, holding the cloth to the back of her neck. "I told Noah I went to the museum. I *work* in a museum."

"What about the journalists? Why did they take you?"

She told him about Noah's delay, about meeting the journalists in the lobby, about the PLO press conference and meeting Dr.

313

Poudrakis and Dr. Khatoubi. "I have notes to prove it, Asher."
She started to get up; he shook his head.

"Don't," he said. "I'm not the one you have to convince. David
doesn't understand your brand of honesty, how easy it is to swin-
dle you. You let people use you as though you have no interests
of your own! Why, Freya, why?"

"Don't speak to me as though I were a child! If I'm a case of
arrested development how did it happen? You have no right . . .
Do you think I *liked* reading out my passport number so I could
come to your house?"

"Your *passport?*" Asher's eyes widened, then he began to
laugh. "Your *passport?* Who asked you?"

"Aviva. Yes. I read the numbers to her twice."

"Aviva. I'll be damned."

Then Freya was silent, embarrassed that she had involved
Aviva, angry that Asher criticized her brand of honesty yet again.
How many times had he accused her of objectivity from the
safety of freedom?

Asher was glancing at her obliquely, his grin vanished into
tight lines. He took the compress and went to freshen it before
he sat down on the bed next to her.

"I'm sorry about the passport," he said softly. "But I have to
know details, Freya. Which hospital was it?"

"Mukhassed. A couple of blocks from Sabra. There wasn't
much of it left."

"Who told you to go to Mukhassed?"

She averted her eyes. "A nurse at Red Crescent. She was so
desperate, I gave her. . ."

"*What*, what did you give her?"

"Money! You ought to understand about getaway money."

Asher bolted up and walked toward the balcony. With his back
to her, the look of him, one hand running through his hair, the
other thrust into his pocket, was of a man trapped in a maze. "Do
you know what you've done, Freya? I wouldn't give a damn about
it, if it weren't one more irony for us. We're desperate to stay
together and every move you make. . ." He stood there staring

out the balcony window while she held the washcloth to the back of her neck, hardly daring to breathe.

He turned to look at her. "You understand, Freya? You took a stroll through the enemy camp and passed out money. If the *Shin Bet* finds out, they'll have you deported."

The washcloth dropped away. "Deported? You're making it up. You want to punish me. Are you making it up? You can't *let* them deport me!"

"*Let* them? You despise my job and you do as you damn well please. Then you rely on my clout to dig you out. You've positioned yourself above it all, Freya, loyalty and self-interest mean nothing to you. My credibility means nothing to you."

It was like an electric shock. "Asher," she murmured. "I love you . . . you can't mean what you say." It couldn't come to this. She *had* been loyal. Wasn't it enough that her loyalty to Asher had fragmented her life?

"David will track it down . . . you can be sure. It's not so difficult to pick you out of a crowd."

Freya was holding the back of her neck. "Why is it that everything I do . . . Uri warned me. But I couldn't stand what I saw. I wanted to patch."

"You don't think any of us understand? None of us see what you see?"

Silence was deepening; she remembered Carole gesturing, *no*, when she pulled out the money; afterward, Carole told her to keep her mouth shut about it. Asher was moving toward her. She reached out for him and when he took her in his arms the horror of the day erupted inside her. She was sobbing against him, clutching his arm. "Oh Sher, what have I done to you? I never wanted anything but to be with you."

"Don't cry. We'll learn." He held her, smoothing her hair. "I must have been crazy to sweep you off to Ramat Eshkol. I left you alone in a country you know nothing about. I never gave you a chance. I put David on your tail so I wouldn't lose you again."

"They won't have to deport me, Sher. You and Noah won't be disgraced. I'll go back to Chicago. I'll take responsibility so

315

they can't blame you. Tell them you were conned. I conned you!"

"Freya stop it! You think being conned by a beautiful woman is a credit to a senior intelligence officer? For God's sake, you're overwrought. Let's go over it again. Who did you give the money to at Sabra?"

"She was Chinese. She was crazy with rumors about what was going to happen. I was so frightened for her, I had to do something. Her name was Miss Wehn."

Astonishment came over Asher's face. "*Wehn?*" he said. "What else did she tell you?"

"She said she could have gotten out but she wouldn't leave Sani. She said someone should have gotten her out and when the journalist asked how, she denied she said it."

Asher's look of astonishment had turned to a wry smile of irony. "You won't be deported," he said. "You were recruited. Recruited by the general himself! My special envoy." His face brightened in a great smile.

"*Recruited...?*"

"Freya, my God, it's cruel to tell you. Miss Wehn is not from an ancient Chinese province. She was an independent volunteer nurse. The third world is where her heart beats."

"What's wrong with that?" Freya cried.

"Nothing's wrong with that. Except that she worked both sides. She had to save them all from themselves. Poor Wehn. Her conscience justified taking things into her own hands, making decisions she had no right to make. We got a lot of information from her."

"I don't believe you! I'm the one who writes stories, not you!"

"Well, Freya," Asher said. "Write a story about a world without order. Where nothing is as it seems. Where you can't read the signals because they're not recognizable."

Freya sat there staring at him, a steady pulse beating in her neck. "I don't believe you. Miss Wehn loves those people."

"She does," Asher said. "What's that got to do with it? I told you once, Freya. You're obsessed with free choice and you're absolutely certain that given free choice, every motive is pure. Any shade of grey is only smog. You don't see how people need to

316

finagle, scramble, not for what they choose to do, but what they must do."

"Not everyone is so cynical. I don't think Miss Wehn is. . ."

"Cynical? No, Wehn isn't cynical. She's like the rest of humanity. You can't accept that because you're stuck on white hats and black hats."

"Miss Wehn wouldn't inform! She warned the photographer not to take nasty pictures. She called those people, heroic!"

"Why wouldn't she? Wehn knew they couldn't win the war. She saw them die. She's a nurse! When their suffering was unbearable, she decided to shorten the process, save as many lives as she could."

"You took advantage of that!"

"We did not. She approached us about ammunition hidden in houses. She let us know, then she warned them that we knew and they dumped the ammunition into trash cans. When she got wind of a rumor she passed it on. She saw how hopeless it was. When someone you love is suffering, about to die with no hope to recover, and you pull the plug, should you be convicted of murder?"

Freya put her face in her hands. "Why didn't you get her out?"

"She wouldn't go! She chose being found out in the camp and getting torn to pieces. *Responsibility*, she called it."

"But she was wild to get out! Why didn't you help her?"

"You think there was time to wait until Wehn made up her mind?"

Freya didn't answer. Asher moved toward her, smiling. "I recruited you. The general himself. I sent you on the sly to give Wehn money, so she could get out if she changed her mind. You won't be deported. You'll get a citation!"

"Why didn't you check her out? Why didn't you make sure?"

Asher smiled. "There was no opportunity until you came along, Dr. Schweitzer."

"David will never believe you!"

"He might. By now, David has a look in his eye when he talks about you. I tell you, Freya, when once a man rescues a woman, he gets possessive." Asher grinned.

317

"I won't stay, Asher."

"You won't stay where?"

"I should go home."

"You will *not* go home," Asher said. "Do you think I'd let you leave when there's a chance we could work out our lives? Aviva will accept a divorce. She wants to build a clientele in Paris or so she's been telling me for years. She may be working on her third husband."

"Because of me? I don't want that!" She stopped for a moment. Aviva was with Asher in everything political, she would never compromise him. But Aviva was working on her third husband? Could love switch like fashions of the season?

"I have to go home, Asher," she said. "If I stay, I'd only be a problem to you."

"Will you?" Asher's smile was tender. "Remember when I said I admired your Pushkin mentality, your grit?"

"I'd be a problem to you," she repeated.

"You're a problem I couldn't live without." He was unbuttoning her blouse, drawing her close to him, pressing his lips to her shoulder. "We've been as far over the brink as we need to go."

She put her arms around him. She understood that he was also vulnerable; he needed the story of recruiting her as much to save his own face as to block her deportation.

"I'm taking you home tomorrow, Freya. To Jerusalem. To our home."

Her mouth was dry. "I love you, Sher. But the story . . . wouldn't we know that lying is okay when it suits us? The next step is lying to each other."

He moved back to look at her. "Don't be ridiculous," he said. "What's wrong with getting out of a fix? You let the cab driver get out. You insisted! Were you so sure he was innocent? Why can't you give us the same chance?"

She didn't answer. She was thinking of all the coverups in her life, pretending to be something she wasn't. A loving daughter, a devoted wife, Berman's obedient student.

"Asher," she murmured. "I won't lie to myself anymore. I'm no agent. I wouldn't be anyone's agent, not even yours."

He was holding her, his lips brushing her face. "Don't you think I know why you helped Miss Wehn? I wouldn't distort your feelings, Freya. I'll think of something else, maybe even the truth. David won't question it."

She thought of telling Asher what David had left out of his report about the cemetery charade. David hadn't told Asher about Noah's wound, either. Deftly, he had passed the responsibility to Noah's commanding officer. How much more was covered up? Still, if she told Asher about the funeral, how David had subtly discouraged her from seeing Asher's doctor friend, David would surely retaliate by pursuing her activities in Beirut. Why did Asher say that David wouldn't question the truth? Slowly, she realized that David's job was to protect Asher, not to endanger him. David wriggled in the same can of worms as she did.

She had to see Uri. She owed him an account of herself and of Noah. He had tried to help her and if his motives were fueled by his need for Noah's return to Gan Zvi, still, he had tried to keep her centered for her own sake in an unpredictable country where she was totally ignorant, worse, even arrogant. Had she actually believed that she could behave as though the rules were the same as in Chicago, Illinois, USA? Had she really thought that she could outflank the Israeli Secret Service, as though she were playing a board game? Uri had tried to make her understand and protect her, but she wasn't able to listen. There were too many cross currents, she was too torn to hear him.

"I want to see Uri, Asher."

As though she had touched a nerve, Asher moved away. He was standing over her and he looked down on her, perplexed, questioning. "Uri," he said flatly. But he only pulled his belt tighter and straightened his shoulders. "Uri then," he said tersely. "I have a meeting with David but I'll be back for dinner. We'll stay here tonight . . . we need the night together. I'll take you to Gan Zvi tomorrow. And talk to your journalists. Tell them to keep their mouths shut."

Then he was at the door. She called him back.

"Sher, I love you!"

He opened the door before he turned back. "Stay put, then.

Don't go anywhere." He hesitated. "Don't worry, I *won't* let them. . ." He had stopped before he said it but his face looked troubled and uncertain.

Asher's roses were still fragrant in the vase as she lay across the bed where Asher had held her. The familiar excitement was kindling inside her; she thought it absurd and inappropriate that now, at this moment, she should fantasize making love. It wasn't normal. Why was it that only when she and Asher made love that she felt the spark of her own identity? She was never herself with Momma or Poppa. She was a good daughter by rote, stuffed like a goose on romantic slogans and parables. *A shoe maker working all day, all night. Who should profit from his labor? Land, Bread, Peace.* How simple and splendid. She had never forgotten that night in the park when the great fart came from the King of Hope. Or Mr. Berman and the little medals he gave to the worthy—the golden haired boy, Lenin as a baby. All the parents were charged for the medals whether one's child got one or not. *Overcharged,* Poppa said, and Marsalka accused Berman of pocketing the profits. Once, Mr. Berman called her back as the *Schule* was letting out. He held her in front of him, his knees gripping her waist, one hand stroking her hair, the other fondling her. He was whispering, "Beautiful, beautiful" . . . Squirming out of his grasp, her face hot, she ran sobbing all the way to Frick Street. Freya kept her secret inside her like wormwood.

She had always felt unworthy. She never even tried to get what she needed. Instead of joining Asher where her life was, she had stayed in Chicago to care for Momma. On the surface it was the fact that she was needed, but down deep she knew that Momma's needing didn't count; it was the need in herself not to let anyone discover what lay in her heart. When she gave up Asher, she gave up everything. She had denied herself for twenty years and no matter how she fought against it, she accepted the prophecy that she would be like the martyred Rosa Luxemburg.

Miss Wehn was like that, tricked by her own conscience. Sani died because Miss Wehn played both sides . . . she had informed about hiding ammunition. Miss Wehn had swallowed the poison and when it was too late she had begged to be saved. Freya began to shiver.

She lay there, cold, staring at the door where Asher and Noah had gone out. She promised Noah that she would stay. And suddenly, she was calm. She had not swallowed the poison. Not yet. She was not the martyred Rosa. She would never leave Asher.

She could breathe again as though someone had taken their hands off her throat, then, after a moment, she bolted up; the calm broke as though she were experiencing a second waking. She could be deported. How much of Asher's clout had she dissipated by playing Lady Bountiful in the enemy camp?

It didn't matter whether or not David informed the *Shin Bet*. He would know. Beni would know and Asher would be beholden to both of them. But she had travelled across Hell's highway with them and they *knew* she was not a security risk. She *had* done *Shin Bet* work for them with Miss Wehn. Still, would it count? From the look on Asher's face before he closed the door, she knew she would have to face some kind of interrogation.

It was only now, sitting there, staring, that Freya understood that she was enmeshed in political tradeoffs. The choice to live with Asher and Noah was out of her hands.

She reached for the phone and stopped, remembering Carole's face when she pulled out the money for Miss Wehn—her shock. For Carole, she was a steppingstone to Asher. Now, she had to tell Carole to deny everything. She could hear her: *You bloody nit! Didn't you know the trouble you'd make if you told the great lover?* If she had made a friend easily she had lost one just as easily. And Layla? How did she bring her the news? The bearer of bad tidings is rejected, shunned. She pulled back from the phone.

She was staring at the door where she had watched Asher leave, where she had stopped herself from crying out that she could never live without him. They had a pact to tough it out no matter what.

She sat rocking back and forth while images coiled in her mind like snakes, the orchards burnt out, blackened geraniums, the

321

baby who was an enemy, a broken crib, the songs of birds, deadened. Noah's scarred face.

Then, suddenly, she stopped rocking and bent over so that her forehead touched her knees. Early streaks of dawn were brightening the sky. It was still hot. There was another day to get through before she and Asher would be together again. Asher would stay the night and they would make love. It was their common language.

Where could she imagine the two people she loved most in the world? On the bleak mountain roads where snipers were heroes? How long did it take for a grenade to explode? How long before the moment passed to the next moment when the world collapsed?

She got up and took three aspirins, then she stood at the open balcony window, looking out. Clouds of black smoke coiled across the lightening sky. There were growls of distant artillery.